Instrument Flying Handbook

Revised Edition

U.S. Department of Transportation
FEDERAL AVIATION ADMINISTRATION
Flight Standards Service

FAA-H-8083-15B

Skyhorse Publishing

All inquiries should be addressed to Skyhorse Publishing, 307 West 36th Street, 11th Floor, New York, NY 10018.

Skyhorse Publishing books may be purchased in bulk at special discounts for sales promotion, corporate gifts, fund-raising, or educational purposes. Special editions can also be created to specifications. For details, contact the Special Sales Department, Skyhorse Publishing, 307 West 36th Street, 11th Floor, New York, NY 10018 or info@skyhorsepublishing.com.

Visit our website at www.skyhorsepublishing.com.

10 9 8 7 6 5 4 3 2 1

Library of Congress Cataloging-in-Publication Data is available on file.

ISBN: 978-1-62636-237-6

Printed in China

Preface

This Instrument Flying Handbook is designed for use by instrument flight instructors and pilots preparing for instrument rating tests. Instructors may find this handbook a valuable training aid as it includes basic reference material for knowledge testing and instrument flight training. Other Federal Aviation Administration (FAA) publications should be consulted for more detailed information on related topics.

This handbook conforms to pilot training and certification concepts established by the FAA. There are different ways of teaching, as well as performing, flight procedures and maneuvers and many variations in the explanations of aerodynamic theories and principles. This handbook adopts selected methods and concepts for instrument flying. The discussion and explanations reflect the most commonly used practices and principles. Occasionally the word "must" or similar language is used where the desired action is deemed critical. The use of such language is not intended to add to, interpret, or relieve a duty imposed by Title 14 of the Code of Federal Regulations (14 CFR).

All of the aeronautical knowledge and skills required to operate in instrument meteorological conditions (IMC) are detailed. Chapters are dedicated to human and aerodynamic factors affecting instrument flight, the flight instruments, attitude instrument flying for airplanes, basic flight maneuvers used in IMC, attitude instrument flying for helicopters, navigation systems, the National Airspace System (NAS), the air traffic control (ATC) system, instrument flight rules (IFR) flight procedures, and IFR emergencies. Clearance shorthand and an integrated instrument lesson guide are also included.

This handbook supersedes FAA-H-8081-15A, Instrument Flying Handbook, dated 2007.

This handbook may be purchased from the Superintendent of Documents, United States Government Printing Office (GPO), Washington, DC 20402-9325, or from GPO's website.

http://bookstore.gpo.gov

This handbook is also available for download, in PDF format, from the Regulatory Support Division's (AFS-600) website.

http://www.faa.gov/about/office_org/headquarters_offices/avs/offices/afs/afs600

This handbook is published by the United States Department of Transportation, Federal Aviation Administration, Airman Testing Standards Branch, AFS-630, P.O. Box 25082, Oklahoma City, OK 73125.

Comments regarding this publication should be sent, in email form, to the following address.

AFS630comments@faa.gov

Acknowledgments

This handbook was produced as a combined Federal Aviation Administration (FAA) and industry effort. The FAA wishes to acknowledge the following contributors:

The laboratory of Dale Purves, M.D. and Mr. Al Seckel in providing imagery (found in Chapter 3) for visual illusions from the book, *The Great Book of Optical Illusions*, Firefly Books, 2004

Sikorsky Aircraft Corporation and Robinson Helicopter Company for imagery provided in Chapter 2

Garmin Ltd. for providing flight system information and multiple display systems to include integrated flight, GPS and communication systems; information and hardware used with WAAS, LAAS; and information concerning encountering emergencies with high-technology systems

Universal Avionics System Corporation for providing background information of the Flight Management System and an overview on Vision–1 and Traffic Alert and Collision Avoidance systems (TCAS)

Meggitt/S-Tec for providing detailed autopilot information regarding installation and use

Cessna Aircraft Company in providing instrument panel layout support and information on the use of onboard systems

Kearfott Guidance and Navigation Corporation in providing background information on the Ring-LASAR gyroscope and its history

Honeywell International Inc., for Terrain Awareness Systems (TAWS) and various communication and radio systems sold under the Bendix-King name

Chelton Flight Systems and Century Flight Systems, Inc., for providing autopilot information relating to Highway in the Sky (Chelton) and HSI displays (Century)

Avidyne Corporation for providing displays with alert systems developed and sold by Ryan International, L3 Communications, and Tectronics

Additional appreciation is extended to the Aircraft Owners and Pilots Association (AOPA), the AOPA Air Safety Institute, and the National Business Aviation Association (NBAA) for their technical support and input.

Introduction

Is an Instrument Rating Necessary?

The answer to this question depends entirely upon individual needs. Pilots may not need an instrument rating if they fly in familiar uncongested areas, stay continually alert to weather developments, and accept an alternative to their original plan. However, some cross-country destinations may take a pilot to unfamiliar airports and/or through high activity areas in marginal visual or instrument meteorological conditions (IMC). Under these conditions, an instrument rating may be an alternative to rerouting, rescheduling, or canceling a flight. Many accidents are the result of pilots who lack the necessary skills or equipment to fly in marginal visual meteorological conditions (VMC) or IMC and attempt flight without outside references.

Pilots originally flew aircraft strictly by sight, sound, and feel while comparing the aircraft's attitude to the natural horizon. As aircraft performance increased, pilots required more inflight information to enhance the safe operation of their aircraft. This information has ranged from a string tied to a wing strut, to development of sophisticated electronic flight information systems (EFIS) and flight management systems (FMS). Interpretation of the instruments and aircraft control have advanced from the "one, two, three" or "needle, ball, and airspeed" system to the use of "attitude instrument flying" techniques.

Navigation began by using ground references with dead reckoning and has led to the development of electronic navigation systems. These include the automatic direction finder (ADF), very-high frequency omnidirectional range (VOR), distance measuring equipment (DME), tactical air navigation (TACAN), long range navigation (LORAN), global positioning system (GPS), instrument landing system (ILS), microwave landing system (MLS), and inertial navigation system (INS).

Perhaps you want an instrument rating for the same basic reason you learned to fly in the first place—because you like flying. Maintaining and extending your proficiency, once you have the rating, means less reliance on chance and more on skill and knowledge. Earn the rating—not because you might need it sometime, but because it represents achievement and provides training you will use continually and build upon as long as you fly. But most importantly it means greater safety in flying.

Instrument Rating Requirements

A private or commercial pilot must have an instrument rating and meet the appropriate currency requirements if that pilot operates an aircraft using an instrument flight rules (IFR) flight plan in conditions less than the minimums prescribed for visual flight rules (VFR), or in any flight in Class A airspace.

You will need to carefully review the aeronautical knowledge and experience requirements for the instrument rating as outlined in Title 14 of the Code of Federal Regulations (14 CFR) part 61. After completing the Federal Aviation Administration (FAA) Knowledge Test issued for the instrument rating, and all the experience requirements have been satisfied, you are eligible to take the practical test. The regulations specify minimum total and pilot-in-command time requirements. This minimum applies to all applicants regardless of ability or previous aviation experience.

Training for the Instrument Rating

A person who wishes to add the instrument rating to his or her pilot certificate must first make commitments of time, money, and quality of training. There are many combinations of training methods available. Independent studies may be adequate preparation to pass the required FAA Knowledge Test for the instrument rating. Occasional periods of ground and flight instruction may provide the skills necessary to pass the required test. Or, individuals may choose a training facility that provides comprehensive aviation education and the training necessary to ensure the pilot will pass all the required tests and operate safely in the National Airspace System (NAS). The aeronautical knowledge may be administered by educational institutions, aviation-oriented schools, correspondence courses, and appropriately rated instructors. Each person must decide for themselves which training program best meets his or her needs and at the same time maintain a high quality of training. Interested persons should make inquiries regarding the available training at nearby airports, training facilities, in aviation publications, and through the FAA Flight Standards District Office (FSDO).

Although the regulations specify minimum requirements, the amount of instructional time needed is determined not by the regulation, but by the individual's ability to achieve a satisfactory level of proficiency. A professional pilot with diversified flying experience may easily attain a satisfactory level of proficiency in the minimum time required by regulation. Your own time requirements will depend upon a variety of factors, including previous flying experience, rate of learning, basic ability, frequency of flight training, type of aircraft flown, quality of ground school training, and quality of flight instruction, to name a few. The total instructional time you will need, the scheduling of such time, is up to the individual most qualified to judge your proficiency—the instructor who supervises your progress and endorses your record of flight training.

You can accelerate and enrich much of your training by informal study. An increasing number of visual aids and programmed instrument courses is available. The best course is one that includes a well-integrated flight and ground school curriculum. The sequential nature of the learning process requires that each element of knowledge and skill be learned and applied in the right manner at the right time.

Part of your instrument training may utilize a flight simulator, flight training device, basic aviation training device (BATD), or an advanced aviation training device (AATD). This ground-based flight training equipment is a valuable tool for developing your instrument cross-check and learning procedures, such as intercepting and tracking, holding patterns, and instrument approaches. Once these concepts are fully understood, you can then continue with inflight training and refine these techniques for full transference of your new knowledge and skills.

Holding the instrument rating does not necessarily make you a competent all-weather pilot. The rating certifies only that you have complied with the minimum experience requirements, that you can plan and execute a flight under IFR, that you can execute basic instrument maneuvers, and that you have shown acceptable skill and judgment in performing these activities. Your instrument rating permits you to fly into instrument weather conditions with no previous instrument weather experience. Your instrument rating is issued on the assumption that you have the good judgment to avoid situations beyond your capabilities. The instrument training program you undertake should help you to develop not only essential flying skills but also the judgment necessary to use the skills within your own limits.

Regardless of the method of training selected, the curriculum in Appendix B, Instrument Training Lesson Guide, provides guidance as to the minimum training required for the addition of an instrument rating to a private or commercial pilot certificate.

Maintaining the Instrument Rating

Once you hold the instrument rating, you may not act as pilot-in-command under IFR or in weather conditions less than the minimums prescribed for VFR, unless you meet the recent flight experience requirements outlined in 14 CFR part 61. These procedures must be accomplished within the preceding 6 months and include six instrument approaches, holding procedures, and intercepting and tracking courses through the use of navigation systems. If you do not meet the experience requirements during these 6 months, you have another 6 months to meet these minimums. If the requirements are still not met, you must pass an instrument proficiency check, which is an inflight evaluation by a qualified instrument flight instructor using tasks outlined in the instrument rating practical test standards (PTS).

The instrument currency requirements must be accomplished under actual or simulated instrument conditions. You may log instrument flight time during the time for which you control the aircraft solely by reference to the instruments. This can be accomplished by wearing a view-limiting device, such as a hood, flying an approved flight-training device, or flying in actual IMC.

It takes only one harrowing experience to clarify the distinction between minimum practical knowledge and a thorough understanding of how to apply the procedures and techniques used in instrument flight. Your instrument training is never complete; it is adequate when you have absorbed every foreseeable detail of knowledge and skill to ensure a solution will be available if and when you need it.

Table of Contents

Chapter 7, Section I
Airplane Basic Flight Maneuvers Using Analog Instrumentation....7-1

Chapter 7, Section II
Airplane Basic Flight Maneuvers Using an Electronic Flight Display....7-33

Chapter 8
Helicopter Attitude Instrument Flying 8-1

Chapter 9
Navigation Systems 9-1

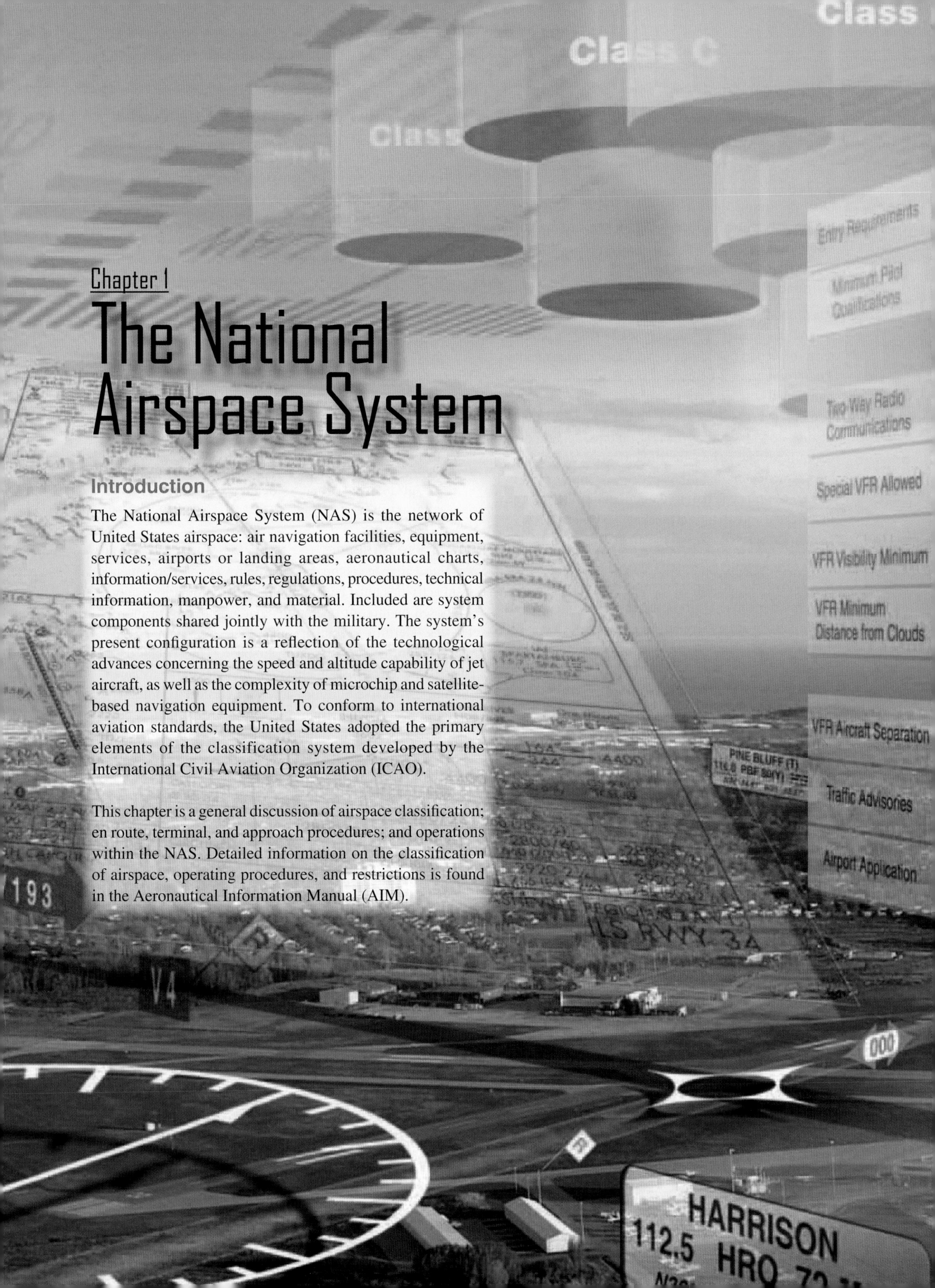

Chapter 1

The National Airspace System

Introduction

The National Airspace System (NAS) is the network of United States airspace: air navigation facilities, equipment, services, airports or landing areas, aeronautical charts, information/services, rules, regulations, procedures, technical information, manpower, and material. Included are system components shared jointly with the military. The system's present configuration is a reflection of the technological advances concerning the speed and altitude capability of jet aircraft, as well as the complexity of microchip and satellite-based navigation equipment. To conform to international aviation standards, the United States adopted the primary elements of the classification system developed by the International Civil Aviation Organization (ICAO).

This chapter is a general discussion of airspace classification; en route, terminal, and approach procedures; and operations within the NAS. Detailed information on the classification of airspace, operating procedures, and restrictions is found in the Aeronautical Information Manual (AIM).

Airspace Classification

Airspace in the United States *[Figure 1-1]* is designated as follows:

1. Class A. Generally, airspace from 18,000 feet mean sea level (MSL) up to and including flight level (FL) 600, including the airspace overlying the waters within 12 nautical miles (NM) of the coast of the 48 contiguous states and Alaska. Unless otherwise authorized, all pilots must operate their aircraft under instrument flight rules (IFR).
2. Class B. Generally, airspace from the surface to 10,000 feet MSL surrounding the nation's busiest airports in terms of airport operations or passenger enplanements. The configuration of each Class B airspace area is individually tailored, consists of a surface area and two or more layers (some Class B airspace areas resemble upside-down wedding cakes), and is designed to contain all published instrument procedures once an aircraft enters the airspace. An air traffic control (ATC) clearance is required for all aircraft to operate in the area, and all aircraft that are so cleared receive separation services within the airspace.
3. Class C. Generally, airspace from the surface to 4,000 feet above the airport elevation (charted in MSL) surrounding those airports that have an operational control tower are serviced by a radar approach control and have a certain number of IFR operations or passenger enplanements. Although the configuration of each Class C area is individually tailored, the airspace usually consists of a surface area with a 5 NM radius, an outer circle with a 10 NM radius that extends from 1,200 feet to 4,000 feet above the airport elevation and an outer area. Each aircraft must establish two-way radio communications with the ATC facility providing air traffic services prior to entering the airspace and thereafter maintain those communications while within the airspace.
4. Class D. Generally, airspace from the surface to 2,500 feet above the airport elevation (charted in MSL) surrounding those airports that have an operational control tower. The configuration of each Class D airspace area is individually tailored and, when instrument procedures are published, the airspace normally designed to contain the procedures. Arrival extensions for instrument approach procedures (IAPs) may be Class D or Class E airspace. Unless otherwise authorized, each aircraft must establish two-way radio communications with the ATC facility providing air traffic services prior to entering the airspace and thereafter maintain those communications while in the airspace.
5. Class E. Generally, if the airspace is not Class A, B, C, or D, and is controlled airspace, then it is Class E airspace. Class E airspace extends upward from either the surface or a designated altitude to the overlying or adjacent controlled airspace. When designated as a surface area, the airspace is configured to contain all instrument procedures. Also in this class are federal airways, airspace beginning at either 700 or 1,200 feet above ground level (AGL) used to transition to and from the terminal or en route environment, and en route domestic and offshore airspace areas designated below 18,000 feet MSL. Unless designated at a lower altitude, Class E airspace begins at 14,500 MSL over the United States, including that airspace overlying the waters within 12 NM of the coast of the 48 contiguous states and Alaska, up to but not including 18,000 feet MSL, and the airspace above FL 600.
6. Class G. Airspace not designated as Class A, B, C, D, or E. Class G airspace is essentially uncontrolled by ATC except when associated with a temporary control tower.

Special Use Airspace

Special use airspace is the designation for airspace in which certain activities must be confined or where limitations may be imposed on aircraft operations that are not part of those activities. Certain special use airspace areas can create limitations on the mixed use of airspace. The special use airspace depicted on instrument charts includes the area name or number, effective altitude, time and weather conditions of operation, the controlling agency, and the chart panel location. On National Aeronautical Navigation Products (AeroNav Products) en route charts, this information is available on one of the end panels.

Prohibited areas contain airspace of defined dimensions within which the flight of aircraft is prohibited. Such areas are established for security or other reasons associated with the national welfare. These areas are published in the Federal Register and are depicted on aeronautical charts. The area is charted as a "P" followed by a number (e.g., "P-123").

Restricted areas are areas where operations are hazardous to nonparticipating aircraft and contain airspace within which the flight of aircraft, while not wholly prohibited, is subject to restrictions. Activities within these areas must be confined because of their nature, or limitations may be imposed upon aircraft operations that are not a part of those activities, or both. Restricted areas denote the existence of unusual, often invisible, hazards to aircraft (e.g., artillery firing, aerial gunnery, or guided missiles). IFR flights may be authorized to transit the airspace and are routed accordingly. Penetration

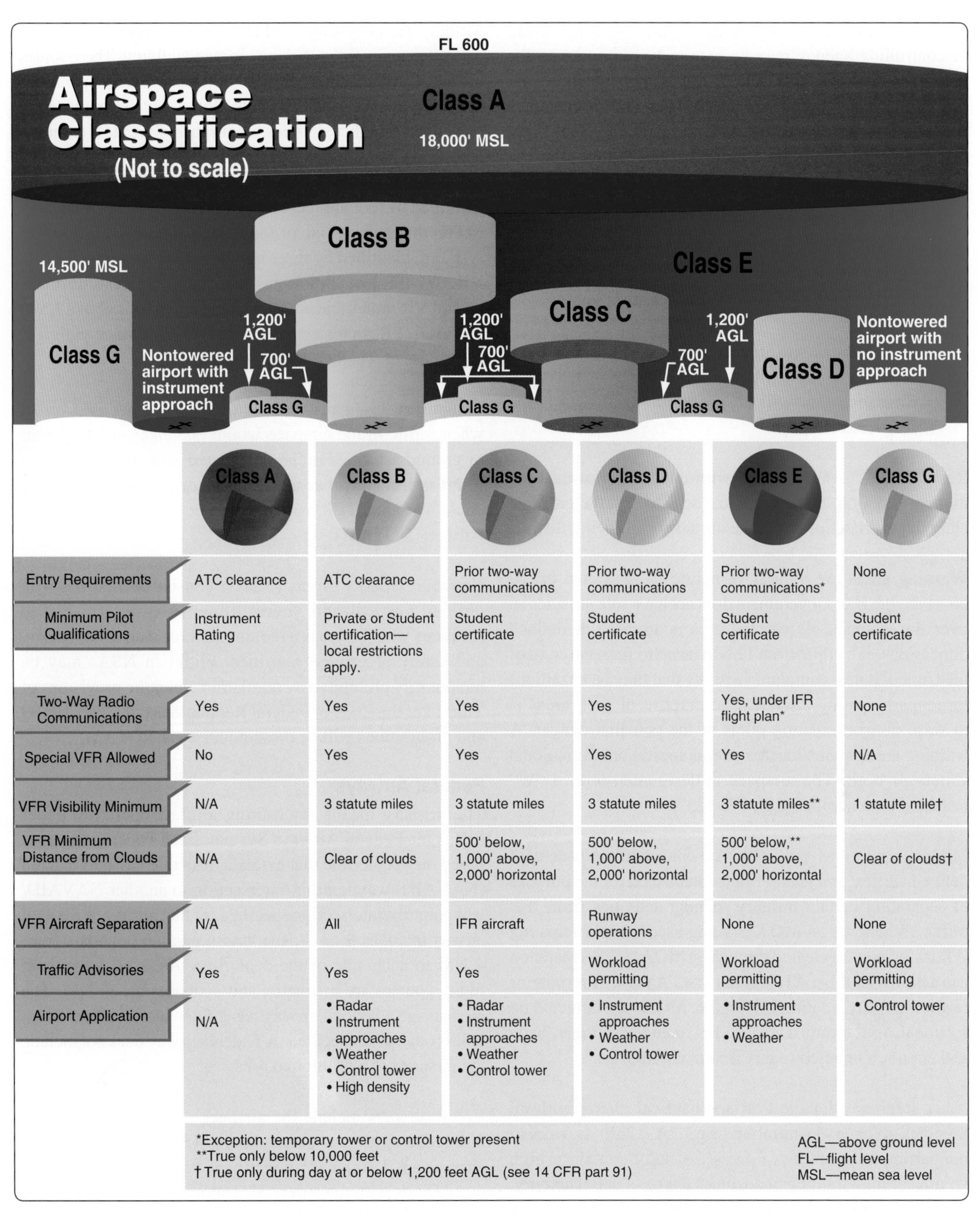

	Class A	Class B	Class C	Class D	Class E	Class G
Entry Requirements	ATC clearance	ATC clearance	Prior two-way communications	Prior two-way communications	Prior two-way communications*	None
Minimum Pilot Qualifications	Instrument Rating	Private or Student certification—local restrictions apply.	Student certificate	Student certificate	Student certificate	Student certificate
Two-Way Radio Communications	Yes	Yes	Yes	Yes	Yes, under IFR flight plan*	None
Special VFR Allowed	No	Yes	Yes	Yes	Yes	N/A
VFR Visibility Minimum	N/A	3 statute miles	3 statute miles	3 statute miles	3 statute miles**	1 statute mile†
VFR Minimum Distance from Clouds	N/A	Clear of clouds	500' below, 1,000' above, 2,000' horizontal	500' below, 1,000' above, 2,000' horizontal	500' below,** 1,000' above, 2,000' horizontal	Clear of clouds†
VFR Aircraft Separation	N/A	All	IFR aircraft	Runway operations	None	None
Traffic Advisories	Yes	Yes	Yes	Workload permitting	Workload permitting	Workload permitting
Airport Application	N/A	• Radar • Instrument approaches • Weather • Control tower • High density	• Radar • Instrument approaches • Weather • Control tower	• Instrument approaches • Weather • Control tower	• Instrument approaches • Weather	• Control tower

*Exception: temporary tower or control tower present
**True only below 10,000 feet
† True only during day at or below 1,200 feet AGL (see 14 CFR part 91)

AGL—above ground level
FL—flight level
MSL—mean sea level

Figure 1-1. *Airspace classifications.*

of restricted areas without authorization from the using or controlling agency may be extremely hazardous to the aircraft and its occupants. ATC facilities apply the following procedures when aircraft are operating on an IFR clearance (including those cleared by ATC to maintain visual flight rules (VFR)-On-Top) via a route that lies within joint-use restricted airspace:

1. If the restricted area is not active and has been released to the Federal Aviation Administration (FAA), the ATC facility will allow the aircraft to operate in the restricted airspace without issuing specific clearance for it to do so.
2. If the restricted area is active and has not been released to the FAA, the ATC facility will issue a clearance that will ensure the aircraft avoids the restricted airspace.

Restricted areas are charted with an "R" followed by a number (e.g., "R-5701") and are depicted on the en route chart appropriate for use at the altitude or FL being flown.

Warning areas are similar in nature to restricted areas; however, the U.S. Government does not have sole jurisdiction over the airspace. A warning area is airspace of defined dimensions, extending from 12 NM outward from the coast of the United States, containing activity that may be hazardous to nonparticipating aircraft. The purpose of such areas is to warn nonparticipating pilots of the potential danger. A warning area may be located over domestic or international waters or both. The airspace is designated with a "W" followed by a number (e.g., "W-123").

Military operations areas (MOAs) consist of airspace with defined vertical and lateral limits established for the purpose of separating certain military training activities from IFR traffic. Whenever an MOA is being used, nonparticipating IFR traffic may be cleared through an MOA if IFR separation can be provided by ATC. Otherwise, ATC will reroute or restrict nonparticipating IFR traffic. MOAs are depicted on sectional, VFR terminal area, and en route low altitude charts and are not numbered (e.g., "Boardman MOA").

Alert areas are depicted on aeronautical charts with an "A" followed by a number (e.g., "A-123") to inform nonparticipating pilots of areas that may contain a high volume of pilot training or an unusual type of aerial activity. Pilots should exercise caution in alert areas. All activity within an alert area shall be conducted in accordance with regulations, without waiver, and pilots of participating aircraft, as well as pilots transiting the area, shall be equally responsible for collision avoidance.

Military Training Routes (MTRs) are routes used by military aircraft to maintain proficiency in tactical flying. These routes are usually established below 10,000 feet MSL for operations at speeds in excess of 250 knots. Some route segments may be defined at higher altitudes for purposes of route continuity. Routes are identified as IFR (IR) and VFR (VR) followed by a number. MTRs with no segment above 1,500 feet AGL are identified by four number characters (e.g., IR1206, VR1207). MTRs that include one or more segments above 1,500 feet AGL are identified by three number characters (e.g., IR206, VR207). IFR low altitude en route charts depict all IR routes and all VR routes that accommodate operations above 1,500 feet AGL. IR routes are conducted in accordance with IFR regardless of weather conditions.

Temporary flight restrictions (TFRs) are put into effect when traffic in the airspace would endanger or hamper air or ground activities in the designated area. For example, a forest fire, chemical accident, flood, or disaster-relief effort could warrant a TFR, which would be issued as a Notice to Airmen (NOTAM).

National Security Areas (NSAs) consist of airspace with defined vertical and lateral dimensions established at locations where there is a requirement for increased security and safety of ground facilities. Flight in NSAs may be temporarily prohibited by regulation under the provisions of Title 14 of the Code of Federal Regulations (14 CFR) part 99 and prohibitions will be disseminated via NOTAM.

Federal Airways

The primary means for routing aircraft operating under IFR is the Federal Airways System. Each Federal airway is based on a centerline that extends from one navigational aid (NAVAID)/waypoint/fix/intersection to another NAVAID/waypoint/fix/intersection specified for that airway. A Federal airway includes the airspace within parallel boundary lines 4 NM to each side of the centerline. As in all instrument flight, courses are magnetic, and distances are in NM. The airspace of a Federal airway has a floor of 1,200 feet AGL, unless otherwise specified. A Federal airway does not include the airspace of a prohibited area.

Victor airways include the airspace extending from 1,200 feet AGL up to, but not including 18,000 feet MSL. The airways are designated on sectional and IFR low altitude en route charts with the letter "V" followed by a number (e.g., "V23"). Typically, Victor airways are given odd numbers when oriented north/south and even numbers when oriented east/west. If more than one airway coincides on a route segment, the numbers are listed serially (e.g., "V287-495-500"). *[Figure 1-2]*

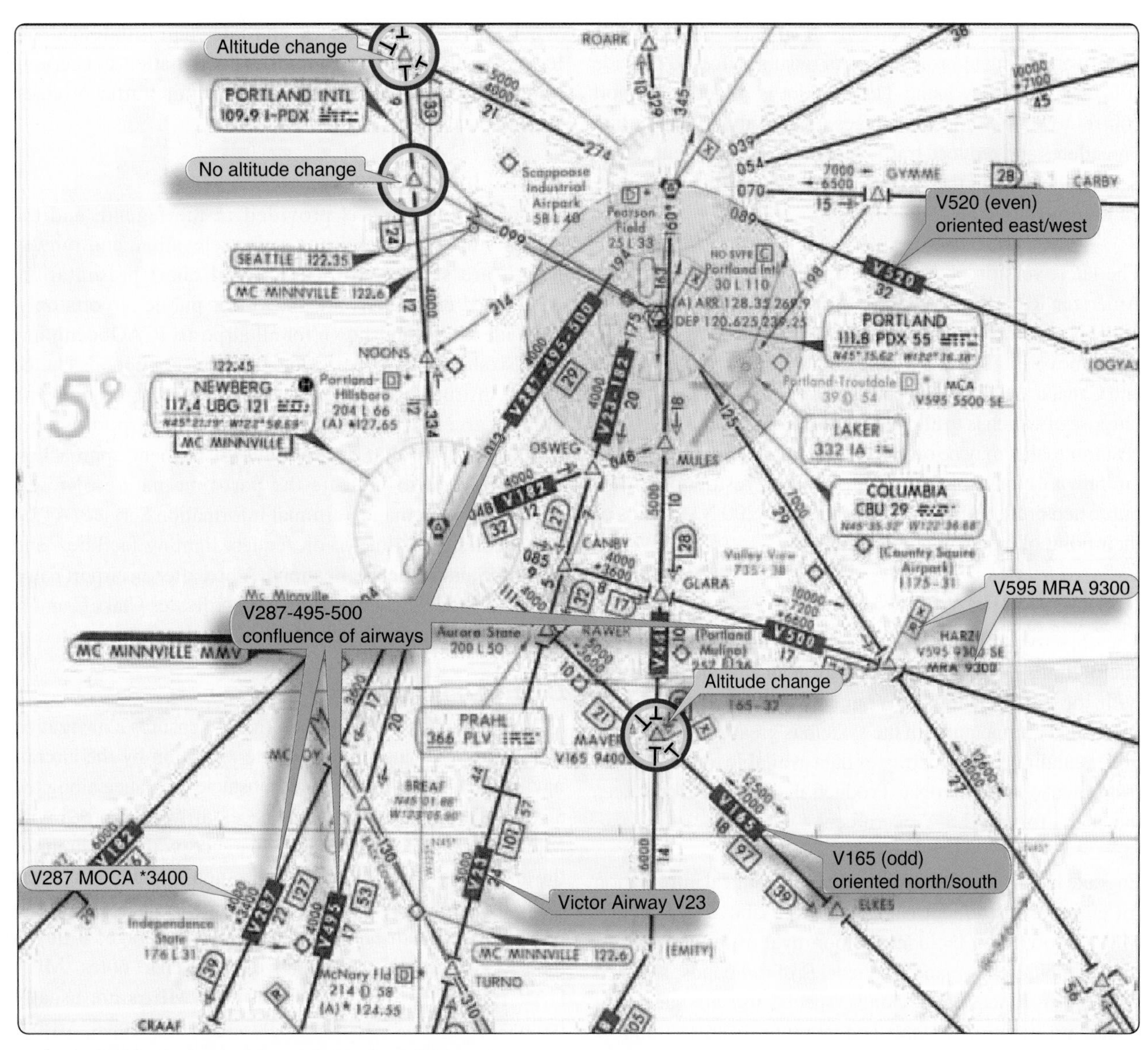

Figure 1-2. *Victor airways and charted IFR altitudes.*

Jet routes exist only in Class A airspace, from 18,000 feet MSL to FL 450, and are depicted on high-altitude en route charts. The letter "J" precedes a number to label the airway (e.g., J12).

Area navigation (RNAV) routes have been established in both the low-altitude and the high-altitude structures in recent years and are depicted on the en route low and high chart series. High altitude RNAV routes are identified with a "Q" prefix (except the Q-routes in the Gulf of Mexico) and low altitude RNAV routes are identified with a "T" prefix. RNAV routes and data are depicted in aeronautical blue.

In addition to the published routes, a random RNAV route may be flown under IFR if it is approved by ATC. Random RNAV routes are direct routes, based on RNAV capability, between waypoints defined in terms of latitude/longitude coordinates, degree-distance fixes, or offsets from established routes/airways at a specified distance and direction.

Radar monitoring by ATC is required on all random RNAV routes. These routes can only be approved in a radar environment. Factors that are considered by ATC in approving random RNAV routes include the capability to provide radar monitoring and compatibility with traffic volume and flow. ATC will radar monitor each flight; however, navigation on the random RNAV route is the responsibility of the pilot.

Other Routing

Preferred IFR routes have been established between major terminals to guide pilots in planning their routes of flight, minimizing route changes, and aiding in the orderly management of air traffic on Federal airways. Low and high altitude preferred routes are listed in the Airport/Facility Directory (A/FD). To use a preferred route, reference the departure and arrival airports; if a routing exists for your flight, then airway instructions are listed.

Tower En Route Control (TEC) is an ATC program that uses overlapping approach control radar services to provide IFR clearances. By using TEC, a pilot is routed by airport control towers. Some advantages include abbreviated filing procedures and reduced traffic separation requirements. TEC is dependent upon the ATC's workload, and the procedure varies among locales.

The latest version of Advisory Circular (AC) 90-91, North American Route Program (NRP), provides guidance to users of the NAS for participation in the NRP. All flights operating at or above FL 290 within the conterminous United States and Canada are eligible to participate in the NRP, the primary purpose of which is to allow operators to plan minimum time/cost routes that may be off the prescribed route structure. NRP aircraft are not subject to route-limiting restrictions (e.g., published preferred IFR routes) beyond a 200 NM radius of their point of departure or destination.

IFR En Route Charts

The objective of IFR en route flight is to navigate within the lateral limits of a designated airway at an altitude consistent with the ATC clearance. Your ability to fly instruments safely and competently in the system is greatly enhanced by understanding the vast array of data available to the pilot on instrument charts. AeroNav Products maintains and produces the charts for the U.S. Government.

En route high-altitude charts provide aeronautical information for en route instrument navigation at or above 18,000 feet MSL. Information includes the portrayal of Jet and RNAV routes, identification and frequencies of radio aids, selected airports, distances, time zones, special use airspace, and related information. Established jet routes from 18,000 feet MSL to FL 450 use NAVAIDs not more than 260 NM apart. The charts are revised every 56 days.

To effectively depart from one airport and navigate en route under instrument conditions, a pilot needs the appropriate IFR en route low-altitude chart(s). The IFR low altitude en route chart is the instrument equivalent of the sectional chart. When folded, the cover of the AeroNav Products en route chart displays an index map of the United States showing the coverage areas. Cities near congested airspace are shown in black type and their associated area chart is listed in the box in the lower left-hand corner of the map coverage box. Also noted is an explanation of the off-route obstruction clearance altitude (OROCA). The effective date of the chart is printed on the other side of the folded chart. Information concerning MTRs is also included on the chart cover. The en route charts are revised every 56 days.

When the AeroNav Products en route chart is unfolded, the legend is displayed and provides information concerning airports, NAVAIDs, communications, air traffic services, and airspace.

Airport Information

Airport information is provided in the legend, and the symbols used for the airport name, elevation, and runway length are similar to the sectional chart presentation. Associated city names are shown for public airports only. FAA identifiers are shown for all airports. ICAO identifiers are also shown for airports outside of the contiguous United States. Instrument approaches can be found at airports with blue or green symbols, while the brown airport symbol denotes airports that do not have instrument approaches. Stars are used to indicate the part-time nature of tower operations, Automatic Terminal Information Service (ATIS) frequencies, part-time or on request lighting facilities, and part-time airspace classifications. A box after an airport name with a "C" or "D" inside (e.g., [D]) indicates Class C and D airspace, respectively, per *Figure 1-3*.

Charted IFR Altitudes

The minimum en route altitude (MEA) ensures a navigation signal strong enough for adequate reception by the aircraft navigation (NAV) receiver and obstacle clearance along the airway. Communication is not necessarily guaranteed with MEA compliance. The obstacle clearance, within the limits of the airway, is typically 1,000 feet in non-mountainous areas and 2,000 feet in designated mountainous areas. MEAs can be authorized with breaks in the signal coverage; if this is the case, the AeroNav Products en route chart notes "MEA GAP" parallel to the affected airway. MEAs are usually bidirectional; however, they can be single-directional. Arrows are used to indicate the direction to which the MEA applies.

The minimum obstruction clearance altitude (MOCA), as the name suggests, provides the same obstruction clearance as an MEA; however, the NAV signal reception is ensured only within 22 NM of the closest NAVAID defining the route. The MOCA is listed below the MEA and indicated on AeroNav Products charts by a leading asterisk (e.g., "*3400"—see *Figure 1-2*, V287 at bottom left).

The minimum reception altitude (MRA) identifies the lowest altitude at which an intersection can be determined from an off-course NAVAID. If the reception is line-of-sight based, signal coverage only extends to the MRA or above. However, if the aircraft is equipped with distance measuring equipment (DME) and the chart indicates the intersection can be identified with such equipment, the pilot could define the

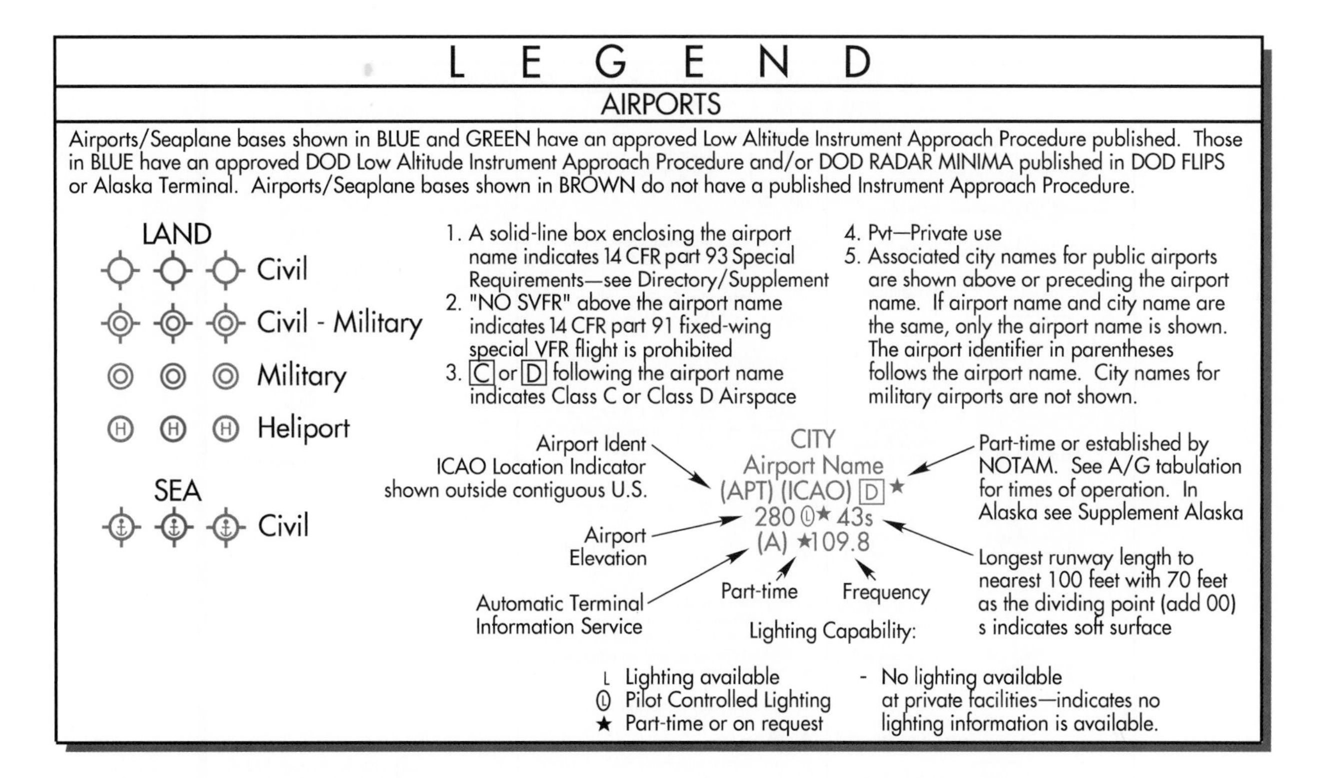

L E G E N D
AIRPORTS
Airports/Seaplane bases shown in BLUE and GREEN have an approved Low Altitude Instrument Approach Procedure published. Those in BLUE have an approved DOD Low Altitude Instrument Approach Procedure and/or DOD RADAR MINIMA published in DOD FLIPS or Alaska Terminal. Airports/Seaplane bases shown in BROWN do not have a published Instrument Approach Procedure.
LAND
Civil
Civil - Military
Military
Heliport
SEA
Civil
1. A solid-line box enclosing the airport name indicates 14 CFR part 93 Special Requirements—see Directory/Supplement
2. "NO SVFR" above the airport name indicates 14 CFR part 91 fixed-wing special VFR flight is prohibited
3. C or D following the airport name indicates Class C or Class D Airspace
4. Pvt—Private use
5. Associated city names for public airports are shown above or preceding the airport name. If airport name and city name are the same, only the airport name is shown. The airport identifier in parentheses follows the airport name. City names for military airports are not shown.
Airport Ident
ICAO Location Indicator shown outside contiguous U.S.
Airport Elevation
Automatic Terminal Information Service
CITY
Airport Name
(APT) (ICAO) D ★
280 Ⓛ★ 43s
(A) ★109.8
Part-time
Frequency
Lighting Capability:
Part-time or established by NOTAM. See A/G tabulation for times of operation. In Alaska see Supplement Alaska
Longest runway length to nearest 100 feet with 70 feet as the dividing point (add 00) s indicates soft surface
L Lighting available
Ⓛ Pilot Controlled Lighting
★ Part-time or on request
- No lighting available at private facilities—indicates no lighting information is available.

Figure 1-3. *En route airport legend.*

fix without attaining the MRA. On AeroNav Products charts, the MRA is indicated by the symbol and the altitude preceded by "MRA" (e.g., "MRA 9300"). *[Figure 1-2]*

The minimum crossing altitude (MCA) is charted when a higher MEA route segment is approached. The MCA is usually indicated when a pilot is approaching steeply rising terrain and obstacle clearance and/or signal reception is compromised. In this case, the pilot is required to initiate a climb so the MCA is reached by the time the intersection is crossed. On AeroNav Products charts, the MCA is indicated by the symbol , and the Victor airway number, altitude, and the direction to which it applies (e.g. "V24 8000 SE").

The maximum authorized altitude (MAA) is the highest altitude at which the airway can be flown with assurance of receiving adequate navigation signals. Chart depictions appear as "MAA-15000."

When an MEA, MOCA, and/or MAA change on a segment other than at a NAVAID, a sideways "T" (—|) is depicted on the chart. If there is an airway break without the symbol, one can assume the altitudes have not changed (see the upper left area of *Figure 1-2*). When a change of MEA to a higher MEA is required, the climb may commence at the break, ensuring obstacle clearance. *[Figure 1-4]*

Navigation Features

Types of NAVAIDs

Very high frequency omnidirectional ranges (VORs) are the principal NAVAIDs that support the Victor and Jet airways. Many other navigation tools are also available to the pilot. For example, nondirectional beacons (NDBs) can broadcast signals accurate enough to provide stand-alone approaches, and DME allows the pilot to pinpoint a reporting point on the airway. Though primarily navigation tools, these NAVAIDs can also transmit voice broadcasts.

Tactical air navigation (TACAN) channels are represented as the two- or three-digit numbers following the three-letter identifier in the NAVAID boxes. The AeroNav Products terminal procedures provide a frequency-pairing table for the TACAN-only sites. On AeroNav Products charts, very-high frequencies and ultra-high frequencies (VHF/UHF) NAVAIDs (e.g., VORs) are depicted in black, while low frequencies and medium frequencies (LF/MF) are depicted as brown. *[Figure 1-5]*

Identifying Intersections

Intersections along the airway route are established by a variety of NAVAIDs. An open triangle △ indicates the location of an ATC reporting point at an intersection. If the triangle is solid ▲, a report is compulsory. *[Figure 1-4]* NDBs, localizers,

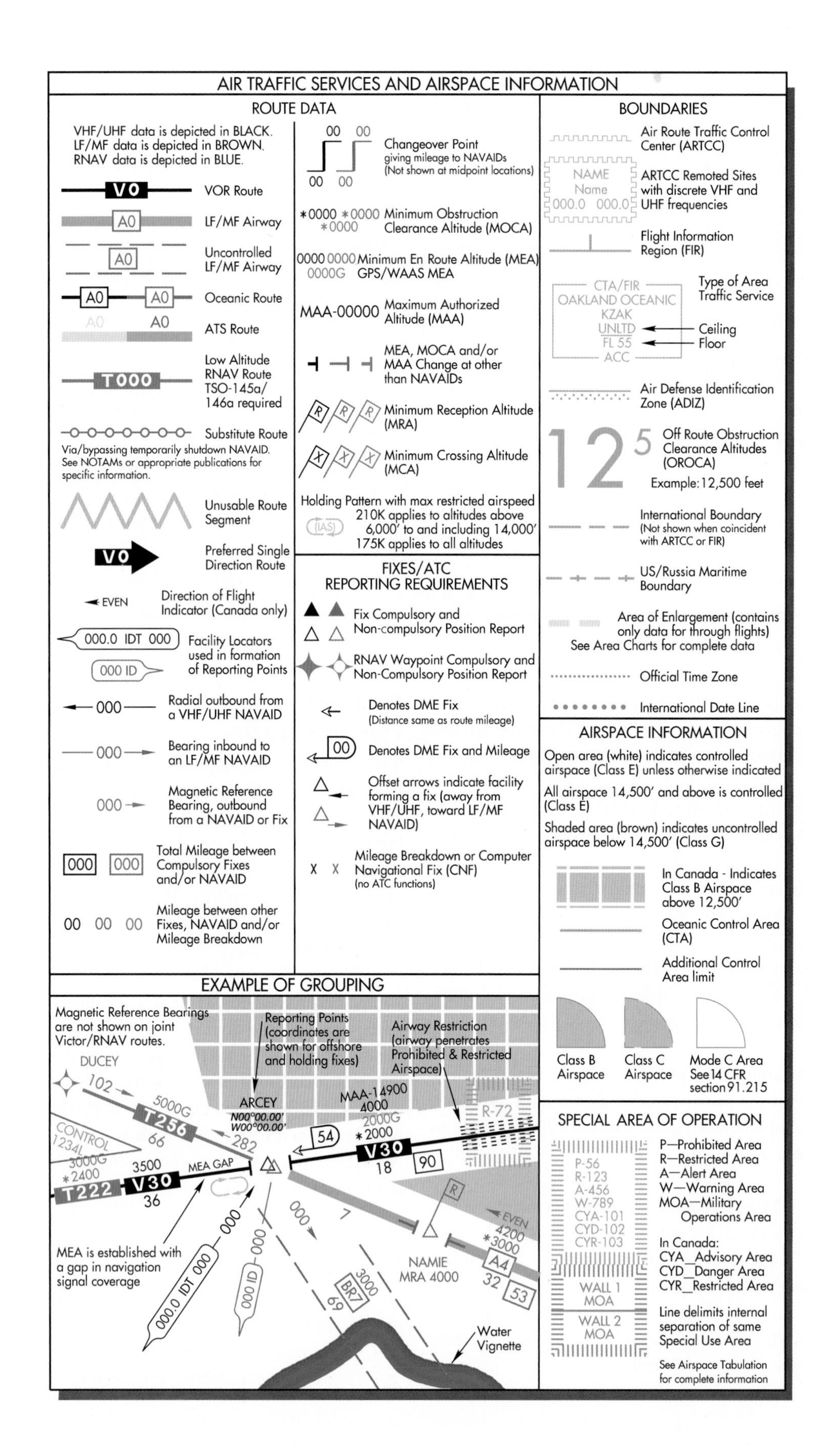

Figure 1-4. *Legend from en route low attitude chart, air traffic services and airspace information section.*

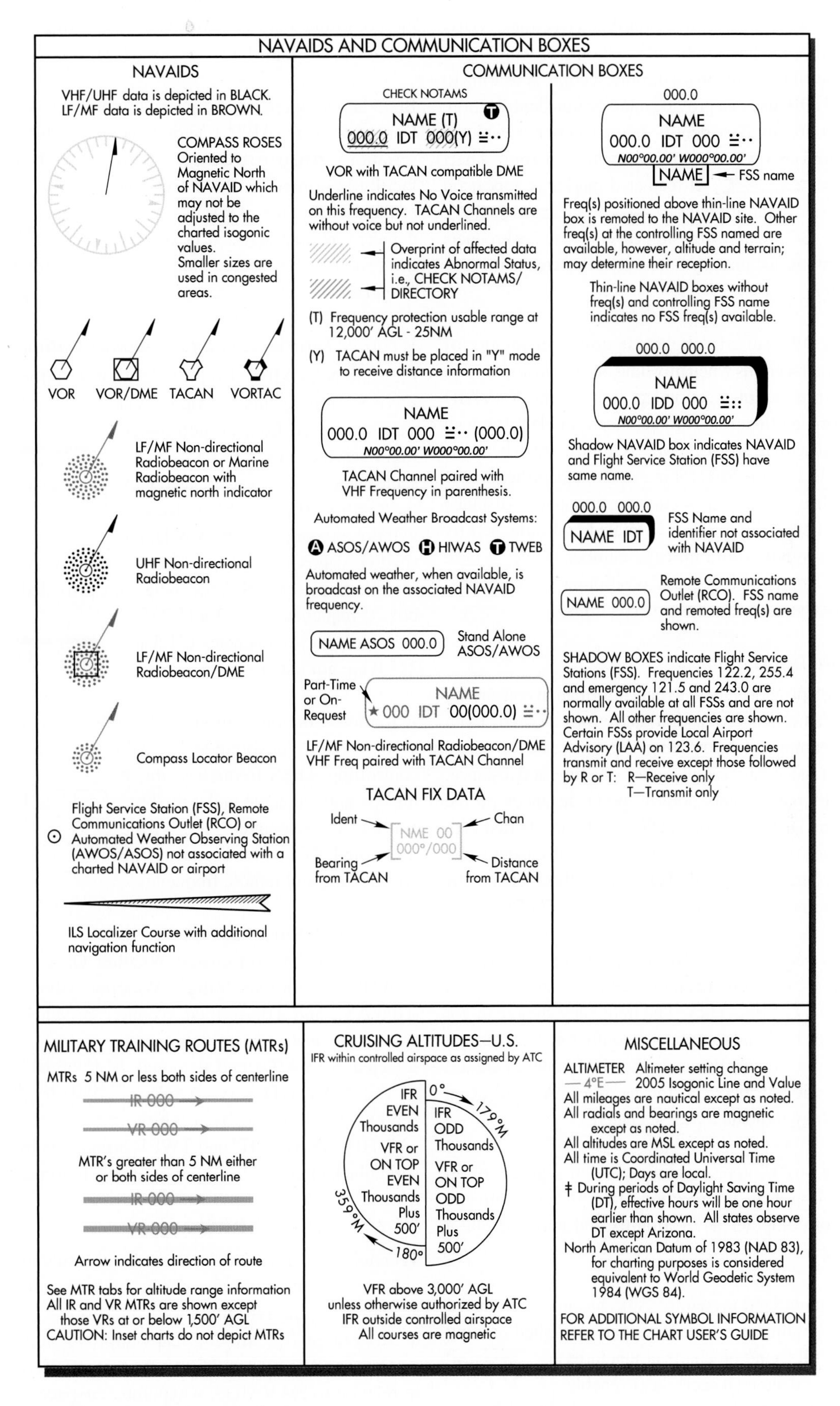

Figure 1-5. *Legend from en route low attitude chart.*

and off-route VORs are used to establish intersections. NDBs are sometimes collocated with intersections, in which case passage of the NDB would mark the intersection. A bearing to an off-route NDB also can provide intersection identification. A localizer course used to identify an intersection is depicted by a feathered arrowhead symbol on the en route chart ( ). If feathered markings appear on the left-hand side of the arrowhead ( ), a back course (BC) signal is transmitted. On AeroNav Products en route charts, the localizer symbol is only depicted to identify an intersection.

Off-route VORs remain the most common means of identifying intersections when traveling on an airway. Arrows depicted next to the intersection indicate the NAVAID to be used for identification. Another means of identifying an intersection is with the use of DME. A hollow arrowhead indicates DME is authorized for intersection identification. If the DME mileage at the intersection is a cumulative distance of route segments, the mileage is totaled and indicated by a D-shaped symbol with a mileage number inside 00. *[Figure 1-4]* Approved IFR global positioning system (GPS) units can also be used to report intersections.

Other Route Information

DME and GPS provide valuable route information concerning such factors as mileage, position, and ground speed. Even without this equipment, information is provided on the charts for making the necessary calculations using time and distance. The en route chart depicts point-to-point distances on the airway system. Distances from VOR to VOR are charted with a number inside of a box 000. To differentiate distances when two airways coincide, the word "TO" with the three-letter VOR identifier appear to the left of the distance boxes TO PDX 97.

VOR changeover points (COPs) are depicted on the charts by this symbol 45/53. The numbers indicate the distance at which to change the VOR frequency. The frequency change might be required due to signal reception or conflicting frequencies. If a COP does not appear on an airway, the frequency should be changed midway between the facilities. A COP at an intersection may indicate a course change.

Occasionally an "x" appears at a separated segment of an airway that is not an intersection. The "x" is a mileage breakdown or computer navigation fix and may indicate a course change.

Today's computerized system of ATC has greatly reduced the need for holding en route. However, published holding patterns are still found on charts at junctures where ATC has deemed it necessary to enable traffic flow. When a holding pattern is charted, the controller may provide the holding direction and the statement "as published." *[Figure 1-4]*

Boundaries separating the jurisdiction of Air Route Traffic Control Centers (ARTCC) are depicted on charts with blue serrations. The name of the controlling facility is printed on the corresponding side of the division line. ARTCC remote sites are depicted as blue serrated boxes and contain the center name, sector name, and the sector frequency. *[Figure 1-4]*

Weather Information and Communication Features

En route NAVAIDs also provide weather information and serve communication functions. When a NAVAID is shown as a shadowed box, an automated flight service station (AFSS) of the same name is directly associated with the facility. If an AFSS is located without an associated NAVAID, the shadowed box is smaller and contains only the name and identifier. The AFSS frequencies are provided above the box. (Frequencies 122.2 and 255.4, and emergency frequencies 121.5 and 243.0 are not listed.)

A Remote Communications Outlet (RCO) associated with a NAVAID is designated by a thin-lined box with the controlling AFSS frequency above the box and the name under the box. Without an associated facility, the thin-lined RCO box contains the AFSS name and remote frequency.

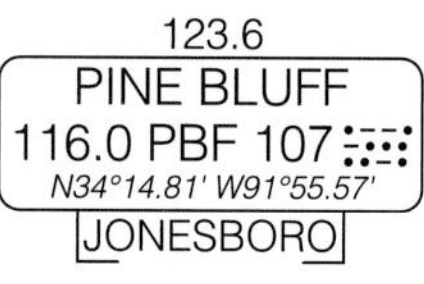

Automated Surface Observing Station (ASOS), Automated Weather Observing Station (AWOS), Hazardous Inflight Weather Advisory Service (HIWAS), and Transcribed Weather Broadcast (TWEB) are continuously transmitted over selected NAVAIDs and depicted in the NAVAID box. ASOS/AWOS are depicted by a white "A", HIWAS by a "H" and TWEB broadcasts by a "T" in a solid black circle in the upper right or left corner.

New Technologies

Technological advances have made multifunction displays and moving maps more common in newer aircraft. Even older aircraft are being retrofitted to include "glass" in the flight deck. *[Figure 1-6]* Moving maps improve pilot situational awareness (SA) by providing a picture of aircraft location in relation to NAVAIDS, waypoints, airspace, terrain, and

Figure 1-6. *Moving map display.*

hazardous weather. GPS systems can be certified for terminal area and en route use as well as approach guidance.

Additional breakthroughs in display technology are the new electronic chart systems or electronic flight bags that facilitate the use of electronic documents in the general aviation flight deck. *[Figure 1-7]* An electronic chart or flight bag is a self-powered electronic library that stores and displays en route charts and other essential documents on a screen. These electronic devices can store the digitized United States terminal procedures, en route charts, the complete A/FD, in addition to 14 CFR and the AIM. Full touch-screen based computers allow pilots to view airport approach and area charts electronically while flying. With FAA approval, an operator may replace paper charts as well as other paper materials including minimum equipment lists (MELs), standard operating procedures (SOPs), standard instrument departures (SIDs), standard terminal arrival routes (STARs), checklists, and flight deck manuals. As with paper flight publications, the electronic database needs to be current to provide accurate information regarding NAVAIDS, waypoints, and terminal procedures. Databases are updated every 28 days and are available from various commercial vendors. Pilots should be familiar with equipment operation, capabilities, and limitations prior to use.

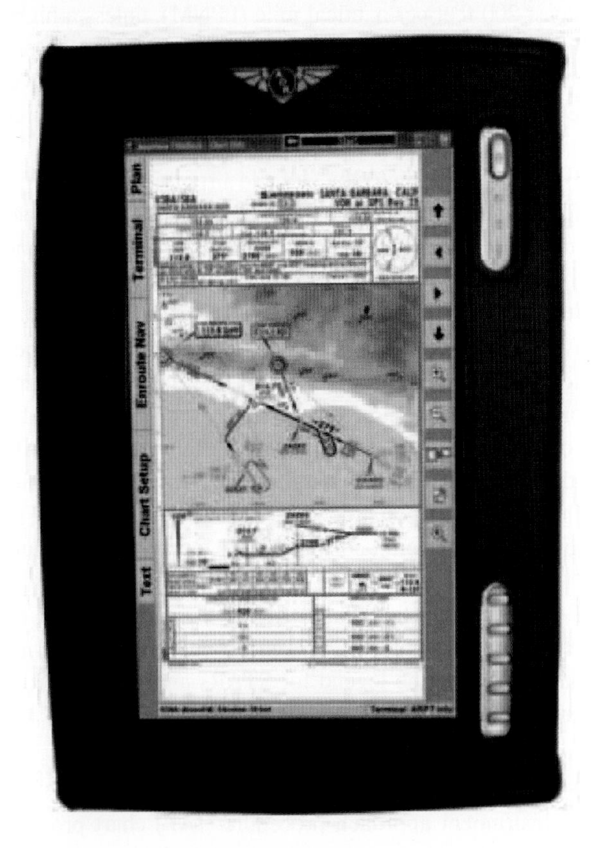

Figure 1-7. *Example of an electronic flight bag.*

Terminal Procedures Publications

While the en route charts provide the information necessary to safely transit broad regions of airspace, the United States Terminal Procedures Publication (TPP) enables pilots to guide their aircraft in the airport area. Whether departing or arriving, these procedures exist to make the controllers' and pilots' jobs safer and more efficient. Available in booklets by region (published by AeroNav Products), the TPP includes approach procedures, STARs, Departure Procedures (DPs), and airport diagrams.

Departure Procedures

There are two types of DPs: Obstacle Departure Procedures (ODP) and SIDs. *[Figure 1-8]* Both types of DPs provide obstacle clearance protection to aircraft in instrument meteorological conditions (IMC), while reducing communications and departure delays. DPs are published in text and/or charted graphic form. Regardless of the format, all DPs provide a way to depart the airport and transition to the en route structure safely. When possible, pilots are strongly encouraged to file and fly a DP at night, during marginal visual meteorological conditions (VMC) and IMC.

All DPs provide obstacle clearance provided the aircraft crosses the end of the runway at least 35 feet AGL; climbs to 400 feet above airport elevation before turning; and climbs at least 200 feet per nautical mile (FPNM), unless a higher climb gradient is specified to the assigned altitude. ATC may vector an aircraft off a previously assigned DP; however, the 200 FPNM or the FPNM specified in the DP is required.

Textual ODPs are listed by city and airport in the IFR Take-Off Minimums and DPs section of the TPP. SIDs are depicted in the TPP following the approach procedures for the airport.

Standard Terminal Arrival Routes

STARs depict prescribed routes to transition the instrument pilot from the en route structure to a fix in the terminal area from which an instrument approach can be conducted. If a pilot does not have the appropriate STAR, write "No STAR" in the flight plan. However, if the controller is busy, the pilot might be cleared along the same route and, if necessary, the controller has the pilot copy the entire text of the procedure.

STARs are listed alphabetically at the beginning of the AeroNav Products booklet. *Figure 1-9* shows an example of a STAR, and the legend for STARs and DPs printed in AeroNav Products booklets.

Instrument Approach Procedure Charts

The instrument approach procedure (IAP) chart provides the method to descend and land safely in low visibility conditions. The FAA establishes an IAP after thorough analyses of obstructions, terrain features, and navigational facilities. Maneuvers, including altitude changes, course corrections, and other limitations, are prescribed in the IAP. The approach charts reflect the criteria associated with the United States Standard for Terminal Instrument Approach Procedures (TERPs), which prescribes standardized methods for use in designing instrument flight procedures.

In addition to the AeroNav Products, other governmental and corporate entities produce approach procedures. The U.S. Military IAPs are established and published by the Department of Defense and are available to the public upon request. Special IAPs are approved by the FAA for individual operators and are not available to the general public. Foreign country standard IAPs are established and published according to the individual country's publication procedures. The information presented in the following sections highlight features of the United States TPP.

The instrument approach chart is divided into six main sections, which include the margin identification, pilot briefing (and notes), plan view, profile view, landing minimums, and airport diagram. *[Figure 1-10]* An examination of each section follows.

Margin Identification

The margin identification, at the top and bottom of the chart, depicts the airport location and procedure identification. The civil approach plates are organized by city, then airport name and state. For example, Orlando Executive in Orlando, Florida, is alphabetically listed under "O" for Orlando. Military approaches are organized by airport name first.

The chart's amendment status appears below the city and state in the bottom margin. The amendment number is followed by the five-digit julian-date of the last chart change."05300" is read, "the 300th day of 2005." At the center of the top margin is the FAA chart reference number and the approving authority. At the bottom center, the airport's latitude and longitude coordinates are provided. If a chart is original, the date of issuance can be used instead of the julian-date.

The procedure chart title (top and bottom margin area of *Figure 1-10*) is derived from the type of navigational facility providing final approach course guidance. A runway number is listed when the approach course is aligned within 30° of the runway centerline. This type of approach allows a straight-in landing under the right conditions. The type of approach followed by a letter identifies approaches that do not have straight-in landing minimums. Examples include procedure titles at the same airport, which have only circling minimums. The first approach of this type created

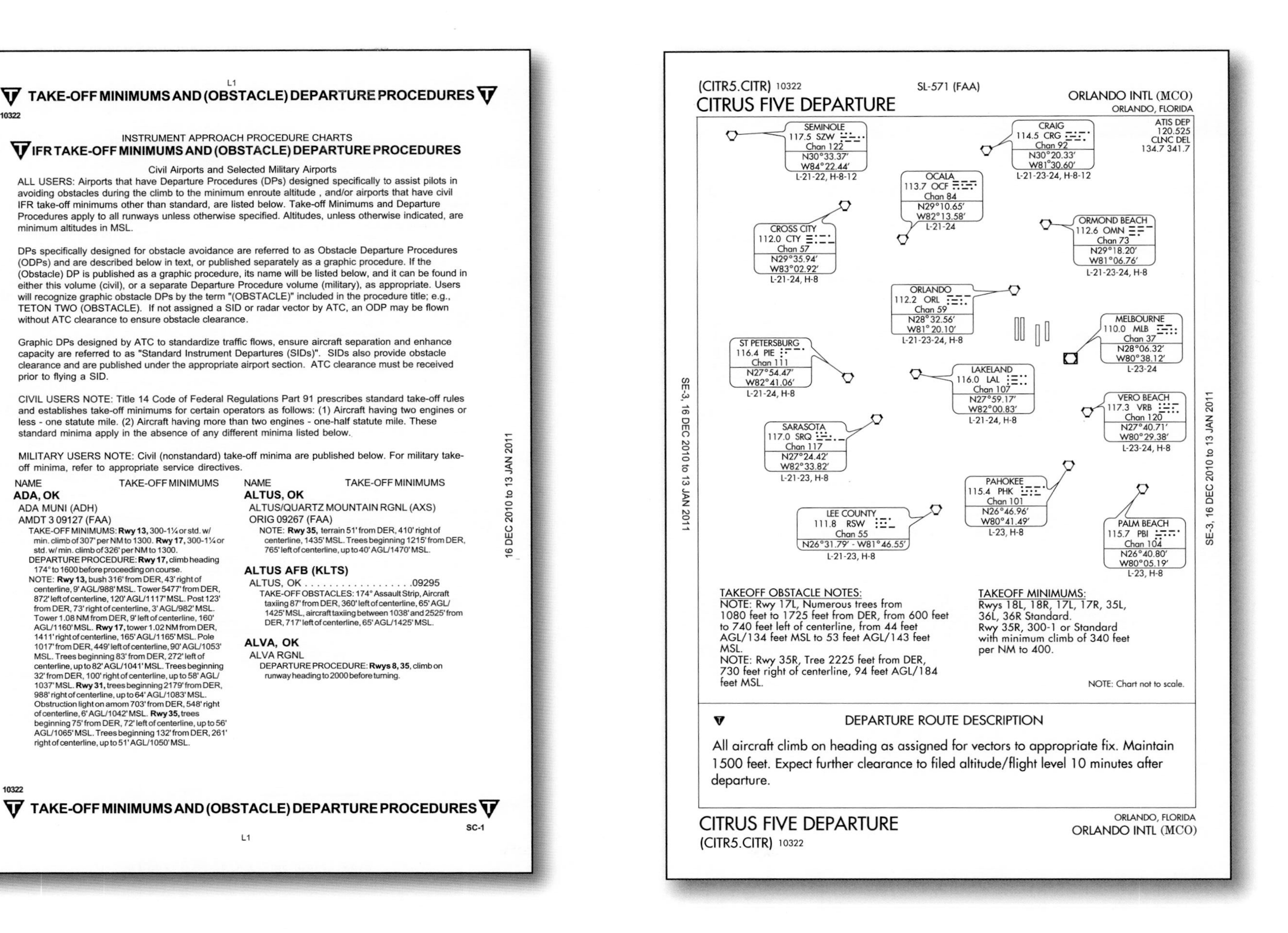

L1

TAKE-OFF MINIMUMS AND (OBSTACLE) DEPARTURE PROCEDURES

10322

INSTRUMENT APPROACH PROCEDURE CHARTS

IFR TAKE-OFF MINIMUMS AND (OBSTACLE) DEPARTURE PROCEDURES

Civil Airports and Selected Military Airports

ALL USERS: Airports that have Departure Procedures (DPs) designed specifically to assist pilots in avoiding obstacles during the climb to the minimum enroute altitude , and/or airports that have civil IFR take-off minimums other than standard, are listed below. Take-off Minimums and Departure Procedures apply to all runways unless otherwise specified. Altitudes, unless otherwise indicated, are minimum altitudes in MSL.

DPs specifically designed for obstacle avoidance are referred to as Obstacle Departure Procedures (ODPs) and are described below in text, or published separately as a graphic procedure. If the (Obstacle) DP is published as a graphic procedure, its name will be listed below, and it can be found in either this volume (civil), or a separate Departure Procedure volume (military), as appropriate. Users will recognize graphic obstacle DPs by the term "(OBSTACLE)" included in the procedure title; e.g., TETON TWO (OBSTACLE). If not assigned a SID or radar vector by ATC, an ODP may be flown without ATC clearance to ensure obstacle clearance.

Graphic DPs designed by ATC to standardize traffic flows, ensure aircraft separation and enhance capacity are referred to as "Standard Instrument Departures (SIDs)". SIDs also provide obstacle clearance and are published under the appropriate airport section. ATC clearance must be received prior to flying a SID.

CIVIL USERS NOTE: Title 14 Code of Federal Regulations Part 91 prescribes standard take-off rules and establishes take-off minimums for certain operators as follows: (1) Aircraft having two engines or less - one statute mile. (2) Aircraft having more than two engines - one-half statute mile. These standard minima apply in the absence of any different minima listed below.

MILITARY USERS NOTE: Civil (nonstandard) take-off minima are published below. For military take-off minima, refer to appropriate service directives.

NAME TAKE-OFF MINIMUMS

ADA, OK

ADA MUNI (ADH)

AMDT 3 09127 (FAA)

TAKE-OFF MINIMUMS: **Rwy 13,** 300-1¼ or std. w/ min. climb of 307' per NM to 1300. **Rwy 17,** 300-1¼ or std. w/ min. climb of 326' per NM to 1300.

DEPARTURE PROCEDURE: **Rwy 17,** climb heading 174° to 1600 before proceeding on course.

NOTE: **Rwy 13,** bush 316' from DER, 43' right of centerline, 9' AGL/988' MSL. Tower 5477' from DER, 872' left of centerline, 120' AGL/1117' MSL. Post 123' from DER, 73' right of centerline, 3' AGL/982' MSL. Tower 1.08 NM from DER, 9' left of centerline, 160' AGL/1160' MSL. **Rwy 17,** tower 1.02 NM from DER, 1411' right of centerline, 165' AGL/1165' MSL. Pole 1017' from DER, 449' left of centerline, 90' AGL/1053' MSL. Trees beginning 83' from DER, 272' left of centerline, up to 82' AGL/1041' MSL. Trees beginning 32' from DER, 100' right of centerline, up to 58' AGL/1037' MSL. **Rwy 31,** trees beginning 2179' from DER, 988' right of centerline, up to 64' AGL/1083' MSL. Obstruction light on amom 703' from DER, 548' right of centerline, 6' AGL/1042' MSL. **Rwy 35,** trees beginning 75' from DER, 72' left of centerline, up to 56' AGL/1065' MSL. Trees beginning 132' from DER, 261' right of centerline, up to 51' AGL/1050' MSL.

NAME TAKE-OFF MINIMUMS

ALTUS, OK

ALTUS/QUARTZ MOUNTAIN RGNL (AXS)

ORIG 09267 (FAA)

NOTE: **Rwy 35,** terrain 51' from DER, 410' right of centerline, 1435' MSL. Trees beginning 1215' from DER, 765' left of centerline, up to 40' AGL/1470' MSL.

ALTUS AFB (KLTS)

ALTUS, OK 09295

TAKE-OFF OBSTACLES: 174° Assault Strip, Aircraft taxiing 87' from DER, 360' left of centerline, 65' AGL/1425' MSL, aircraft taxiing between 1038' and 2525' from DER, 717' left of centerline, 65' AGL/1425' MSL.

ALVA, OK

ALVA RGNL

DEPARTURE PROCEDURE: **Rwys 8, 35**, climb on runway heading to 2000 before turning.

10322

TAKE-OFF MINIMUMS AND (OBSTACLE) DEPARTURE PROCEDURES

SC-1

L1

16 DEC 2010 to 13 JAN 2011

Figure 1-8. *Obstacle departure procedures (ODP) and standard instrument departures (SID).*

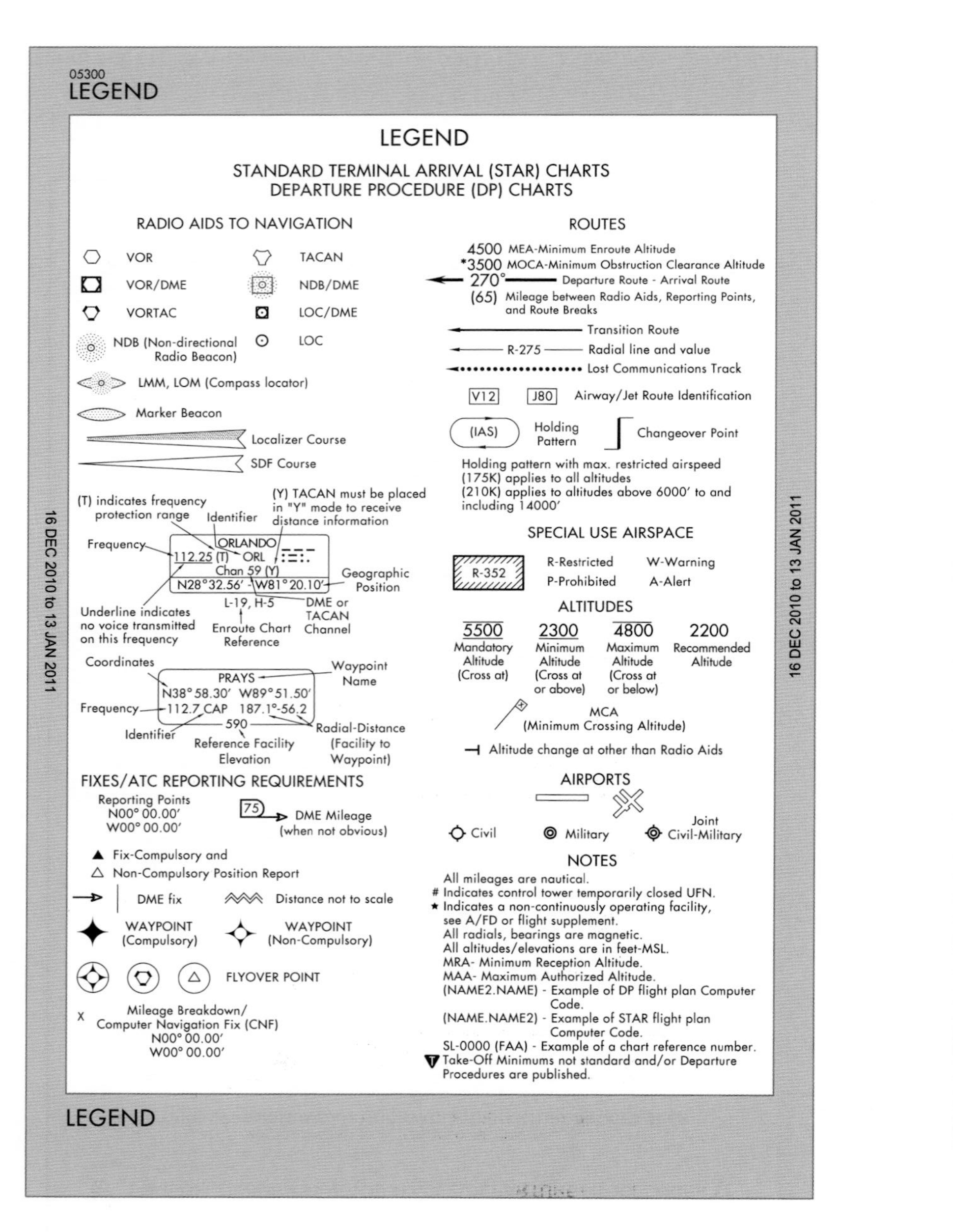

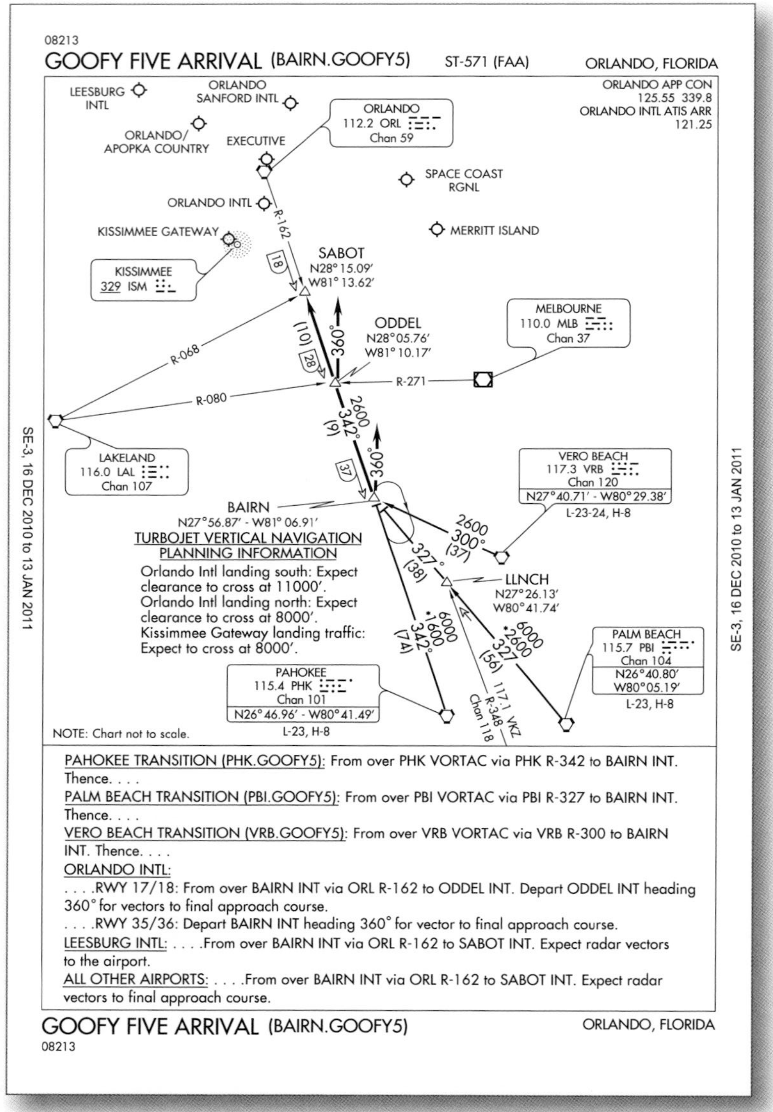

Figure 1-9. *DP chart legend and STAR.*

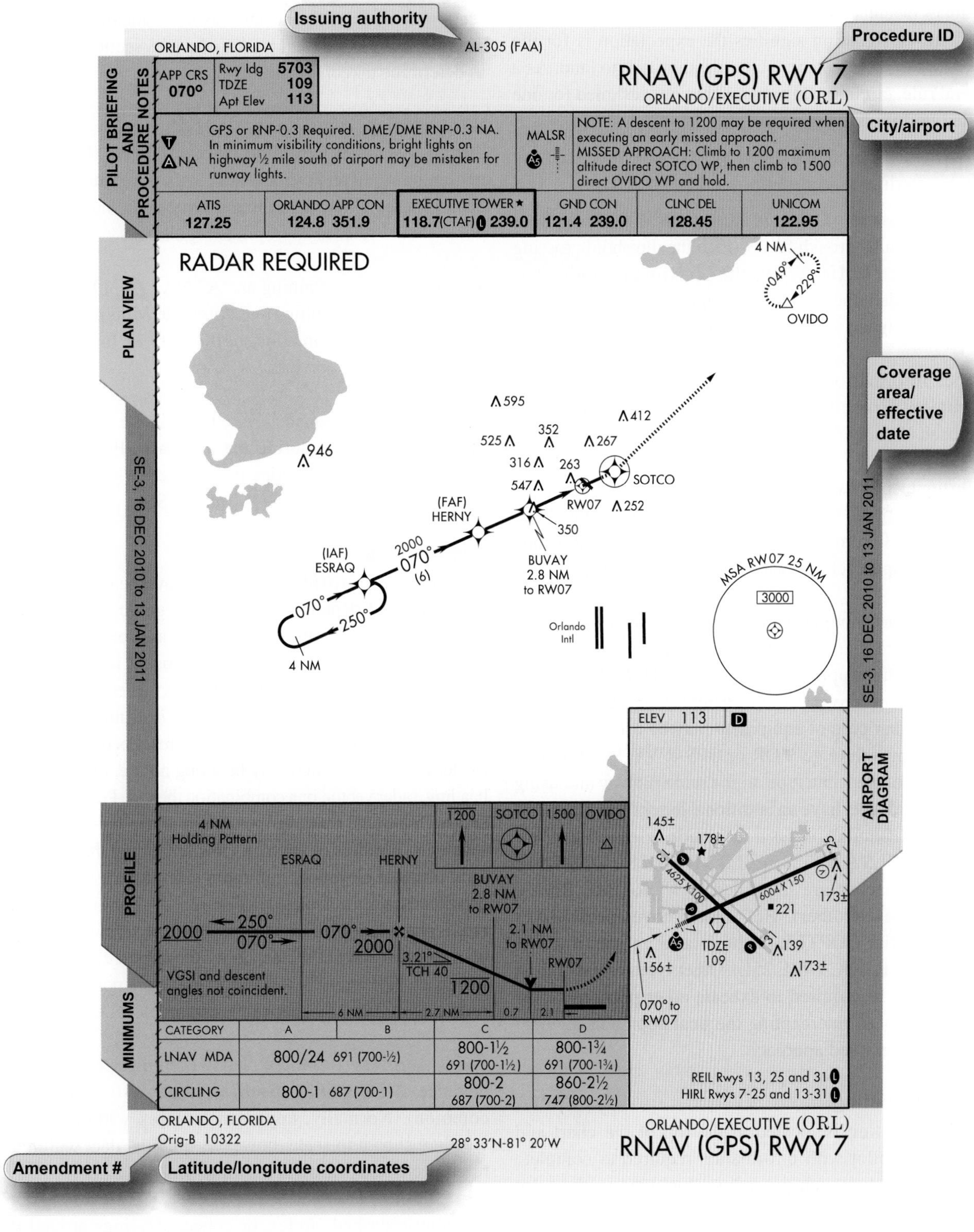

CATEGORY	A	B	C	D
LNAV MDA	800/24 691 (700-½)		800-1½ 691 (700-1½)	800-1¾ 691 (700-1¾)
CIRCLING	800-1 687 (700-1)		800-2 687 (700-2)	860-2½ 747 (800-2½)

Figure 1-10. *Instrument approach chart.*

at the airport is labeled with the letter A, and the lettering continues in alphabetical order (e.g., "VOR-A or "LDA-B"). The letter designation signifies the expectation is for the procedure to culminate in a circling approach to land. As a general rule, circling-only approaches are designed for one of the two following reasons:

- The final approach course alignment with the runway centerline exceeds 30°.
- The descent gradient is greater than 400 FPNM from the final approach fix (FAF) to the threshold crossing height (TCH). When this maximum gradient is exceeded, the circling-only approach procedure may be designed to meet the gradient criteria limits.

Further information on this topic can be found in the Instrument Procedures Handbook, Chapter 4, under Approach Naming Chart Conventions.

To distinguish between the left, right, and center runways, an "L," "R," or "C" follows the runway number (e.g., "ILS RWY 16R"). In some cases, an airport might have more than one circling approach, shown as VOR-A, VOR/DME-B, etc.

More than one navigational system separated by a slash indicates more than one type of equipment is required to execute the final approach (e.g., VOR/DME RWY 31). More than one navigational system separated by "or" indicates either type of equipment may be used to execute the final approach (e.g., VOR or GPS RWY 15). Multiple approaches of the same type, to the same runway and using the same guidance, have an additional letter from the end of the alphabet, number, or term in the title (e.g., ILS Z RWY 28, SILVER ILS RWY 28, or ILS 2 RWY 28). VOR/DME RNAV approaches are identified as VOR/DME RNAV RWY (runway number). Helicopters have special IAPs designated with COPTER in the procedure identification (e.g., COPTER LOC/DME 25L). Other types of navigation systems may be required to execute other portions of the approach prior to intercepting the final approach segment or during the missed approach.

The Pilot Briefing

The pilot briefing is located at the top of the chart and provides the pilot with information required to complete the published approach procedure. Included in the pilot briefing are the NAVAID providing approach guidance, its frequency, the final approach course, and runway information. A notes section contains additional procedural information. For example, a procedural note might indicate restrictions for circling maneuvers. Some other notes might concern a local altimeter setting and the resulting change in the minimums. The use of RADAR may also be noted in this section. Additional notes may be found in the plan view.

When a triangle containing a "T" (T) appears in the notes section, it signifies the airport has nonstandard IFR takeoff minimums. Pilots should refer to the DPs section of the TPP to determine takeoff minimums.

When a triangle containing an "A" (A) appears in the notes section, it signifies the airport has nonstandard IFR alternate minimums. Civil pilots should refer to the Alternate Minimums Section of the TPP to determine alternate minimums. Military pilots should refer to appropriate regulations.

When a triangle containing an "A" NA (A NA) appears in the notes area, it signifies that Alternate Minimums are Not Authorized due to unmonitored facility or the absence of weather reporting service.

Communication frequencies are listed in the order in which they would be used during the approach. Frequencies for weather and related facilities are included, where applicable, such as ATIS, ASOS, AWOS, and AFSSs.

The Plan View

The plan view provides a graphical overhead view of the procedure and depicts the routes that guide the pilot from the en route segments to the initial approach fix (IAF). *[Figure 1-10]* During the initial approach, the aircraft has departed the en route phase of flight and is maneuvering to enter an intermediate or final segment of the instrument approach. An initial approach can be made along prescribed routes within the terminal area, which may be along an arc, radial, course, heading, radar vector, or a combination thereof. Procedure turns and high-altitude teardrop penetrations are initial approach segments. Features of the plan view, including the procedure turn, obstacle elevation, minimum safe altitude (MSA), and procedure track are depicted in *Figure 1-11*. Terrain is depicted in the plan view portion of all IAPs if the terrain within the plan view exceeds 4,000 feet above the airport elevation, or if within a 6 NM radius of the airport reference point the terrain rises at least 2,000 feet above the airport elevation.

Some AeroNav Products charts contain a reference or distance circle with a specified radius (10 NM is most common). Normally, approach features within the plan view are shown to scale; however, only the data within the reference circle is always drawn to scale.

Concentric dashed circles, or concentric rings around the distance circle, are used when the information necessary to the procedure will not fit to scale within the limits of the plan view area. They serve as a means to systematically arrange this information in its relative position outside and beyond the reference circle. These concentric rings are labeled en route facilities and feeder facilities.

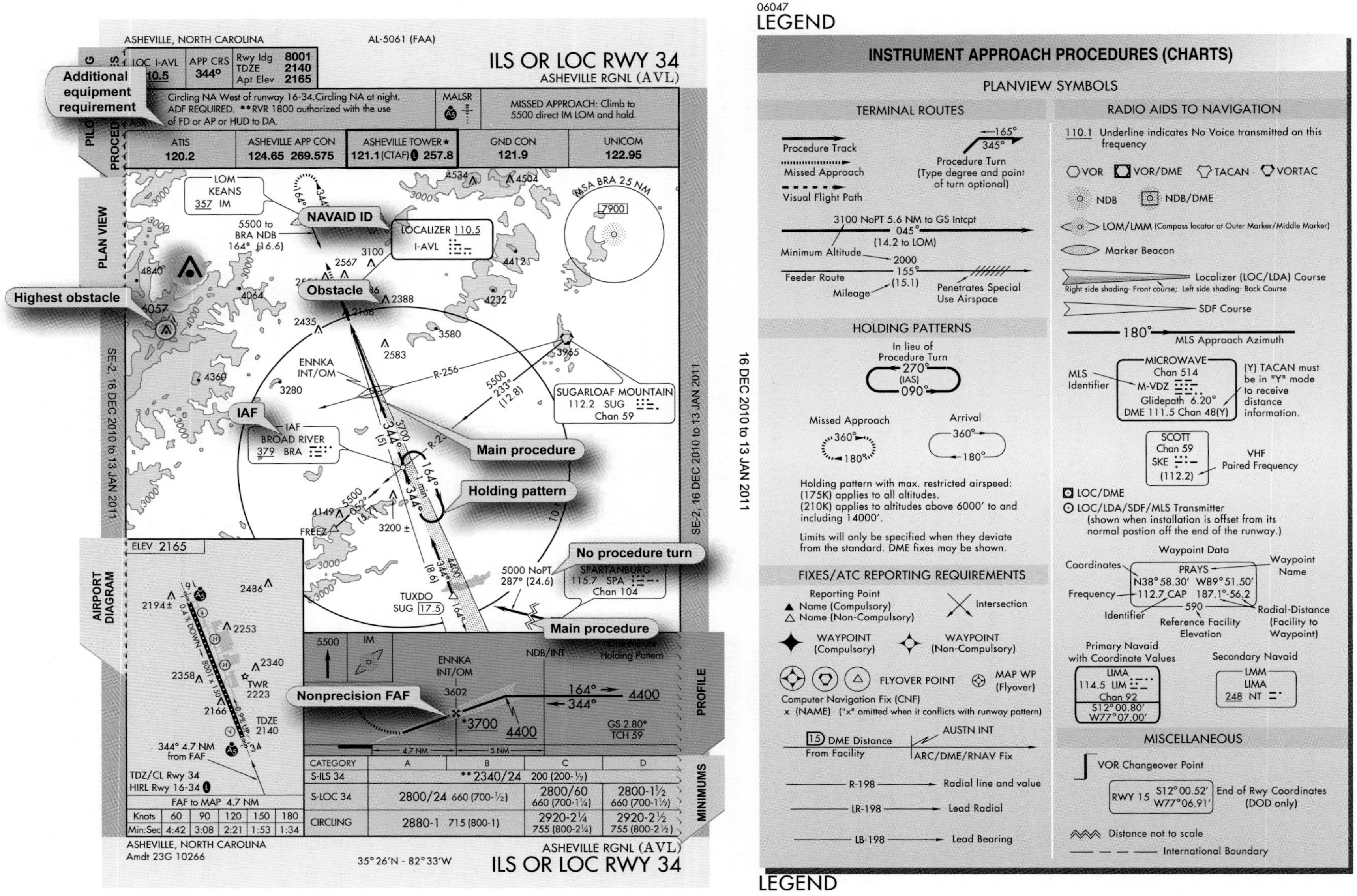

Figure 1-11. *IAP plan view and symbol legends.*

The primary airport depicted in the plan view is drawn with enough detail to show the runway orientation and final approach course alignment. Airports other than the primary approach airport are not normally depicted in the AeroNav Products plan view.

Known spot elevations are indicated on the plan view with a dot in MSL altitude. The largest dot and number combination indicates the highest elevation. An inverted "V" with a dot in the center depicts an obstacle (Λ). The highest obstacle is indicated with a bolder, larger version of the same symbol. *[Figure 1-11]*

The MSA circle appears in the plan view, except in approaches for which the Terminal Arrival Area (TAA) format is used or appropriate NAVAIDs (e.g., VOR or NDB) are unavailable. The MSA is provided for emergency purposes only and guarantees 1,000 feet obstruction clearance in the sector indicated with reference to the bearings in the circle. For conventional navigation systems, the MSA is normally based on the primary omnidirectional facility (NAVAID) on which the IAP is predicated. The MSA depiction on the approach chart contains the facility identifier of the NAVAID used to determine the MSA altitudes. For RNAV approaches, the MSA is based on the runway waypoint for straight-in approaches or the airport waypoint for circling approaches. For GPS approaches, the MSA center header is the missed approach waypoint. The MSL altitudes appear in boxes within the circle, which is typically a 25 NM radius unless otherwise indicated. The MSA circle header refers to the letter identifier of the NAVAID or waypoint that describes the center of the circle.

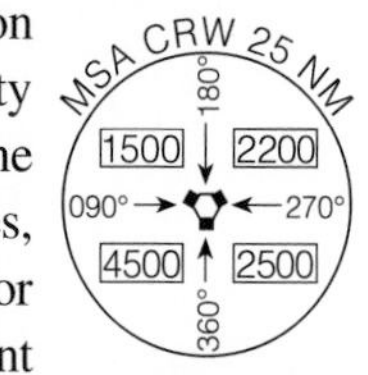

NAVAIDs necessary for the completion of the instrument procedure include the facility name, letter identifier, and Morse code sequence. They may also furnish the frequency, Morse code, and channel. A heavy-lined NAVAID box depicts the primary NAVAID used for the approach. An "I" in front of the NAVAID identifier (in *Figure 1-11*, "I-AVL") listed in the NAVAID box indicates a localizer. The requirement for an ADF, DME, or RADAR in the approach is noted in the plan view.

Intersections, fixes, radials, and course lines describe route and approach sequencing information. The main procedure or final approach course is a thick, solid line (▬▬▬). A DME arc, which is part of the main procedure course, is also represented as a thick, solid line (▬▬▬). A feeder route is depicted with a medium line (———) and provides heading, altitude, and distance information. (All three components must be designated on the chart to provide a navigable course.) Radials, such as lead radials, are shown by thin lines (______). The missed approach track is drawn using a thin, hash marked line with a directional arrow (ıııııııııııııııııııııııııı►). A visual flightpath segment appears as a thick dashed line with a directional arrow (- - - - - - - -> Visual Flightpath). IAFs are charted IAF when associated with a NAVAID or when freestanding.

The missed approach holding pattern track is represented with a thin, dashed line. When collocated, the missed approach holding pattern and procedure turn holding pattern are indicated as a solid, black line. Arrival holding patterns are depicted as thin, solid lines.

Terminal Arrival Area (TAA)

The design objective of the TAA procedure is to provide a transition method for arriving aircraft with GPS/RNAV equipment. TAAs also eliminate or reduce the need for feeder routes, departure extensions, and procedure turns or course reversal. The TAA is controlled airspace established in conjunction with the standard or modified RNAV approach configurations.

The standard TAA has three areas: straight-in, left base, and right base. The arc boundaries of the three areas of the TAA are published portions of the approach and allow aircraft to transition from the en route structure direct to the nearest IAF. When crossing the boundary of each of these areas or when released by ATC within the area, the pilot is expected to proceed direct to the appropriate waypoint IAF for the approach area being flown. A pilot has the option in all areas of proceeding directly to the holding pattern.

The TAA has a "T" structure that normally provides a No Procedure Turn (NoPT) for aircraft using the approach. *[Figure 1-12]* The TAA provides the pilot and air traffic controller with an efficient method for routing traffic from the en route to the terminal structure. The basic "T" contained in the TAA normally aligns the procedure on runway centerline with the missed approach point (MAP) located at the threshold, the FAF 5 NM from the threshold, and the intermediate fix (IF) 5 NM from the FAF.

In order to accommodate descent from a high en route altitude to the initial segment altitude, a hold in lieu of a procedure turn provides the aircraft with an extended distance for the necessary descent gradient. The holding pattern constructed for this purpose is always established on the center IAF waypoint. Other modifications may be required for parallel runways or special operational requirements. When published, the RNAV chart depicts the TAA through the use of icons representing each TAA associated with the RNAV procedure. These icons are depicted in the plan view of the approach, generally arranged on the chart in accordance with their position relative to the aircraft's arrival from the en route structure.

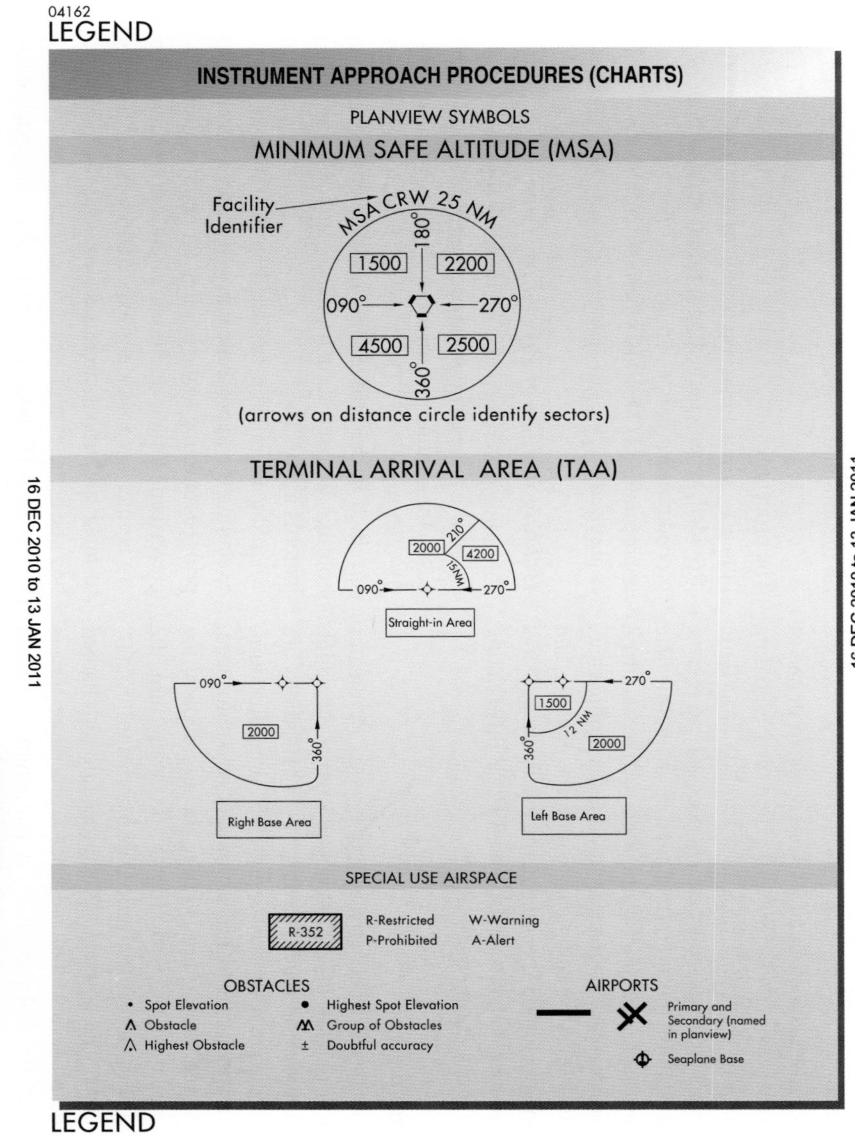

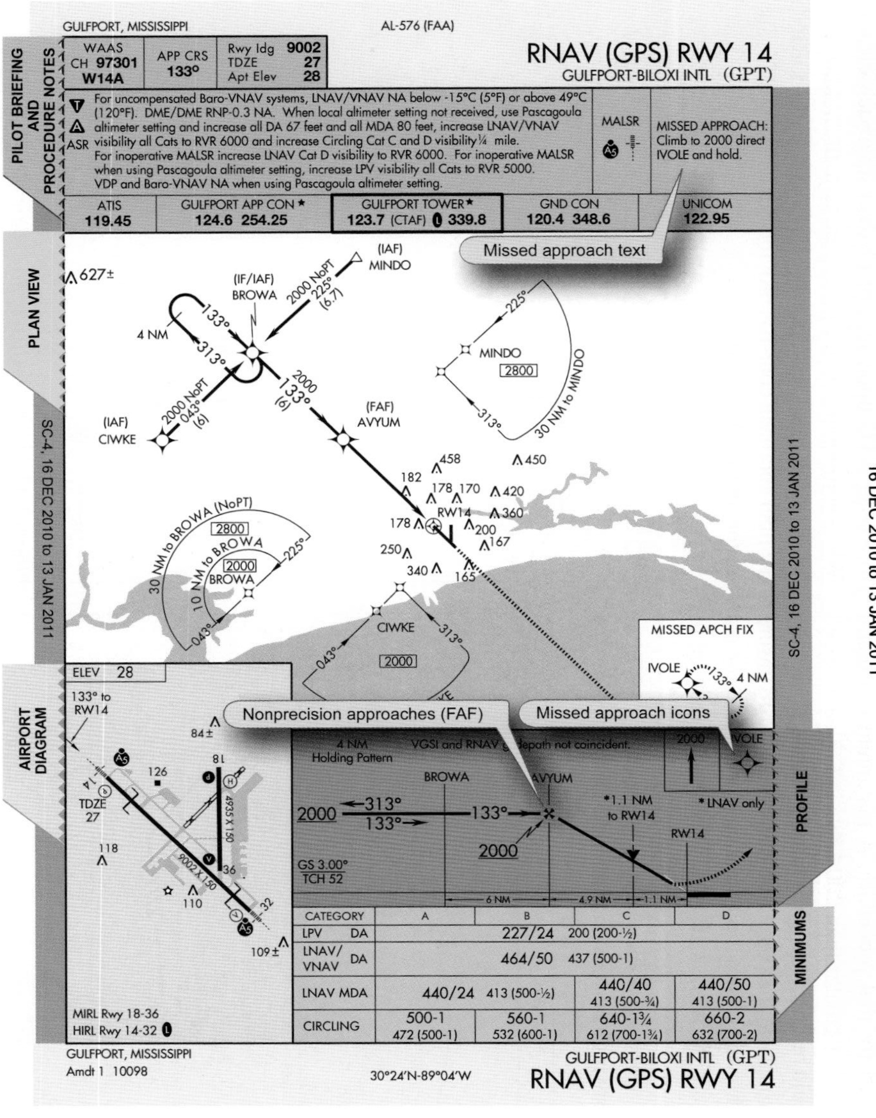

Figure 1-12. *Basic "T" design of terminal arrival area (TAA) and legend.*

Course Reversal Elements in Plan View and Profile View

Course reversals included in an IAP are depicted in one of three different ways: a 45°/180° procedure turn, a holding pattern in lieu of procedure turn, or a teardrop procedure. The maneuvers are required when it is necessary to reverse direction to establish the aircraft inbound on an intermediate or final approach course. Components of the required procedure are depicted in the plan view and the profile view. The maneuver must be completed within the distance and at the minimum altitude specified in the profile view. Pilots should coordinate with the appropriate ATC facility relating to course reversal during the IAP.

Procedure Turns

A procedure turn barbed arrow (045° 225° 090°) indicates the direction or side of the outbound course on which the procedure turn is made. *[Figure 1-13]* Headings are provided for course reversal using the 45° procedure turn. However, the point at which the turn may be commenced, and the type and rate of turn is left to the discretion of the pilot. Some of the options are the 45° procedure turn, the racetrack pattern, the teardrop procedure turn, or the 80°/260° course reversal. The absence of the procedure turn barbed arrow in the plan view indicates that a procedure turn is not authorized for that procedure. A maximum procedure turn speed of not greater than 200 knots indicated airspeed (KIAS) should be observed when turning outbound over the IAF and throughout the procedure turn maneuver to ensure staying within the obstruction clearance area. The normal procedure turn distance is 10 NM. This may be reduced to a minimum of 5 NM where only Category A or helicopter aircraft are operated, or increased to as much as 15 NM to accommodate high performance aircraft. Descent below the procedure turn altitude begins after the aircraft is established on the inbound course.

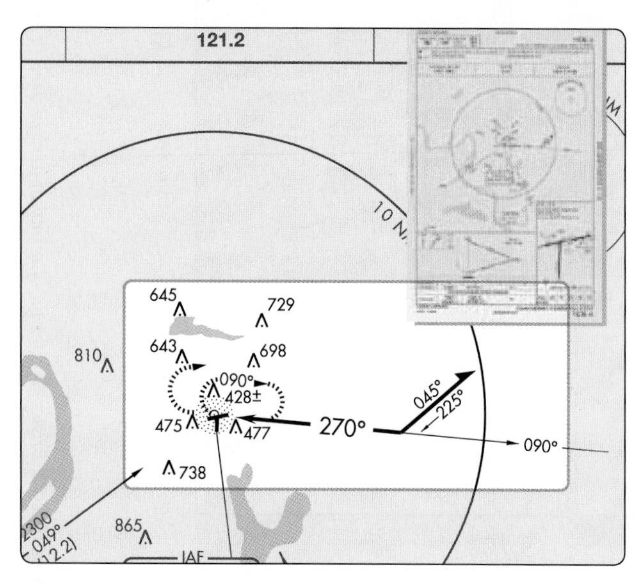

Figure 1-13. *45° procedure turn.*

The procedure turn is *not* required when the symbol "NoPT" appears, when radar vectoring to the final approach is provided, when conducting a timed approach, or when the procedure turn is not authorized. Pilots should contact the appropriate ATC facility when in doubt if a procedure turn is required.

Holding in Lieu of Procedure Turn

A holding pattern in lieu of a procedure turn may be specified for course reversal in some procedures. *[Figure 1-14]* In such cases, the holding pattern is established over an intermediate fix (IF) or a FAF. The holding pattern distance or time specified in the profile view must be observed. Maximum holding airspeed limitations as set forth for all holding patterns apply. The holding pattern maneuver is completed when the aircraft is established on the inbound course after executing the appropriate entry. If cleared for the approach prior to returning to the holding fix and the aircraft is at the prescribed altitude, additional circuits of the holding pattern are neither necessary nor expected by ATC. If pilots elect to make additional circuits to lose excessive altitude or to become better established on course, it is their responsibility to advise ATC upon receipt of their approach clearance. When holding in lieu of a procedure turn, the holding pattern must be followed, except when RADAR VECTORING to the final approach course is provided or when NoPT is shown on the approach course.

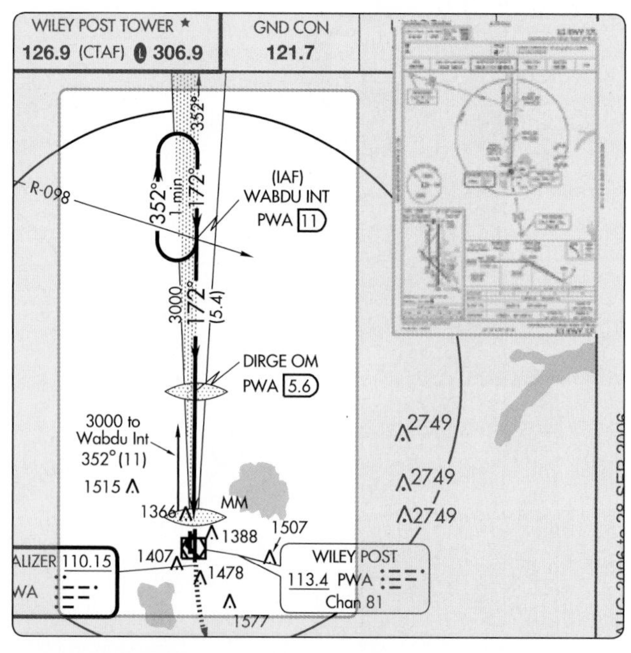

Figure 1-14. *Holding in lieu of procedure turn.*

Teardrop Procedure

When a teardrop procedure turn is depicted and a course reversal is required, unless otherwise authorized by ATC, this type of procedure must be executed. *[Figure 1-15]* The teardrop procedure consists of departure from an IAF on the published outbound course followed by a turn toward and intercepting the inbound course at or prior to the intermediate fix or point. Its purpose is to permit an aircraft to reverse direction and lose considerable altitude within reasonably limited airspace. Where no fix is available to mark the beginning of the intermediate segment, it shall be assumed to commence at a point 10 NM prior to the FAF. When the facility is located on the airport, an aircraft is considered to be on final approach upon completion of the penetration turn. However, the final approach segment begins on the final approach course 10 NM from the facility.

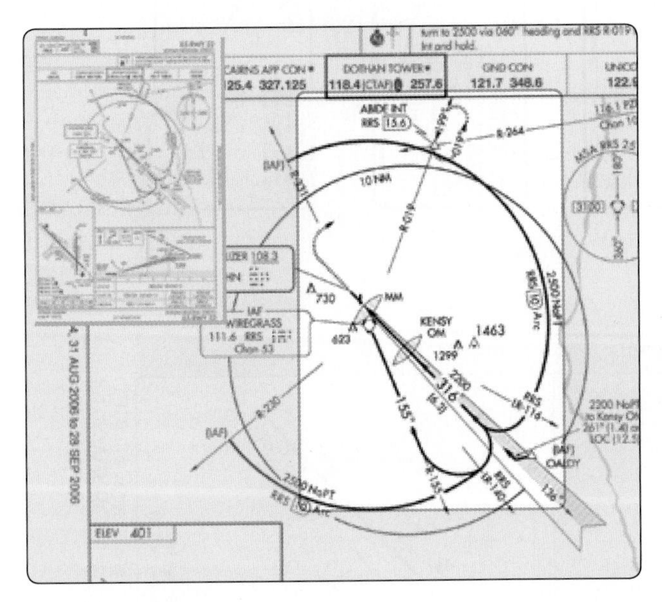

Figure 1-15. *Teardrop procedure.*

The Profile View

The profile view is a depiction of the procedure from the side and illustrates the vertical approach path altitudes, headings, distances, and fixes. *[Figures 1-10, 1-11,* and *1-12]* The view includes the minimum altitude and the maximum distance for the procedure turn, altitudes over prescribed fixes, distances between fixes, and the missed approach procedure. The profile view aids in the pilot's interpretation of the IAP. The profile view is not drawn to scale. *[Figures 1-10, 1-11, 1-12,* and *1-16]*

The precision approach glideslope (GS) intercept altitude is a minimum altitude for GS interception after completion of the procedure turn, illustrated by an altitude number and "zigzag" line. It applies to precision approaches, and except where otherwise prescribed, also applies as a minimum altitude for crossing the FAF when the GS is inoperative or not used. Precision approach profiles also depict the GS angle of descent, threshold crossing height (TCH), and GS altitude at the outer marker (OM).

For nonprecision approaches, a final descent is initiated and the final segment begins at either the FAF or the final approach point (FAP). The FAF is identified by use of the Maltese cross symbol in the profile view (✖). *[Figure 1-11]* When no FAF is depicted, the final approach point is the point at which the aircraft is established inbound on the final approach course. *[Figure 1-16]*

Stepdown fixes in nonprecision procedures are provided between the FAF and the airport for authorizing a lower minimum descent altitude (MDA) after passing an obstruction. Stepdown fixes can be identified by NAVAID, NAVAID fix, waypoint, or radar and are depicted by a hash marked line (¦). Normally, there is only one stepdown fix between the FAF and the MAP, but there can be several. If the stepdown fix cannot be identified for any reason, the minimum altitude at the stepdown fix becomes the MDA for the approach. However, circling minimums apply if they are higher than the stepdown fix minimum altitude, and a circling approach is required.

The visual descent point (VDP) is a defined point on the final approach course of a nonprecision straight-in approach procedure. A normal descent from the MDA to the runway touchdown point may be commenced, provided visual reference is established. The VDP is identified on the profile view of the approach chart by the symbol "V." *[Figure 1-12]*

The MAP varies depending upon the approach flown. For the ILS, the MAP is at the decision altitude/decision height (DA/DH). For nonprecision procedures, the pilot determines the MAP by timing from FAF when the approach aid is away from the airport, by a fix or NAVAID when the navigation facility is located on the field, or by waypoints as defined by GPS or VOR/DME RNAV. The pilot may execute the MAP early, but pilots should, unless otherwise cleared by ATC, fly the IAP as specified on the approach plate to the MAP at or above the MDA or DA/DH before executing a turning maneuver.

A complete description of the MAP appears in the pilot briefing section. *[Figure 1-16]* Icons indicating what is to be accomplished at the MAP are located in the profile view. When initiating a missed approach, the pilot is directed to climb straight ahead (e.g., "Climb to 2,000") or commence a turning climb to a specified altitude (e.g., "Climbing right turn to 2,000."). In some cases, the procedure directs the pilot to climb straight ahead to an initial altitude, then turn or enter

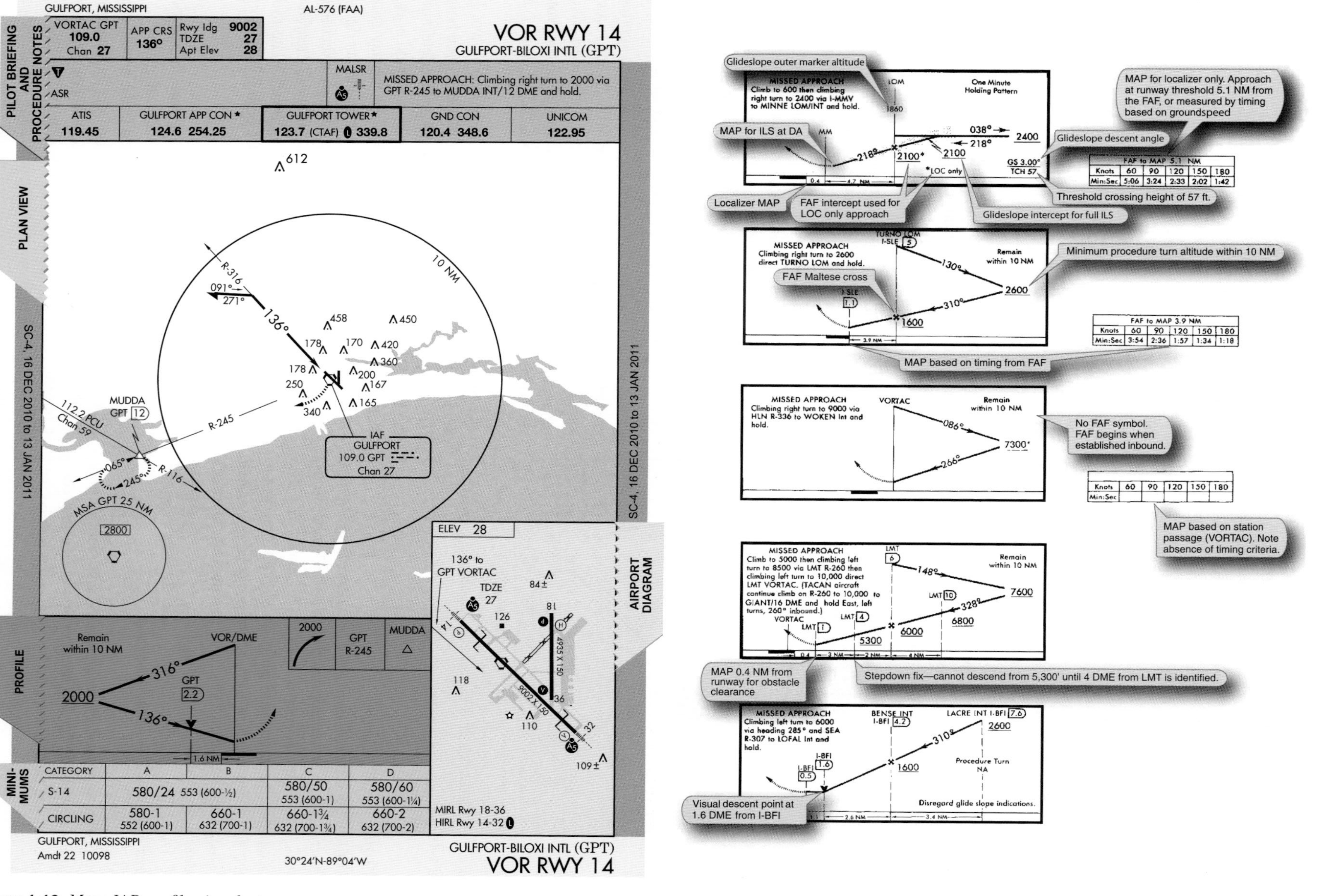

Figure 1-16. *More IAP profile view features.*

a climbing turn to the holding altitude (e.g., "Climb to 900, then climbing right turn to 2,500 direct ABC VOR and hold.")

When the MAP specifies holding at a facility or fix, the pilot proceeds according to the missed approach track and pattern depicted on the plan view. An alternate MAP may also be issued by ATC. The textual description also specifies the NAVAID(s) or radials that identify the holding fix.

The profile view also depicts minimum, maximum, recommended, and mandatory block altitudes used in approaches. The minimum altitude is depicted with the altitude underscored (2500). On final approach, aircraft are required to maintain an altitude at or above the depicted altitude until reaching the subsequent fix. The maximum altitude is depicted with the altitude overscored (4300), and aircraft must remain at or below the depicted altitude. Mandatory altitudes are depicted with the altitude both underscored and overscored (5500), and altitude is to be maintained at the depicted value. Recommended altitudes are advisory altitudes and are neither over- nor underscored. When an over- or underscore spans two numbers, a mandatory block altitude is indicated, and aircraft are required to maintain altitude within the range of the two numbers. *[Figures 1-11* and *1-12]*

The Vertical Descent Angle (VDA) found on nonprecision approach charts provides the pilot with information required to establish a stabilized approach descent from the FAF or stepdown fix to the TCH. *[Figure 1-17]* Pilots can use the published angle and estimated or actual groundspeed to find a target rate of descent using the rate of descent table in the back of the TPP.

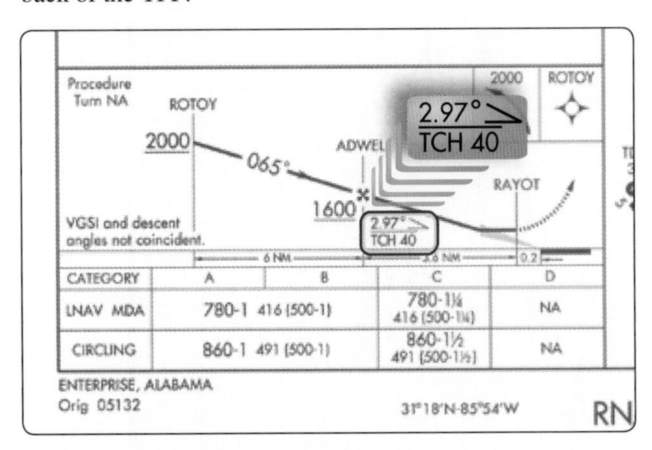

Figure 1-17. *Vertical descent angle (VDA).*

Landing Minimums

The minimums section sets forth the lowest altitude and visibility requirements for the approach, whether precision or nonprecision, straight-in or circling, or radar vectored. When a fix is incorporated in a nonprecision final segment, two sets of minimums may be published depending upon how the fix can be identified. Two sets of minimums may also be published when a second altimeter source is used in the procedure. The minimums ensure that final approach obstacle clearance is provided from the start of the final segment to the runway or MAP, whichever occurs last. The same minimums apply to both day and night operations unless different minimums are specified in the notes section of the pilot briefing. Published circling minimums provide obstacle clearance when pilots remain within the appropriate area of protection. *[Figure 1-18]*

Minimums are specified for various aircraft approach categories based upon a value 1.3 times the stalling speed of the aircraft in the landing configuration at maximum certified gross landing weight. If it is necessary to maneuver at speeds in excess of the upper limit of a speed range for a category, the minimums for the next higher category should be used. For example, an aircraft that falls into category A, but is circling to land at a speed in excess of 91 knots, should use approach category B minimums when circling to land. *[Figure 1-19]*

The minimums for straight-in and circling appear directly under each aircraft category. *[Figure 1-19]* When there is no solid division line between minimums for each category on the rows for straight-in or circling, the minimums apply to the two or more categories.

The terms used to describe the minimum approach altitudes differ between precision and nonprecision approaches. Precision approaches use DH, which is referenced to the height above threshold elevation (HAT). Nonprecision approaches use MDA, referenced to "feet MSL." The MDA is also referenced to HAT for straight-in approaches, or height above airport (HAA) for circling approaches. On AeroNav Products charts, the figures listed parenthetically are for military operations and are not used in civil aviation.

Visibility figures are provided in statute miles or runway visual range (RVR), which is reported in hundreds of feet. RVR is measured by a transmissometer, which represents the horizontal distance measured at points along the runway. It is based on the sighting of either high intensity runway lights or on the visual contrast of other targets, whichever yields the greater visual range. RVR is horizontal visual range, not slant visual range, and is used in lieu of prevailing visibility in determining minimums for a particular runway. It is illustrated in hundreds of feet if less than a mile (i.e., "24" is an RVR of 2,400 feet). *[Figures 1-19* and *1-20]*

Visibility figures are depicted after the DA/DH or MDA in the minimums section. If visibility in statute miles is indicated,

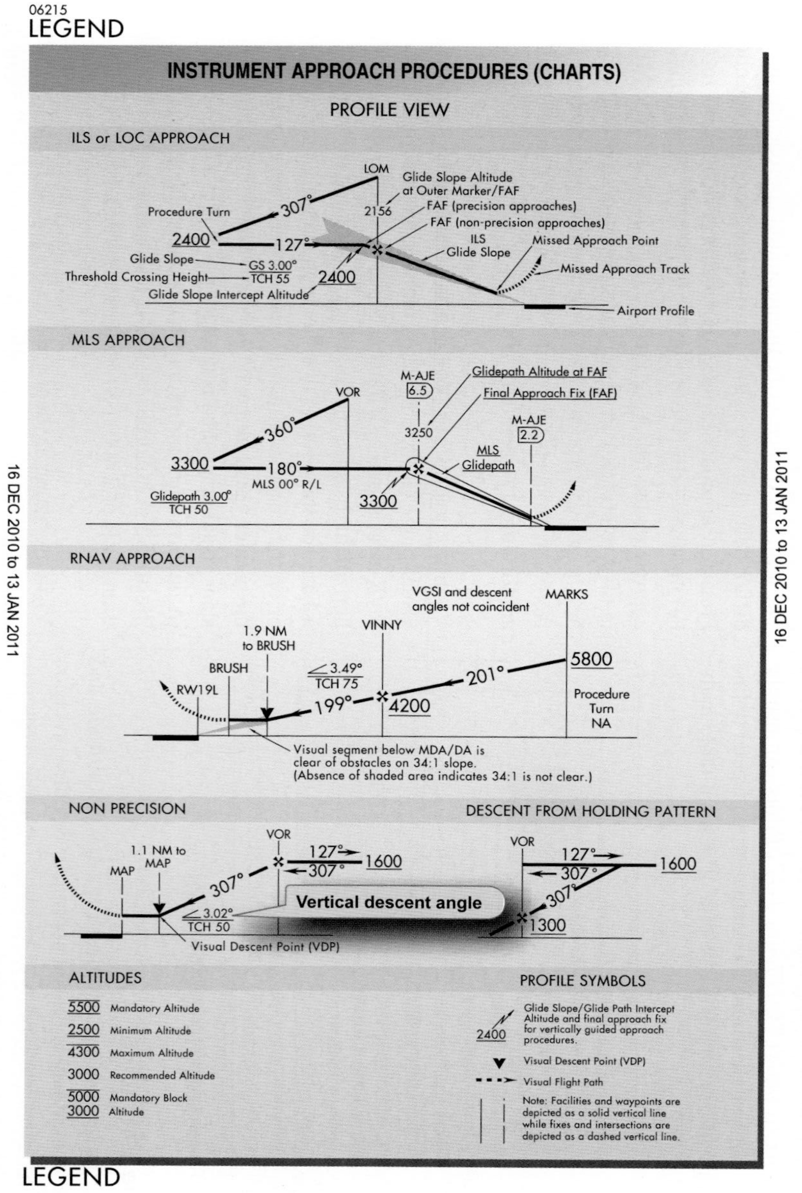

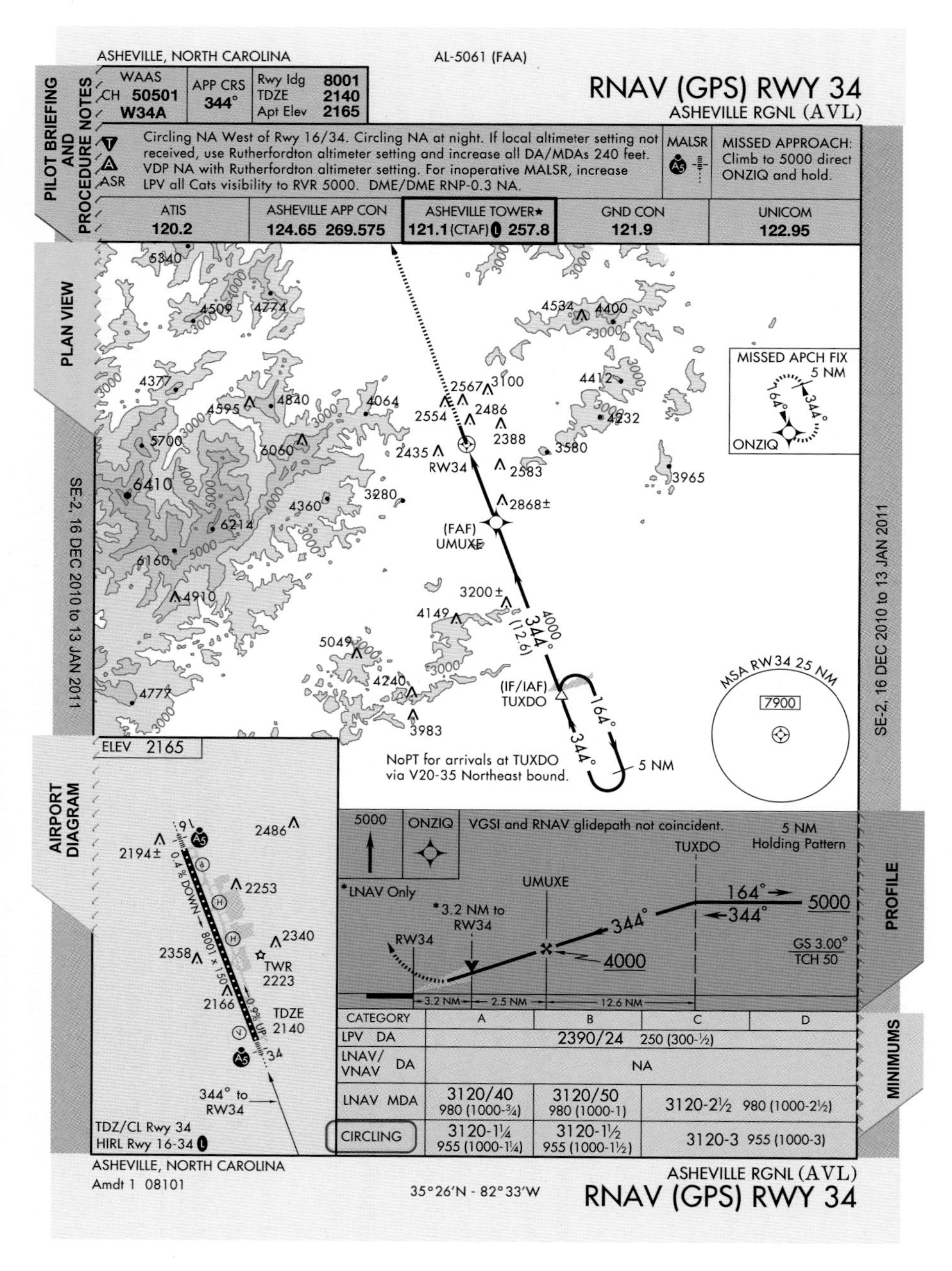

Figure 1-18. *IAP profile legend.*

DESCENT TABLE 99028

RATE OF DESCENT TABLE

A rate of descent table is provided for use in planning and executing precision descents under known or approximate ground speed conditions. It will be especially useful for approaches when the localizer only is used for course guidance. A best speed, power, altitude combination can be programmed which will result in a stable glide rate and altitude favorable for executing a landing if minimums exist upon breakout. Care should always be exercised so that minimum descent altitude and missed approach point are not exceeded.

ANGLE OF DESCENT (degrees and tenths)	FEET /NM	GROUND SPEED (knots)										
		30	45	60	75	90	105	120	135	150	165	180
2.0	210	105	160	210	265	320	370	425	475	530	585	635
2.5	265	130	200	265	330	395	465	530	595	665	730	795
VERTICAL PATH ANGLE 2.7	287	143	215	287	358	430	501	573	645	716	788	860
2.8	297	149	223	297	371	446	520	594	669	743	817	891
2.9	308	154	231	308	385	462	539	616	693	769	846	923
3.0	318	159	239	318	398	478	557	637	716	796	876	955
3.1	329	165	247	329	411	494	576	658	740	823	905	987
3.2	340	170	255	340	425	510	594	679	764	849	934	1019
3.3	350	175	263	350	438	526	613	701	788	876	963	1051
3.4	361	180	271	361	451	541	632	722	812	902	993	1083
3.5	370	185	280	370	465	555	650	740	835	925	1020	1110
4.0	425	210	315	425	530	635	740	845	955	1060	1165	1270
4.5	475	240	355	475	595	715	835	955	1075	1190	1310	1430
5.0	530	265	395	530	660	795	925	1060	1190	1325	1455	1590
5.5	580	290	435	580	730	875	1020	1165	1310	1455	1600	1745
6.0	635	315	475	635	795	955	1110	1270	1430	1590	1745	1950
6.5	690	345	515	690	860	1030	1205	1375	1550	1720	1890	2065
7.0	740	370	555	740	925	1110	1295	1480	1665	1850	2035	2220
7.5	795	395	595	795	990	1190	1390	1585	1785	1985	2180	2380
8.0	845	425	635	845	1055	1270	1480	1690	1905	2115	2325	2540
8.5	900	450	675	900	1120	1345	1570	1795	2020	2245	2470	2695
9.0	950	475	715	950	1190	1425	1665	1900	2140	2375	2615	2855
9.5	1005	500	750	1005	1255	1505	1755	2005	2255	2510	2760	3010
10.0	1055	530	790	1055	1320	1585	1845	2110	2375	2640	2900	3165
10.5	1105	555	830	1105	1385	1660	1940	2215	2490	2770	3045	3320
11.0	1160	580	870	1160	1450	1740	2030	2320	2610	2900	3190	3480
11.5	1210	605	910	1210	1515	1820	2120	2425	2725	3030	3335	3635
12.0	1260	630	945	1260	1575	1890	2205	2520	2835	3150	3465	3780

16 DEC 2010 to 13 JAN 2011

DESCENT TABLE 99028

16 DEC 2010 to 13 JAN 2011

ORLANDO, FLORIDA

AL-305 (FAA)

RNAV (GPS) RWY 7

ORLANDO/EXECUTIVE (ORL)

PILOT BRIEFING AND PROCEDURE NOTES

APP CRS 070°	Rwy ldg 5703 TDZE 109 Apt Elev 113

GPS or RNP-0.3 Required. DME/DME RNP-0.3 NA. In minimum visibility conditions, bright lights on highway ½ mile south of airport may be mistaken for runway lights.

MALSR

NOTE: A descent to 1200 may be required when executing an early missed approach. MISSED APPROACH: Climb to 1200 maximum altitude direct SOTCO WP, then climb to 1500 direct OVIDO WP and hold.

ATIS	ORLANDO APP CON	EXECUTIVE TOWER ★	GND CON	CLNC DEL	UNICOM
127.25	124.8 351.9	118.7(CTAF) 239.0	121.4 239.0	128.45	122.95

PLAN VIEW

RADAR REQUIRED

4 NM 049° 229° OVIDO 595 412 352 525 267 316 263 547 SOTCO RW07 252 946 (FAF) HERNY 350 (IAF) ESRAQ 2000 070° (6) BUVAY 2.8 NM to RW07 070° 250° 4 NM Orlando Intl MSA RW07 25 NM 3000

SE-3, 16 DEC 2010 to 13 JAN 2011

ELEV 113

AIRPORT DIAGRAM

145± 178± 4625 X 100 6004 X 150 221 173± 139 173± 156± TDZE 109 070° to RW07

REIL Rwys 13, 25 and 31
HIRL Rwys 7-25 and 13-31

PROFILE

4 NM Holding Pattern ESRAQ HERNY 1200 SOTCO 1500 OVIDO BUVAY 2.8 NM to RW07 2.1 NM to RW07 RW07 2000 250° 070° 070° 2000 3.21° TCH 40 1200 VGSI and descent angles not coincident. 24 2.7 NM 0.7 2.1

MINIMUMS

CATEGORY	A	B	C	D
LNAV MDA	800 (24) 691 (700-½)		800-1½ 691 (700-1½)	800-1¾ 691 (700-1¾)
CIRCLING	800-1 687 (700-1)		800-2 687 (700-2)	860-2½ 747 (800-2½)

RVR

ORLANDO, FLORIDA
Orig-B 10322

28° 33'N-81° 20'W

ORLANDO/EXECUTIVE (ORL)
RNAV (GPS) RWY 7

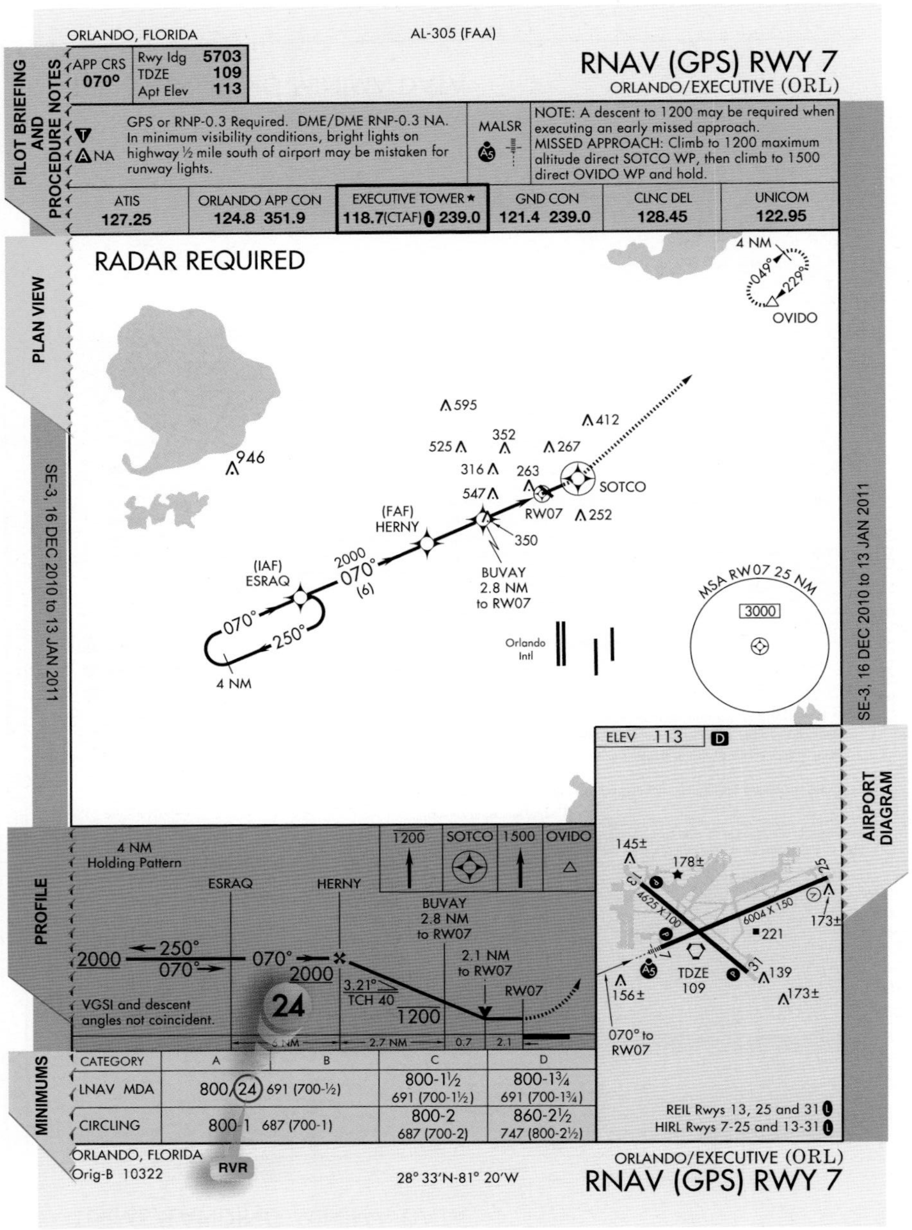

Figure 1-19. *Descent rate table.*

TERMS/LANDING MINIMA DATA

IFR LANDING MINIMA

The United States Standard for Terminal Instrument Procedures (TERPS) is the approved criteria for formulating instrument approach procedures. Landing minima are established for six aircraft approach categories (ABCDE and COPTER). In the absence of COPTER MINIMA, helicopters may use the CAT A minimums of other procedures. The standard format for RNAV minima and landing minima portrayal follows:

RNAV (GPS) MINIMA

CATEGORY	A	B	C	D
LPV DA	1540/24 258 (300-½)			
LNAV/VNAV DA	1600/24 318 (400-½)			1600/40 318 (400-¾)
LNAV MDA	1840/24 558 (600-½)		1840/50 558 (600-1)	1840/60 558 (600-1¼)
CIRCLING	1840-1 545 (600-1)		1840-1½ 545 (600-1½)	1860-2 565 (600-2)

NOTE: The W symbol indicates outages of the WAAS vertical guidance may occur daily at this location due to initial system limitations. WAAS NOTAMS for vertical outages are not provided for this approach. Use LNAV minima for flight planning at these locations, whether as a destination or alternate. For flight operations at these locations, when the WAAS avionics indicate that LNAV/VNAV or LPV service is available, then vertical guidance may be used to complete the approach using the displayed level of service. Should an outage occur during the procedure, reversion to LNAV minima may be required. As the WAAS coverage is expanded, the W will be removed.

RNAV minimums are dependent on navigation equipment capability, as stated in the applicable AFM, AFMS, or other FAA approved document, and as outlined below.

GLS (GLobal Navigation Satellite System (GNSS) Landing System)

The GLS (NA) minima line will be removed from existing RNAV (GPS) approach charts when LPV minima is published.

LPV (An Approach Procedure with Vertical Guidance (APV) based on WAAS lateral and vertical guidance)

Must have WAAS avionics approved for LPV approach.

LNAV/VNAV (Lateral navigation/Vertical navigation)

Must have either:
a.) WAAS avionics approved for LNAV/VNAV approach, or
b.) A certified Baro-VNAV system with an IFR approach approved GPS,or
c.) A certified Baro-VNAV system with an IFR approach approved WAAS, or
d.) An approach certified RNP-0.3 system with barometric vertical guidance (Baro-VNAV).
Other RNAV systems require special approval.

NOTES:
1. LNAV-VNAV minima not applicable for Baro-VNAV equipment if chart is annotated "Baro-VNAV NA" or when below the minimum published temperature, e.g., Baro-VNAV NA below -17°C(2°F).
2. DME/DME based RNP-0.3 systems may be used only when a chart note indicates DME/DME availability; e.g., "DME/DME RNP-0.3 Authorized." Specific DME facilities may be required; e.g., "DME/DME RNP-0.3 Authorized. ABC, XYZ required."

LNAV (Lateral navigation)

Must have IFR approach approved GPS, WAAS, or RNP-0.3 system.Other RNAV systems require special approval.
NOTE: DME/DME based RNP-0.3 systems may be used only when a chart note indicates DME/DME availability; e.g., "DME/DME RNP-0.3 Authorized." Specific DME facilities may be required; e.g., "DME/DME RNP-0.3 Authorized. ABC, XYZ required."

LANDING MINIMA FORMAT

In this example airport elevation is 1179, and runway touchdown zone elevation is 1152.

CATEGORY	A	B	C	D
S-ILS 27	1352/24 200 (200-½)			
S-LOC 27	1440/24 288 (300-½)			1440/50 288 (300-1)
CIRCLING	1540-1 361 (400-1)	1640-1 461 (500-1)	1640-1½ 461 (500-1½)	1740-2 561 (600-2)

16 DEC 2010 to 13 JAN 2011

TERMS/LANDING MINIMA DATA

COPTER MINIMA ONLY

CATEGORY	COPTER
H-176°	680-½ 363 (400-½)

No circling minimums are provided

RADAR MINIMA

PAR (c)	10 2.5°/42/1000	ABCDE	**195**/16	100	(100-¼)				
(d)	28 2.5°/48/1068	ABCDE	**187**/16	100	(100-¼)				
ASR	10	ABC	**560**/40	463	(500-¾)	D	**560**/50	463	(500-1)
		E	**580**/60	463	(500-1¼)				
	28	AB	**600**/50	513	(600-1)	C	**600**/60	513	(600-1¼)
		DE	**600**-1½	513	(600-1½)				
CIR (b)	10	AB	**560**-1¼	463	(500-1¼)	C	**560**-1½	463	(500-1½)
	28	AB	**600**-1¼	503	(600-1¼)	C	**600**-1½	503	(600-1½)
	10, 28	DE	**660**-2	563	(600-2)				

Radar Minima:
1. Minima shown are the lowest permitted by established criteria. Pilots should consult applicable directives for their category of aircraft.
2. The circling MDA and weather minima to be used are those for the runway to which the final approach is flown - not the landing runway. In the above RADAR MINIMA example, a category C aircraft flying a radar approach to runway 10, circling to land on runway 28, must use an MDA of 560 feet with weather minima of 500-1½.

▲ Alternate Minimums not standard. Civil users refer to tabulation. USA/USN/USAF pilots refer to appropriate regulations.
▲ NA Alternate minimums are Not Authorized due to unmonitored facility or absence of weather reporting service.
▼ Take-off Minimums not standard and/or Departure Procedures are published. Refer to tabulation.

AIRCRAFT APPROACH CATEGORIES

Aircraft approach category indicates a grouping of aircraft based on a speed of VREF, if specified, or if VREF not specified, 1.3 VSO at the maximum certificated landing weight. VREF, VSO, and the maximum certificated landing weight are those values as established for the aircraft by the certification authority of the country of registry. Helicopters are Category A aircraft. An aircraft shall fit in only one category. However, if it is necessary to operate at a speed in excess of the upper limit of the speed range for an aircraft's category, the minimums for the category for that speed shall be used. For example, an airplane which fits into Category B, but is circling to land at a speed of 145 knots, shall use the approach Category D minimums. As an additional example, a Category A airplane (or helicopter) which is operating at 130 knots on a straight-in approach shall use the approach Category C minimums. See following category limits:

MANEUVERING TABLE

Approach Category	A	B	C	D	E
Speed (Knots)	0-90	91-120	121-140	141-165	Abv 165

Comparable Values of RVR and Visibility

The following table shall be used for converting RVR to ground or flight visibility. For converting RVR values that fall between listed values, use the next higher RVR value; do not interpolate. For example, when converting 1800 RVR, use 2400 RVR with the resultant visibility of 1/2 mile.

RVR (feet)	Visibility (statute miles)	RVR (feet)	Visibility (statute miles)
1600	¼	4500	⅞
2400	½	5000	1
3200	⅝	6000	1¼
4000	¾		

16 DEC 2010 to 13 JAN 2011

Figure 1-20. *Terms/landing minima data.*

an altitude number, hyphen, and a whole or fractional number appear; for example, 530-1, which indicates "530 feet MSL" and 1 statute mile visibility. This is the descent minimum for the approach. The RVR value is separated from the minimum altitude with a slash, such as "1065/24," which indicates 1,065 feet MSL and an RVR of 2,400 feet. If RVR is prescribed for the procedure, but not available, a conversion table is used to provide the equivalent visibility in this case, of ½ statute mile visibility. *[Figure 1-20]* The conversion table is also available in the TPP.

When an alternate airport is required, standard IFR alternate minimums apply. For aircraft other than helicopters, precision approach procedures require a 600-feet ceiling and 2 statute miles visibility; nonprecision approaches require an 800-feet ceiling and 2 statute miles visibility. Helicopter alternate minimums are a ceiling that is 200 feet above the minimum for the approach to be flown and visibility of at least 1 statute mile, but not less than the minimum visibility for the approach to be flown. When a black triangle with a white "A" appears in the notes section of the pilot briefing, it indicates non-standard IFR alternate minimums exist for the airport. If an "NA" appears after the "A" (▲NA), then alternate minimums are not authorized. This information is found in the beginning of the TPP.

In addition to the COPTER approaches, instrument-equipped helicopters may fly standard approach procedures. The required visibility minimum may be reduced to one-half the published visibility minimum for category A aircraft, but in no case may it be reduced to less than ¼ mile or 1,200 feet RVR.

Two terms are specific to helicopters. Height above landing (HAL) means height above a designated helicopter landing area used for helicopter IAPs. "Point in space approach" refers to a helicopter IAP to a MAP more than 2,600 feet from an associated helicopter landing area.

Airport Sketch /Airport Diagram

Prior to all flights, pilots should take the time and study the airport layout for all of the airports that they intend to land, including those that may be used as an alternate. During the flight planning phase, study the taxi procedures for the departure airport and landing procedures for the arrival airport. The expected taxi route should be checked against the airport diagram or taxi chart, and special attention should be given to the unique or complex intersections along the taxi route. Pilots should identify critical times and locations on the taxi route (e.g., transitioning through complex intersections, crossing intervening runways, entering and lining up on the runway for takeoff, and approaching and lining up on the runway for landing).

By knowing the layout of the airport and their particular procedures, pilots are able to anticipate, understand, and safely execute all ATC directives and procedures. A major contributor to runway incursions is pilots not knowing the airport layout and procedures. This lack of situational awareness causes unnecessary accidents that can be avoided by proper flight planning. The FAA believes that following the aircraft's progress on the airport diagram to be sure that the instructions received from ATC are being followed is one of the key procedures in reducing runway incursions. To do this, pilots must take the time prior to the flight to study all procedures so that they are not trying to learn about the airport while they are receiving ATC instructions.

The airport sketch, located on the bottom of the chart, includes many helpful features. IAPs for some of the larger airports devote an entire page to an airport diagram. Airport sketch information concerning runway orientation, lighting, final approach bearings, airport beacon, and obstacles all serve to guide the pilot in the final phases of flight. See *Figure 1-21* for a legend of airport diagram/airport sketch features (see also *Figure 1-10* for an example of an airport diagram).

The airport elevation is indicated in a separate box at the top left of the airport sketch. The touchdown zone elevation (TDZE), which is the highest elevation within the first 3,000 feet of the runway, is designated at the approach end of the procedure's runway.

Beneath the airport sketch is a time and speed table when applicable. The table provides the distance and the amount of time required to transit the distance from the FAF to the MAP for selected groundspeeds.

The approach lighting systems and the visual approach lights are depicted on the airport sketch. White on black symbols (Ⓥ) are used for identifying pilot-controlled lighting (PCL). Runway lighting aids are also noted (e.g., REIL, HIRL), as is the runway centerline lighting (RCL). *[Figure 1-22]*

The airport diagram shows the paved runway configuration in solid black, while the taxiways and aprons are shaded gray. Other runway environment features are shown, such as the runway identification, dimensions, magnetic heading, displaced threshold, arresting gear, usable length, and slope.

Inoperative Components

Certain procedures can be flown with inoperative components. According to the Inoperative Components Table, for example, an instrument landing system (ILS) approach with a malfunctioning Medium Intensity Approach Lighting System with Runway Alignment Indicator Lights (MALSR = MALS with RAIL) can be flown if the minimum visibility is

06215
LEGEND

16 DEC 2010 to 13 JAN 2011

INSTRUMENT APPROACH PROCEDURES (CHARTS)

AIRPORT DIAGRAM/AIRPORT SKETCH

Runways

Hard Surface · Other Than Hard Surface · Stopways, Taxiways, Parking Areas, Water Runways · Displaced Threshold

Closed Runway · Closed Taxiway · Under Construction · Metal Surface

ARRESTING GEAR: Specific arresting gear systems; e.g., BAK12, MA-1A etc., shown on airport diagrams, not applicable to Civil Pilots. Military Pilots refer to appropriate DOD publications.

uni-directional · bi-directional · Jet Barrier

REFERENCE FEATURES

Buildings........................
Tanks........................
Obstructions........................
Airport Beacon #........................
Runway Radar Reflectors........................
Control Tower #........................

When Control Tower and Rotating Beacon are co-located, Beacon symbol will be used and further identified as TWR.

Runway length depicted is the physical length of the runway (end-to-end, including displaced thresholds if any) but excluding areas designated as stopways. Where a displaced threshold is shown and/or part of the runway is otherwise not available for landing, an annotation is added to indicate the landing length of the runway; e.g., Rwy 13 ldg 5000'.

Helicopter Alighting Areas

Negative Symbols used to identify Copter Procedures landing point........................

Runway TDZ elevation........................TDZE 123

Runway Slope........................0.3% DOWN / 0.8% UP
(shown when runway slope exceeds 0.3%)

NOTE:
Runway Slope measured to midpoint on runways 8000 feet or longer.

U.S. Navy Optical Landing System (OLS) "OLS" location is shown because of its height of approximately 7 feet and proximity to edge of runway may create an obstruction for some types of aircraft.

Approach light symbols are shown in the Flight Information Handbook.

Airport diagram scales are variable.

True/magnetic North orientation may vary from diagram to diagram

Coordinate values are shown in 1 or ½ minute increments. They are further broken down into 6 second ticks, within each 1 minute increments.

Positional accuracy within ±600 feet unless otherwise noted on the chart.

NOTE:
All new and revised airport diagrams are shown referenced to the World Geodetic System (WGS) (noted on appropriate diagram), and may not be compatible with local coordinates published in FLIP. (Foreign Only)

Runway Weight Bearing Capacity/or PCN Pavement Classification Number is shown as a codified expression.
Refer to the appropriate Supplement/Directory for applicable codes e.g.,
RWY 14-32 S75, T185, ST175, TT325
PCN 80 F/D/X/U

SCOPE

Airport diagrams are specifically designed to assist in the movement of ground traffic at locations with complex runway/taxiway configurations and provide information for updating Computer Based Navigation Systems (I.E., INS, GPS) aboard aircraft. Airport diagrams are not intended to be used for approach and landing or departure operations. For revisions to Airport Diagrams: Consult FAA Order 7910.4.

LEGEND

16 DEC 2010 to 13 JAN 2011

CATEGORY	A	B	C	D
S-5	1540-1¼ 1180 (1200-1¼)	1540-1½ 1180 (1200-1½)	1540-3 1180 (1200-3)	NA
CIRCLING	1540-1¼ 1179 (1200-1¼)	1540-1½ 1179 (1200-1½)	1540-3 1179 (1200-3)	NA
CESVA INT/RADAR MINIMUMS				
S-5	820-1 460 (500-1)		820-1¼ 460 (500-1¼)	NA
CIRCLING	860-1 499 (500-1)		860-1½ 499 (500-1½)	NA

Figure 1-21. *Airport legend and diagram.*

04330
LEGEND

INSTRUMENT APPROACH PROCEDURES (CHARTS)

APPROACH LIGHTING SYSTEM - UNITED STATES

Approach lighting and visual glide slope systems are indicated on the airport sketch by an identifier, e.g., (A2), (V), etc.

A dot "•" portrayed with approach lighting letter identifier indicates sequenced flashing lights (F) installed with the approach lighting system e.g., (A1). Negative symbology, e.g., (A1), (V) indicates Pilot Controlled Lighting (PCL).

RUNWAY TOUCHDOWN ZONE AND CENTERLINE LIGHTING SYSTEMS

TDZ/CL

RUNWAY CENTERLINE LIGHTS

CL

TDZL

TDZL

AVAILABILITY of TDZ/CL will be shown by NOTE in SKETCH e.g. "TDZ/CL Rwy 15"

APPROACH LIGHTING SYSTEM

(A) **ALSF-2**

500'

1000'

2400'/3000'

GREEN

WHITE

RED

RED

WHITE

SEQUENCED FLASHING LIGHTS

NOTE: CIVIL ALSF-2 MAY BE OPERATED AS SSALR DURING FAVORABLE WEATHER CONDITIONS

(High Intensity)
LENGTH 2400/3000 FEET

APPROACH LIGHTING SYSTEM

(A1) **ALSF-1**

RED

GREEN

1000'

2400'/3000'

WHITE

SEQUENCED FLASHING LIGHTS

(High Intensity)
LENGTH 2400/3000 FEET

SHORT APPROACH LIGHTING SYSTEM

(A2)

SALS/SALSF
(High Intensity)

SAME AS INNER 1500' OF ALSF-1

SIMPLIFIED SHORT APPROACH LIGHTING SYSTEM
with Runway Alignment Indicator Lights

(A3) **SSALR**

1000'

2400'/3000'

GREEN

WHITE

SEQUENCED FLASHING LIGHTS

(High Intensity)
LENGTH 2400/3000 FEET

MEDIUM INTENSITY (MALS and MALSF) OR SIMPLIFIED SHORT (SSALS and SSALF) APPROACH LIGHTING SYSTEMS

(A4)

GREEN

1000'

1400'

400'

WHITE

SEQUENCED FLASHING LIGHTS FOR MALSF/SSALF ONLY

LENGTH 1400 FEET

MEDIUM INTENSITY APPROACH LIGHTING SYSTEM
with Runway Alignment Indicator Lights

(A5) **MALSR**

SAME LIGHT CONFIGURATION AS SSALR.

OMNIDIRECTIONAL APPROACH LIGHTING SYSTEM

ODALS

36

THRESHOLD

1500'

SEQUENCED FLASHING LIGHTS

LENGTH 1500 FEET

(V) **VISUAL APPROACH SLOPE INDICATOR**

VASI

VISUAL APPROACH SLOPE INDICATOR WITH STANDARD THRESHOLD CLEARANCE PROVIDED.

ALL LIGHTS WHITE — TOO HIGH
FAR LIGHTS RED / NEAR LIGHTS WHITE — ON GLIDE SLOPE
ALL LIGHTS RED — TOO LOW

VASI 2 — VASI 4 — VASI 12

700' 800'

36 THRESHOLD — 36 THRESHOLD — 36 THRESHOLD

(V3) **VISUAL APPROACH SLOPE INDICATOR**

VASI

VISUAL APPROACH SLOPE INDICATOR WITH A THRESHOLD CROSSING HEIGHT TO ACCOMODATE LONG BODIED OR JUMBO AIRCRAFT.

VASI 6 — VASI 16

36 THRESHOLD — 36 THRESHOLD

LEGEND

04330
LEGEND

INSTRUMENT APPROACH PROCEDURES (CHARTS)

APPROACH LIGHTING SYSTEM - UNITED STATES

Approach lighting and visual glide slope systems are indicated on the airport sketch by an identifier, e.g., (A2), (V), etc.

A dot "•" portrayed with approach lighting letter identifier indicates sequenced flashing lights (F) installed with the approach lighting system e.g., (A1). Negative symbology, e.g., (A1), (V) indicates Pilot Controlled Lighting (PCL).

(P) **PRECISION APPROACH PATH INDICATOR**

PAPI

Too low — Slightly low

On correct approach path

Slightly high — Too high

Legend: □ White ■ Red

(V1) **"T"-VISUAL APPROACH SLOPE INDICATOR**

"T"-VASI

"T" ON BOTH SIDES OF RWY
ALL LIGHTS VARIABLE WHITE.
CORRECT APPROACH SLOPE-ONLY CROSS BAR VISIBLE.
UPRIGHT "T"- FLY UP.
INVERTED "T"- FLY DOWN.
RED "T"- GROSS UNDERSHOOT.

(V2) **PULSATING VISUAL APPROACH SLOPE INDICATOR**

PVASI

Above Glide Path — Pulsating White
On Glide Path — Steady, White or Alternating Red/White
Below Glide Path — Pulsating Red

Threshold

CAUTION: When viewing the pulsating visual approach slope indicators in the pulsating white or pulsating red sectors, it is possible to mistake this lighting aid for another aircraft or a ground vehicle. Pilots should exercise caution when using this type of system.

(V4) **TRI-COLOR VISUAL APPROACH SLOPE INDICATOR**

TRCV

Above Glide Path — Amber
On Glide Path — Green
Below Glide Path — Red
Amber

CAUTION: When the aircraft descends from green to red, the pilot may see a dark amber color during the transition from green to red.

(V5) **ALIGNMENT OF ELEMENTS SYSTEMS**

APAP

Above glide path — On Glide Path — Below Glide Path

Painted panels which may be lighted at night. To use the system the pilot positions the aircraft so the elements are in alignment.

LEGEND

16 DEC 2010 to 13 JAN 2011

Figure 1-22. *Approach lighting legend.*

increased by ¼ mile. *[Figure 1-23]* A note in this section might read, "Inoperative Table does not apply to ALS or HIRL Runway 13L."

RNAV Instrument Approach Charts

To avoid unnecessary duplication and proliferation of approach charts, approach minimums for unaugmented GPS, Wide Area Augmentation System (WAAS), Local Area Augmentation System (LAAS) are published on the same approach chart as lateral navigation/vertical navigation (LNAV/VNAV). Other types of equipment may be authorized to conduct the approach based on the minima notes in the front of the TPP approach chart books. Approach charts titled "RNAV RWY XX" may be used by aircraft with navigation systems that meet the required navigational performance (RNP) values for each segment of the approach. *[Figure 1-24]*

The chart may contain as many as four lines of approach minimums: global landing system (GLS), WAAS and LAAS, LNAV/VNAV, LNAV, and circling. LNAV/VNAV is an instrument approach with lateral and vertical guidance with integrity limits similar to barometric vertical navigation (BARO VNAV).

RNAV procedures that incorporate a final approach stepdown fix may be published without vertical navigation on a separate chart also titled RNAV. During a transition period when GPS procedures are undergoing revision to a new title, both RNAV and GPS approach charts and formats are published. ATC clearance for the RNAV procedure authorizes a properly certificated pilot to utilize any landing minimums for which the aircraft is certified.

Chart terminology changes slightly to support the new procedure types:

1. DA replaces the term DH. DA conforms to the international convention where altitudes relate to MSL and heights relate to AGL. DA will eventually be published for other types of IAPs with vertical guidance, as well. DA indicates to the pilot that the published descent profile is flown to the DA (MSL), where a missed approach is initiated if visual references for landing are not established. Obstacle clearance is provided to allow a momentary descent below DA while transitioning from the final approach to the missed approach. The aircraft is expected to follow the missed approach instructions while continuing along the published final approach course to at least the published runway threshold waypoint or MAP (if not at the threshold) before executing any turns.
2. MDA continues to be used only for the LNAV and circling procedures.
3. TCH has been traditionally used in precision approaches as the height of the GS above threshold. With publication of LNAV/VNAV minimums and RNAV descent angles, including graphically depicted descent profiles, TCH also applies to the height of the "descent angle," or glidepath, at the threshold. Unless otherwise required for larger type aircraft that may be using the IAP, the typical TCH is 30 to 50 feet.

The minima format changes slightly:

1. Each line of minima on the RNAV IAP is titled to reflect the RNAV system applicable (e.g., LPV, LNAV/VNAV, and LNAV). Circling minima is also provided.
2. The minima title box also indicates the nature of the minimum altitude for the IAP. For example: DA is published next to the minima line title for minimums supporting vertical guidance, and MDA is published where the minima line supports only lateral guidance. During an approach where an MDA is used, descent below MDA is not authorized.
3. Where two or more systems share the same minima, each line of minima is displayed separately.

For more information concerning government charts, the AeroNav Products can be contacted by telephone or via their internet address at:

National Aeronautical Navigation Products (AeroNav Products)
Telephone 800-626-3677
www.aeronav.faa.gov

11237

INOP COMPONENTS

08 MAR 2012 to 05 APR 2012

INOPERATIVE COMPONENTS OR VISUAL AIDS TABLE

Landing minimums published on instrument approach procedure charts are based upon full operation of all components and visual aids associated with the particular instrument approach chart being used. Higher minimums are required with inoperative components or visual aids as indicated below. If more than one component is inoperative, each minimum is raised to the highest minimum required by any single component that is inoperative. ILS glide slope inoperative minimums are published on the instrument approach charts as localizer minimums. This table may be amended by notes on the approach chart. Such notes apply only to the particular approach category(ies) as stated. See legend page for description of components indicated below.

(1) ILS, MLS, PAR and RNAV (LPV line of minima)

Inoperative Component or Aid	Approach Category	Increase Visibility
ALSF 1 & 2, MALSR, & SSALR	ABCD	¼ mile

(2) ILS with visibility minimum of 1,800 RVR

ALSF 1 & 2, MALSR, & SSALR	ABCD	To 4000 RVR
TDZL RCLS	ABCD	To 2400 RVR*
RVR	ABCD	To ½ mile

*1800 RVR authorized with the use of FD or AP or HUD to DA.

(3) VOR, VOR/DME, TACAN, LOC, LOC/DME, LDA, LDA/DME, SDF, SDF/DME, GPS, ASR and RNAV (LNAV/VNAV, LNAV and LP lines of minima)

Inoperative Visual Aid	Approach Category	Increase Visibility
ALSF 1 & 2, MALSR, & SSALR	ABCD	½ mile
SSALS,MALS, & ODALS	ABC	¼ mile

(4) NDB

ALSF 1 & 2, MALSR, & SSALR	C	½ mile
	ABD	¼ mile
MALS, SSALS, ODALS	ABC	¼ mile

08 MAR 2012 to 05 APR 2012

INOP COMPONENTS

Figure 1-23. *IAP inoperative components table.*

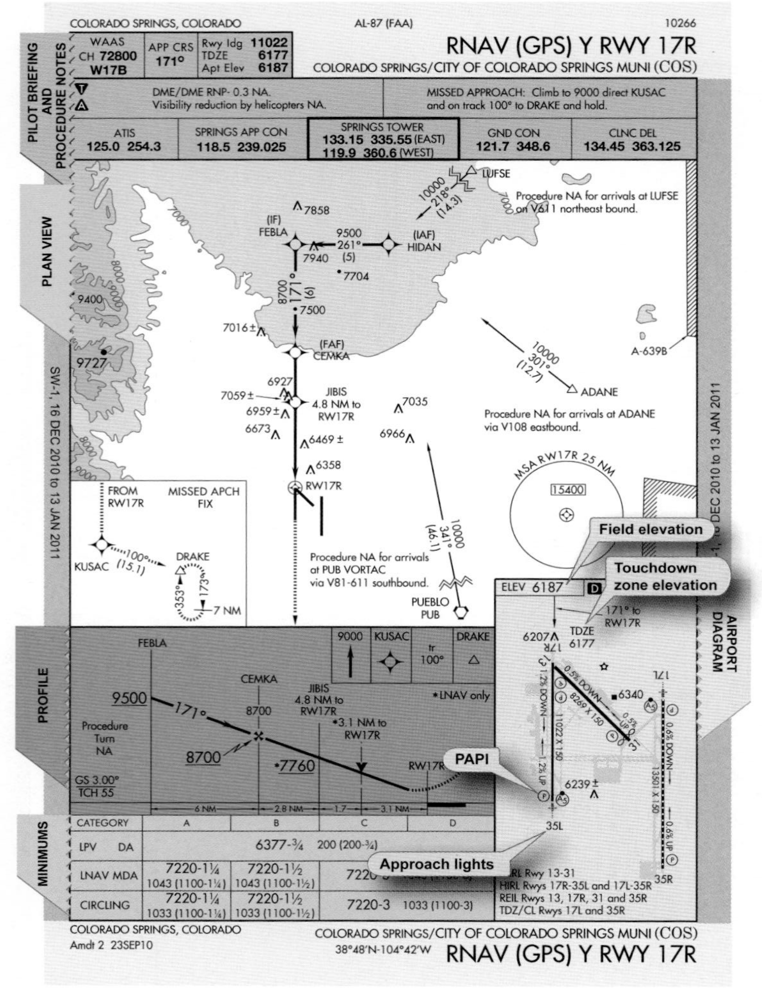

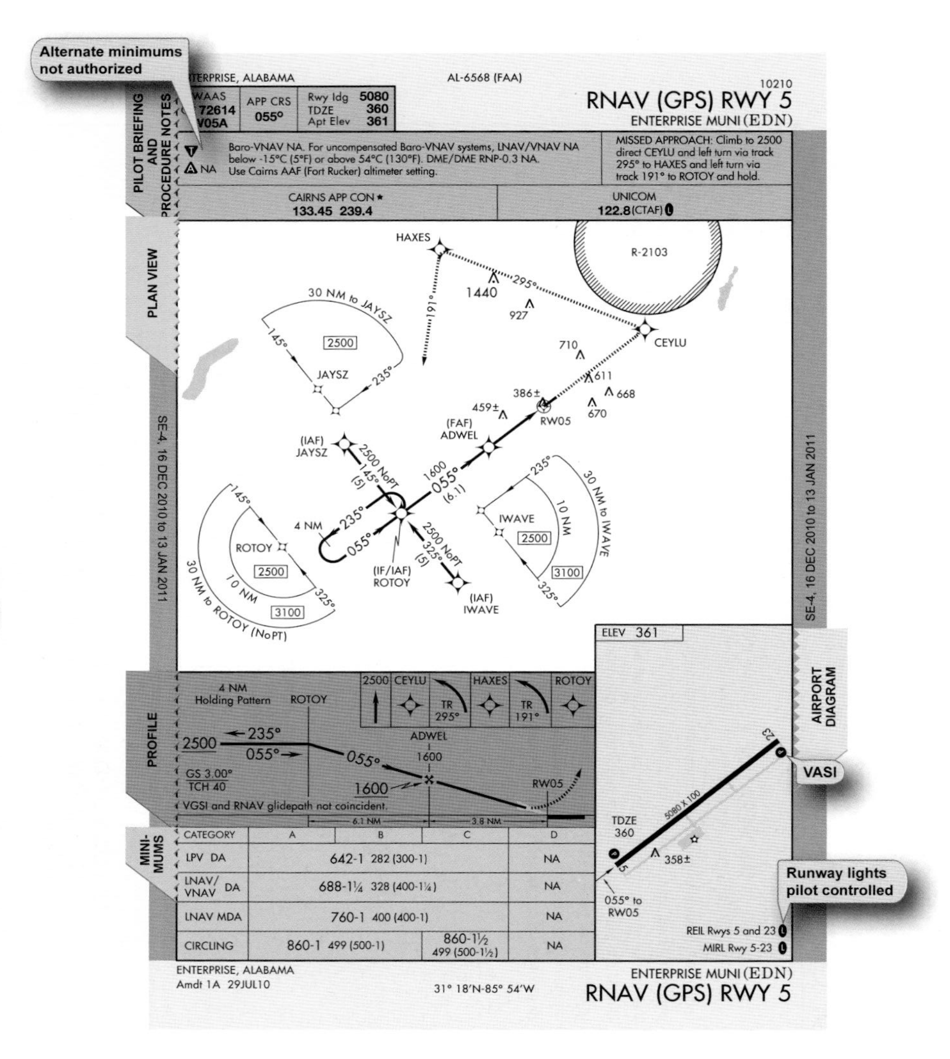

Figure 1-24. *RNAV instrument approach charts.*

Chapter 2

The Air Traffic Control System

Introduction

This chapter covers the communication equipment, communication procedures, and air traffic control (ATC) facilities and services available for a flight under instrument flight rules (IFR) in the National Airspace System (NAS).

Communication Equipment

Navigation/Communication Equipment

Civilian pilots communicate with ATC on frequencies in the very high frequency (VHF) range between 118.000 and 136.975 MHz. To derive full benefit from the ATC system, radios capable of 25 kHz spacing are required (e.g., 134.500, 134.575, 134.600). If ATC assigns a frequency that cannot be selected, ask for an alternative frequency.

Figure 2-1 illustrates a typical radio panel installation consisting of a communications transceiver on the left and a navigational receiver on the right. Many radios allow the pilot to have one or more frequencies stored in memory and one frequency active for transmitting and receiving (called simplex operation). It is possible to communicate with some flight service stations (FSS) by transmitting on 122.1 MHz (selected on the communication radio) and receiving on a VHF omnidirectional range (VOR) frequency (selected on the navigation radio). This is called duplex operation.

An audio panel allows a pilot to adjust the volume of the selected receiver(s) and to select the desired transmitter. *[Figure 2-2]* The audio panel has two positions for receiver selection, cabin speaker, and headphone (some units might have a center "OFF" position). Use of a hand-held microphone and the cabin speaker introduces the distraction of reaching for and hanging up the microphone. A headset with a boom microphone is recommended for clear communications. The microphone should be positioned close to the lips to reduce

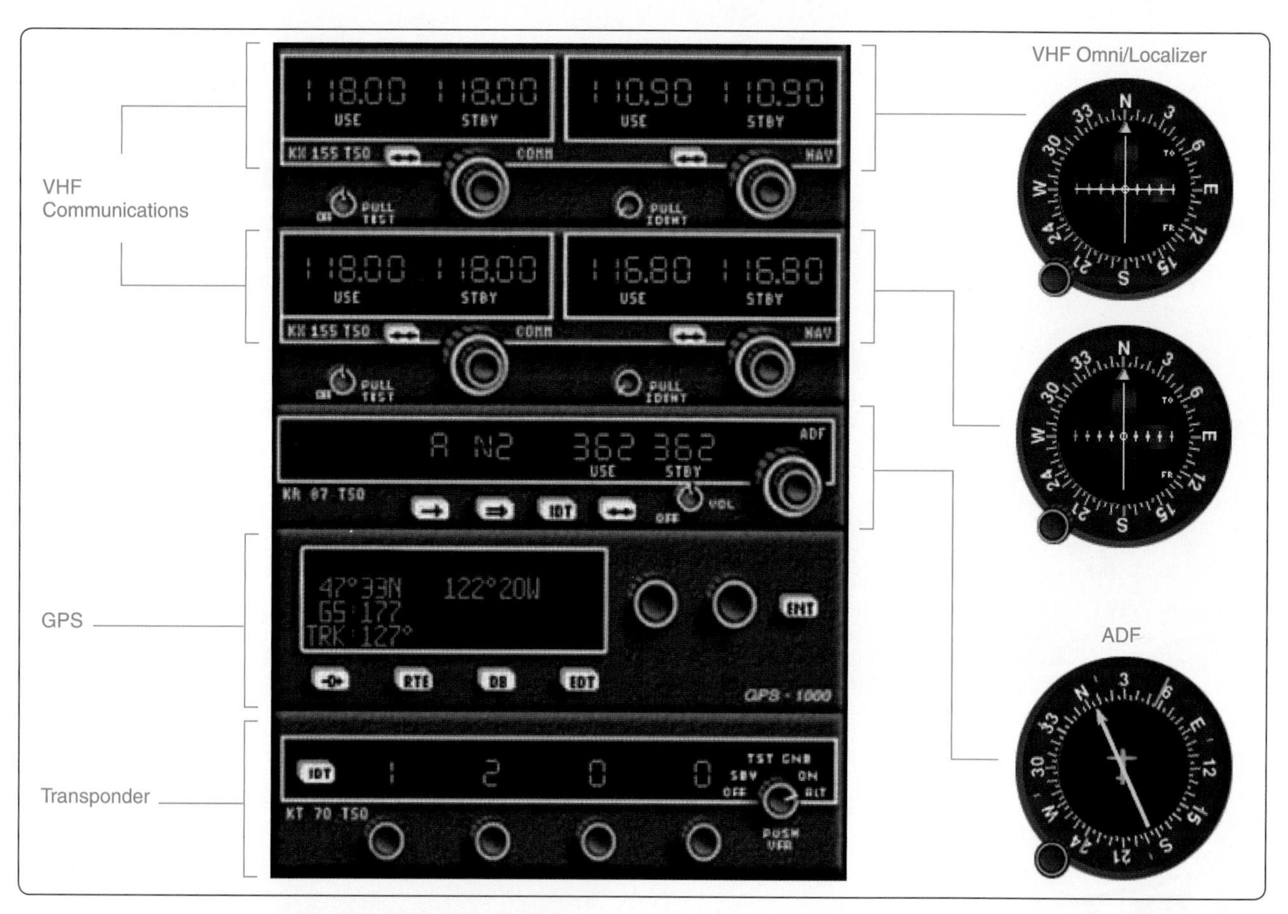

Figure 2-1. *Typical navigation/communication installation.*

Figure 2-2. *Audio panel.*

the possibility of ambient flight deck noise interfering with transmissions to the controller. Headphones deliver the received signal directly to the ears; therefore, ambient noise does not interfere with the pilot's ability to understand the transmission. *[Figure 2-3]*

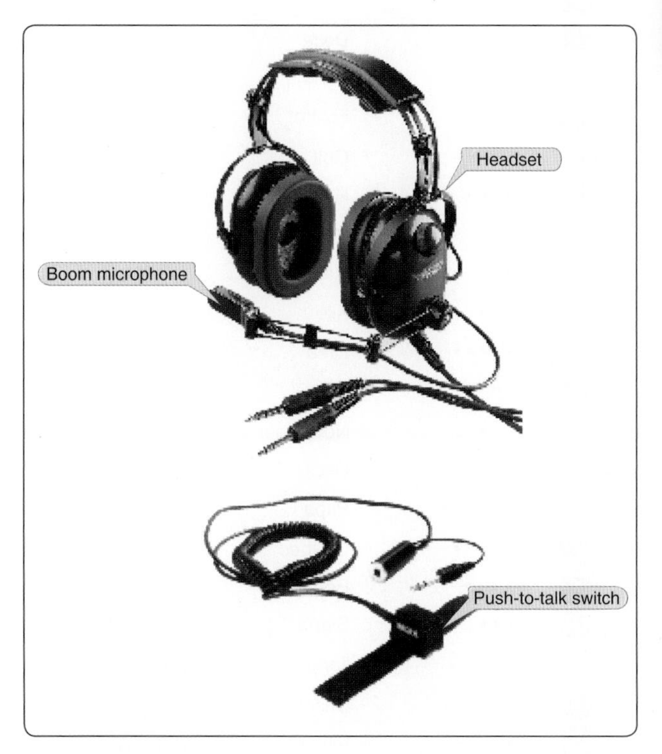

Figure 2-3. *Boom microphone, headset, and push-to-talk switch.*

Switching the transmitter selector between COM1 and COM2 changes both transmitter and receiver frequencies. It is necessary only when a pilot wants to monitor one frequency while transmitting on another. One example is listening to Automatic Terminal Information Service (ATIS) on one receiver while communicating with ATC on the other. Monitoring a navigation receiver to check for proper identification is another reason to use the switch panel.

Most audio switch panels also include a marker beacon receiver. All marker beacons transmit on 75 MHz, so there is no frequency selector.

Figure 2-4 illustrates an increasingly popular form of navigation/communication radio; it contains a global positioning system (GPS) receiver and a communications transceiver. Using its navigational capability, this unit can determine when a flight crosses an airspace boundary or fix and can automatically select the appropriate communications frequency for that location in the communications radio.

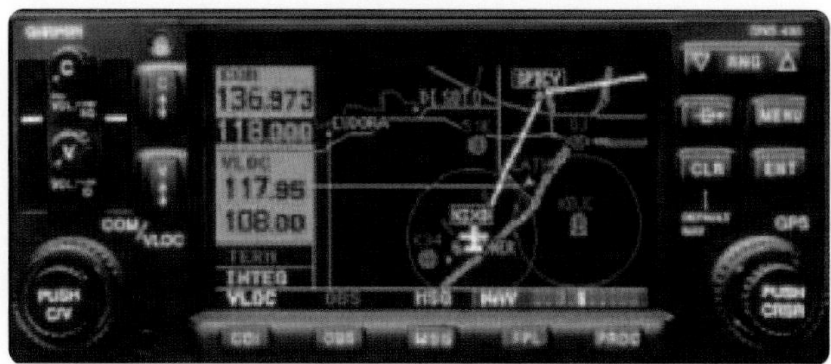

Figure 2-4. *Combination GPS-com unit.*

Radar and Transponders

ATC radars have a limited ability to display primary returns, which is energy reflected from an aircraft's metallic structure. Their ability to display secondary returns (transponder replies to ground interrogation signals) makes possible the many advantages of automation.

A transponder is a radar beacon transmitter/receiver installed in the instrument panel. ATC beacon transmitters send out interrogation signals continuously as the radar antenna rotates. When an interrogation is received by a transponder, a coded reply is sent to the ground station where it is displayed on the controller's scope. A reply light on the transponder panel flickers every time it receives and replies to a radar interrogation. Transponder codes are assigned by ATC.

When a controller asks a pilot to "ident" and the ident button is pushed, the return on the controller's scope is intensified for precise identification of a flight. When requested, briefly push the ident button to activate this feature. It is good practice for pilots to verbally confirm that they have changed codes or pushed the ident button.

Mode C (Altitude Reporting)

Primary radar returns indicate only range and bearing from the radar antenna to the target; secondary radar returns can display altitude, Mode C, on the control scope if the aircraft is equipped with an encoding altimeter or blind encoder. In either case, when the transponder's function switch is in the ALT position, the aircraft's pressure altitude is sent to the controller. Adjusting the altimeter's Kollsman window has no effect on the altitude read by the controller.

Transponders, when installed, must be ON at all times when operating in controlled airspace; altitude reporting is required by regulation in Class B and Class C airspace and inside a 30-mile circle surrounding the primary airport in Class B airspace. Altitude reporting should also be ON at all times.

Communication Procedures

Clarity in communication is essential for a safe instrument flight. This requires pilots and controllers to use terms that are understood by both—the Pilot/Controller Glossary in the Aeronautical Information Manual (AIM) is the best source of terms and definitions. The AIM is revised twice a year and new definitions are added, so the glossary should be reviewed frequently. Because clearances and instructions are comprised largely of letters and numbers, a phonetic pronunciation guide has been developed for both. *[Figure 2-5]*

ATC must follow the guidance of the Air Traffic Control Manual when communicating with pilots. The manual presents the controller with different situations and prescribes precise terminology that must be used. This is advantageous for pilots because once they have recognized a pattern or format, they can expect future controller transmissions to follow that format. Controllers are faced with a wide variety of communication styles based on pilot experience, proficiency, and professionalism.

Pilots should study the examples in the AIM, listen to other pilots communicate, and apply the lessons learned to their own communications with ATC. Pilots should ask for clarification of a clearance or instruction. If necessary, use plain English to ensure understanding, and expect the controller to reply in the same way. A safe instrument flight is the result of cooperation between controller and pilot.

Character	Morse Code	Telephony	Phonic (Pronunciation)
A	•—	Alfa	(AL-FAH)
B	—•••	Bravo	(BRAH-VOH)
C	—•—•	Charlie	(CHAR-LEE) or (SHAR-LEE)
D	—••	Delta	(DELL-TAH)
E	•	Echo	(ECK-OH)
F	••—•	Foxtrot	(FOKS-TROT)
G	——•	Golf	(GOLF)
H	••••	Hotel	(HOH-TEL)
I	••	India	(IN-DEE-AH)
J	•———	Juliett	(JEW-LEE-ETT)
K	—•—	Kilo	(KEY-LOH)
L	•—••	Lima	(LEE-MAH)
M	——	Mike	(MIKE)
N	—•	November	(NO-VEM-BER)
O	———	Oscar	(OSS-CAH)
P	•——•	Papa	(PAH-PAH)
Q	——•—	Quebec	(KEH-BECK)
R	•—•	Romeo	(ROW-ME-OH)
S	•••	Sierra	(SEE-AIR-RAH)
T	—	Tango	(TANG-GO)
U	••—	Uniform	(YOU-NEE-FORM) or (OO-NEE-FORM)
V	•••—	Victor	(VIK-TAH)
W	•——	Whiskey	(WISS-KEY)
X	—••—	Xray	(ECKS-RAY)
Y	—•——	Yankee	(YANG-KEY)
Z	——••	Zulu	(ZOO-LOO)
1	•————	One	(WUN)
2	••———	Two	(TOO)
3	•••——	Three	(TREE)
4	••••—	Four	(FOW-ER)
5	•••••	Five	(FIFE)
6	—••••	Six	(SIX)
7	——•••	Seven	(SEV-EN)
8	———••	Eight	(AIT)
9	————•	Nine	(NI-NER)
0	—————	Zero	(ZEE-RO)

Figure 2-5. *Phonetic pronunciation guide.*

Communication Facilities

The controller's primary responsibility is separation of aircraft operating under IFR. This is accomplished with ATC facilities, to include the FSS, airport traffic control tower (ATCT), terminal radar approach control (TRACON), and air route traffic control center (ARTCC).

Flight Service Stations (FSS)

A pilot's first contact with ATC is usually through FSS, either by radio or telephone. FSSs provide pilot briefings, receive and process flight plans, relay ATC clearances, originate Notices to Airmen (NOTAMs), and broadcast aviation weather. Some facilities provide En Route Flight Advisory Service (EFAS), take weather observations, and advise United States Customs and Immigration of international flights.

Telephone contact with Flight Service can be obtained by dialing 1-800-WX-BRIEF. This number can be used anywhere in the United States and connects to the nearest FSS based on the area code from which the call originates. There are a variety of methods of making radio contact: direct transmission, remote communication outlets (RCOs), ground communication outlets (GCOs), and by using duplex transmissions through navigational aids (NAVAIDs). The best source of information on frequency usage is the Airport/Facility Directory (A/FD) and the legend panel on sectional charts also contains contact information.

The briefer sends a flight plan to the host computer at the ARTCC (Center). After processing the flight plan, the computer sends flight strips to the tower, to the radar facility that handles the departure route, and to the Center controller whose sector the flight first enters. *Figure 2-6* shows a typical strip. These strips are delivered approximately 30 minutes prior to the proposed departure time. Strips are delivered to en route facilities 30 minutes before the flight is expected to enter their airspace. If a flight plan is not opened, it will "time out" 2 hours after the proposed departure time.

When departing an airport in Class G airspace, a pilot receives an IFR clearance from the FSS by radio or telephone. It contains either a clearance void time, in which case an aircraft must be airborne prior to that time, or a release time. Pilots should not take off prior to the release time. Pilots can help the controller by stating how soon they expect to be airborne. If the void time is, for example, 10 minutes past the hour and an aircraft is airborne at exactly 10 minutes past the hour, the clearance is void—a pilot must take off prior to the void time. A specific void time may be requested when filing a flight plan.

ATC Towers

Several controllers in the tower cab are involved in handling an instrument flight. Where there is a dedicated clearance delivery position, that frequency is found in the A/FD and on the instrument approach chart for the departure airport. Where there is no clearance delivery position, the ground controller performs this function. At the busiest airports, pre-taxi clearance is required; the frequency for pre-taxi clearance can be found in the A/FD. Taxi clearance should be requested not more than 10 minutes before proposed taxi time.

It is recommended that pilots read their IFR clearance back to the clearance delivery controller. Instrument clearances can be overwhelming when attempting to copy them verbatim, but they follow a format that allows a pilot to be prepared when responding "Ready to copy." The format is: clearance limit (usually the destination airport); route, including any departure procedure; initial altitude; frequency (for departure control); and transponder code. With the exception of the transponder code, a pilot knows most of these items before engine start. One technique for clearance copying is writing C-R-A-F-T.

Assume an IFR flight plan has been filed from Seattle, Washington to Sacramento, California via V-23 at 7,000 feet. Traffic is taking off to the north from Seattle-Tacoma (Sea-Tac) airport and, by monitoring the clearance delivery frequency, a pilot can determine the departure procedure being assigned to southbound flights. The clearance limit is the destination airport, so write "SAC" after the letter C. Write "SEATTLE TWO – V23" after R for Route because departure control issued this departure to other flights. Write "70" after the A, the departure control frequency printed on the approach charts for Sea-Tac after F, and leave the space after the letter T blank—the transponder code is generated by computer and can seldom be determined in advance. Then, call clearance delivery and report "Ready to copy."

As the controller reads the clearance, check it against what is already written down; if there is a change, draw a line through that item and write in the changed item. Chances are the changes are minimal, and most of the clearance is copied before keying the microphone. Still, it is worthwhile to develop clearance shorthand to decrease the verbiage that must be copied (see Appendix 1).

Pilots are required to have either the text of a departure procedure (DP) or a graphic representation (if one is available), and should review it before accepting a clearance. This is another reason to find out ahead of time which DP is in use. If the DP includes an altitude or a departure control frequency, those items are not included in the clearance.

The last clearance received supersedes all previous clearances. For example, if the DP says "Climb and maintain 2,000 feet, expect higher in 6 miles," but upon contacting the departure controller a new clearance is received: "Climb and maintain 8,000 feet," the 2,000 feet restriction has been canceled. This rule applies in both terminal and Center airspace.

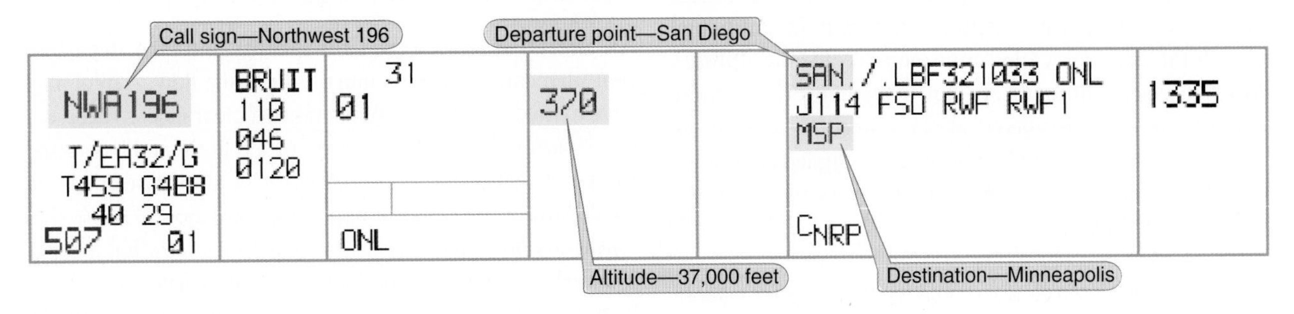

Figure 2-6. *Flight strip.*

When reporting "ready to copy" an IFR clearance before the strip has been received from the Center computer, pilots are advised "clearance on request." The controller initiates contact when it has been received. This time can be used for taxi and pre-takeoff checks.

The local controller is responsible for operations in the Class D airspace and on the active runways. At some towers, designated as IFR towers, the local controller has vectoring authority. At visual flight rules (VFR) towers, the local controller accepts inbound IFR flights from the terminal radar facility and cannot provide vectors. The local controller also coordinates flights in the local area with radar controllers. Although Class D airspace normally extends 2,500 feet above field elevation, towers frequently release the top 500 feet to the radar controllers to facilitate overflights. Accordingly, when a flight is vectored over an airport at an altitude that appears to enter the tower controller's airspace, there is no need to contact the tower controller—all coordination is handled by ATC.

The departure radar controller may be in the same building as the control tower, but it is more likely that the departure radar position is remotely located. The tower controller will not issue a takeoff clearance until the departure controller issues a release.

Terminal Radar Approach Control (TRACON)

TRACONs are considered terminal facilities because they provide the link between the departure airport and the en route structure of the NAS. Terminal airspace normally extends 30 nautical miles (NM) from the facility with a vertical extent of 10,000 feet; however, dimensions vary widely. Class B and Class C airspace dimensions are provided on aeronautical charts. At terminal radar facilities, the airspace is divided into sectors, each with one or more controllers, and each sector is assigned a discrete radio frequency. All terminal facilities are approach controls and should be addressed as "Approach" except when directed to do otherwise (e.g., "Contact departure on 120.4.").

Terminal radar antennas are located on or adjacent to the airport. *Figure 2-7* shows a typical configuration. Terminal controllers can assign altitudes lower than published procedural altitudes called minimum vectoring altitudes (MVAs). These altitudes are not published or accessible to pilots, but are displayed at the controller's position. *[Figure 2-8]* However, when pilots are assigned an altitude that seems to be too low, they should query the controller before descending.

When a pilot accepts a clearance and reports ready for takeoff, a controller in the tower contacts the TRACON for a release.

Figure 2-7. *Combined radar and beacon antenna.*

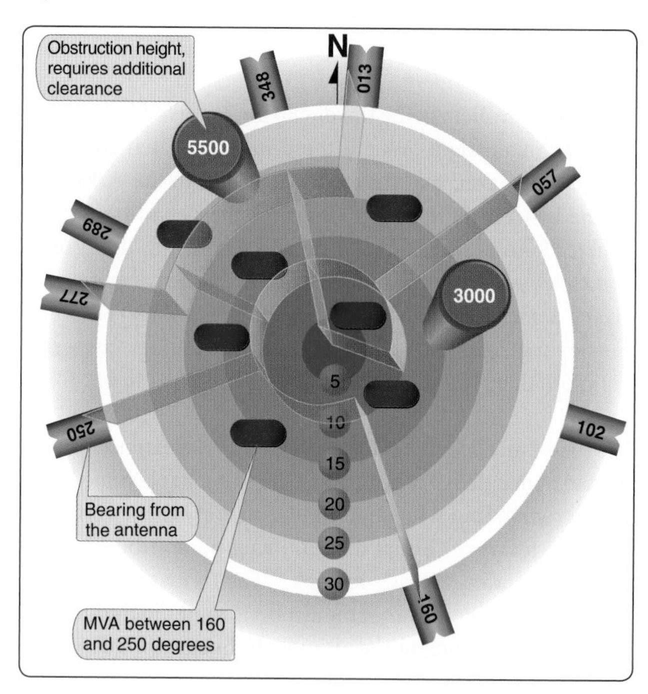

Figure 2-8. *Minimum vectoring altitude (MVA) chart.*

An aircraft is not cleared for takeoff until the departure controller can fit the flight into the departure flow. A pilot may have to hold for release. When takeoff clearance is received, the departure controller is aware of the flight and is waiting for a call. All of the information the controller needs is on the departure strip or the computer screen; there is no need to repeat any portion of the clearance to that controller. Simply establish contact with the facility when instructed to do so by the tower controller. The terminal facility computer picks up the transponder and initiates tracking as soon as it detects the assigned code. For this reason, the transponder should remain on standby until takeoff clearance has been received.

The aircraft appears on the controller's radar display as a target with an associated data block that moves as the aircraft moves through the airspace. The data block includes aircraft identification, aircraft type, altitude, and airspeed.

A TRACON controller uses Airport Surveillance Radar (ASR) to detect primary targets and Automated Radar Terminal Systems (ARTS) to receive transponder signals; the two are combined on the controller's scope. *[Figure 2-9]*

At facilities with ASR-3 equipment, radar returns from precipitation are not displayed as varying levels of intensity, and controllers must rely on pilot reports and experience to provide weather avoidance information. With ASR-9 equipment, the controller can select up to six levels of intensity. Light precipitation does not require avoidance tactics but precipitation levels of moderate, heavy, or extreme should cause pilots to plan accordingly. Along with precipitation, the pilot must additionally consider the temperature, which if between –20° and +5 °C causes icing even during light precipitation. The returns from higher levels of intensity may obscure aircraft data blocks, and controllers may select the higher levels only on pilot request. When uncertainty exists about the weather ahead, ask the controller if the facility can display intensity levels—pilots of small aircraft should avoid intensity levels 3 or higher.

Tower En Route Control (TEC)

At many locations, instrument flights can be conducted entirely in terminal airspace. These tower en route control (TEC) routes are generally for aircraft operating below 10,000 feet, and they can be found in the A/FD. Pilots desiring to use TEC should include that designation in the remarks section of the flight plan.

Pilots are not limited to the major airports at the city pairs listed in the A/FD. For example, a tower en route flight from an airport in New York (NYC) airspace could terminate at any airport within approximately 30 miles of Bradley International (BDL) airspace, such as Hartford (HFD). *[Figure 2-10]*

A valuable service provided by the automated radar equipment at terminal radar facilities is the Minimum Safe Altitude Warnings (MSAW). This equipment predicts an aircraft's position in 2 minutes based on present path of flight—the controller issues a safety alert if the projected path encounters terrain or an obstruction. An unusually rapid descent rate on a nonprecision approach can trigger such an alert.

Air Route Traffic Control Center (ARTCC)

ARTCC facilities are responsible for maintaining separation between IFR flights in the en route structure. Center radars (Air Route Surveillance Radar (ARSR)) acquire and track transponder returns using the same basic technology as terminal radars. *[Figure 2-11]*

Earlier Center radars display weather as an area of slashes (light precipitation) and Hs (moderate rainfall), as illustrated in *Figure 2-12*. Because the controller cannot detect higher levels of precipitation, pilots should be wary of areas showing moderate rainfall. Newer radar displays show weather as three levels of blue. Controllers can select the level of weather to be displayed. Weather displays of higher levels of intensity can make it difficult for controllers to see aircraft data blocks, so pilots should not expect ATC to keep weather displayed continuously.

Center airspace is divided into sectors in the same manner as terminal airspace; additionally, most Center airspace is divided by altitudes into high and low sectors. Each sector has a dedicated team of controllers and a selection of radio frequencies because each Center has a network of remote transmitter/receiver sites. All Center frequencies can be found in the back of the A/FD in the format shown in *Figure 2-13;* they are also found on en route charts.

Each ARTCC's area of responsibility covers several states; when flying from the vicinity of one remote communication site toward another, expect to hear the same controller on different frequencies.

Center Approach/Departure Control

The majority of airports with instrument approaches do not lie within terminal radar airspace and, when operating to or from these airports, pilots communicate directly with the Center controller. Departing from a tower-controlled airport, the tower controller provides instructions for contacting the appropriate Center controller. When departing an airport without an operating control tower, the clearance includes instructions such as "Upon entering controlled airspace, contact Houston Center on 126.5." Pilots are responsible for terrain clearance until reaching the controller's MVA. Simply hearing "Radar contact" does not relieve a pilot of this responsibility.

If obstacles in the departure path require a steeper-than-standard climb gradient (200 feet per nautical mile (FPNM)), then the controller advises the pilot. However, it is the pilot's responsibility to check the departure airport listing in the A/FD to determine if there are trees or wires in the departure path. When in doubt, ask the controller for the required climb gradient.

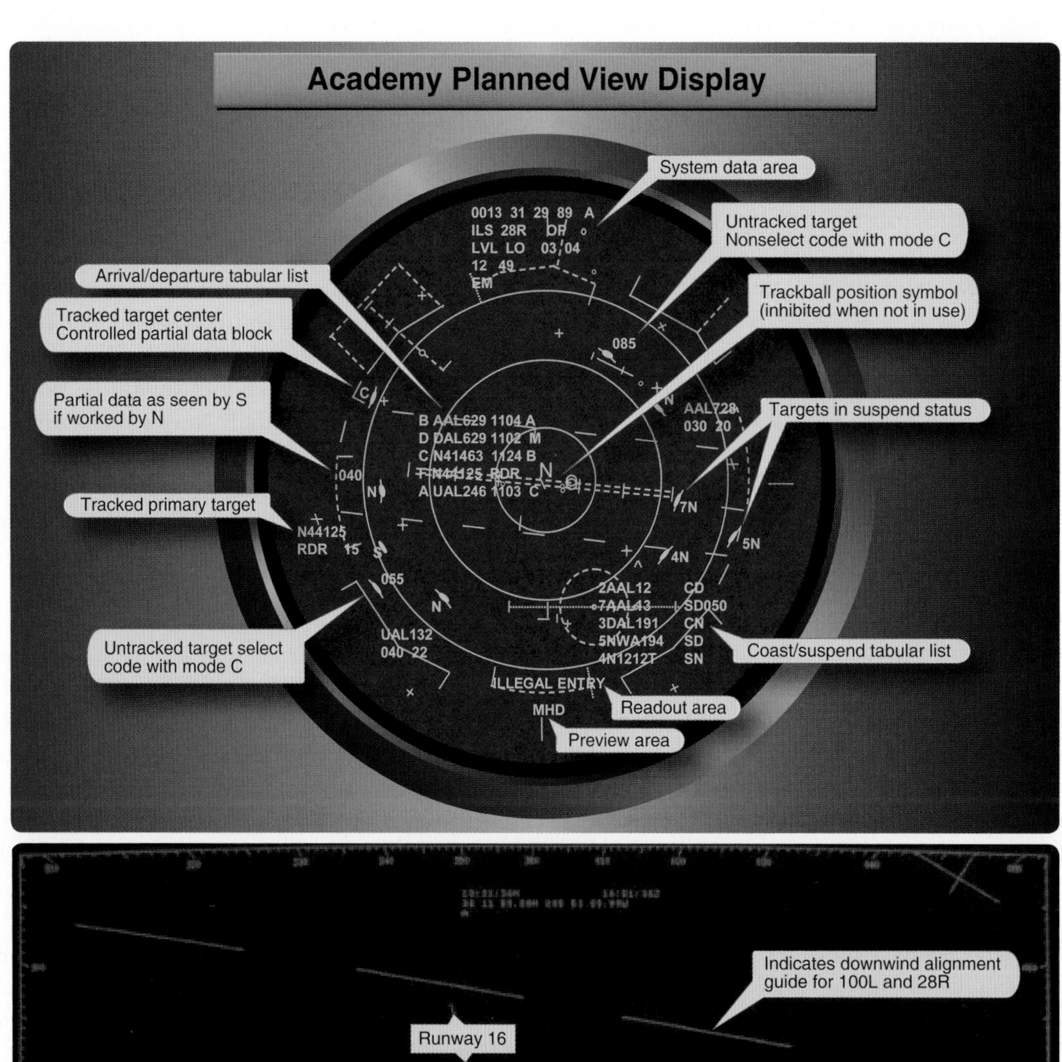

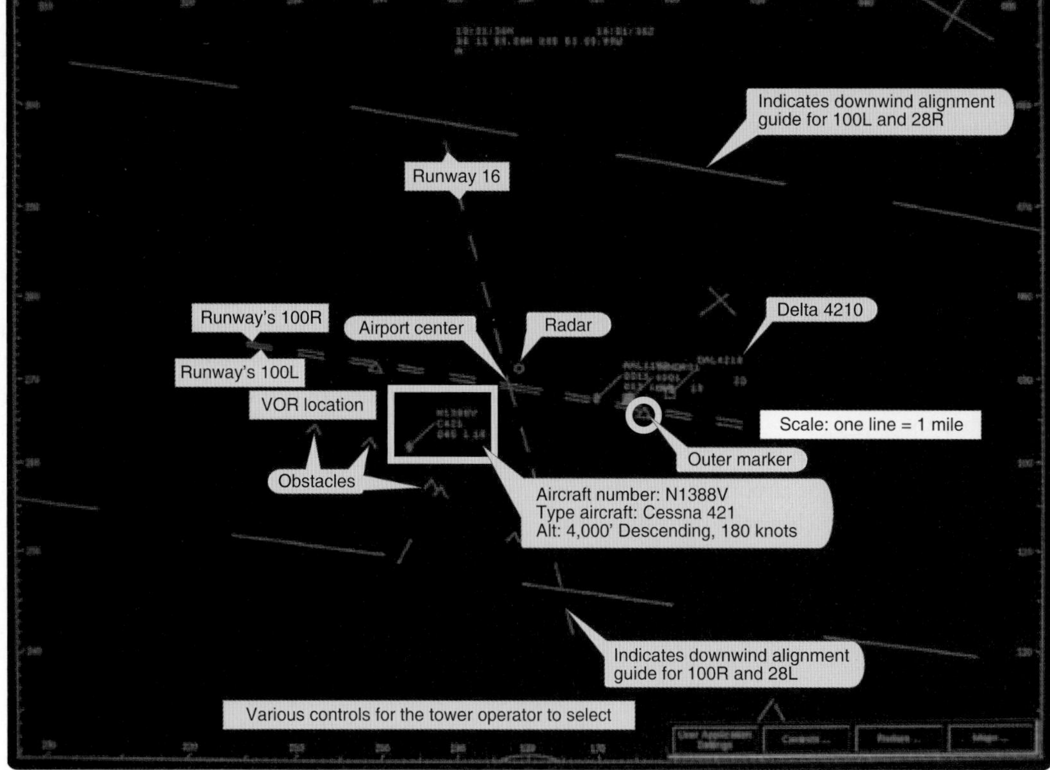

Figure 2-9. *The top image is a display as seen by controllers in an air traffic facility. It is an ARTS III (Automated Radar Terminal System). The display shown provides an explanation of the symbols in the graphic. The lower figure is an example of the Digital Bright Radar Indicator Tower Equipment (DBRITE) screen as seen by tower personnel. It provides tower controllers with a visual display of the airport surveillance radar, beacon signals, and data received from ARTS III. The display shown provides an explanation of the symbols in the graphic.*

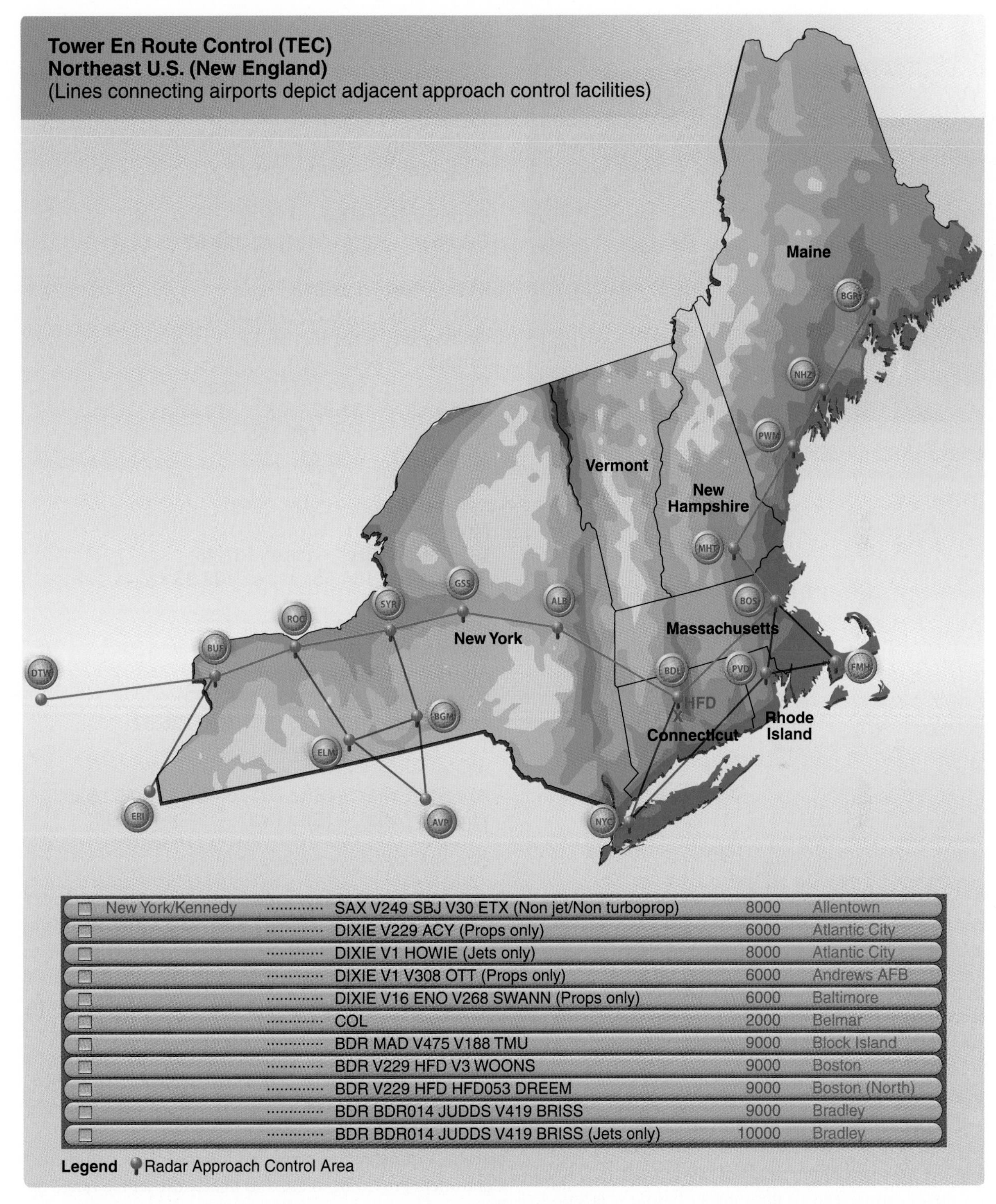

☐ New York/Kennedy	....	SAX V249 SBJ V30 ETX (Non jet/Non turboprop)	8000	Allentown
☐	....	DIXIE V229 ACY (Props only)	6000	Atlantic City
☐	....	DIXIE V1 HOWIE (Jets only)	8000	Atlantic City
☐	....	DIXIE V1 V308 OTT (Props only)	6000	Andrews AFB
☐	....	DIXIE V16 ENO V268 SWANN (Props only)	6000	Baltimore
☐	....	COL	2000	Belmar
☐	....	BDR MAD V475 V188 TMU	9000	Block Island
☐	....	BDR V229 HFD V3 WOONS	9000	Boston
☐	....	BDR V229 HFD HFD053 DREEM	9000	Boston (North)
☐	....	BDR BDR014 JUDDS V419 BRISS	9000	Bradley
☐	....	BDR BDR014 JUDDS V419 BRISS (Jets only)	10000	Bradley

Figure 2-10. *A portion of the New York area tower en route list (from the A/FD).*

Figure 2-11. *Center radar displays.*

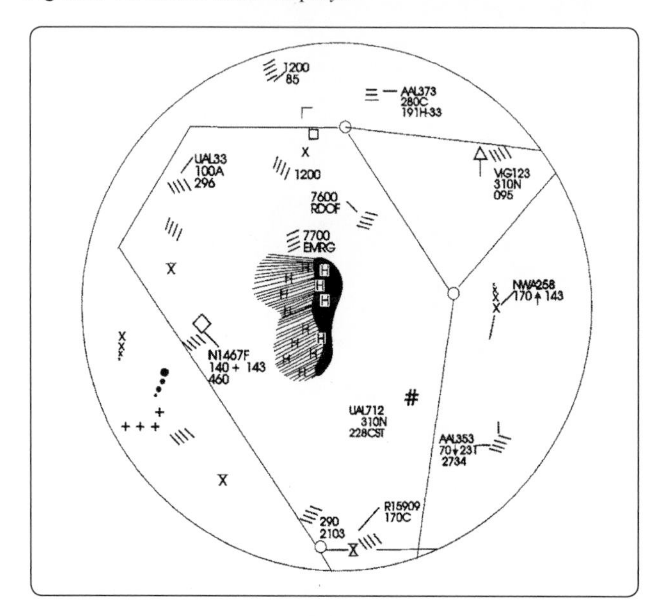

Figure 2-12. *A center controller's scope.*

A common clearance in these situations is "When able, proceed direct to the Astoria VOR..." The words "when able" mean to proceed to the waypoint, intersection, or NAVAID when the pilot is able to navigate directly to that point using onboard available systems providing proper guidance, usable signal, etc. If provided such guidance while flying VFR, the pilot remains responsible for terrain and obstacle clearance. Using the standard climb gradient, an aircraft is 2 miles from the departure end of the runway before it is safe to turn (400 feet above ground level (AGL)). When a Center controller issues a heading, a direct route, or says "direct when able," the controller becomes responsible for terrain and obstruction clearance.

®FORT WORTH CENTER	
H-2-3-4, L-4-6-13-14-15-17 (KZFW)	134.4
Abilene	–134.25 127.45
Ardmore	–132.975 128.1
Big Spring	–133.7
Blue Ridge	–127.6 124.87
Brownwood	–127.45
Clinton-Sherman	–132.45 128.4 126.3
Cumby	–132.85 132.02 126.57
Dublin	–135.375 128.32 127.15
El Dorado	–133.875 128.2
Frankston	–135.25 134.025
Gainsville	–134.15 126.77
Hobbs	–133.1
Keller	–135.275 134.15 133.25
Lubbock	–133.35 127.7 126.45
Marshall	–135.1 128.125
McAlester	–135.45 132.2
Midland (Site A)	–133.1 132.075
Mineral Wells	–135.6 127.0
Monroe	–135.1
Oklahoma City	–133.9 132.45
Paducah	–134.55 133.5 133.35 126.45
Paris	–127.6
Plainview	–126.45
San Angelo	–132.075 126.15
Scurry	–135.75 126.725
Shreveport	–135.1 132.275
Texarkana	–134.475 133.95 126.57
Tyler	–135.25 134.025
Waco	–133.3
Wichita Falls	–(Site Nr1) - 134.55 132.925
Wichita Falls	–(Site Nr2) - 133.5 127.95

Figure 2-13. *A/FD center frequencies listing.*

Another common Center clearance is "Leaving (altitude) fly (heading) or proceed direct when able." This keeps the terrain/obstruction clearance responsibility in the flight deck until above the minimum IFR altitude. A controller cannot issue an IFR clearance until an aircraft is above the minimum IFR altitude unless it is able to climb in VFR conditions.

On a Center controller's scope, 1 NM is about ⅟28 of an inch. When a Center controller is providing Approach/Departure control services at an airport many miles from the radar antenna, estimating headings and distances is very difficult. Controllers providing vectors to final must set the range on their scopes to not more than 125 NM to provide the greatest possible accuracy for intercept headings. Accordingly, at locations more distant from a Center radar antenna, pilots should expect a minimum of vectoring.

ATC Inflight Weather Avoidance Assistance

ATC Radar Weather Displays

ATC radar systems are able to display areas of precipitation by sending out a beam of radio energy that is reflected back to the radar antenna when it strikes an object or moisture, which may be in the form of rain drops, hail, or snow. The larger the object, or the denser its reflective surface, the stronger the return. Radar weather processors indicate the intensity of reflective returns in terms of decibels with respect to the radar reflectively factor (dBZ).

ATC systems cannot detect the presence or absence of clouds. ATC radar systems can often determine the intensity of a precipitation area, but the specific character of that area (snow, rain, hail, VIRGA, etc.) cannot be determined. For this reason, ATC refers to all weather areas displayed on ATC radar scopes as "precipitation."

All ATC facilities using radar weather processors with the ability to determine precipitation intensity describes the intensity to pilots as:

1. "LIGHT" (< 30 dBZ)
2. "MODERATE" (30 to 40 dBZ)
3. "HEAVY" (>40 to 50 dBZ)
4. "EXTREME" (>50 dBZ)

ARTCC controllers do not use the term "LIGHT" because their systems do not display "LIGHT" precipitation intensities. ATC facilities that, due to equipment limitations, cannot display the intensity levels of precipitation, describe the location of the precipitation area by geographic position or position relative to the aircraft. Since the intensity level is not available, the controller states, "INTENSITY UNKNOWN."

ARTCC facilities normally use a Weather and Radar Processor (WARP) to display a mosaic of data obtained from multiple NEXRAD sites. The WARP processor is only used in ARTCC facilities.

There is a time delay between actual conditions and those displayed to the controller. For example, the precipitation data on the ARTCC controller's display could be up to 6 minutes old. When the WARP is not available, a secondary system, the narrowband ARSR is utilized. The ARSR system can display two distinct levels of precipitation intensity that is described to pilots as "MODERATE" (30 to 40 dBZ) and "HEAVY to EXTREME" (>40 dBZ).

ATC radar systems cannot detect turbulence. Generally, turbulence can be expected to occur as the rate of rainfall or intensity of precipitation increases. Turbulence associated with greater rates of rainfall/precipitation is normally more severe than any associated with lesser rates of rainfall/precipitation. Turbulence should be expected to occur near convective activity, even in clear air. Thunderstorms are a form of convective activity that implies severe or greater turbulence. Operation within 20 miles of thunderstorms should be approached with great caution, as the severity of turbulence can be markedly greater than the precipitation intensity might indicate.

Weather Avoidance Assistance

ATC's first duty priority is to separate aircraft and issue safety alerts. ATC provides additional services to the extent possible, contingent upon higher priority duties and other factors including limitations of radar, volume of traffic, frequency congestion, and workload. Subject to the above factors/limitations, controllers issue pertinent information on weather or chaff areas; and if requested, assist pilots, to the extent possible, in avoiding areas of precipitation. Pilots should respond to a weather advisory by acknowledging the advisory and, if desired, requesting an alternate course of action, such as:

1. Request to deviate off course by stating the direction and number of degrees or miles needed to deviate from the original course;
2. Request a change of altitude; or
3. Request routing assistance to avoid the affected area. Because ATC radar systems cannot detect the presence or absence of clouds and turbulence, such assistance conveys no guarantee that the pilot will not encounter hazards associated with convective activity. Pilots wishing to circumnavigate precipitation areas by a specific distance should make their desires clearly known to ATC at the time of the request for services. Pilots must advise ATC when they can resume normal navigation.

IFR pilots shall not deviate from their assigned course or altitude without an ATC clearance. Plan ahead for possible course deviations because hazardous convective conditions can develop quite rapidly. This is important to consider because the precipitation data displayed on ARTCC radar scopes can be up to 6 minutes old, and thunderstorms can develop at rates exceeding 6,000 feet per minute (fpm). When encountering weather conditions that threaten the safety of the aircraft, the pilot may exercise emergency authority as stated in 14 CFR part 91, section 91.3 should an immediate deviation from the assigned clearance be necessary and time does not permit approval by ATC.

Generally, when weather disrupts the flow of air traffic, greater workload demands are placed on the controller. Requests for deviations from course and other services should be made as far in advance as possible to better assure the controller's ability to approve these requests promptly. When requesting approval to detour around weather activity, include the following information to facilitate the request:

1. The proposed point where detour commences;
2. The proposed route and extent of detour (direction and distance);
3. The point where original route will be resumed;
4. Flight conditions (instrument meteorological conditions (IMC) or visual meteorological conditions (VMC);
5. Whether the aircraft is equipped with functioning airborne radar; and
6. Any further deviation that may become necessary.

To a large degree, the assistance that might be rendered by ATC depends upon the weather information available to controllers. Due to the extremely transitory nature of hazardous weather, the controller's displayed precipitation information may be of limited value.

Obtaining IFR clearance or approval to circumnavigate hazardous weather can often be accommodated more readily in the en route areas away from terminals because there is usually less congestion and, therefore, greater freedom of action. In terminal areas, the problem is more acute because of traffic density, ATC coordination requirements, complex departure and arrival routes, and adjacent airports. As a consequence, controllers are less likely to be able to accommodate all requests for weather detours in a terminal area. Nevertheless, pilots should not hesitate to advise controllers of any observed hazardous weather and should specifically advise controllers if they desire circumnavigation of observed weather.

Pilot reports (PIREPs) of flight conditions help define the nature and extent of weather conditions in a particular area. These reports are disseminated by radio and electronic means to other pilots. Provide PIREP information to ATC regarding pertinent flight conditions, such as:

1. Turbulence;
2. Visibility;
3. Cloud tops and bases; and
4. The presence of hazards such as ice, hail, and lightning.

Approach Control Facility

An approach control facility is a terminal ATC facility that provides approach control service in the terminal area. Services are provided for arriving and departing VFR and IFR aircraft and, on occasion, en route aircraft. In addition, for airports with parallel runways with ILS or LDA approaches, the approach control facility provides monitoring of the approaches.

Approach Control Advances

Precision Runway Monitor (PRM)

Over the past few years, a new technology has been installed at airports that permits a decreased separation distance between parallel runways. The system is called a Precision Runway Monitor (PRM) and is comprised of high-update radar, high-resolution ATC displays, and PRM-certified controllers. *[Figure 2-14]*

Figure 2-14. *High-resolution ATC displays used in PRM.*

PRM Radar

The PRM uses a Monopulse Secondary Surveillance Radar (MSSR) that employs electronically-scanned antennas. Because the PRM has no scan rate restrictions, it is capable of providing a faster update rate (up to 1.0 second) over conventional systems, thereby providing better target presentation in terms of accuracy, resolution, and track prediction. The system is designed to search, track, process, and display SSR-equipped aircraft within airspace of over 30 miles in range and over 15,000 feet in elevation. Visual and audible alerts are generated to warn controllers to take corrective actions.

PRM Benefits

Typically, PRM is used with dual approaches with centerlines separated less than 4,300 feet but not less than 3,000 feet (under most conditions). *[Figure 2-15]* Separating the two final approach courses is a No Transgression Zone (NTZ) with surveillance of that zone provided by two controllers, one for each active approach. The system tracking software provides PRM monitor controllers with aircraft identification, position, speed, projected position, as well as visual and aural alerts.

Control Sequence

The IFR system is flexible and accommodating if pilots do their homework, have as many frequencies as possible written down before they are needed, and have an alternate in mind if the flight cannot be completed as planned. Pilots should familiarize themselves with all the facilities and services available along the planned route of flight. *[Figure 2-16]* Always know where the nearest VFR conditions can be found, and be prepared to head in that direction if the situation deteriorates.

A typical IFR flight, with departure and arrival at airports with control towers, would use the ATC facilities and services in the following sequence:

1. FSS: Obtain a weather briefing for a departure, destination and alternate airports, and en route conditions, and then file a flight plan by calling 1-800-WX-BRIEF.
2. ATIS: Preflight complete, listen for present conditions and the approach in use.
3. Clearance Delivery: Prior to taxiing, obtain a departure clearance.

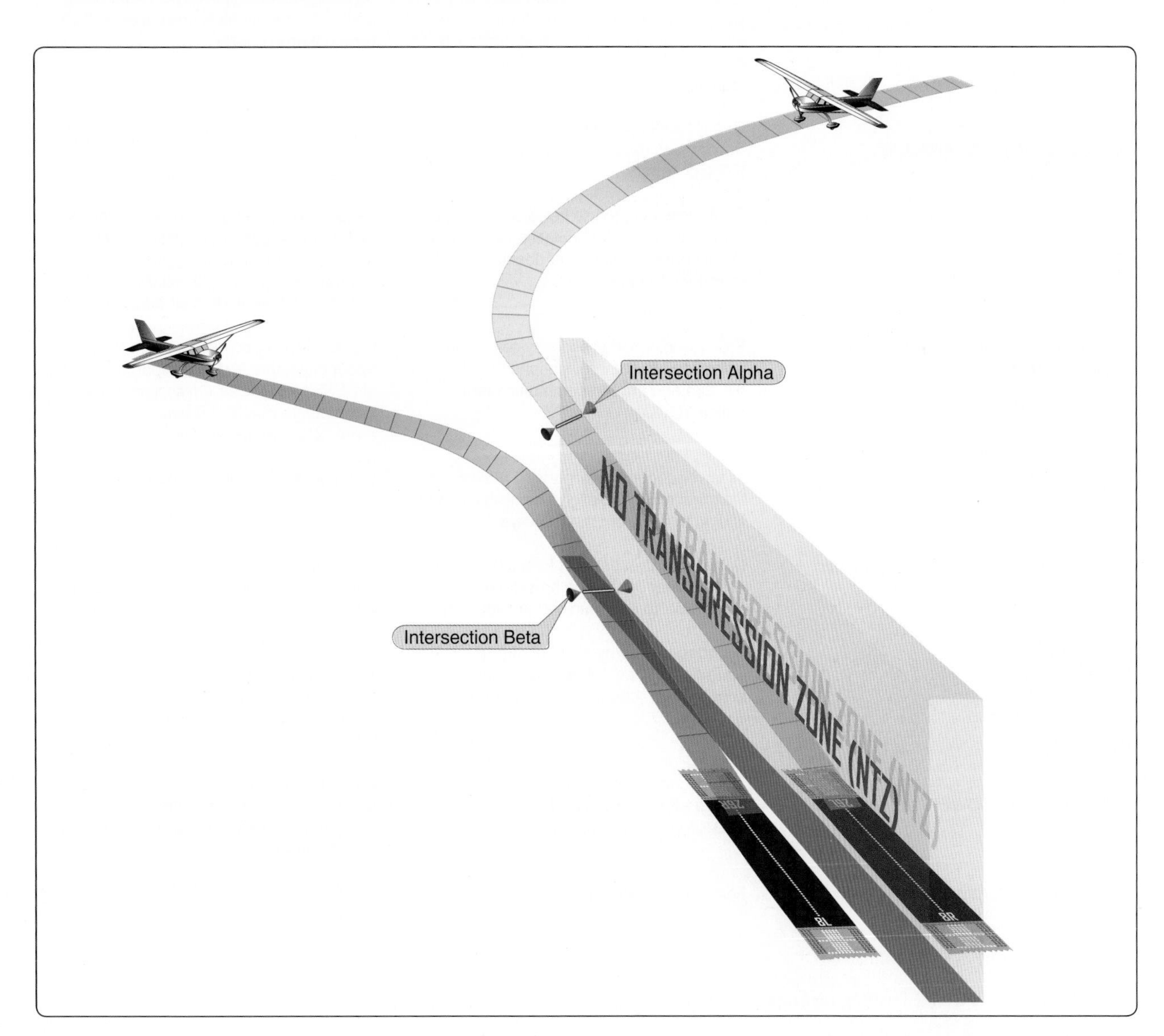

Figure 2-15. *Aircraft management using PRM. (Note the no transgression zone (NTZ) and how the aircraft are separated.)*

Communications Facility	Description	Frequency
Airport Advisory Area "[AFSS name] RADIO"	AFSS personnel provide traffic advisories to pilots operating within 10 miles of the airport.	123.6 MHz.
UNICOM "[airport name] UNICOM"	Airport advisories from an airport without an operating control tower or AFSS.	Listed in A/FD under the city name; also on sectional charts in airport data block.
Air Route Traffic Control Center (ARTCC) "CENTER"	En route radar facilities that maintain separation between IFR flights, and between IFR flights and known VFR flights. Centers provide VFR traffic advisories on a workload permitting basis.	Listed in A/FD and on instrument enroute charts.
Approach/Departure Control "[airport name] APPROACH" (unless otherwise advised)	Positions at a terminal radar facility responsible for handling of IFR flights to and from the primary airport (where Class B airspace exists).	Listed in A/FD; also on sectional charts in the communications panel and on terminal area charts.
Automatic Terminal Information Service (ATIS)	Continuous broadcast of audio tape prepared by ATC controller containing wind direction and velocity, temperature, altimeter setting, runway and approach in use, and other information of interest to pilots.	Listed in A/FD under the city name; also on sectional charts in airport data block and in the communications panel, and on terminal area charts.
Clearance Delivery "[airport name] CLEARANCE"	Control tower position responsible for transmitting departure clearances to IFR flights.	Listed on instrument approach procedure charts.
Common Traffic Advisory Frequency (CTAF)	CTAF provides a single frequency for pilots in the area to use for contacting the facility and/or broadcasting their position and intentions to other pilots.	Listed in A/FD; also on sectional charts in the airport data block (followed by a white C on a blue or magenta background). At airports with no tower, CTAF is 122.9, the "MULTICOM" frequency.
Automated Flight Service Station (AFSS) "[facility name] RADIO"	Provides information and services to pilots, using remote communications outlets (RCOs) and ground communications outlets (GCOs).	Listed in A/FD and sectional charts, both under city name and in a separate listing of AFSS frequencies. On sectional charts, listed above the VOR boxes, or in separate boxes when remote.
Ground Control "[airport name] GROUND"	At tower-controlled airports, a position in the tower responsible for controlling aircraft taxiing to and from the runways.	Listed in A/FD under city name.
Hazardous Inflight Weather Advisory Service (HIWAS)	Continuous broadcast of forecast hazardous weather conditions on selected NAVAIDs. No communication capability.	Black circle with white "H" in VOR frequency box; notation in A/FD airport listing under "Radio Aids to Navigation."
MULTICOM "[airport name] TRAFFIC"	Intended for use by pilots at airports with no radio facilities. Pilots should use self-announce procedures given in the AIM.	122.9 MHz. A/FD shows 122.9 as CTAF; also on sectional charts 122.9 is followed by a white C on a dark background, indicating CTAF.
Tower "[airport name] TOWER"	"Local" controller responsible for operations on the runways and in Class B, C, or D airspace surrounding the airport.	Listed in A/FD under city name; also on sectional and terminal control area charts in the airport data block and communications panel.
En Route Flight Advisory Service (EFAS) "FLIGHT WATCH"	For inflight weather information.	122.0 MHz (0600-2200 local time)

Figure 2-16. *ATC facilities, services, and radio call signs.*

4. Ground Control: Noting that the flight is IFR, receive taxi instructions.
5. Tower: Pre-takeoff checks complete, receive clearance to takeoff.
6. Departure Control: Once the transponder "tags up" with the ARTS, the tower controller instructs the pilot to contact Departure to establish radar contact.
7. ARTCC: After departing the departure controller's airspace, aircraft is handed off to Center, who coordinates the flight while en route. Pilots may be in contact with multiple ARTCC facilities; they coordinate the hand-offs.
8. EFAS/ Hazardous Inflight Weather Advisory Service (HIWAS): Coordinate with ATC before leaving their frequency to obtain inflight weather information.
9. ATIS: Coordinate with ATC before leaving their frequency to obtain ATIS information.
10. Approach Control: Center hands off to approach control where pilots receive additional information and clearances.
11. Tower: Once cleared for the approach, pilots are instructed to contact tower control; the flight plan is canceled by the tower controller upon landing.

A typical IFR flight, with departure and arrival at airports without operating control towers, would use the ATC facilities and services in the following sequence:

1. FSS: Obtain a weather briefing for departure, destination, and alternate airports, and en route conditions, and then file a flight plan by calling 1-800-WX-BRIEF. Provide the latitude/longitude description for small airports to ensure that Center is able to locate departure and arrival locations.
2. FSS or UNICOM: ATC clearances can be filed and received on the UNICOM frequency if the licensee has made arrangements with the controlling ARTCC; otherwise, file with FSS via telephone. Be sure all preflight preparations are complete before filing. The clearance includes a clearance void time. Pilots must be airborne prior to the void time.
3. ARTCC: After takeoff, establish contact with Center. During the flight, pilots may be in contact with multiple ARTCC facilities; ATC coordinates the hand-offs.
4. EFAS/HIWAS: Coordinate with ATC before leaving their frequency to obtain inflight weather information.
5. Approach Control: Center hands off to approach control where pilots receive additional information and clearances. If a landing under VMC is possible, pilots may cancel their IFR clearance before landing.

Letters of Agreement (LOA)

The ATC system is indeed a system and very little happens by chance. As a flight progresses, controllers in adjoining sectors or adjoining Centers coordinate its handling by telephone or by computer. Where there is a boundary between the airspace controlled by different facilities, the location and altitude for hand-off is determined by Letters of Agreement (LOA) negotiated between the two facility managers. This information is not available to pilots in any Federal Aviation Administration (FAA) publication. For this reason, it is good practice to note on the en route chart the points at which hand-offs occur. Each time a flight is handed off to a different facility, the controller knows the altitude and location—this was part of the hand-off procedure.

Chapter 3

Human Factors

Introduction

Human factors is a broad field that examines the interaction between people, machines, and the environment for the purpose of improving performance and reducing errors. As aircraft became more reliable and less prone to mechanical failure, the percentage of accidents related to human factors increased. Some aspect of human factors now accounts for over 80 percent of all accidents. Pilots, who have a good understanding of human factors, are better equipped to plan and execute a safe and uneventful flight.

Flying in instrument meteorological conditions (IMC) can result in sensations that are misleading to the body's sensory system. A safe pilot needs to understand these sensations and effectively counteract them. Instrument flying requires a pilot to make decisions using all available resources.

The elements of human factors covered in this chapter include sensory systems used for orientation and illusions in flight. For more information about physiological and psychological factors, medical factors, aeronautical decision-making (ADM), and crew resource management (CRM), refer to the Pilot's Handbook of Aeronautical Knowledge.

Sensory Systems for Orientation

Orientation is the awareness of the position of the aircraft and of oneself in relation to a specific reference point. Disorientation is the lack of orientation, and spatial disorientation specifically refers to the lack of orientation with regard to position in space and to other objects.

Orientation is maintained through the body's sensory organs in three areas: visual, vestibular, and postural. The eyes maintain visual orientation. The motion sensing system in the inner ear maintains vestibular orientation. The nerves in the skin, joints, and muscles of the body maintain postural orientation. When healthy human beings are in their natural environment, these three systems work well. When the human body is subjected to the forces of flight, these senses can provide misleading information. It is this misleading information that causes pilots to become disoriented.

Eyes

Of all the senses, vision is most important in providing information to maintain safe flight. Even though the human eye is optimized for day vision, it is also capable of vision in very low light environments. During the day, the eye uses receptors called cones, while at night, vision is facilitated by the use of rods. Both of these provide a level of vision optimized for the lighting conditions that they were intended. That is, cones are ineffective at night and rods are ineffective during the day.

Rods, which contain rhodopsin (called visual purple), are especially sensitive to light and increased light washes out the rhodopsin compromising the night vision. Hence, when strong light is momentarily introduced at night, vision may be totally ineffective as the rods take time to become effective again in darkness. Smoking, alcohol, oxygen deprivation, and age affect vision, especially at night. It should be noted that at night, oxygen deprivation, such as one caused from a climb to a high altitude, causes a significant reduction in vision. A return back to the lower altitude does not restore a pilot's vision in the same transitory period used at the climb altitude.

The eye also has two blind spots. The day blind spot is the location on the light sensitive retina where the optic nerve fiber bundle (which carries messages from the eye to the brain) passes through. This location has no light receptors, and a message cannot be created there to be sent to the brain. The night blind spot is due to a concentration of cones in an area surrounding the fovea on the retina. Because there are no rods in this area, direct vision on an object at night will disappear. As a result, off-center viewing and scanning at night is best for both obstacle avoidance and to maximize situational awareness (SA). (See the Pilot's Handbook of Aeronautical Knowledge and the Aeronautical Information Manual (AIM) for detailed reading.)

The brain also processes visual information based upon color, relationship of colors, and vision from objects around us. *Figure 3-1* demonstrates the visual processing of information. The brain assigns color based on many items, to include an object's surroundings. In the figure below, the orange square on the shaded side of the cube is actually the same color as the brown square in the center of the cube's top face.

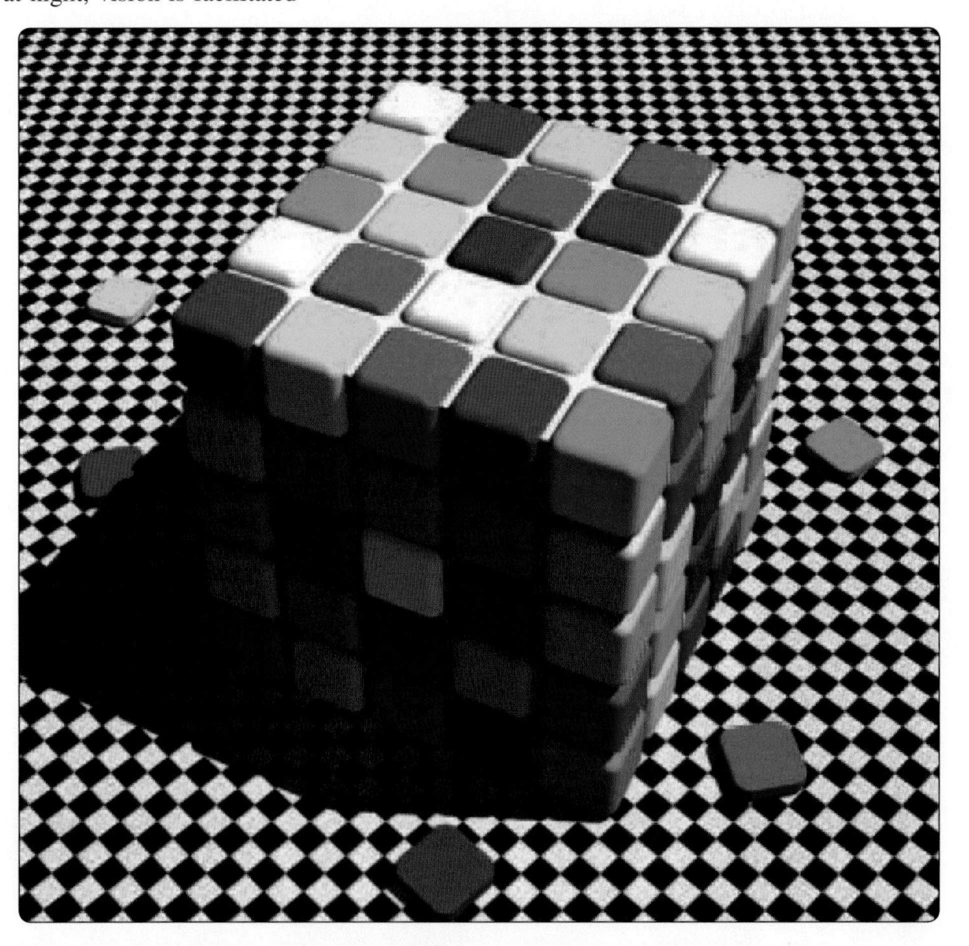

Figure 3-1. *Rubik's cube graphic depicting the visual processing of information.*

Isolating the orange square from surrounding influences will reveal that it is actually brown. The application to a real environment is evident when processing visual information that is influenced by surroundings. The ability to pick out an airport in varied terrain or another aircraft in a light haze are examples of problems with interpretation that make vigilance all the more necessary.

Figure 3-2 illustrates problems with perception. Both tables are the same lengths. Objects are easily misinterpreted in size to include both length and width. Being accustomed to a 75-foot-wide runway on flat terrain is most likely going to influence a pilot's perception of a wider runway on uneven terrain simply because of the inherent processing experience.

Vision Under Dim and Bright Illumination

Under conditions of dim illumination, aeronautical charts and aircraft instruments can become unreadable unless adequate flight deck lighting is available. In darkness, vision becomes more sensitive to light. This process is called dark adaptation. Although exposure to total darkness for at least 30 minutes is required for complete dark adaptation, a pilot can achieve a moderate degree of dark adaptation within 20 minutes under dim red flight deck lighting.

Red light distorts colors (filters the red spectrum), especially on aeronautical charts, and makes it very difficult for the eyes to focus on objects inside the aircraft. Pilots should use it only where optimum outside night vision capability is necessary. White flight deck lighting (dim lighting) should be available when needed for map and instrument reading, especially under IMC conditions.

Since any degree of dark adaptation is lost within a few seconds of viewing a bright light, pilots should close one eye when using a light to preserve some degree of night vision. During night flights in the vicinity of lightning, flight deck lights should be turned up to help prevent loss of night vision due to the bright flashes. Dark adaptation is also impaired by exposure to cabin pressure altitudes above 5,000 feet, carbon monoxide inhaled through smoking, deficiency of Vitamin A in the diet, and prolonged exposure to bright sunlight.

During flight in visual meteorological conditions (VMC), the eyes are the major orientation source and usually provide accurate and reliable information. Visual cues usually prevail over false sensations from other sensory systems. When these visual cues are taken away, as they are in IMC, false sensations can cause the pilot to quickly become disoriented.

An effective way to counter these false sensations is to recognize the problem, disregard the false sensations, rely on the flight instruments, and use the eyes to determine the aircraft attitude. The pilot must have an understanding of the problem and the skill to control the aircraft using only instrument indications.

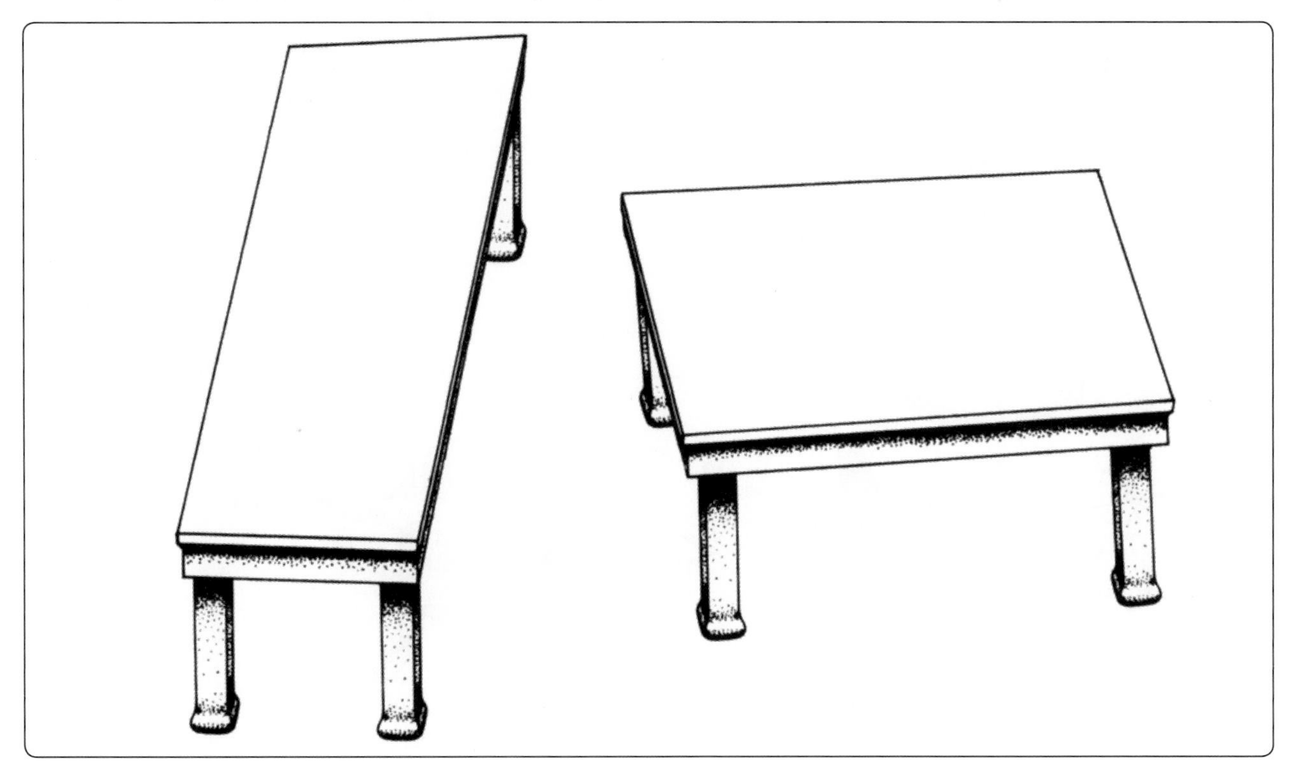

Figure 3-2. *Shepard's tables illustrating problems with perception as both tables are the same length.*

Ears

The inner ear has two major parts concerned with orientation: the semicircular canals and the otolith organs. *[Figure 3-3]* The semicircular canals detect angular acceleration of the body, while the otolith organs detect linear acceleration and gravity. The semicircular canals consist of three tubes at approximate right angles to each other, each located on one of three axes: pitch, roll, or yaw as illustrated in *Figure 3-4*. Each canal is filled with a fluid called endolymph fluid. In the center of the canal is the cupola, a gelatinous structure that rests upon sensory hairs located at the end of the vestibular nerves. It is the movement of these hairs within the fluid that causes sensations of motion.

Because of the friction between the fluid and the canal, it may take about 15–20 seconds for the fluid in the ear canal to reach the same speed as the canal's motion.

To illustrate what happens during a turn, visualize the aircraft in straight-and-level flight. With no acceleration of the aircraft, the hair cells are upright, and the body senses that no turn has occurred. Therefore, the position of the hair cells and the actual sensation correspond.

Placing the aircraft into a turn puts the semicircular canal and its fluid into motion, with the fluid within the semicircular canal lagging behind the accelerated canal walls. *[Figure 3-5]* This lag creates a relative movement of the fluid within the canal. The canal wall and the cupula move in the opposite direction from the motion of the fluid.

The brain interprets the movement of the hairs to be a turn in the same direction as the canal wall. The body correctly senses that a turn is being made. If the turn continues at a constant rate for several seconds or longer, the motion of the fluid in

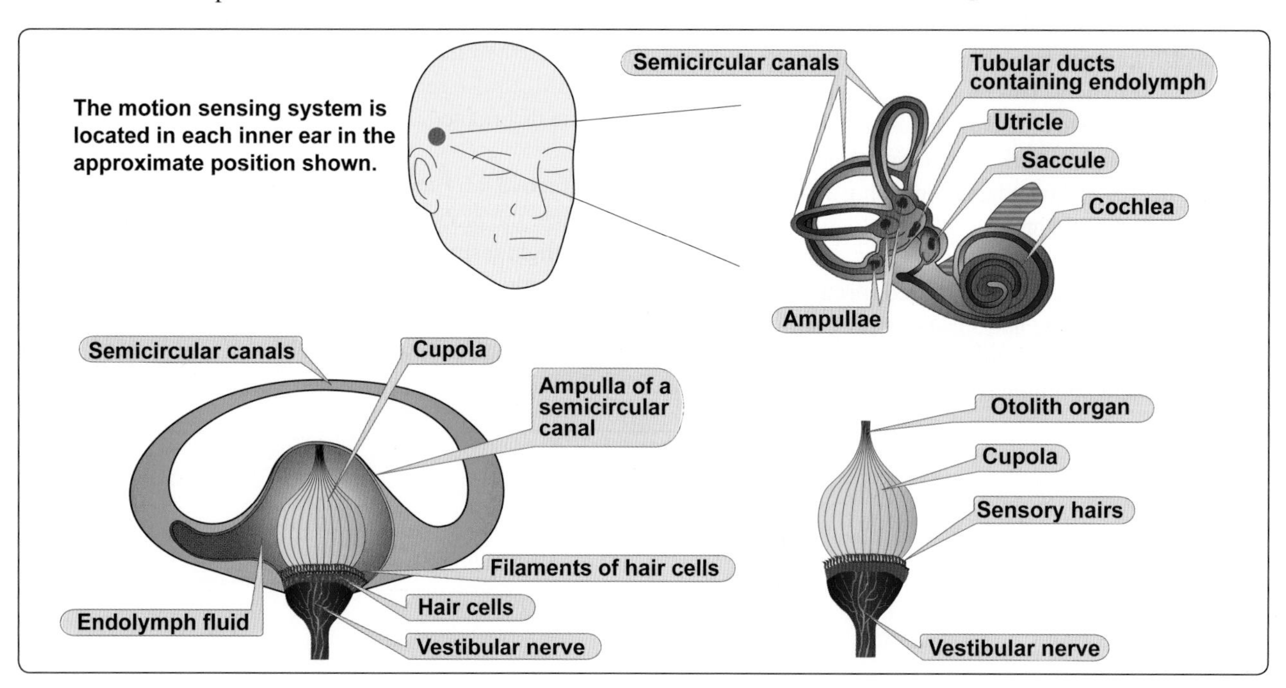

Figure 3-3. *Inner ear orientation.*

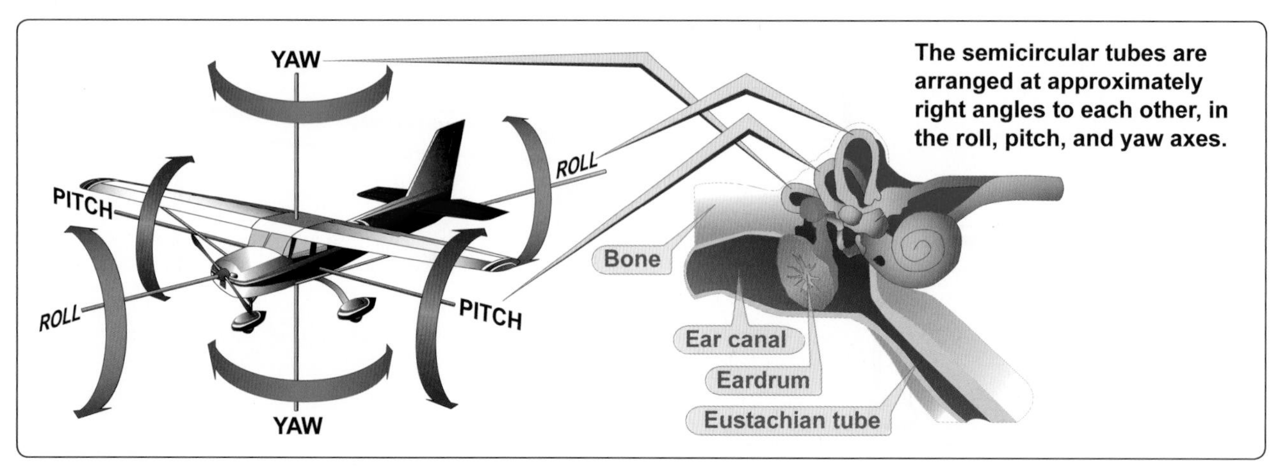

Figure 3-4. *Angular acceleration and the semicircular tubes.*

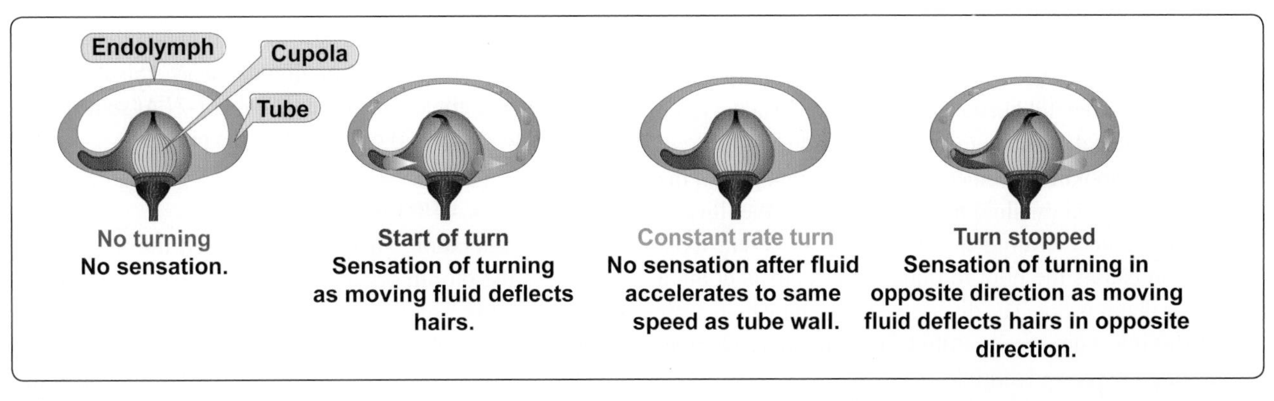
Endolymph
Cupola
Tube
No turning
No sensation.
Start of turn
Sensation of turning as moving fluid deflects hairs.
Constant rate turn
No sensation after fluid accelerates to same speed as tube wall.
Turn stopped
Sensation of turning in opposite direction as moving fluid deflects hairs in opposite direction.

Figure 3-5. *Angular acceleration.*

the canals catches up with the canal walls. The hairs are no longer bent, and the brain receives the false impression that turning has stopped. Thus, the position of the hair cells and the resulting sensation during a prolonged, constant turn in either direction results in the false sensation of no turn.

When the aircraft returns to straight-and-level flight, the fluid in the canal moves briefly in the opposite direction. This sends a signal to the brain that is falsely interpreted as movement in the opposite direction. In an attempt to correct the falsely perceived turn, the pilot may reenter the turn placing the aircraft in an out-of-control situation.

The otolith organs detect linear acceleration and gravity in a similar way. Instead of being filled with a fluid, a gelatinous membrane containing chalk-like crystals covers the sensory hairs. When the pilot tilts his or her head, the weight of these crystals causes this membrane to shift due to gravity, and the sensory hairs detect this shift. The brain orients this new position to what it perceives as vertical. Acceleration and deceleration also cause the membrane to shift in a similar manner. Forward acceleration gives the illusion of the head tilting backward. *[Figure 3-6]* As a result, during takeoff and while accelerating, the pilot may sense a steeper than normal climb resulting in a tendency to nose-down.

Nerves

Nerves in the body's skin, muscles, and joints constantly send signals to the brain, which signals the body's relation to gravity. These signals tell the pilot his or her current position. Acceleration is felt as the pilot is pushed back into the seat. Forces, created in turns, can lead to false sensations of the true direction of gravity and may give the pilot a false sense of which way is up.

Uncoordinated turns, especially climbing turns, can cause misleading signals to be sent to the brain. Skids and slips give the sensation of banking or tilting. Turbulence can create motions that confuse the brain as well. Pilots need to be aware that fatigue or illness can exacerbate these sensations and ultimately lead to subtle incapacitation.

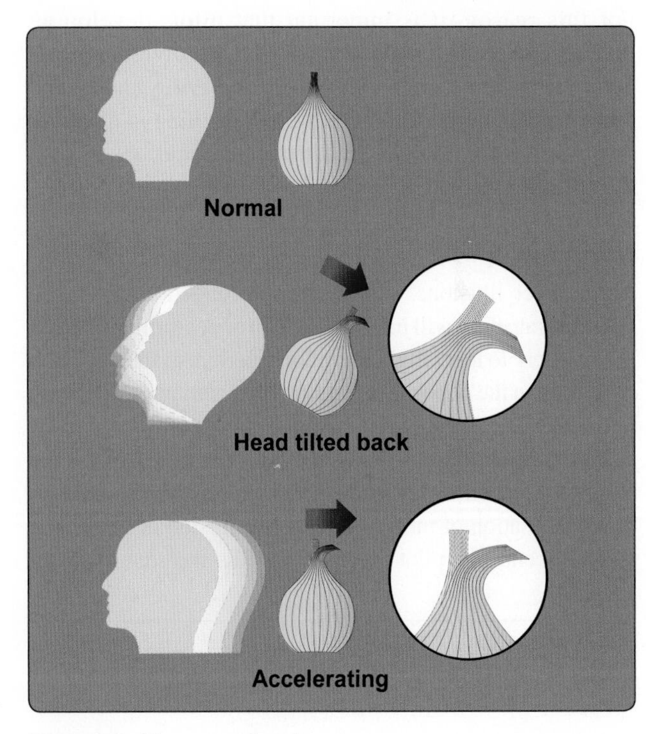
Normal
Head tilted back
Accelerating

Figure 3-6. *Linear acceleration.*

Illusions Leading to Spatial Disorientation

The sensory system responsible for most of the illusions leading to spatial disorientation is the vestibular system. Visual illusions can also cause spatial disorientation.

Vestibular Illusions

The Leans

A condition called "the leans" can result when a banked attitude, to the left for example, may be entered too slowly to set in motion the fluid in the "roll" semicircular tubes. *[Figure 3-5]* An abrupt correction of this attitude sets the fluid in motion, creating the illusion of a banked attitude to the right. The disoriented pilot may make the error of rolling the aircraft into the original left banked attitude, or if level flight is maintained, feel compelled to lean in the perceived vertical plane until this illusion subsides.

Coriolis Illusion

The coriolis illusion occurs when a pilot has been in a turn long enough for the fluid in the ear canal to move at the same speed as the canal. A movement of the head in a different plane, such as looking at something in a different part of the flight deck, may set the fluid moving and create the illusion of turning or accelerating on an entirely different axis. This action causes the pilot to think the aircraft is doing a maneuver that it is not. The disoriented pilot may maneuver the aircraft into a dangerous attitude in an attempt to correct the aircraft's perceived attitude.

For this reason, it is important that pilots develop an instrument cross-check or scan that involves minimal head movement. Take care when retrieving charts and other objects in the flight deck—if something is dropped, retrieve it with minimal head movement and be alert for the coriolis illusion.

Graveyard Spiral

As in other illusions, a pilot in a prolonged coordinated, constant-rate turn, will have the illusion of not turning. During the recovery to level flight, the pilot experiences the sensation of turning in the opposite direction. The disoriented pilot may return the aircraft to its original turn. Because an aircraft tends to lose altitude in turns unless the pilot compensates for the loss in lift, the pilot may notice a loss of altitude. The absence of any sensation of turning creates the illusion of being in a level descent. The pilot may pull back on the controls in an attempt to climb or stop the descent. This action tightens the spiral and increases the loss of altitude; hence, this illusion is referred to as a graveyard spiral. *[Figure 3-7]* At some point, this could lead to a loss of control by the pilot.

Somatogravic Illusion

A rapid acceleration, such as experienced during takeoff, stimulates the otolith organs in the same way as tilting the head backwards. This action creates the somatogravic illusion of being in a nose-up attitude, especially in situations without good visual references. The disoriented pilot may push the aircraft into a nose-low or dive attitude. A rapid deceleration by quick reduction of the throttle(s) can have the opposite effect with the disoriented pilot pulling the aircraft into a nose-up or stall attitude.

Inversion Illusion

An abrupt change from climb to straight-and-level flight can stimulate the otolith organs enough to create the illusion of tumbling backwards or inversion illusion. The disoriented pilot may push the aircraft abruptly into a nose-low attitude, possibly intensifying this illusion.

Elevator Illusion

An abrupt upward vertical acceleration, as can occur in an updraft, can stimulate the otolith organs to create the illusion of being in a climb. This is called elevator illusion. The disoriented pilot may push the aircraft into a nose-low

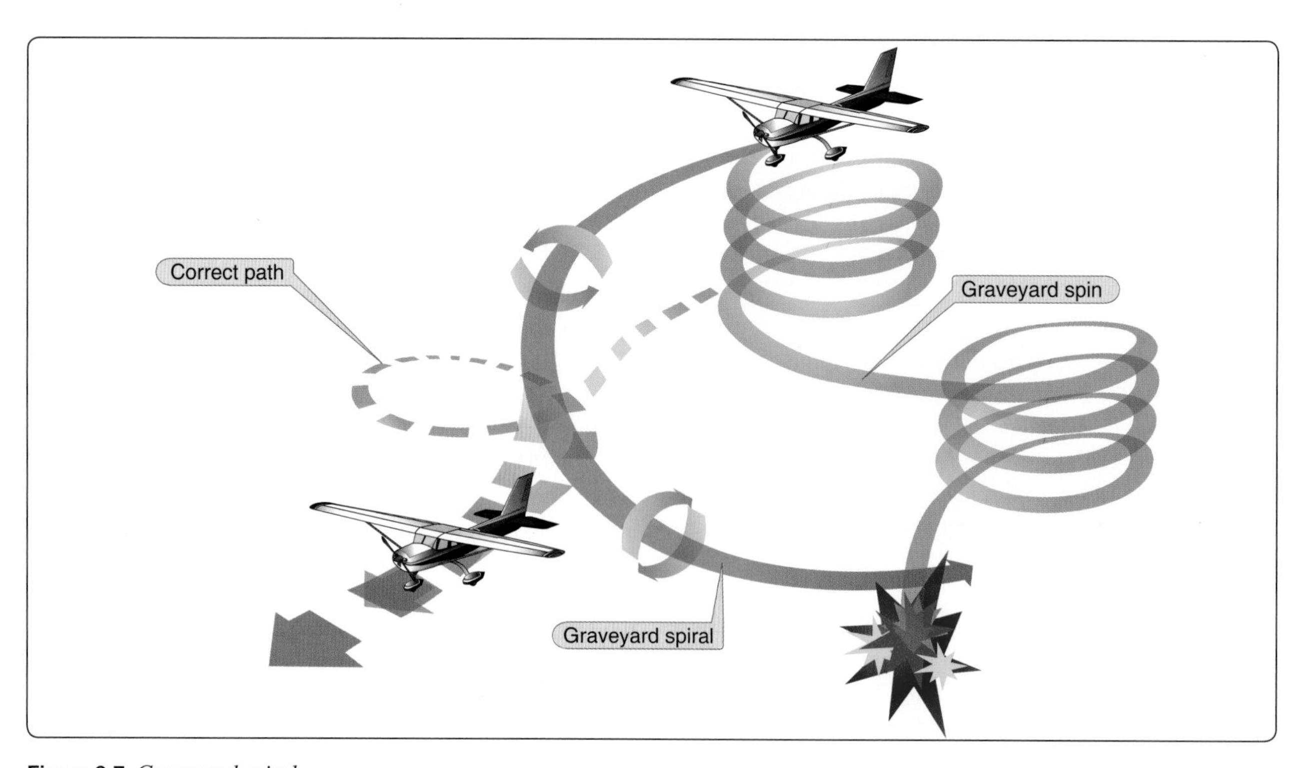

Figure 3-7. *Graveyard spiral.*

attitude. An abrupt downward vertical acceleration, usually in a downdraft, has the opposite effect with the disoriented pilot pulling the aircraft into a nose-up attitude.

Visual Illusions

Visual illusions are especially hazardous because pilots rely on their eyes for correct information. Two illusions that lead to spatial disorientation, false horizon and autokinesis, are concerned with only the visual system.

False Horizon

A sloping cloud formation, an obscured horizon, an aurora borealis, a dark scene spread with ground lights and stars, and certain geometric patterns of ground lights can provide inaccurate visual information, or false horizon, for aligning the aircraft correctly with the actual horizon. The disoriented pilot may place the aircraft in a dangerous attitude.

Autokinesis

In the dark, a stationary light will appear to move about when stared at for many seconds. The disoriented pilot could lose control of the aircraft in attempting to align it with the false movements of this light called autokinesis.

Postural Considerations

The postural system sends signals from the skin, joints, and muscles to the brain that are interpreted in relation to the Earth's gravitational pull. These signals determine posture. Inputs from each movement update the body's position to the brain on a constant basis. "Seat of the pants" flying is largely dependent upon these signals. Used in conjunction with visual and vestibular clues, these sensations can be fairly reliable. However, because of the forces acting upon the body in certain flight situations, many false sensations can occur due to acceleration forces overpowering gravity. *[Figure 3-8]* These situations include uncoordinated turns, climbing turns, and turbulence.

Demonstration of Spatial Disorientation

There are a number of controlled aircraft maneuvers a pilot can perform to experiment with spatial disorientation. While each maneuver normally creates a specific illusion, any false sensation is an effective demonstration of disorientation. Thus, even if there is no sensation during any of these maneuvers, the absence of sensation is still an effective demonstration in that it shows the inability to detect bank or roll. There are several objectives in demonstrating these various maneuvers.

1. They teach pilots to understand the susceptibility of the human system to spatial disorientation.
2. They demonstrate that judgments of aircraft attitude based on bodily sensations are frequently false.
3. They help lessen the occurrence and degree of disorientation through a better understanding of the relationship between aircraft motion, head movements, and resulting disorientation.
4. They help instill a greater confidence in relying on flight instruments for assessing true aircraft attitude.

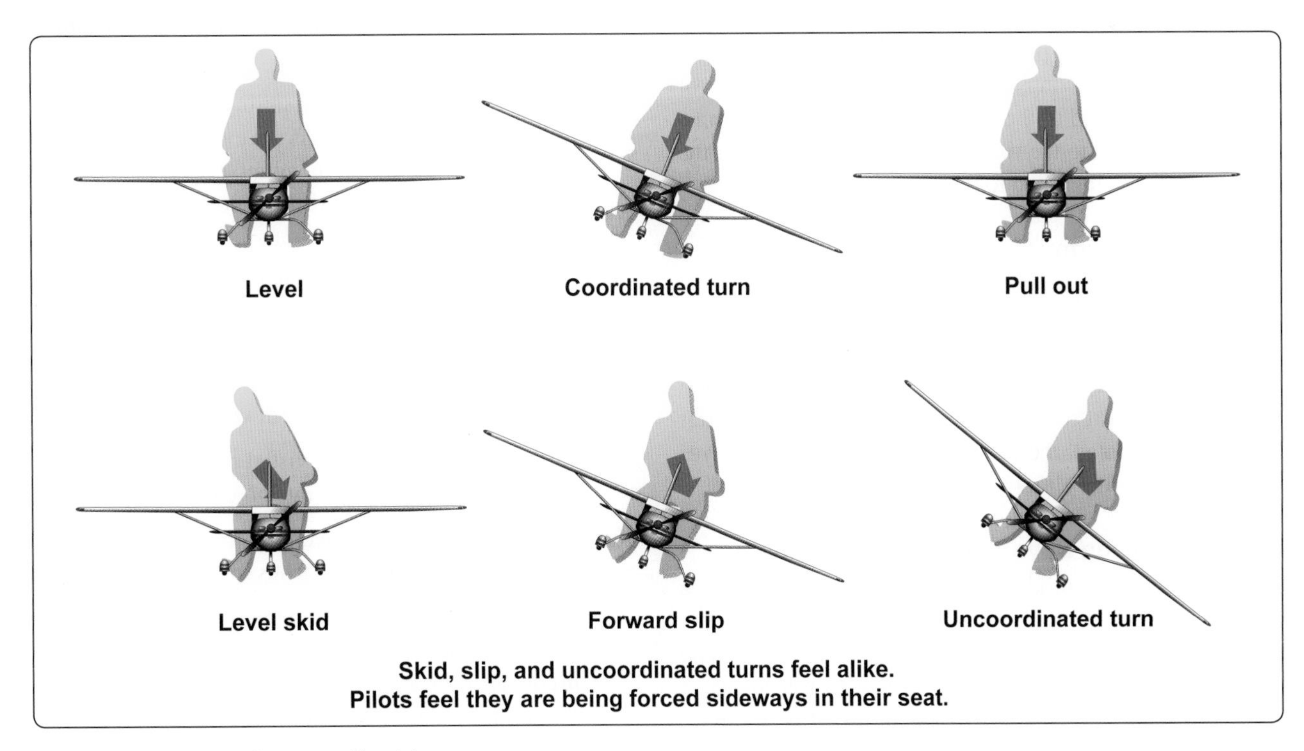

Figure 3-8. *Sensations from centrifugal force.*

A pilot should not attempt any of these maneuvers at low altitudes or in the absence of an instructor pilot or an appropriate safety pilot.

Climbing While Accelerating

With the pilot's eyes closed, the instructor pilot maintains approach airspeed in a straight-and-level attitude for several seconds, and then accelerates while maintaining straight-and-level attitude. The usual illusion during this maneuver, without visual references, is that the aircraft is climbing.

Climbing While Turning

With the pilot's eyes still closed and the aircraft in a straight-and-level attitude, the instructor pilot now executes, with a relatively slow entry, a well-coordinated turn of about 1.5 positive G (approximately 50° bank) for 90°. While in the turn, without outside visual references and under the effect of the slight positive G, the usual illusion produced is that of a climb. Upon sensing the climb, the pilot should immediately open the eyes and see that a slowly established, coordinated turn produces the same feeling as a climb.

Diving While Turning

Repeating the previous procedure, with the exception that the pilot's eyes should be kept closed until recovery from the turn is approximately one-half completed can create this sensation. With the eyes closed, the usual illusion is that the aircraft is diving.

Tilting to Right or Left

While in a straight-and-level attitude, with the pilot's eyes closed, the instructor pilot executes a moderate or slight skid to the left with wings level. This creates the illusion of the body being tilted to the right. The same illusion can be sensed with a skid to the right with wings level, except the body feels it is being tilted to the left.

Reversal of Motion

This illusion can be demonstrated in any of the three planes of motion. While straight and level, with the pilot's eyes closed, the instructor pilot smoothly and positively rolls the aircraft to approximately a 45° bank attitude. This creates the illusion of a strong sense of rotation in the opposite direction. After this illusion is noted, the pilot should open his or her eyes and observe that the aircraft is in a banked attitude.

Diving or Rolling Beyond the Vertical Plane

This maneuver may produce extreme disorientation. While in straight-and-level flight, the pilot should sit normally, either with eyes closed or gaze lowered to the floor. The instructor pilot starts a positive, coordinated roll toward a 30° or 40° angle of bank. As this is in progress, the pilot tilts his or her head forward, looks to the right or left, then immediately returns his or her head to an upright position. The instructor pilot should time the maneuver so the roll is stopped as the pilot returns his or her head upright. An intense disorientation is usually produced by this maneuver, and the pilot experiences the sensation of falling downward into the direction of the roll.

In the descriptions of these maneuvers, the instructor pilot is doing the flying, but having the pilot do the flying can also be a very effective demonstration. The pilot should close his or her eyes and tilt their head to one side. The instructor pilot tells the pilot what control inputs to perform. The pilot then attempts to establish the correct attitude or control input with eyes closed and head tilted. While it is clear the pilot has no idea of the actual attitude, he or she will react to what the senses are saying. After a short time, the pilot will become disoriented, and the instructor pilot then tells the pilot to look up and recover. The benefit of this exercise is the pilot experiences the disorientation while flying the aircraft.

Coping with Spatial Disorientation

To prevent illusions and their potentially disastrous consequences, pilots can:

1. Understand the causes of these illusions and remain constantly alert for them. Take the opportunity to understand and then experience spatial disorientation illusions in a device, such as a Barany chair, a Vertigon, or a Virtual Reality Spatial Disorientation Demonstrator.
2. Always obtain and understand preflight weather briefings.
3. Before flying in marginal visibility (less than 3 miles) or where a visible horizon is not evident such as flight over open water during the night, obtain training and maintain proficiency in airplane control by reference to instruments.
4. Do not continue flight into adverse weather conditions or into dusk or darkness unless proficient in the use of flight instruments. If intending to fly at night, maintain night-flight currency and proficiency. Include cross-country and local operations at various airfields.
5. Ensure that when outside visual references are used, they are reliable, fixed points on the Earth's surface.
6. Avoid sudden head movement, particularly during takeoffs, turns, and approaches to landing.
7. Be physically tuned for flight into reduced visibility. Ensure proper rest, adequate diet, and, if flying at night, allow for night adaptation. Remember that illness, medication, alcohol, fatigue, sleep loss, and

mild hypoxia are likely to increase susceptibility to spatial disorientation.

8. Most importantly, become proficient in the use of flight instruments and rely upon them. Trust the instruments and disregard your sensory perceptions.

The sensations that lead to illusions during instrument flight conditions are normal perceptions experienced by pilots. These undesirable sensations cannot be completely prevented, but through training and awareness, pilots can ignore or suppress them by developing absolute reliance on the flight instruments. As pilots gain proficiency in instrument flying, they become less susceptible to these illusions and their effects.

Optical Illusions

Of the senses, vision is the most important for safe flight. However, various terrain features and atmospheric conditions can create optical illusions. These illusions are primarily associated with landing. Since pilots must transition from reliance on instruments to visual cues outside the flight deck for landing at the end of an instrument approach, it is imperative they be aware of the potential problems associated with these illusions and take appropriate corrective action. The major illusions leading to landing errors are described below.

Runway Width Illusion

A narrower-than-usual runway can create an illusion the aircraft is at a higher altitude than it actually is, especially when runway length-to-width relationships are comparable. *[Figure 3-9A]* The pilot who does not recognize this illusion will fly a lower approach with the risk of striking objects along the approach path or landing short. A wider-than-usual runway can have the opposite effect with the risk of leveling out high and landing hard or overshooting the runway.

Runway and Terrain Slopes Illusion

An upsloping runway, upsloping terrain, or both can create an illusion the aircraft is at a higher altitude than it actually is. *[Figure 3-9B]* The pilot who does not recognize this illusion will fly a lower approach. Downsloping runways and downsloping approach terrain can have the opposite effect.

Featureless Terrain Illusion

An absence of surrounding ground features, as in an overwater approach, over darkened areas, or terrain made featureless by snow, can create an illusion the aircraft is at a higher altitude than it actually is. This illusion, sometimes referred to as the "black hole approach," causes pilots to fly a lower approach than is desired.

Water Refraction

Rain on the windscreen can create an illusion of being at a higher altitude due to the horizon appearing lower than it is. This can result in the pilot flying a lower approach.

Haze

Atmospheric haze can create an illusion of being at a greater distance and height from the runway. As a result, the pilot has a tendency to be low on the approach. Conversely, extremely clear air (clear bright conditions of a high attitude airport) can give the pilot the illusion of being closer than he or she actually is, resulting in a high approach that may cause an overshoot or go around. The diffusion of light due to water particles on the windshield can adversely affect depth perception. The lights and terrain features normally used to gauge height during landing become less effective for the pilot.

Fog

Flying into fog can create an illusion of pitching up. Pilots who do not recognize this illusion often steepen the approach quite abruptly.

Ground Lighting Illusions

Lights along a straight path, such as a road or lights on moving trains, can be mistaken for runway and approach lights. Bright runway and approach lighting systems, especially where few lights illuminate the surrounding terrain, may create the illusion of less distance to the runway. The pilot who does not recognize this illusion will often fly a higher approach.

How to Prevent Landing Errors Due to Optical Illusions

To prevent these illusions and their potentially hazardous consequences, pilots can:

1. Anticipate the possibility of visual illusions during approaches to unfamiliar airports, particularly at night or in adverse weather conditions. Consult airport diagrams and the Airport/Facility Directory (A/FD) for information on runway slope, terrain, and lighting.
2. Make frequent reference to the altimeter, especially during all approaches, day and night.
3. If possible, conduct aerial visual inspection of unfamiliar airports before landing.
4. Use Visual Approach Slope Indicator (VASI) or Precision Approach Path Indicator (PAPI) systems for a visual reference or an electronic glideslope, whenever they are available.

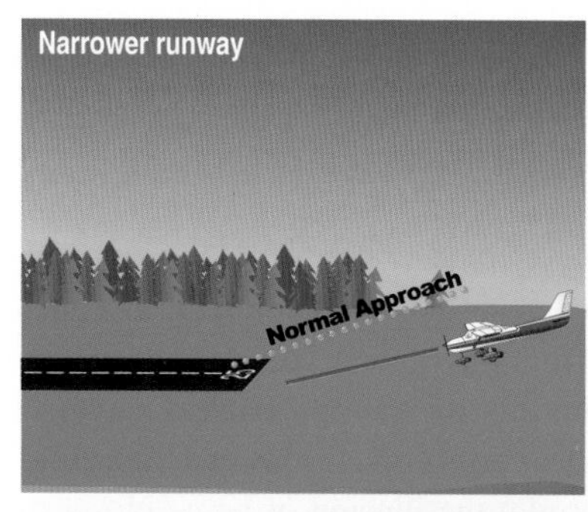

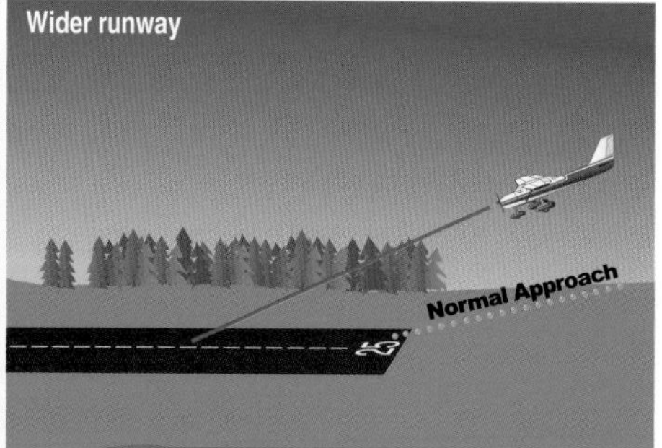

Figure 3-9A
Runway width illusion

- **A narrower-than-usual runway can created an illusion that the aircraft is higher than it actually is, leading to a lower approach.**
- **A wider-than-usual runway can create an illusion that the aircraft is lower than it actually is, leading to a higher approach.**

Downsloping runway
Upsloping runway
Normal Approach
Normal Approach

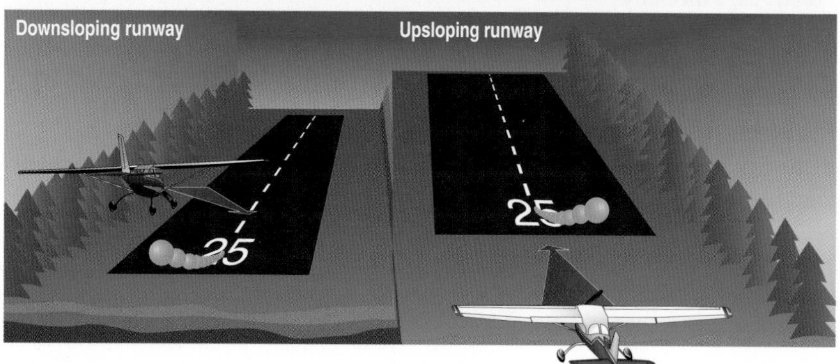

Figure 3-9B
Runway slope illusion

- **A downsloping runway can create the illusion that the aircraft is lower than it actually is, leading to a higher approach.**
- **An upsloping runway can create the illusion that the aircraft is higher than it actually is, leading to a lower approach.**

Normal approach
Approach due to illusion

Figure 3-9. *Runway width and slope illusions.*

5. Utilize the visual descent point (VDP) found on many nonprecision instrument approach procedure charts.
6. Recognize that the chances of being involved in an approach accident increase when some emergency or other activity distracts from usual procedures.
7. Maintain optimum proficiency in landing procedures.

Chapter 4

Aerodynamic Factors

Introduction

Several factors affect aircraft performance including the atmosphere, aerodynamics, and aircraft icing. Pilots need an understanding of these factors for a sound basis for prediction of aircraft response to control inputs, especially with regard to instrument approaches, while holding, and when operating at reduced airspeed in instrument meteorological conditions (IMC). Although these factors are important to the pilot flying visual flight rules (VFR), they must be even more thoroughly understood by the pilot operating under instrument flight rules (IFR). Instrument pilots rely strictly on instrument indications to precisely control the aircraft; therefore, they must have a solid understanding of basic aerodynamic principles in order to make accurate judgments regarding aircraft control inputs.

The Wing

To understand aerodynamic forces, a pilot needs to understand basic terminology associated with airfoils. *Figure 4-1* illustrates a typical airfoil.

The chord line is the straight line intersecting the leading and trailing edges of the airfoil, and the term chord refers to the chord line longitudinal length (length as viewed from the side).

The mean camber is a line located halfway between the upper and lower surfaces. Viewing the wing edgewise, the mean camber connects with the chord line at each end. The mean camber is important because it assists in determining aerodynamic qualities of an airfoil. The measurement of the maximum camber; inclusive of both the displacement of the mean camber line and its linear measurement from the end of the chord line, provide properties useful in evaluating airfoils.

Review of Basic Aerodynamics

The instrument pilot must understand the relationship and differences between several factors that affect the performance of an aircraft in flight. Also, it is crucial to understand how the aircraft reacts to various control and power changes, because the environment in which instrument pilots fly has inherent hazards not found in visual flying. The basis for this understanding is found in the four forces acting on an aircraft and Newton's Three Laws of Motion.

Relative Wind is the direction of the airflow with respect to an airfoil.

Angle of Attack (AOA) is the acute angle measured between the relative wind, or flightpath and the chord of the airfoil. *[Figure 4-2]*

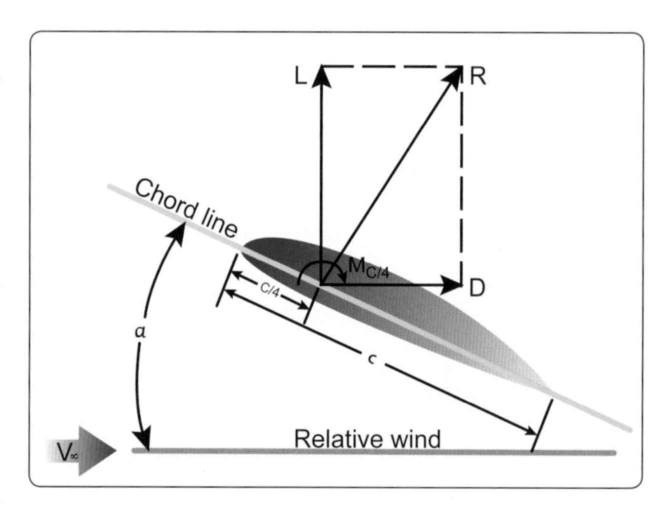

Figure 4-2. *Angle of attack and relative wind.*

Flightpath is the course or track along which the aircraft is flying or is intended to be flown.

The Four Forces

The four basic forces *[Figure 4-3]* acting upon an aircraft in flight are lift, weight, thrust, and drag.

Lift

Lift is a component of the total aerodynamic force on an airfoil and acts perpendicular to the relative wind. Relative wind is the direction of the airflow with respect to an airfoil. This force acts straight up from the average (called mean) center of pressure (CP), which is called the center of lift. It should be noted that it is a point along the chord line of an airfoil through which all aerodynamic forces are considered to act. The magnitude of lift varies proportionately with speed, air density, shape and size of the airfoil, and AOA. During straight-and-level flight, lift and weight are equal.

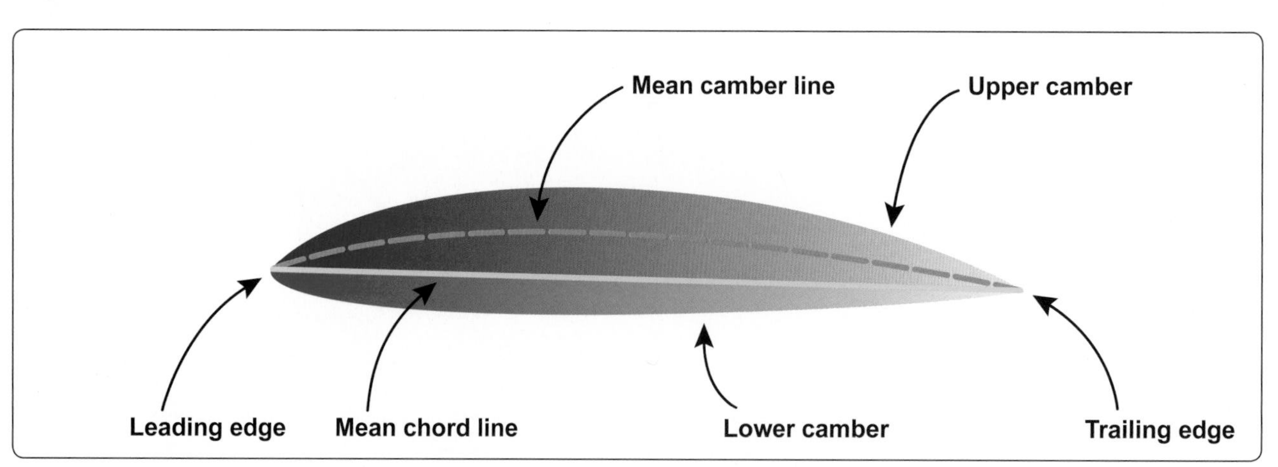

Figure 4-1. *The airfoil.*

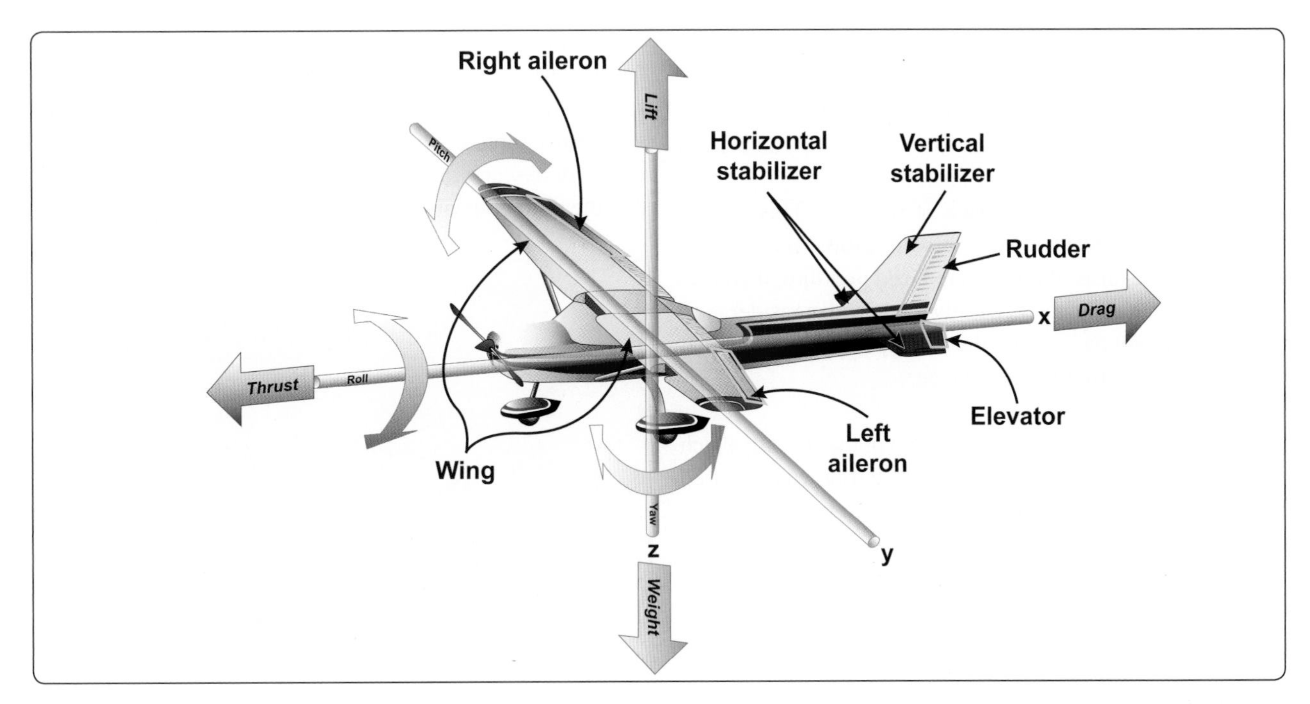

Figure 4-3. *The four forces and three axes of rotation.*

Weight

Weight is the force exerted by an aircraft from the pull of gravity. It acts on an aircraft through its center of gravity (CG) and is straight down. This should not be confused with the center of lift, which can be significantly different from the CG. As an aircraft is descending, weight is greater than lift.

Thrust

Thrust is the forward force produced by the powerplant/propeller or rotor. It opposes or overcomes the force of drag. As a general rule, it acts parallel to the longitudinal axis.

Drag

Drag is the net aerodynamic force parallel to the relative wind and is generally a sum of two components: induced drag and parasite drag.

Induced Drag

Induced drag is caused from the creation of lift and increases with AOA. Therefore, if the wing is not producing lift, induced drag is zero. Conversely, induced drag decreases with airspeed.

Parasite Drag

Parasite drag is all drag not caused from the production of lift. Parasite drag is created by displacement of air by the aircraft, turbulence generated by the airfoil, and the hindrance of airflow as it passes over the surface of the aircraft or components. All of these forces create drag not from the production of lift but the movement of an object through an air mass. Parasite drag increases with speed and includes skin friction drag, interference drag, and form drag.

• Skin Friction Drag

Covering the entire "wetted" surface of the aircraft is a thin layer of air called a boundary layer. The air molecules on the surface have zero velocity in relation to the surface; however, the layer just above moves over the stagnant molecules below because it is pulled along by a third layer close to the free stream of air. The velocities of the layers increase as the distance from the surface increases until free stream velocity is reached, but all are affected by the free stream. The distance (total) between the skin surface and where free stream velocity is reached is called the boundary layer. At subsonic levels the cumulative layers are about the thickness of a playing card, yet their motion sliding over one another creates a drag force. This force retards motion due to the viscosity of the air and is called skin friction drag. Because skin friction drag is related to a large surface area its affect on smaller aircraft is small versus large transport aircraft where skin friction drag may be considerable.

• Interference Drag

Interference drag is generated by the collision of airstreams creating eddy currents, turbulence, or restrictions to smooth flow. For instance, the airflow around a fuselage and around the wing meet at some point, usually near the wing's root. These airflows interfere with each other causing a greater drag than the individual values. This is often the case when external items are placed on an aircraft. That is, the drag of each item individually, added to that of the aircraft, are less than that of the two items when allowed to interfere with one another.

• Form Drag

Form drag is the drag created because of the shape of a component or the aircraft. If one were to place a circular disk in an air stream, the pressure on both the top and bottom would be equal. However, the airflow starts to break down as the air flows around the back of the disk. This creates turbulence and hence a lower pressure results. Because the total pressure is affected by this reduced pressure, it creates a drag. Newer aircraft are generally made with consideration to this by fairing parts along the fuselage (teardrop) so that turbulence and form drag is reduced.

Total lift must overcome the total weight of the aircraft, which is comprised of the actual weight and the tail-down force used to control the aircraft's pitch attitude. Thrust must overcome total drag in order to provide forward speed with which to produce lift. Understanding how the aircraft's relationship between these elements and the environment provide proper interpretation of the aircraft's instruments.

Newton's First Law, the Law of Inertia

Newton's First Law of Motion is the Law of Inertia. It states that a body at rest will remain at rest, and a body in motion will remain in motion, at the same speed and in the same direction until affected by an outside force. The force with which a body offers resistance to change is called the force of inertia. Two outside forces are always present on an aircraft in flight: gravity and drag. The pilot uses pitch and thrust controls to counter or change these forces to maintain the desired flightpath. If a pilot reduces power while in straight-and-level flight, the aircraft will slow due to drag. However, as the aircraft slows there is a reduction of lift, which causes the aircraft to begin a descent due to gravity. *[Figure 4-4]*

Newton's Second Law, the Law of Momentum

Newton's Second Law of Motion is the Law of Momentum, which states that a body will accelerate in the same direction as the force acting upon that body, and the acceleration will be directly proportional to the net force and inversely proportional to the mass of the body. Acceleration refers either to an increase or decrease in velocity, although deceleration is commonly used to indicate a decrease. This law governs the aircraft's ability to change flightpath and speed, which are controlled by attitude (both pitch and bank) and thrust inputs. Speeding up, slowing down, entering climbs or descents, and turning are examples of accelerations that the pilot controls in everyday flight. *[Figure 4-5]*

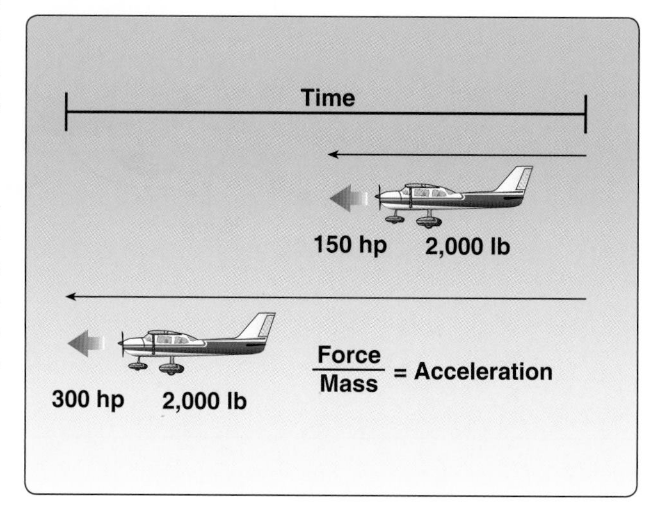

Figure 4-5. *Newton's Second Law of Motion: the Law of Momentum.*

Newton's Third Law, the Law of Reaction

Newton's Third Law of Motion is the Law of Reaction, which states that for every action there is an equal and opposite reaction. As shown in *Figure 4-6*, the action of the jet engine's thrust or the pull of the propeller lead to the reaction of the aircraft's forward motion. This law is also responsible for a portion of the lift that is produced by a wing, from the downward deflection of the airflow around it. This downward force of the relative wind results in an equal but opposite (upward) lifting force created by the airflow over the wing. *[Figure 4-6]*

Atmosphere

The atmosphere is the envelope of air which surrounds the Earth. A given volume of dry air contains about 78 percent nitrogen, 21 percent oxygen, and about 1 percent other gases such as argon, carbon dioxide, and others to a lesser degree.

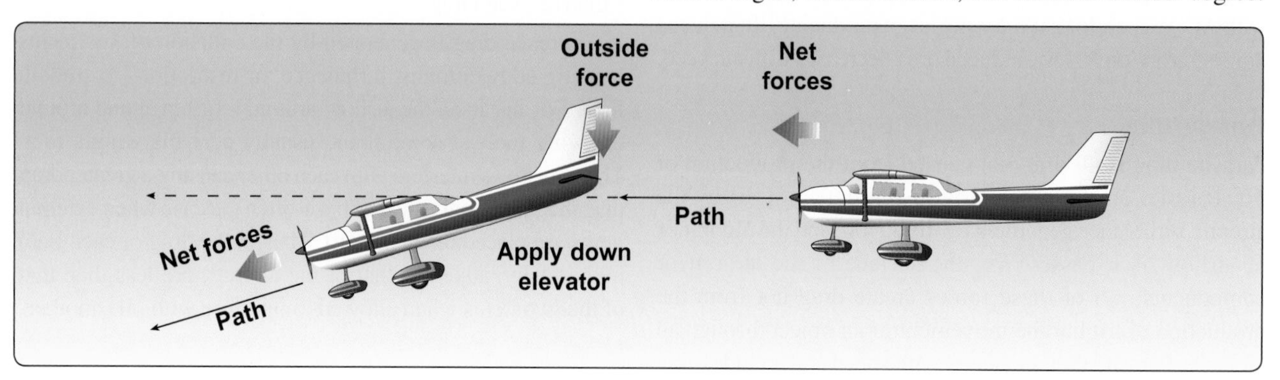

Figure 4-4. *Newton's First Law of Motion: the Law of Inertia.*

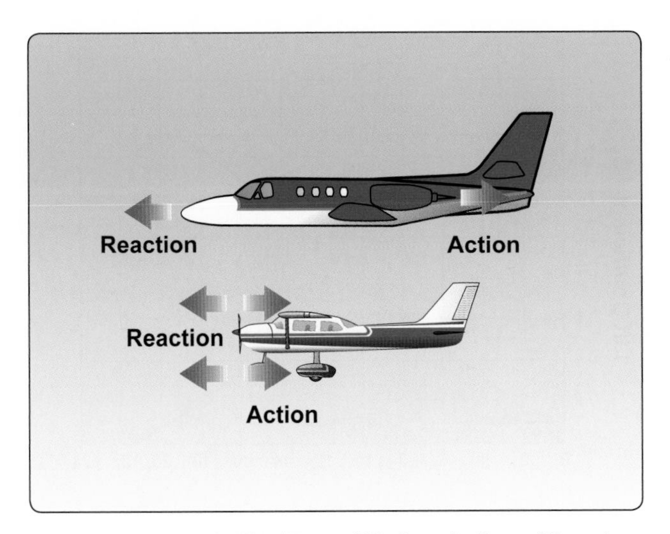

Figure 4-6. *Newton's Third Law of Motion: the Law of Reaction.*

Although seemingly light, air does have weight and a one square inch column of the atmosphere at sea level weighs approximately 14.7 pounds. About one-half of the air by weight is within the first 18,000 feet. The remainder of the air is spread over a vertical distance in excess of 1,000 miles.

Air density is a result of the relationship between temperature and pressure. Air density is inversely related to temperature and directly related to pressure. For a constant pressure to be maintained as temperature increases, density must decrease, and vice versa. For a constant temperature to be maintained as pressure increases, density must increase, and vice versa. These relationships provide a basis for understanding instrument indications and aircraft performance.

Layers of the Atmosphere

There are several layers to the atmosphere with the troposphere being closest to the Earth's surface extending to about 60,000 feet at the equator. Following is the stratosphere, mesosphere, ionosphere, thermosphere, and finally the exosphere. The tropopause is the thin layer between the troposphere and the stratosphere. It varies in both thickness and altitude but is generally defined where the standard lapse (generally accepted at 2 °C per 1,000 feet) decreases significantly (usually down to 1 °C or less).

International Standard Atmosphere (ISA)

The International Civil Aviation Organization (ICAO) established the ICAO Standard Atmosphere as a way of creating an international standard for reference and performance computations. Instrument indications and aircraft performance specifications are derived using this standard as a reference. Because the standard atmosphere is a derived set of conditions that rarely exist in reality, pilots need to understand how deviations from the standard affect both instrument indications and aircraft performance.

In the standard atmosphere, sea level pressure is 29.92 inches of mercury ("Hg) and the temperature is 15 °C (59 °F). The standard lapse rate for pressure is approximately a 1 "Hg decrease per 1,000 feet increase in altitude. The standard lapse rate for temperature is a 2 °C (3.6 °F) decrease per 1,000 feet increase, up to the top of the stratosphere. Since all aircraft performance is compared and evaluated in the environment of the standard atmosphere, all aircraft performance instrumentation is calibrated for the standard atmosphere. Because the actual operating conditions rarely, if ever, fit the standard atmosphere, certain corrections must apply to the instrumentation and aircraft performance. For instance, at 10,000 ISA predicts that the air pressure should be 19.92 "Hg (29.92 "Hg – 10 "Hg = 19.92 "Hg) and the outside temperature at –5 °C (15 °C – 20 °C). If the temperature or the pressure is different than the International Standard Atmosphere (ISA) prediction an adjustment must be made to performance predictions and various instrument indications.

Pressure Altitude

Pressure altitude is the height above the standard datum plane (SDP). The aircraft altimeter is essentially a sensitive barometer calibrated to indicate altitude in the standard atmosphere. If the altimeter is set for 29.92 "Hg SDP, the altitude indicated is the pressure altitude-the altitude in the standard atmosphere corresponding to the sensed pressure.

The SDP is a theoretical level where the pressure of the atmosphere is 29.92 "Hg and the weight of air is 14.7 psi. As atmospheric pressure changes, the SDP may be below, at, or above sea level. Pressure altitude is important as a basis for determining aircraft performance, as well as for assigning flight levels to aircraft operating at or above 18,000 feet. The pressure altitude can be determined by either of two methods: (1) by setting the barometric scale of the altimeter to 29.92 "Hg and reading the indicated altitude, or (2) by applying a correction factor to the indicated altitude according to the reported altimeter setting.

Density Altitude

Density altitude is pressure altitude corrected for nonstandard temperature. As the density of the air increases (lower density altitude), aircraft performance increases. Conversely, as air density decreases (higher density altitude), aircraft performance decreases. A decrease in air density means a high density altitude; an increase in air density means a lower density altitude. Density altitude is used in calculating aircraft performance. Under standard atmospheric conditions, air at each level in the atmosphere has a specific density; under standard conditions, pressure altitude and density altitude identify the same level. Density altitude, then, is the vertical distance above sea level in the standard atmosphere at which a given density is to be found. It can be computed using a Koch Chart or a flight computer with a density altitude function. *[Figure 4-7]*

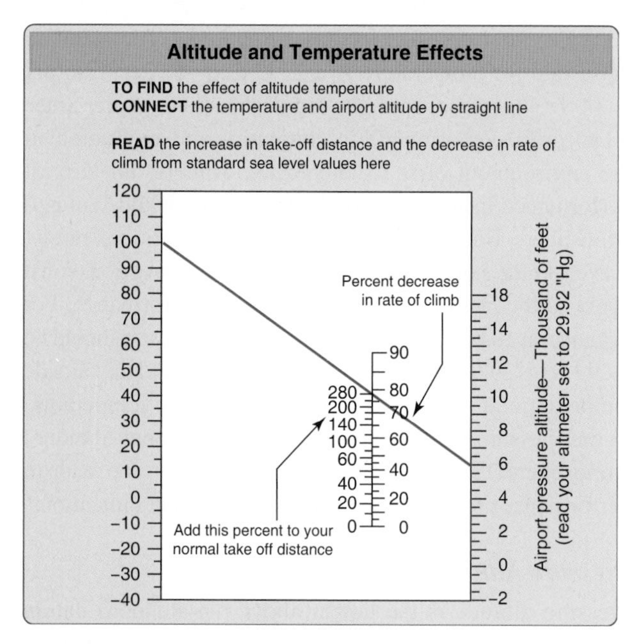

Figure 4-7. *Koch chart sample.*

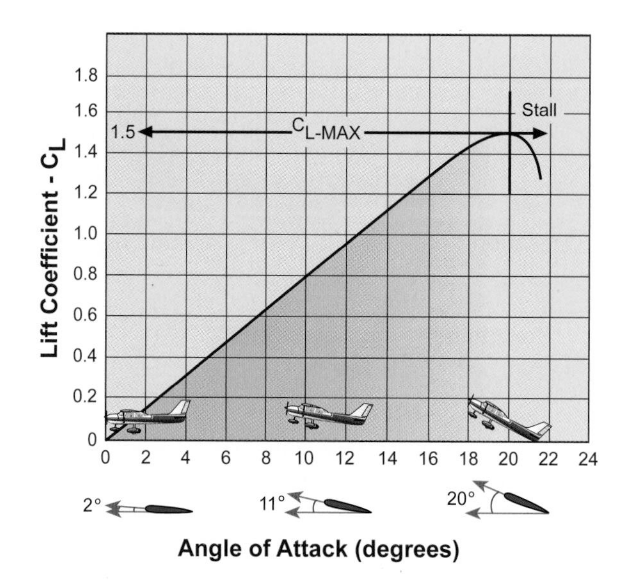

Figure 4-8. *Relationship of lift to AOA.*

If a chart is not available, the density altitude can be estimated by adding 120 feet for every degree Celsius above the ISA. For example, at 3,000 feet PA, the ISA prediction is 9 °C (15 °C – [lapse rate of 2 °C per 1,000 feet x 3 = 6 °C]). However, if the actual temperature is 20 °C (11 °C more than that predicted by ISA) then the difference of 11 °C is multiplied by 120 feet equaling 1,320. Adding this figure to the original 3,000 feet provides a density altitude of 4,320 feet (3,000 feet + 1,320 feet).

Lift

Lift always acts in a direction perpendicular to the relative wind and to the lateral axis of the aircraft. The fact that lift is referenced to the wing, not to the Earth's surface, is the source of many errors in learning flight control. Lift is not always "up." Its direction relative to the Earth's surface changes as the pilot maneuvers the aircraft.

The magnitude of the force of lift is directly proportional to the density of the air, the area of the wings, and the airspeed. It also depends upon the type of wing and the AOA. Lift increases with an increase in AOA up to the stalling angle, at which point it decreases with any further increase in AOA. In conventional aircraft, lift is therefore controlled by varying the AOA and speed.

Pitch/Power Relationship

An examination of *Figure 4-8* illustrates the relationship between pitch and power while controlling flightpath and airspeed. In order to maintain a constant lift, as airspeed is reduced, pitch must be increased. The pilot controls pitch through the elevators, which control the AOA. When back pressure is applied on the elevator control, the tail lowers and the nose rises, thus increasing the wing's AOA and lift. Under most conditions the elevator is placing downward pressure on the tail. This pressure requires energy that is taken from aircraft performance (speed). Therefore, when the CG is closer to the aft portion of the aircraft the elevator downward forces are less. This results in less energy used for downward forces, in turn resulting in more energy applied to aircraft performance.

Thrust is controlled by using the throttle to establish or maintain desired airspeeds. The most precise method of controlling flightpath is to use pitch control while simultaneously using power (thrust) to control airspeed. In order to maintain a constant lift, a change in pitch requires a change in power, and vice versa.

If the pilot wants the aircraft to accelerate while maintaining altitude, thrust must be increased to overcome drag. As the aircraft speeds up, lift is increased. To prevent gaining altitude, the pitch angle must be lowered to reduce the AOA and maintain altitude. To decelerate while maintaining altitude, thrust must be decreased to less than the value of drag. As the aircraft slows down, lift is reduced. To prevent losing altitude, the pitch angle must be increased in order to increase the AOA and maintain altitude.

Drag Curves

When induced drag and parasite drag are plotted on a graph, the total drag on the aircraft appears in the form of a "drag curve." Graph A of *Figure 4-9* shows a curve based on thrust versus drag, which is primarily used for jet aircraft. Graph B

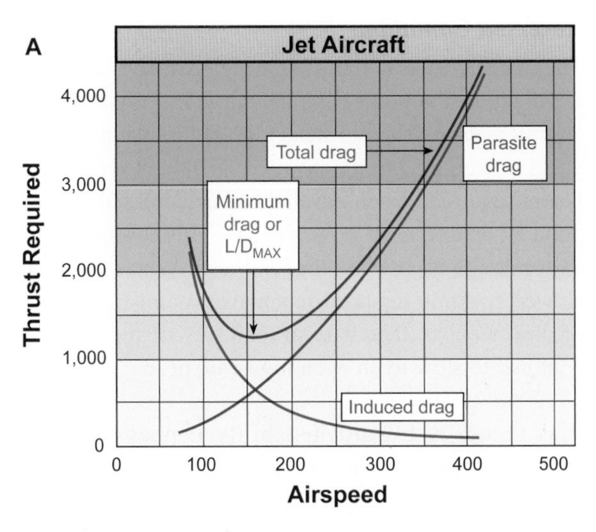

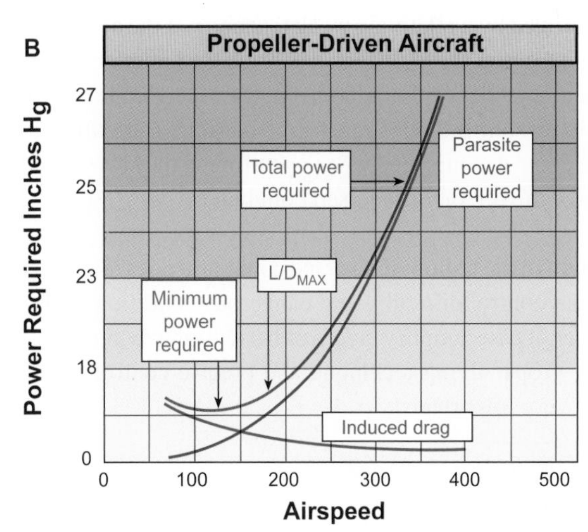

Figure 4-9. *Thrust and power required curves.*

of *Figure 4-9* is based on power versus drag, and it is used for propeller-driven aircraft. This chapter focuses on power versus drag charts for propeller-driven aircraft.

Understanding the drag curve can provide valuable insight into the various performance parameters and limitations of the aircraft. Because power must equal drag to maintain a steady airspeed, the curve can be either a drag curve or a power required curve. The power required curve represents the amount of power needed to overcome drag in order to maintain a steady speed in level flight.

The propellers used on most reciprocating engines achieve peak propeller efficiencies in the range of 80 to 88 percent. As airspeed increases, the propeller efficiency increases until it reaches its maximum. Any airspeed above this maximum point causes a reduction in propeller efficiency. An engine that produces 160 horsepower will have only about 80 percent of that power converted into available horsepower, approximately 128 horsepower. The remainder is lost energy. This is the reason the thrust and power available curves change with speed.

Regions of Command

The drag curve also illustrates the two regions of command: the region of normal command, and the region of reversed command. The term "region of command" refers to the relationship between speed and the power required to maintain or change that speed. "Command" refers to the input the pilot must give in terms of power or thrust to maintain a new speed once reached.

The "region of normal command" occurs where power must be added to increase speed. This region exists at speeds higher than the minimum drag point primarily as a result of parasite drag. The "region of reversed command" occurs where additional power is needed to maintain a slower airspeed. This region exists at speeds slower than the minimum drag point (L/D_{MAX} on the thrust required curve, *Figure 4-9*) and is primarily due to induced drag. *Figure 4-10* shows how one power setting can yield two speeds, points 1 and 2. This is because at point 1 there is high induced drag and low parasite drag, while at point 2 there is high parasite drag and low induced drag.

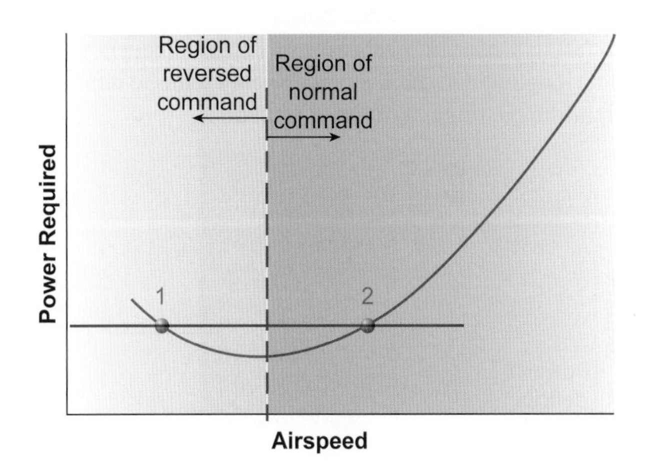

Figure 4-10. *Regions of command.*

Control Characteristics

Most flying is conducted in the region of normal command: for example, cruise, climb, and maneuvers. The region of reversed command may be encountered in the slow-speed phases of flight during takeoff and landing; however, for most general aviation aircraft, this region is very small and is below normal approach speeds.

Flight in the region of normal command is characterized by a relatively strong tendency of the aircraft to maintain the trim speed. Flight in the region of reversed command is characterized by a relatively weak tendency of the aircraft to maintain the trim speed. In fact, it is likely the aircraft exhibits no inherent tendency to maintain the trim speed in this area. For this reason, the pilot must give particular attention to precise control of airspeed when operating in the slow-speed phases of the region of reversed command.

Operation in the region of reversed command does not imply that great control difficulty and dangerous conditions exist. However, it does amplify errors of basic flying technique—making proper flying technique and precise control of the aircraft very important.

Speed Stability

Normal Command

The characteristics of flight in the region of normal command are illustrated at point A on the curve in *Figure 4-11*. If the aircraft is established in steady, level flight at point A, lift is equal to weight, and the power available is set equal to the power required. If the airspeed is increased with no changes to the power setting, a power deficiency exists. The aircraft has a natural tendency to return to the initial speed to balance power and drag. If the airspeed is reduced with no changes to the power setting, an excess of power exists. The aircraft has a natural tendency to speed up to regain the balance between power and drag. Keeping the aircraft in proper trim enhances this natural tendency. The static longitudinal stability of the aircraft tends to return the aircraft to the original trimmed condition.

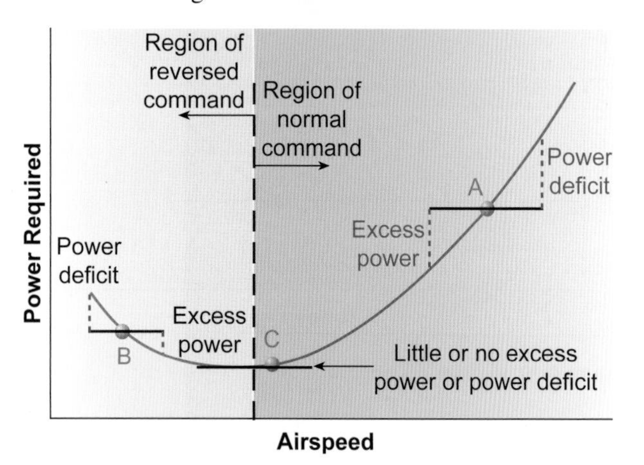

Figure 4-11. *Region of speed stability.*

An aircraft flying in steady, level flight at point C is in equilibrium. *[Figure 4-11]* If the speed were increased or decreased slightly, the aircraft would tend to remain at that speed. This is because the curve is relatively flat and a slight change in speed does not produce any significant excess or deficiency in power. It has the characteristic of neutral stability (i.e., the aircraft's tendency is to remain at the new speed).

Reversed Command

The characteristics of flight in the region of reversed command are illustrated at point B on the curve in *Figure 4-10*. If the aircraft is established in steady, level flight at point B, lift is equal to weight, and the power available is set equal to the power required. When the airspeed is increased greater than point B, an excess of power exists. This causes the aircraft to accelerate to an even higher speed. When the aircraft is slowed to some airspeed lower than point B, a deficiency of power exists. The natural tendency of the aircraft is to continue to slow to an even lower airspeed.

This tendency toward instability happens because the variation of excess power to either side of point B magnifies the original change in speed. Although the static longitudinal stability of the aircraft tries to maintain the original trimmed condition, this instability is more of an influence because of the increased induced drag due to the higher AOA in slow-speed flight.

Trim

The term trim refers to employing adjustable aerodynamic devices on the aircraft to adjust forces so the pilot does not have to manually hold pressure on the controls. One means is to employ trim tabs. A trim tab is a small, adjustable hinged surface, located on the trailing edge of the elevator, aileron, or rudder control surfaces. (Some aircraft use adjustable stabilizers instead of trim tabs for pitch trim.) Trimming is accomplished by deflecting the tab in the direction opposite to that in which the primary control surface must be held. The force of the airflow striking the tab causes the main control surface to be deflected to a position that corrects the unbalanced condition of the aircraft.

Because the trim tabs use airflow to function, trim is a function of speed. Any change in speed results in the need to re-trim the aircraft. An aircraft properly trimmed in pitch seeks to return to the original speed before the change. It is very important for instrument pilots to keep the aircraft in constant trim. This reduces the pilot's workload significantly, allowing attention to other duties without compromising aircraft control.

Slow-Speed Flight

Anytime an aircraft is flying near the stalling speed or the region of reversed command, such as in final approach for a normal landing, the initial part of a go around, or maneuvering in slow flight, it is operating in what is called slow-speed flight. If the aircraft weighs 4,000 pounds, the lift produced by the aircraft must be 4,000 pounds. When lift is less than 4,000 pounds, the aircraft is no longer able to sustain level flight, and consequently descends. During intentional descents, this is an important factor and is used in the total control of the aircraft.

However, because lift is required during low speed flight and is characterized by high AOA, flaps or other high lift devices are needed to either change the camber of the airfoil, or delay the boundary level separation. Plain and split flaps *[Figure 4-12]* are most commonly used to change the camber of an airfoil. It should be noted that with the application of flaps, the aircraft will stall at a lower AOA. For example, if the basic wing stalls at 18° without flaps, then with the addition of flaps to the $C_{L\text{-}MAX}$ position, the new AOA that the wing will stall is 15°. However, the value of lift (flaps extended to the $C_{L\text{-}MAX}$ position) produces more lift than lift at 18° on the basic wing.

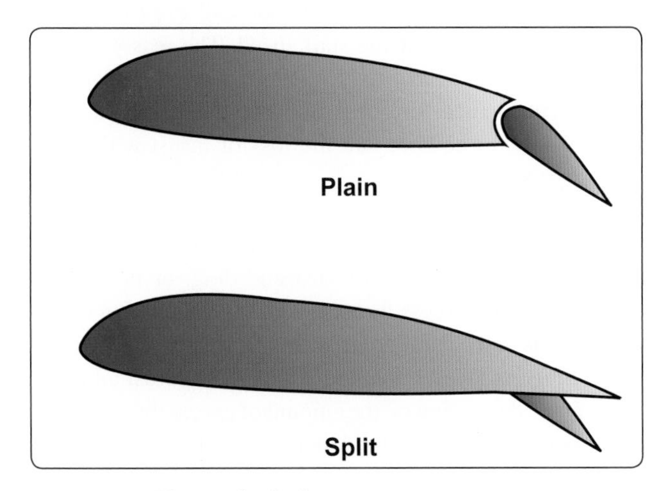

Figure 4-12. *Plain and split flaps.*

Delaying the boundary layer separation is another way to increase $C_{L\text{-}MAX}$. Several methods are employed (such as suction and use of a blowing boundary layer control), but the most common device used on general aviation light aircraft is the vortex generator. Small strips of metal placed along the wing (usually in front of the control surfaces) create turbulence. The turbulence in turn mixes high energy air from outside the boundary layer with boundary layer air. The effect is similar to other boundary layer devices. *[Figure 4-13]*

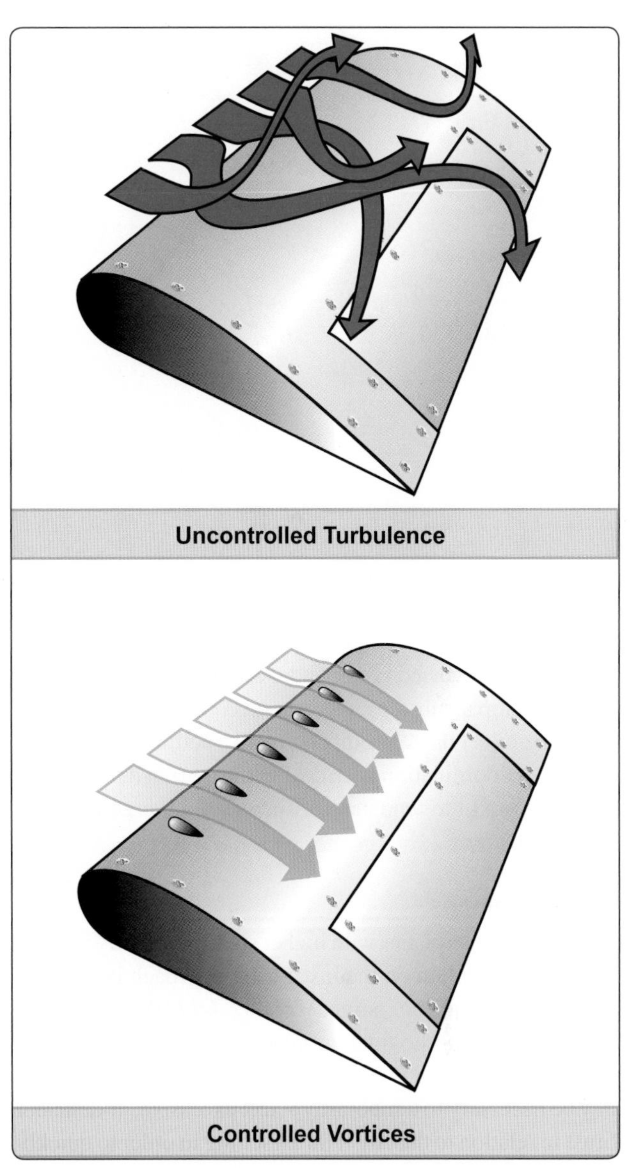

Figure 4-13. *Vortex generators.*

Small Airplanes

Most small airplanes maintain a speed well in excess of 1.3 times V_{SO} on an instrument approach. An airplane with a stall speed of 50 knots (V_{SO}) has a normal approach speed of 65 knots. However, this same airplane may maintain 90 knots (1.8 V_{SO}) while on the final segment of an instrument approach. The landing gear will most likely be extended at the beginning of the descent to the minimum descent altitude, or upon intercepting the glideslope of the instrument landing system. The pilot may also select an intermediate flap setting for this phase of the approach. The airplane at this speed has good positive speed stability, as represented by point A on *Figure 4-11*. Flying in this regime permits the pilot to make slight pitch changes without changing power settings, and accept minor speed changes knowing that when the pitch is returned to the initial setting, the speed returns to the original setting. This reduces the pilot's workload.

Aircraft are usually slowed to a normal landing speed when on the final approach just prior to landing. When slowed to 65 knots, (1.3 V_{SO}), the airplane will be close to point C. *[Figure 4-14]* At this point, precise control of the pitch and power becomes more crucial for maintaining the correct speed. Pitch and power coordination is necessary because the speed stability is relatively neutral since the speed tends to remain at the new value and not return to the original setting. In addition to the need for more precise airspeed control, the pilot normally changes the aircraft's configuration by extending landing flaps. This configuration change means the pilot must be alert to unwanted pitch changes at a low altitude.

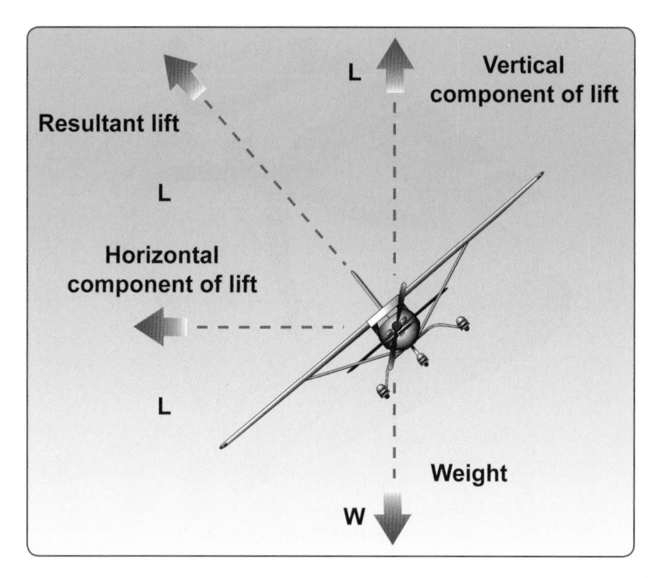

Figure 4-14. *Forces in a turn.*

If allowed to slow several knots, the airplane could enter the region of reversed command. At this point, the airplane could develop an unsafe sink rate and continue to lose speed unless the pilot takes a prompt corrective action. Proper pitch and power coordination is critical in this region due to speed instability and the tendency of increased divergence from the desired speed.

Large Airplanes

Pilots of larger airplanes with higher stall speeds may find the speed they maintain on the instrument approach is near 1.3 V_{SO}, putting them near point C *[Figure 4-11]* the entire time the airplane is on the final approach segment. In this case, precise speed control is necessary throughout the approach. It may be necessary to temporarily select excessive, or deficient thrust in relation to the target thrust setting in order to quickly correct for airspeed deviations.

For example, a pilot is on an instrument approach at 1.3 V_{SO}, a speed near L/D_{MAX}, and knows that a certain power setting maintains that speed. The airplane slows several knots below the desired speed because of a slight reduction in the power setting. The pilot increases the power slightly, and the airplane begins to accelerate, but at a slow rate. Because the airplane is still in the "flat part" of the drag curve, this slight increase in power will not cause a rapid return to the desired speed. The pilot may need to increase the power higher than normally needed to maintain the new speed, allow the airplane to accelerate, then reduce the power to the setting that maintains the desired speed.

Climbs

The ability for an aircraft to climb depends upon an excess power or thrust over what it takes to maintain equilibrium. Excess power is the available power over and above that required to maintain horizontal flight at a given speed. Although the terms power and thrust are sometimes used interchangeably (erroneously implying they are synonymous), distinguishing between the two is important when considering climb performance. Work is the product of a force moving through a distance and is usually independent of time. Power implies work rate or units of work per unit of time, and as such is a function of the speed at which the force is developed. Thrust, also a function of work, means the force which imparts a change in the velocity of a mass.

During takeoff, the aircraft does not stall even though it may be in a climb near the stall speed. The reason is that excess power (used to produce thrust) is used during this flight regime. Therefore, it is important if an engine fails after takeoff, to compensate the loss of thrust with pitch and airspeed.

For a given weight of the aircraft, the angle of climb depends on the difference between thrust and drag, or the excess thrust. When the excess thrust is zero, the inclination of the flightpath is zero, and the aircraft is in steady, level flight. When thrust is greater than drag, the excess thrust allows a climb angle depending on the amount of excess thrust. When thrust is less than drag, the deficiency of thrust induces an angle of descent.

Acceleration in Cruise Flight

Aircraft accelerate in level flight because of an excess of power over what is required to maintain a steady speed. This is the same excess power used to climb. Upon reaching the desired altitude with pitch being lowered to maintain that altitude, the excess power now accelerates the aircraft to its cruise speed. However, reducing power too soon after level off results in a longer period of time to accelerate.

Turns

Like any moving object, an aircraft requires a sideward force to make it turn. In a normal turn, this force is supplied by banking the aircraft in order to exert lift inward, as well as upward. The force of lift is separated into two components at right angles to each other. *[Figure 4-14]* The upward acting lift together with the opposing weight becomes the vertical lift component. The horizontally acting lift and its opposing centrifugal force are the horizontal lift component, or centripetal force. This horizontal lift component is the sideward force that causes an aircraft to turn. The equal and opposite reaction to this sideward force is centrifugal force, which is merely an apparent force as a result of inertia.

The relationship between the aircraft's speed and bank angle to the rate and radius of turns is important for instrument pilots to understand. The pilot can use this knowledge to properly estimate bank angles needed for certain rates of turn, or to determine how much to lead when intercepting a course.

Rate of Turn

The rate of turn, normally measured in degrees per second, is based upon a set bank angle at a set speed. If either one of these elements changes, the rate of turn changes. If the aircraft increases its speed without changing the bank angle, the rate of turn decreases. Likewise, if the speed decreases without changing the bank angle, the rate of turn increases.

Changing the bank angle without changing speed also causes the rate of turn to change. Increasing the bank angle without changing speed increases the rate of turn, while decreasing the bank angle reduces the rate of turn.

The standard rate of turn, 3° per second, is used as the main reference for bank angle. Therefore, the pilot must understand how the angle of bank varies with speed changes, such as slowing down for holding or an instrument approach. *Figure 4-15* shows the turn relationship with reference to a constant bank angle or a constant airspeed, and the effects on rate of turn and radius of turn. A rule of thumb for determining the standard rate turn is to divide the airspeed by ten and add 7. An aircraft with an airspeed of 90 knots takes a bank angle of 16° to maintain a standard rate turn (90 divided by 10 plus 7 equals 16°).

Radius of Turn

The radius of turn varies with changes in either speed or bank. If the speed is increased without changing the bank angle, the radius of turn increases, and vice versa. If the speed is constant, increasing the bank angle reduces the radius of turn, while decreasing the bank angle increases the radius of turn. This means that intercepting a course at a higher speed requires more distance, and therefore, requires a longer lead. If the speed is slowed considerably in preparation for holding or an approach, a shorter lead is needed than that required for cruise flight.

Coordination of Rudder and Aileron Controls

Any time ailerons are used, adverse yaw is produced. Adverse yaw is caused when the ailerons are deflected as a roll motion (as in turn) is initiated. In a right turn, the right aileron is deflected upward while the left is deflected downward. Lift is increased on the left side and reduced on the right, resulting in a bank to the right. However, as a result of producing lift on the left, induced drag is also increased on the left side. The drag causes the left wing to slow down, in turn causing the nose of the aircraft to initially move (left) in the direction opposite of the turn. Correcting for this yaw with rudder, when entering and exiting turns, is necessary for precise control of the airplane when flying on instruments. The pilot can tell if the turn is coordinated by checking the ball in the turn-and-slip indicator or the turn coordinator. *[Figure 4-16]*

As the aircraft banks to enter a turn, a portion of the wing's vertical lift becomes the horizontal component; therefore, without an increase in back pressure, the aircraft loses altitude during the turn. The loss of vertical lift can be offset by increasing the pitch in one-half bar width increments. Trim may be used to relieve the control pressures; however, if used, it has to be removed once the turn is complete.

In a slipping turn, the aircraft is not turning at the rate appropriate to the bank being used, and the aircraft falls to the inside of the turn. The aircraft is banked too much for the rate of turn, so the horizontal lift component is greater than the centrifugal force. A skidding turn results from excess of

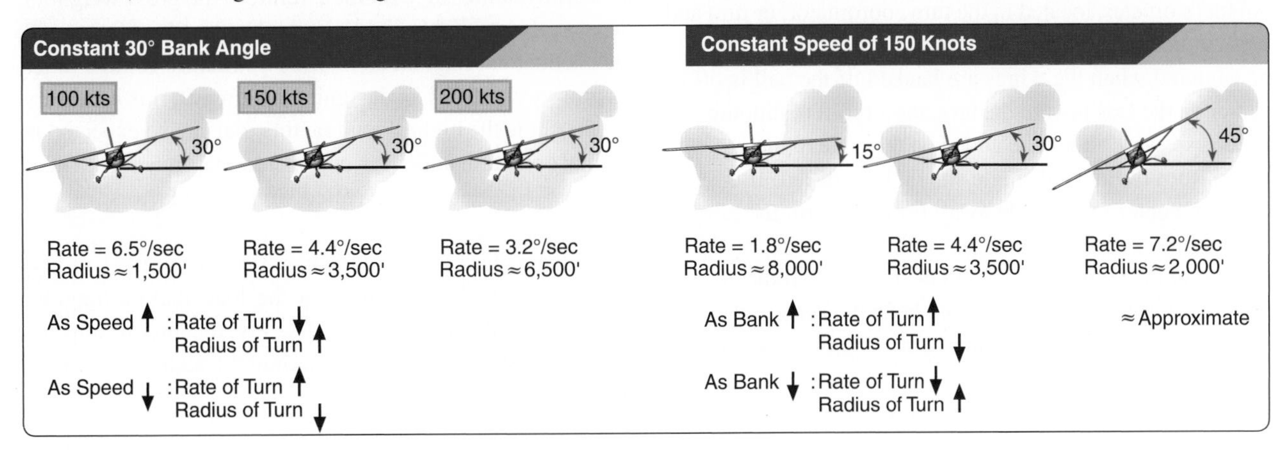

Figure 4-15. *Turns.*

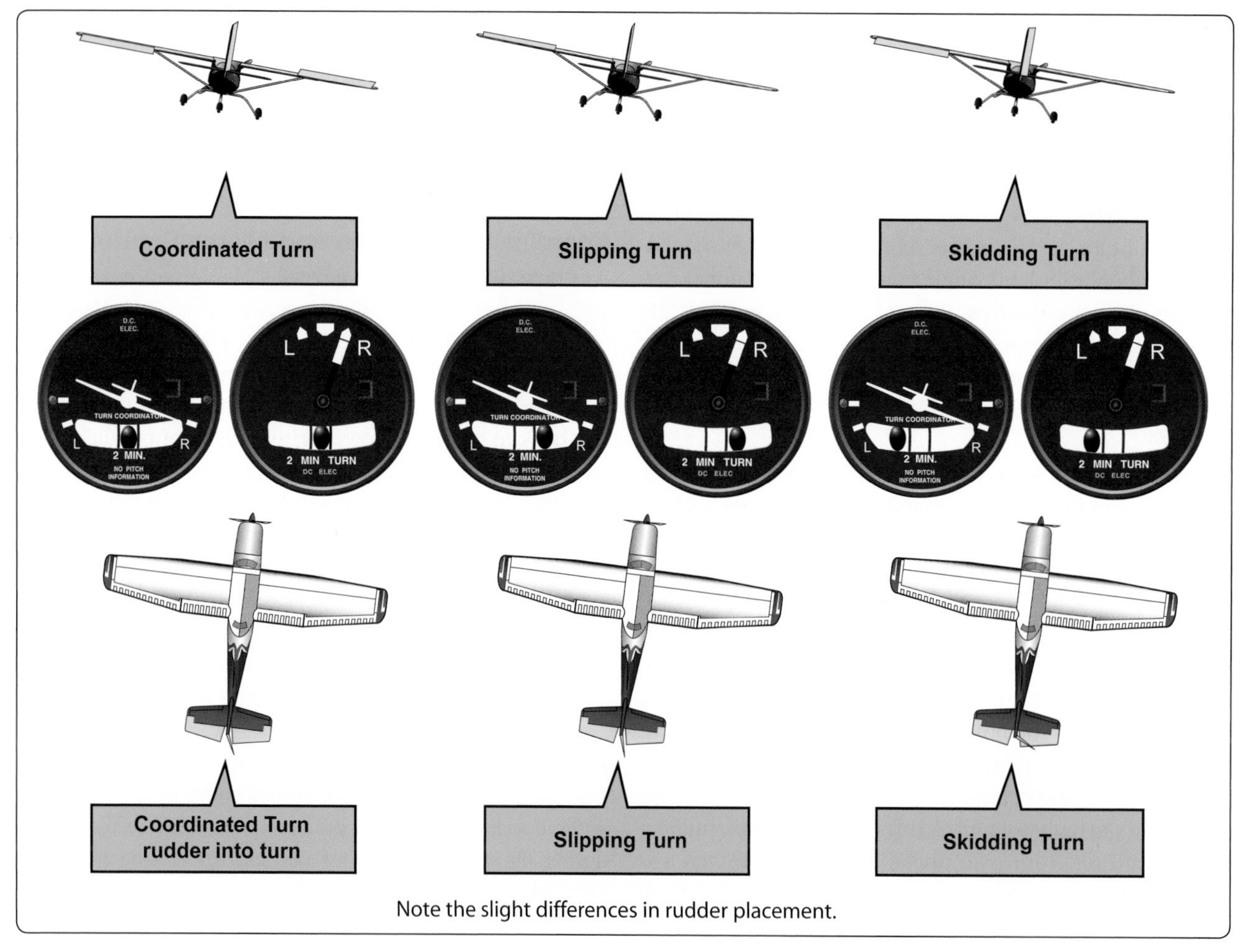

Figure 4-16. *Adverse yaw.*

centrifugal force over the horizontal lift component, pulling the aircraft toward the outside of the turn. The rate of turn is too great for the angle of bank, so the horizontal lift component is less than the centrifugal force.

An inclinometer, located in the turn coordinator, or turn and bank indicator indicates the quality of the turn, and should be centered when the wings are banked. If the ball is off of center on the side toward the turn, the aircraft is slipping and rudder pressure should be added on that side to increase the rate of turn or the bank angle should be reduced. If the ball is off of center on the side away from the turn, the aircraft is skidding and rudder pressure toward the turn should be relaxed or the bank angle should be increased. If the aircraft is properly rigged, the ball should be in the center when the wings are level; use rudder and/or aileron trim if available.

The increase in induced drag (caused by the increase in AOA necessary to maintain altitude) results in a minor loss of airspeed if the power setting is not changed.

Load Factor

Any force applied to an aircraft to deflect its flight from a straight line produces a stress on its structure; the amount of this force is termed load factor. A load factor is the ratio of the aerodynamic force on the aircraft to the gross weight of the aircraft (e.g., lift/weight). For example, a load factor of 3 means the total load on an aircraft's structure is three times its gross weight. When designing an aircraft, it is necessary to determine the highest load factors that can be expected in normal operation under various operational situations. These "highest" load factors are called "limit load factors."

Aircraft are placed in various categories (i.e., normal, utility, and acrobatic) depending upon the load factors they are designed to take. For reasons of safety, the aircraft must be designed to withstand certain maximum load factors without any structural damage.

The specified load may be expected in terms of aerodynamic forces, as in turns. In level flight in undisturbed air, the wings are supporting not only the weight of the aircraft, but centrifugal force as well. As the bank steepens, the horizontal lift component increases, centrifugal force increases, and the load factor increases. If the load factor becomes so great that an increase in AOA cannot provide enough lift to support the load, the wing stalls. Since the stalling speed increases directly with the square root of the load factor, the pilot should be aware of the flight conditions during which the load factor can become critical. Steep turns at slow airspeed, structural ice accumulation, and vertical gusts in turbulent air can increase the load factor to a critical level.

Icing

One of the greatest hazards to flight is aircraft icing. The instrument pilot must be aware of the conditions conducive to aircraft icing. These conditions include the types of icing, the effects of icing on aircraft control and performance, effects of icing on aircraft systems, and the use and limitations of aircraft deice and anti-ice equipment. Coping with the hazards of icing begins with preflight planning to determine where icing may occur during a flight and ensuring the aircraft is free of ice and frost prior to takeoff. This attention to detail extends to managing deice and anti-ice systems properly during the flight, because weather conditions may change rapidly, and the pilot must be able to recognize when a change of flight plan is required.

Types of Icing

Structural Icing

Structural icing refers to the accumulation of ice on the exterior of the aircraft. Ice forms on aircraft structures and surfaces when super-cooled droplets impinge on them and freeze. Small and/or narrow objects are the best collectors of droplets and ice up most rapidly. This is why a small protuberance within sight of the pilot can be used as an "ice evidence probe." It is generally one of the first parts of the airplane on which an appreciable amount of ice forms. An aircraft's tailplane is a better collector than its wings, because the tailplane presents a thinner surface to the airstream.

Induction Icing

Ice in the induction system can reduce the amount of air available for combustion. The most common example of reciprocating engine induction icing is carburetor ice. Most pilots are familiar with this phenomenon, which occurs when moist air passes through a carburetor venturi and is cooled. As a result of this process, ice may form on the venturi walls and throttle plate, restricting airflow to the engine. This may occur at temperatures between 20 °F (–7 °C) and 70 °F (21 °C). The problem is remedied by applying carburetor heat, which uses the engine's own exhaust as a heat source to melt the ice or prevent its formation. On the other hand, fuel-injected aircraft engines usually are less vulnerable to icing but still can be affected if the engine's air source becomes blocked with ice. Manufacturers provide an alternate air source that may be selected in case the normal system malfunctions.

In turbojet aircraft, air that is drawn into the engines creates an area of reduced pressure at the inlet, which lowers the temperature below that of the surrounding air. In marginal icing conditions (i.e., conditions where icing is possible), this reduction in temperature may be sufficient to cause ice to form on the engine inlet, disrupting the airflow into the engine. Another hazard occurs when ice breaks off and is ingested into a running engine, which can cause damage to fan blades, engine compressor stall, or combustor flameout. When anti-icing systems are used, runback water also can refreeze on unprotected surfaces of the inlet and, if excessive, reduce airflow into the engine or distort the airflow pattern in such a manner as to cause compressor or fan blades to vibrate, possibly damaging the engine. Another problem in turbine engines is the icing of engine probes used to set power levels (for example, engine inlet temperature or engine pressure ratio (EPR) probes), which can lead to erroneous readings of engine instrumentation operational difficulties or total power loss.

The type of ice that forms can be classified as clear, rime, or mixed, based on the structure and appearance of the ice. The type of ice that forms varies depending on the atmospheric and flight conditions in which it forms. Significant structural icing on an aircraft can cause serious aircraft control and performance problems.

Clear Ice

A glossy, transparent ice formed by the relatively slow freezing of super cooled water is referred to as clear ice. *[Figure 4-17]* The terms "clear" and "glaze" have been used

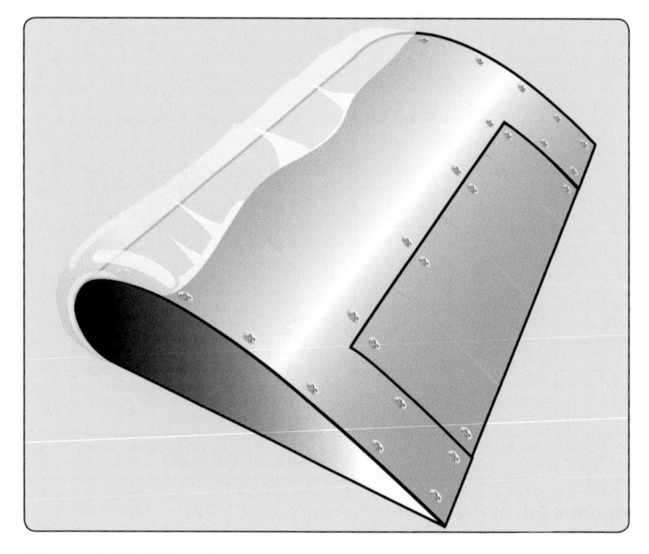

Figure 4-17. *Clear ice.*

for essentially the same type of ice accretion. This type of ice is denser, harder, and sometimes more transparent than rime ice. With larger accretions, clear ice may form "horns." *[Figure 4-18]* Temperatures close to the freezing point, large amounts of liquid water, high aircraft velocities, and large droplets are conducive to the formation of clear ice.

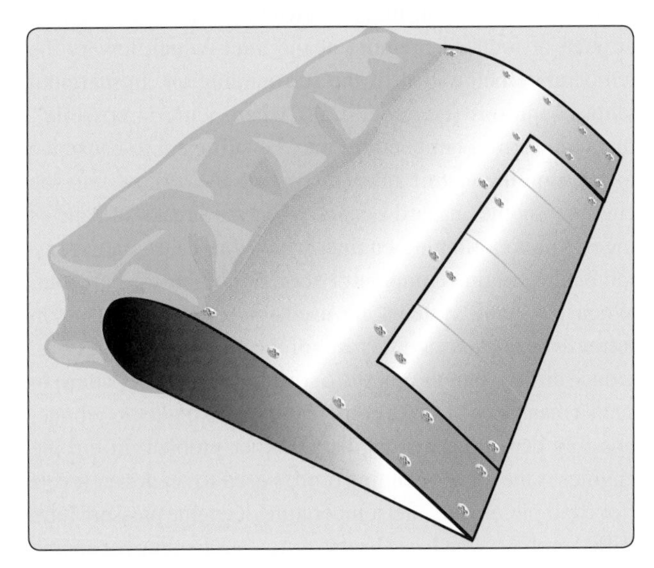

Figure 4-18. *Clear ice buildup with horns.*

Rime Ice

A rough, milky, opaque ice formed by the instantaneous or very rapid freezing of super cooled droplets as they strike the aircraft is known as rime ice. *[Figure 4-19]* The rapid freezing results in the formation of air pockets in the ice, giving it an opaque appearance and making it porous and brittle. For larger accretions, rime ice may form a streamlined extension of the wing. Low temperatures, lesser amounts of liquid water, low velocities, and small droplets are conducive to the formation of rime ice.

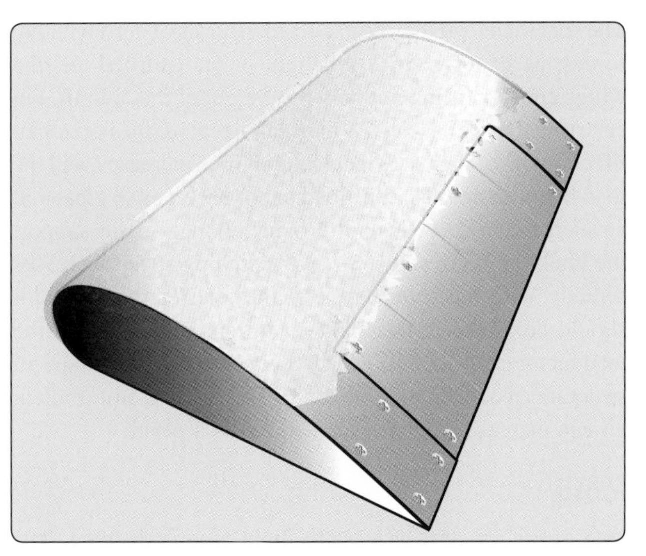

Figure 4-19. *Rime ice.*

Mixed Ice

Mixed ice is a combination of clear and rime ice formed on the same surface. It is the shape and roughness of the ice that is most important from an aerodynamic point of view.

General Effects of Icing on Airfoils

The most hazardous aspect of structural icing is its aerodynamic effects. *[Figure 4-20]* Ice alters the shape of an airfoil, reducing the maximum coefficient of lift and AOA at which the aircraft stalls. Note that at very low AOAs, there may be little or no effect of the ice on the coefficient of lift. Therefore, when cruising at a low AOA, ice on the wing may have little effect on the lift. However, note that the ice significantly reduces the C_{L-MAX}, and the AOA at which it occurs (the stall angle) is much lower. Thus, when slowing down and increasing the AOA for approach, the pilot may find that ice on the wing, which had little effect on lift in cruise now, causes stall to

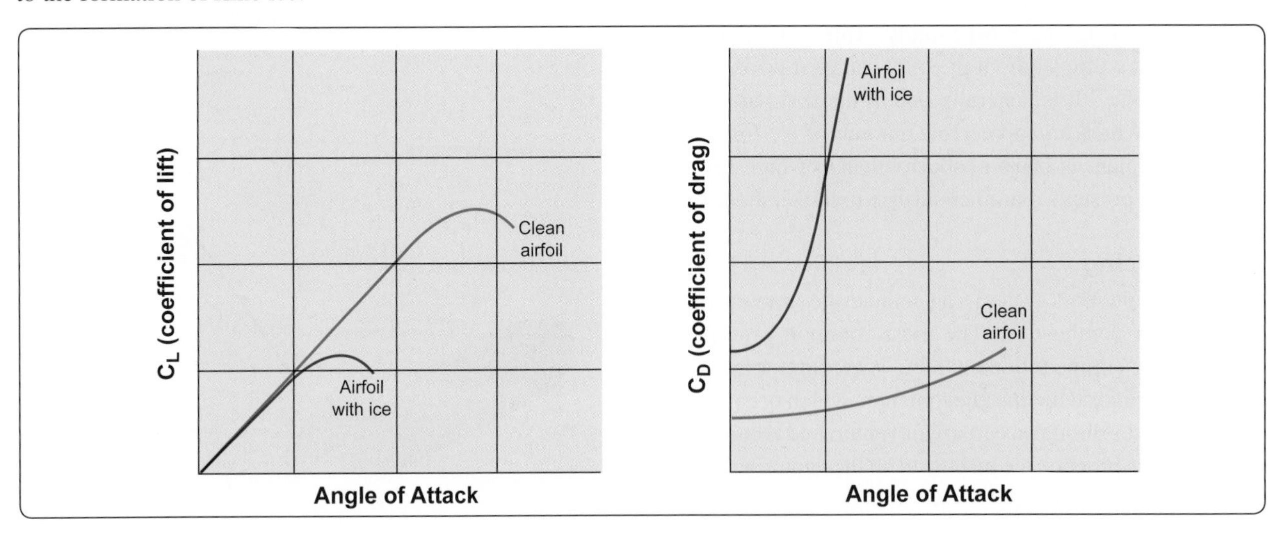

Figure 4-20. *Aerodynamic effects of icing.*

occur at a lower AOA and higher speed. Even a thin layer of ice at the leading edge of a wing, especially if it is rough, can have a significant effect in increasing stall speed. For large ice shapes, especially those with horns, the lift may also be reduced at a lower AOA. The accumulation of ice affects the coefficient of drag of the airfoil. *[Figure 4-20]* Note that the effect is significant even at very small AOAs.

A significant reduction in $C_{L\text{-}MAX}$ and a reduction in the AOA where stall occurs can result from a relatively small ice accretion. A reduction of $C_{L\text{-}MAX}$ by 30 percent is not unusual, and a large horn ice accretion can result in reductions of 40 percent to 50 percent. Drag tends to increase steadily as ice accretes. An airfoil drag increase of 100 percent is not unusual, and for large horn ice accretions, the increase can be 200 percent or even higher.

Ice on an airfoil can have other effects not depicted in these curves. Even before airfoil stall, there can be changes in the pressure over the airfoil that may affect a control surface at the trailing edge. Furthermore, on takeoff, approach, and landing, the wings of many aircraft are multi-element airfoils with three or more elements. Ice may affect the different elements in different ways. Ice may also affect the way in which the air streams interact over the elements.

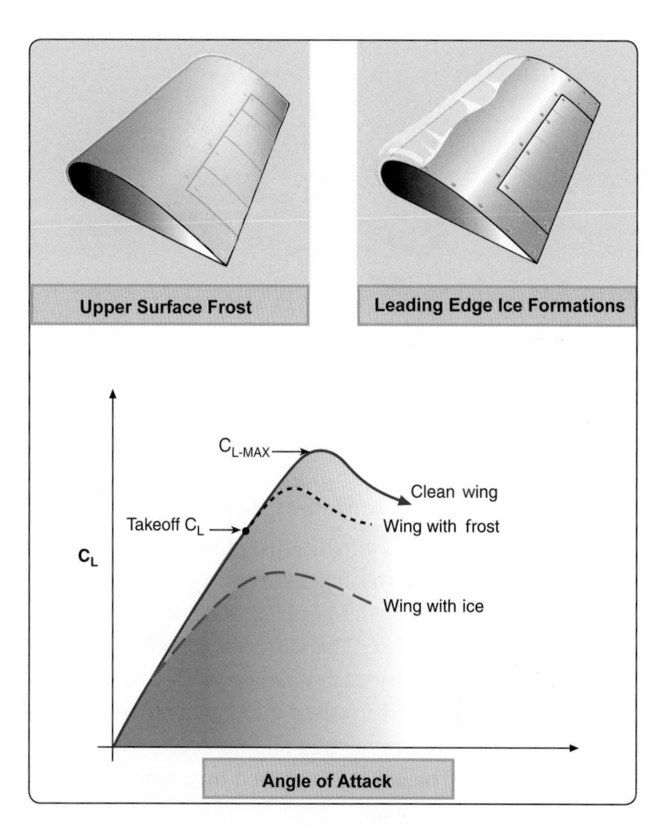

Figure 4-21. *Effect of ice and frost on lift.*

Ice can partially block or limit control surfaces, which limits or makes control movements ineffective. Also, if the extra weight caused by ice accumulation is too great, the aircraft may not be able to become airborne and, if in flight, the aircraft may not be able to maintain altitude. Therefore any accumulation of ice or frost should be removed before attempting flight.

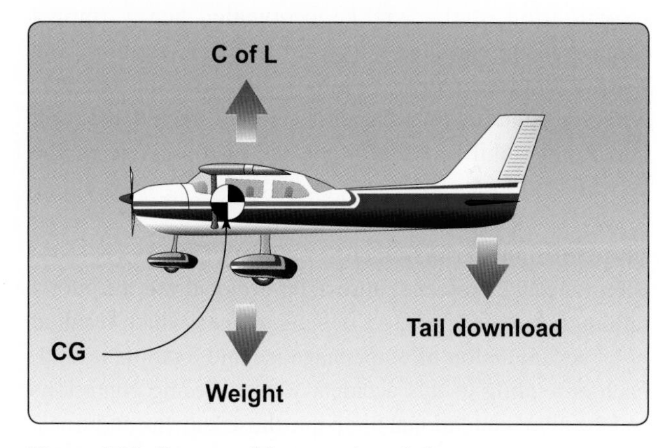

Figure 4-22. *Downward force on the tailplane.*

Another hazard of structural icing is the possible uncommanded and uncontrolled roll phenomenon, referred to as roll upset, associated with severe inflight icing. Pilots flying aircraft certificated for flight in known icing conditions should be aware that severe icing is a condition outside of the aircraft's certification icing envelope. Roll upset may be caused by airflow separation (aerodynamic stall), which induces self-deflection of the ailerons and loss of or degraded roll handling characteristics *[Figure 4-21]*. These phenomena can result from severe icing conditions without the usual symptoms of ice accumulation or a perceived aerodynamic stall.

Most aircraft have a nose-down pitching moment from the wings because the CG is ahead of the CP. It is the role of the tailplane to counteract this moment by providing a downward force. *[Figure 4-22]* The result of this configuration is that actions which move the wing away from stall, such as deployment of flaps or increasing speed, may increase the negative AOA of the tail. With ice on the tailplane, it may stall after full or partial deployment of flaps. *[Figure 4-23]*

Since the tailplane is ordinarily thinner than the wing, it is a more efficient collector of ice. On most aircraft the tailplane is not visible to the pilot, who therefore cannot observe how well it has been cleared of ice by any deicing system. Thus, it is important that the pilot be alert to the possibility of tailplane stall, particularly on approach and landing.

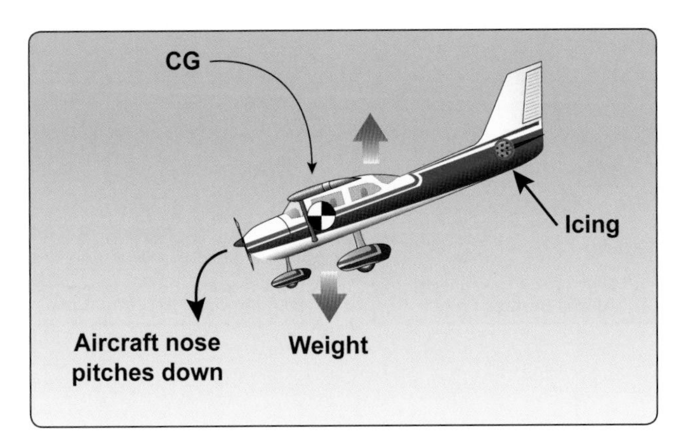

Figure 4-23. *Ice on the tailplane.*

Piper PA-34-200T (Des Moines, Iowa)

The pilot of this flight, which took place on January 9, 1996, said that upon crossing the runway threshold and lowering the flaps 25°, "the airplane pitched down." The pilot "immediately released the flaps and added power, but the airplane was basically uncontrollable at this point." The pilot reduced power and lowered the flaps before striking the runway on its centerline and sliding 1,000 feet before coming to a stop. The accident resulted in serious injury to the pilot, the sole occupant.

Examination of the wreckage revealed heavy impact damage to the airplane's forward fuselage, engines, and wings. Approximately one-half inch of rime ice was observed adhering to the leading edges of the left and right horizontal stabilizers and along the leading edge of the vertical stabilizer.

The National Transportation Safety Board (NTSB) determined the probable cause of the accident was the pilot's failure to use the airplane's deicing system, which resulted in an accumulation of empennage ice and a tailplane stall. Factors relating to this accident were the icing conditions and the pilot's intentional flight into those known conditions.

Tailplane Stall Symptoms

Any of the following symptoms, occurring singly or in combination, may be a warning of tailplane icing:

- Elevator control pulsing, oscillations, or vibrations;
- Abnormal nose-down trim change;
- Any other unusual or abnormal pitch anomalies (possibly resulting in pilot induced oscillations);
- Reduction or loss of elevator effectiveness;
- Sudden change in elevator force (control would move nose-down if unrestrained); and
- Sudden uncommanded nose-down pitch.

If any of the above symptoms occur, the pilot should:

- Immediately retract the flaps to the previous setting and apply appropriate nose-up elevator pressure;
- Increase airspeed appropriately for the reduced flap extension setting;
- Apply sufficient power for aircraft configuration and conditions. (High engine power settings may adversely impact response to tailplane stall conditions at high airspeed in some aircraft designs. Observe the manufacturer's recommendations regarding power settings.);
- Make nose-down pitch changes slowly, even in gusting conditions, if circumstances allow; and
- If a pneumatic deicing system is used, operate the system several times in an attempt to clear the tailplane of ice.

Once a tailplane stall is encountered, the stall condition tends to worsen with increased airspeed and possibly may worsen with increased power settings at the same flap setting. Airspeed, at any flap setting, in excess of the airplane manufacturer's recommendations, accompanied by uncleared ice contaminating the tailplane, may result in a tailplane stall and uncommanded pitch down from which recovery may not be possible. A tailplane stall may occur at speeds less than the maximum flap extended speed (V_{FE}).

Propeller Icing

Ice buildup on propeller blades reduces thrust for the same aerodynamic reasons that wings tend to lose lift and increase drag when ice accumulates on them. The greatest quantity of ice normally collects on the spinner and inner radius of the propeller. Propeller areas on which ice may accumulate and be ingested into the engine normally are anti-iced rather than deiced to reduce the probability of ice being shed into the engine.

Effects of Icing on Critical Aircraft Systems

In addition to the hazards of structural and induction icing, the pilot must be aware of other aircraft systems susceptible to icing. The effects of icing do not produce the performance loss of structural icing or the power loss of induction icing but can present serious problems to the instrument pilot. Examples of such systems are flight instruments, stall warning systems, and windshields.

Flight Instruments

Various aircraft instruments including the airspeed indicator, altimeter, and rate-of-climb indicator utilize pressures sensed by pitot tubes and static ports for normal operation.

When covered by ice these instruments display incorrect information thereby presenting serious hazard to instrument flight. Detailed information on the operation of these instruments and the specific effects of icing is presented in Chapter 5, Flight Instruments.

Stall Warning Systems

Stall warning systems provide essential information to pilots. These systems range from a sophisticated stall warning vane to a simple stall warning switch. Icing affects these systems in several ways resulting in possible loss of stall warning to the pilot. The loss of these systems can exacerbate an already hazardous situation. Even when an aircraft's stall warning system remains operational during icing conditions, it may be ineffective because the wing stalls at a lower AOA due to ice on the airfoil.

Windshields

Accumulation of ice on flight deck windows can severely restrict the pilot's visibility outside of the aircraft. Aircraft equipped for flight into known icing conditions typically have some form of windshield anti-icing to enable the pilot to see outside the aircraft in case icing is encountered in flight. One system consists of an electrically heated plate installed onto the airplane's windshield to give the pilot a narrow band of clear visibility. Another system uses a bar at the lower end of the windshield to spray deicing fluid onto it and prevent ice from forming. On high performance aircraft that require complex windshields to protect against bird strikes and withstand pressurization loads, the heating element often is a layer of conductive film or thin wire strands through which electric current is run to heat the windshield and prevent ice from forming.

Antenna Icing

Because of their small size and shape, antennas that do not lay flush with the aircraft's skin tend to accumulate ice rapidly. Furthermore, they often are devoid of internal anti-icing or deicing capability for protection. During flight in icing conditions, ice accumulations on an antenna may cause it to begin to vibrate or cause radio signals to become distorted and it may cause damage to the antenna. If a frozen antenna breaks off, it can damage other areas of the aircraft in addition to causing a communication or navigation system failure.

Summary

Ice-contaminated aircraft have been involved in many accidents. Takeoff accidents have usually been due to failure to deice or anti-ice critical surfaces properly on the ground. Proper deicing and anti-icing procedures are addressed in two other pilot guides, Advisory Circular (AC) 120-58, Pilot Guide: Large Aircraft Ground Deicing and AC 135-17, Pilot Guide: Small Aircraft Ground Deicing.

The pilot of an aircraft, which is not certificated or equipped for flight in icing conditions, should avoid all icing conditions. The aforementioned guides provide direction on how to do this, and on how to exit icing conditions promptly and safely should they be inadvertently encountered.

The pilot of an aircraft, which is certificated for flight in icing conditions can safely operate in the conditions for which the aircraft was evaluated during the certification process but should never become complacent about icing. Even short encounters with small amounts of rough icing can be very hazardous. The pilot should be familiar with all information in the Aircraft Flight Manual (AFM) or Pilot's Operating Handbook (POH) concerning flight in icing conditions and follow it carefully. Of particular importance are proper operation of ice protection systems and any airspeed minimums to be observed during or after flight in icing conditions. There are some icing conditions for which no aircraft is evaluated in the certification process, such as super-cooled large drops (SLD). These subfreezing water droplets, with diameters greater than 50 microns, occur within or below clouds and sustained flight in these conditions can be very hazardous. The pilot should be familiar with any information in the AFM or POH relating to these conditions, including aircraft-specific cues for recognizing these hazardous conditions within clouds.

The information in this chapter is an overview of the hazards of aircraft icing. For more detailed information refer to AC 91-74, Pilot Guide: Flight in Icing Conditions, AC 91-51, Effect of Icing on Aircraft Control and Airplane Deice and Anti-Ice Systems, AC 20-73, Aircraft Ice Protection and AC 23.143-1, Ice Contaminated Tailplane Stall (ICTS).

Chapter 5

Flight Instruments

Introduction

Aircraft became a practical means of transportation when accurate flight instruments freed the pilot from the necessity of maintaining visual contact with the ground. Flight instruments are crucial to conducting safe flight operations and it is important that the pilot have a basic understanding of their operation. The basic flight instruments required for operation under visual flight rules (VFR) are airspeed indicator (ASI), altimeter, and magnetic direction indicator. In addition to these, operation under instrument flight rules (IFR) requires a gyroscopic rate-of-turn indicator, slip-skid indicator, sensitive altimeter adjustable for barometric pressure, clock displaying hours, minutes, and seconds with a sweep-second pointer or digital presentation, gyroscopic pitch-and-bank indicator (artificial horizon), and gyroscopic direction indicator (directional gyro or equivalent).

Aircraft that are flown in instrument meteorological conditions (IMC) are equipped with instruments that provide attitude and direction reference, as well as navigation instruments that allow precision flight from takeoff to landing with limited or no outside visual reference.

The instruments discussed in this chapter are those required by Title 14 of the Code of Federal Regulations (14 CFR) part 91, and are organized into three groups: pitot-static instruments, compass systems, and gyroscopic instruments. The chapter concludes with a discussion of how to preflight these systems for IFR flight. This chapter addresses additional avionics systems such as Electronic Flight Information Systems (EFIS), Ground Proximity Warning System (GPWS), Terrain Awareness and Warning System (TAWS), Traffic Alert and Collision Avoidance System (TCAS), Head Up Display (HUD), etc., that are increasingly being incorporated into general aviation aircraft.

Pitot/Static Systems

Pitot pressure, or impact air pressure, is sensed through an open-end tube pointed directly into the relative wind flowing around the aircraft. The pitot tube connects to the ASI or an air data computer depending on your aircraft's configuration.

Static Pressure

Static pressure is also used by the ASI as well as the other pitot static instruments for determining altitude and vertical speed. Static pressure may be sensed at one or more locations on an aircraft. Some may be flush mounted on the fuselage or integrated into the electrically heated pitot tube. *[Figure 5-1]* These ports are in locations proven by flight tests to be in undisturbed air, and they may be paired, one on either side of the aircraft. This dual location prevents lateral movement of the aircraft from giving erroneous static pressure indications. The areas around the static ports may be heated with electric heater elements to prevent ice forming over the port and blocking the entry of the static air.

Three basic pressure-operated instruments are found in aircraft instrument panels flown under IFR. These are the ASI, sensitive altimeter, and vertical speed indicator (VSI). All three instruments receive static air pressure for operation with only the ASI receiving both pitot and static pressure. *[Figure 5-2]*

Blockage of the Pitot-Static System

Errors in the ASI and VSI almost always indicate a blockage of the pitot tube, the static port(s), or both. Moisture (including ice), dirt, or even insects can cause a blockage in both systems. During preflight, it is very important to make sure the pitot tube cover is removed and that static port openings are checked for blockage and damage.

Blocked Pitot System

If the pitot tube drain hole becomes obstructed, the pitot system can become partially or completely blocked. When dynamic pressure cannot enter the pitot tube opening, the ASI no longer operates. If the drain hole is open, static pressure equalizes on both sides of the diaphram in the ASI and the

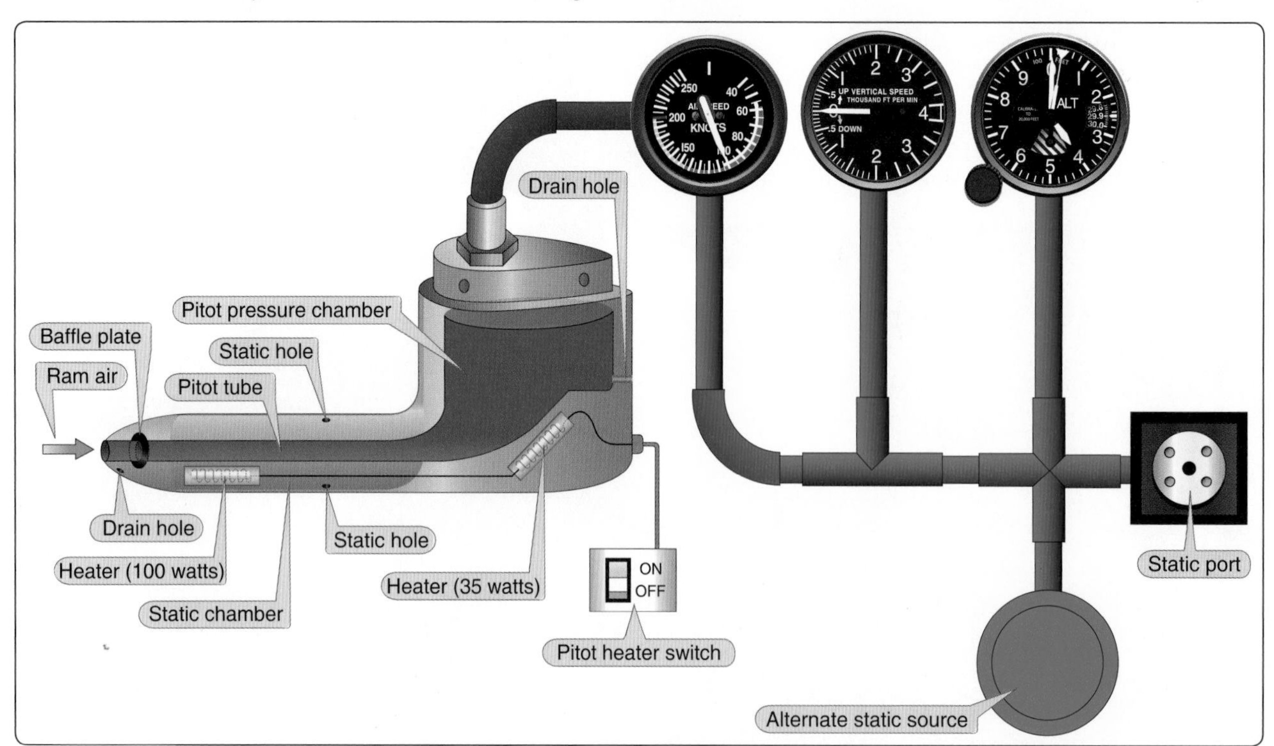

Figure 5-1. *A typical electrically heated pitot-static head.*

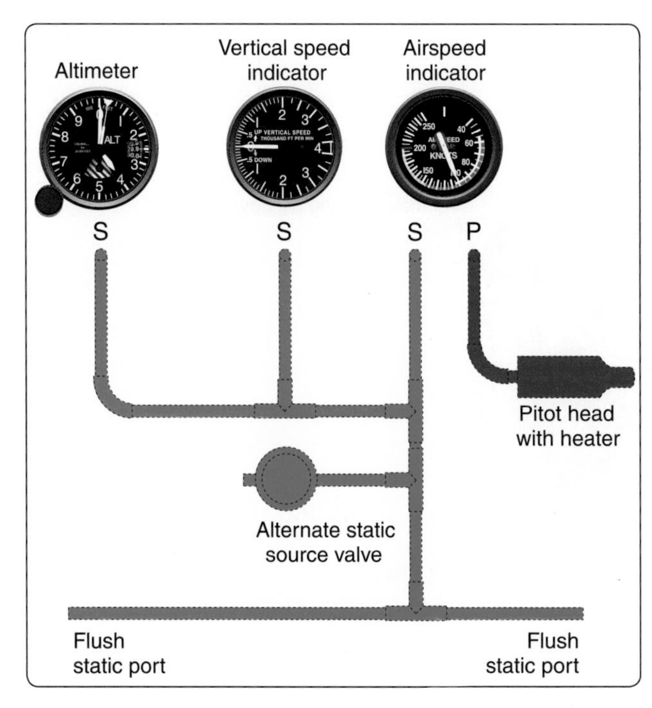

Figure 5-2. *A typical pitot-static system.*

indicated airspeed slowly drops to zero. If the pitot tube ram pressure hole and drain hole become obstructed, the ASI operates like an altimeter as the aircraft climbs and descends. Refer to the Pilot's Handbook of Aeronautical Knowledge (FAA-H-8083-25A) for more in depth information on blocked pitot systems along with different scenarios and how they effect the ASI.

Blocked Static System

When a static system becomes blocked but the pitot tube remains clear the ASI continues to operate but is inaccurate. When the aircraft is operated above the altitude where the static ports became blocked the airspeed indicates lower than the actual airspeed because the trapped static pressure is higher than normal for that altitude. The opposite holds true for operations at lower altitudes; a faster than actual airspeed is displayed due to the relatively low static pressure trapped in the system.

A blockage of the static system can also affect the altimeter and VSI. Trapped static pressure causes the altimeter to freeze at the altitude where the blockage occurred. In the case of the VSI, a blocked static system produces a continuous zero indication.

An alternate static source is provided in some aircraft to provide static pressure should the primary static source become blocked. The alternate static source is normally found inside of the flight deck. Due to the venturi effect of the air flowing around the fuselage, the air pressure inside the flight deck is lower than the exterior pressure.

When the alternate static source pressure is used, the following instrument indications are observed:

1. The altimeter indicates a slightly higher altitude than actual.
2. The ASI indicates an airspeed greater than the actual airspeed.
3. The VSI shows a momentary climb and then stabilizes if the altitude is held constant.

For more information on static system blockages and how to best react to such situations, refer to the Pilot's Handbook of Aeronautical Knowledge (FAA-H-8083-25A).

Effects of Flight Conditions

The static ports are located in a position where the air at their surface is as undisturbed as possible. But under some flight conditions, particularly at a high angle of attack with the landing gear and flaps down, the air around the static port may be disturbed to the extent that it can cause an error in the indication of the altimeter and ASI. Because of the importance of accuracy in these instruments, part of the certification tests for an aircraft is a check of position error in the static system.

The Pilot's Operating Handbook (POH)/Aircraft Flight Manual (AFM) contains any corrections that must be applied to the airspeed for the various configurations of flaps and landing gear.

Pitot/Static Instruments

Sensitive Altimeter

A sensitive altimeter is an aneroid barometer that measures the absolute pressure of the ambient air and displays it in terms of feet or meters above a selected pressure level.

Principle of Operation

The sensitive element in a sensitive altimeter is a stack of evacuated, corrugated bronze aneroid capsules. *[Figure 5-3]* The air pressure acting on these aneroids tries to compress them against their natural springiness, which tries to expand them. The result is that their thickness changes as the air pressure changes. Stacking several aneroids increases the dimension change as the pressure varies over the usable range of the instrument.

Below 10,000 feet, a striped segment is visible. Above this altitude, a mask begins to cover it, and above 15,000 feet, all of the stripes are covered. *[Figure 5-4]*

Another configuration of the altimeter is the drum-type. *[Figure 5-5]* These instruments have only one pointer that

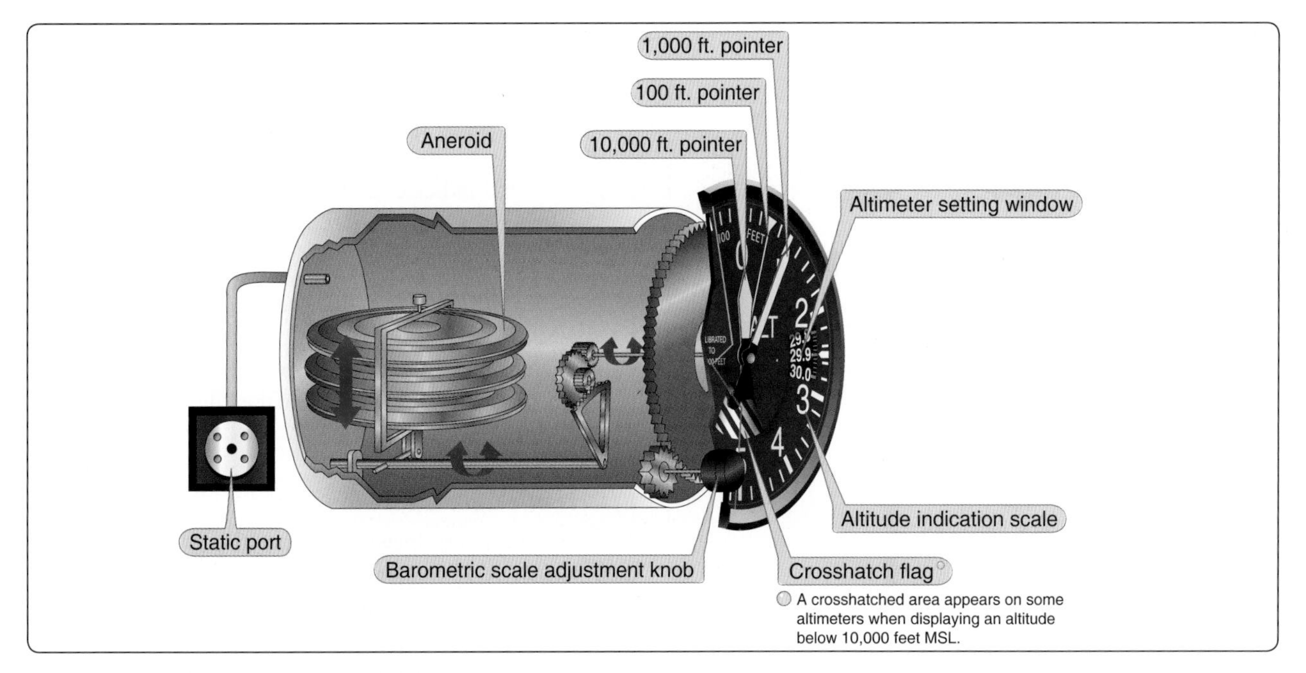

Figure 5-3. *Sensitive altimeter components.*

Figure 5-4. *Three-pointer altimeter.*

Figure 5-5. *Drum-type altimeter.*

makes one revolution for every 1,000 feet. Each number represents 100 feet and each mark represents 20 feet. A drum, marked in thousands of feet, is geared to the mechanism that drives the pointer. To read this type of altimeter, first look at the drum to get the thousands of feet, and then at the pointer to get the feet and hundreds of feet.

A sensitive altimeter is one with an adjustable barometric scale allowing the pilot to set the reference pressure from which the altitude is measured. This scale is visible in a small window called the Kollsman window. A knob on the instrument adjusts the scale. The range of the scale is from 28.00 to 31.00 inches of mercury ("Hg), or 948 to 1,050 millibars.

Rotating the knob changes both the barometric scale and the altimeter pointers in such a way that a change in the barometric scale of 1 "Hg changes the pointer indication by 1,000 feet. This is the standard pressure lapse rate below 5,000 feet. When the barometric scale is adjusted to 29.92 "Hg or 1,013.2 millibars, the pointers indicate the pressure altitude. The pilot displays indicate altitude by adjusting the barometric scale to the local altimeter setting. The altimeter then indicates the height above the existing sea level pressure.

Altimeter Errors

A sensitive altimeter is designed to indicate standard changes from standard conditions, but most flying involves errors caused by nonstandard conditions and the pilot must be able to modify the indications to correct for these errors. There are two types of errors: mechanical and inherent.

Mechanical Altimeter Errors

A preflight check to determine the condition of an altimeter consists of setting the barometric scale to the local altimeter setting. The altimeter should indicate the surveyed elevation of the airport. If the indication is off by more than 75 feet from the surveyed elevation, the instrument should be referred to a certificated instrument repair station for recalibration. Differences between ambient temperature and/or pressure causes an erroneous indication on the altimeter.

Inherent Altimeter Error

When the aircraft is flying in air that is warmer than standard, the air is less dense and the pressure levels are farther apart. When the aircraft is flying at an indicated altitude of 5,000 feet, the pressure level for that altitude is higher than it would be in air at standard temperature, and the aircraft is higher than it would be if the air were cooler. If the air is colder than standard, it is denser and the pressure levels are closer together. When the aircraft is flying at an indicated altitude of 5,000 feet, its true altitude is lower than it would be if the air were warmer. *[Figure 5-6]*

Cold Weather Altimeter Errors

A correctly calibrated pressure altimeter indicates true altitude above mean sea level (MSL) when operating within the International Standard Atmosphere (ISA) parameters of pressure and temperature. Nonstandard pressure conditions are corrected by applying the correct local area altimeter setting.

Temperature errors from ISA result in true altitude being higher than indicated altitude whenever the temperature is warmer than ISA and true altitude being lower than indicated altitude whenever the temperature is colder than ISA. True altitude variance under conditions of colder than ISA temperatures poses the risk of inadequate obstacle clearance.

Under extremely cold conditions, pilots may need to add an appropriate temperature correction determined from the chart in *Figure 5-7* to charted IFR altitudes to ensure terrain and obstacle clearance with the following restrictions:

- Altitudes specifically assigned by Air Traffic Control (ATC), such as "maintain 5,000 feet" shall not be corrected. Assigned altitudes may be rejected if the pilot decides that low temperatures pose a risk of inadequate terrain or obstacle clearance.
- If temperature corrections are applied to charted IFR altitudes (such as procedure turn altitudes, final approach fix crossing altitudes, etc.), the pilot must advise ATC of the applied correction.

ICAO Cold Temperature Error Table

The cold temperature induced altimeter error may be significant when considering obstacle clearances when temperatures are well below standard. Pilots may wish to increase their minimum terrain clearance altitudes with a corresponding increase in ceiling from the normal minimum when flying in extreme cold temperature conditions. Higher altitudes may need to be selected when flying at low terrain clearances. Most flight management systems (FMS) with air data computers implement a capability to compensate for cold temperature errors. Pilots flying with these systems should ensure they are aware of the conditions under which the system automatically compensates. If compensation is applied by the FMS or manually, ATC must be informed that the aircraft is not flying the assigned altitude. Otherwise, vertical separation from other aircraft may be reduced creating a potentially hazardous situation. The table in *Figure 5-7*, derived from International Civil Aviation Organization

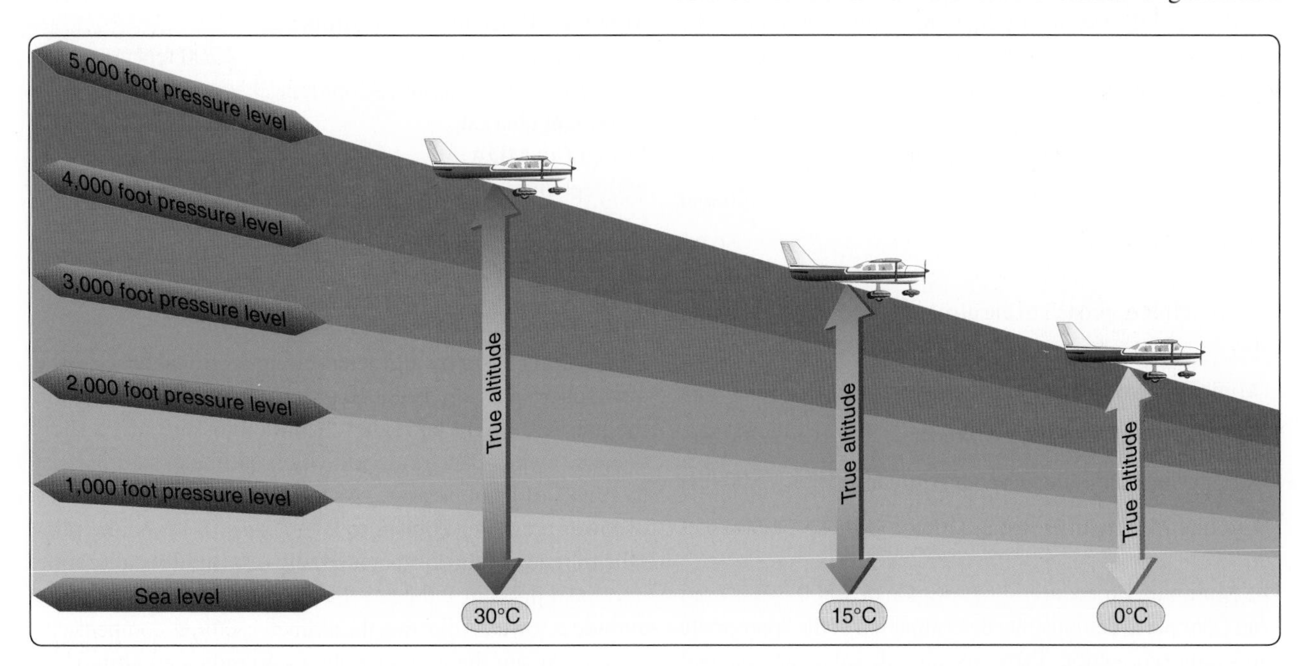

Figure 5-6. *The loss of altitude experienced when flying into an area where the air is colder (more dense) than standard.*

Reported Temp °C	Height Above Airport in Feet													
	200	300	400	500	600	700	800	900	1,000	1,500	2,000	3,000	4,000	5,000
+10	10	10	10	10	20	20	20	20	20	30	40	60	80	90
0	20	20	30	30	40	40	50	50	60	90	120	170	230	280
-10	20	30	40	50	60	70	80	90	100	150	200	290	390	490
-20	30	50	60	70	90	100	120	130	140	210	280	420	570	710
-30	40	60	80	100	120	130	150	170	190	280	380	570	760	950
-40	50	80	100	120	150	170	190	220	240	360	480	720	970	1,210
-50	60	90	120	150	180	210	240	270	300	450	590	890	1,190	1,500

Figure 5-7. *International Civil Aviation Organization (ICAO) cold temperature error table.*

(ICAO) standard formulas, shows how much error can exist when the temperature is extremely cold. To use the table, find the reported temperature in the left column, and then read across the top row to the height above the airport/reporting station. Subtract the airport elevation from the altitude of the final approach fix (FAF). The intersection of the column and row is the amount of possible error.

Example: The reported temperature is –10 degrees Celsius (°C) and the FAF is 500 feet above the airport elevation. The reported current altimeter setting may place the aircraft as much as 50 feet below the altitude indicated by the altimeter.

When using the cold temperature error table, the altitude error is proportional to both the height above the reporting station elevation and the temperature at the reporting station. For IFR approach procedures, the reporting station elevation is assumed to be airport elevation. It is important to understand that corrections are based upon the temperature at the reporting station, not the temperature observed at the aircraft's current altitude and height above the reporting station and not the charted IFR altitude.

To see how corrections are applied, note the following example:

Airport Elevation	496 feet
Airport Temperature	–50 °C

A charted IFR approach to the airport provides the following data:

Minimum Procedure Turn Altitude	1,800 feet
Minimum FAF Crossing Altitude	1,200 feet
Straight-in Minimum Descent Altitude	800 feet
Circling Minimum Descent Altitude (MDA)	1,000 feet

The Minimum Procedure Turn Altitude of 1,800 feet is used as an example to demonstrate determination of the appropriate temperature correction. Typically, altitude values are rounded up to the nearest 100-foot level. The charted procedure turn altitude of 1,800 feet minus the airport elevation of 500 feet equals 1,300 feet. The altitude difference of 1,300 feet falls between the correction chart elevations of 1,000 feet and 1,500 feet. At the station temperature of –50 °C, the correction falls between 300 feet and 450 feet. Dividing the difference in compensation values by the difference in altitude above the airport gives the error value per foot.

In this case, 150 feet divided by 500 feet = 0.33 feet for each additional foot of altitude above 1,000 feet. This provides a correction of 300 feet for the first 1,000 feet and an additional value of 0.33 times 300 feet, or 99 feet, which is rounded to 100 feet. 300 feet + 100 feet = total temperature correction of 400 feet. For the given conditions, correcting the charted value of 1,800 feet above MSL (equal to a height above the reporting station of 1,300 feet) requires the addition of 400 feet. Thus, when flying at an indicated altitude of 2,200 feet, the aircraft is actually flying a true altitude of 1,800 feet.

Minimum Procedure Turn Altitude
1,800 feet charted = 2,200 feet corrected
Minimum FAF Crossing Altitude
1,200 feet charted = 1,500 feet corrected
Straight-in MDA
800 feet charted = 900 feet corrected
Circling MDA
1,000 feet charted = 1,200 feet corrected

Nonstandard Pressure on an Altimeter

Maintaining a current altimeter setting is critical because the atmosphere pressure is not constant. That is, in one location the pressure might be higher than the pressure just a short distance away. Take an aircraft whose altimeter setting is set to 29.92" of local pressure. As the aircraft moves to an area of lower pressure (Point A to B in *Figure 5-8*) and the pilot fails to readjust the altimeter setting (essentially calibrating it to local pressure), then as the pressure decreases, the true altitude is lower. Adjusting the altimeter settings compensates for this. When the altimeter shows an indicated altitude of 5,000 feet, the true altitude at Point A (the height above mean sea level) is only 3,500 feet at Point B. The fact that the altitude indication is not always true lends itself to the memory aid, "When flying from hot to cold or from a high to a low, look out below." *[Figure 5-8]*

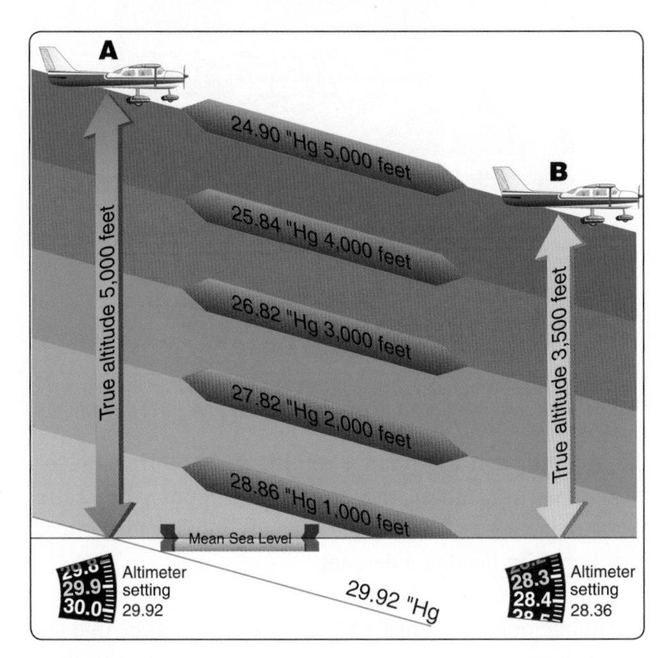

Figure 5-8. *Effects of nonstandard pressure on an altimeter of an aircraft flown into air of lower than standard pressure (air is less dense).*

Altimeter Enhancements (Encoding)

It is not sufficient in the airspace system for only the pilot to have an indication of the aircraft's altitude; the air traffic controller on the ground must also know the altitude of the aircraft. To provide this information, the aircraft is typically equipped with an encoding altimeter.

When the ATC transponder is set to Mode C, the encoding altimeter supplies the transponder with a series of pulses identifying the flight level (in increments of 100 feet) at which the aircraft is flying. This series of pulses is transmitted to the ground radar where they appear on the controller's scope as an alphanumeric display around the return for the aircraft. The transponder allows the ground controller to identify the aircraft and determine the pressure altitude at which it is flying.

A computer inside the encoding altimeter measures the pressure referenced from 29.92 "Hg and delivers this data to the transponder. When the pilot adjusts the barometric scale to the local altimeter setting, the data sent to the transponder is not affected. This is to ensure that all Mode C aircraft are transmitting data referenced to a common pressure level. ATC equipment adjusts the displayed altitudes to compensate for local pressure differences allowing display of targets at correct altitudes. 14 CFR part 91 requires the altitude transmitted by the transponder to be within 125 feet of the altitude indicated on the instrument used to maintain flight altitude.

Reduced Vertical Separation Minimum (RVSM)

Below 31,000 feet, a 1,000 foot separation is the minimum required between usable flight levels. Flight levels (FLs) generally start at 18,000 feet where the local pressure is 29.92 "Hg or greater. All aircraft 18,000 feet and above use a standard altimeter setting of 29.92 "Hg, and the altitudes are in reference to a standard hence termed FL. Between FL 180 and FL 290, the minimum altitude separation is 1,000 feet between aircraft. However, for flight above FL 290 (primarily due to aircraft equipage and reporting capability; potential error) ATC applied the requirement of 2,000 feet of separation. FL 290, an altitude appropriate for an eastbound aircraft, would be followed by FL 310 for a westbound aircraft, and so on to FL 410, or seven FLs available for flight. With 1,000-foot separation, or a reduction of the vertical separation between FL 290 and FL 410, an additional six FLs become available. This results in normal flight level and direction management being maintained from FL 180 through FL 410. Hence the name is Reduced Vertical Separation Minimum (RVSM). Because it is applied domestically, it is called United States Domestic Reduced Vertical Separation Minimum (DRVSM).

However, there is a cost to participate in the DRVSM program which relates to both aircraft equipage and pilot training. For example, altimetry error must be reduced significantly and operators using RVSM must receive authorization from the appropriate civil aviation authority. RVSM aircraft must meet required altitude-keeping performance standards. Additionally, operators must operate in accordance with RVSM policies/procedures applicable to the airspace where they are flying.

The aircraft must be equipped with at least one automatic altitude control—

- Within a tolerance band of ±65 feet about an acquired altitude when the aircraft is operated in straight-and-level flight.
- Within a tolerance band of ±130 feet under no turbulent, conditions for aircraft for which application for type certification occurred on or before April 9, 1997 that are equipped with an automatic altitude control system with flight management/performance system inputs.

That aircraft must be equipped with an altitude alert system that signals an alert when the altitude displayed to the flight crew deviates from the selected altitude by more than (in most cases) 200 feet. For each condition in the full RVSM flight envelope, the largest combined absolute value for residual static source error plus the avionics error may not exceed 200 feet. Aircraft with TCAS must have compatibility with RVSM Operations. *Figure 5-9* illustrates the increase in aircraft permitted between FL 180 and FL 410. Most noteworthy, however, is the economization that aircraft can take advantage of by the higher FLs being available to more aircraft.

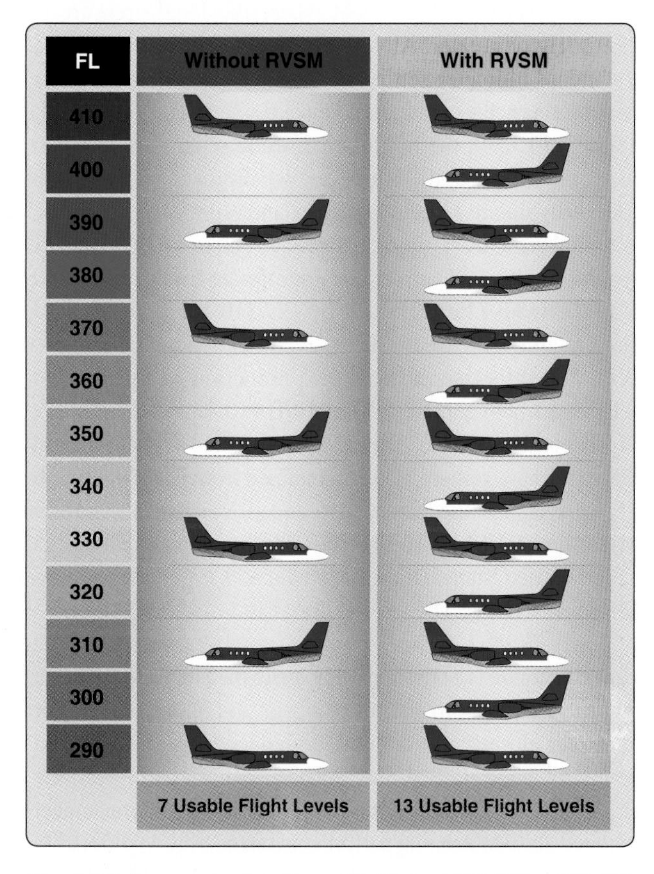

Figure 5-9. *Increase in aircraft permitted between FL 180 and FL 410.*

Vertical Speed Indicator (VSI)

The VSI in *Figure 5-10* is also called a vertical velocity indicator (VVI), and was formerly known as a rate-of-climb indicator. It is a rate-of-pressure change instrument that gives an indication of any deviation from a constant pressure level.

Inside the instrument case is an aneroid very much like the one in an ASI. Both the inside of this aneroid and the inside of the instrument case are vented to the static system, but the case is vented through a calibrated orifice that causes the pressure inside the case to change more slowly than the pressure inside the aneroid. As the aircraft ascends, the static pressure becomes lower. The pressure inside the case compresses the aneroid, moving the pointer upward, showing a climb and indicating the rate of ascent in number of feet per minute (fpm).

When the aircraft levels off, the pressure no longer changes. The pressure inside the case becomes equal to that inside

Figure 5-10. *Rate of climb or descent in thousands of feet per minute.*

the aneroid, and the pointer returns to its horizontal, or zero, position. When the aircraft descends, the static pressure increases. The aneroid expands, moving the pointer downward, indicating a descent.

The pointer indication in a VSI lags a few seconds behind the actual change in pressure. However, it is more sensitive than an altimeter and is useful in alerting the pilot of an upward or downward trend, thereby helping maintain a constant altitude.

Some of the more complex VSIs, called instantaneous vertical speed indicators (IVSI), have two accelerometer-actuated air pumps that sense an upward or downward pitch of the aircraft and instantaneously create a pressure differential. By the time the pressure caused by the pitch acceleration dissipates, the altitude pressure change is effective.

Dynamic Pressure Type Instruments

Airspeed Indicator (ASI)

An ASI is a differential pressure gauge that measures the dynamic pressure of the air through which the aircraft is flying. Dynamic pressure is the difference in the ambient static air pressure and the total, or ram, pressure caused by the motion of the aircraft through the air. These two pressures are taken from the pitot-static system.

The mechanism of the ASI in *Figure 5-11* consists of a thin, corrugated phosphor bronze aneroid, or diaphragm, that receives its pressure from the pitot tube. The instrument case is sealed and connected to the static ports. As the pitot pressure increases or the static pressure decreases, the diaphragm expands. This dimensional change is measured by a rocking shaft and a set of gears that drives a pointer across the instrument dial. Most ASIs are calibrated in knots, or nautical miles per hour; some instruments show statute miles per hour, and some instruments show both.

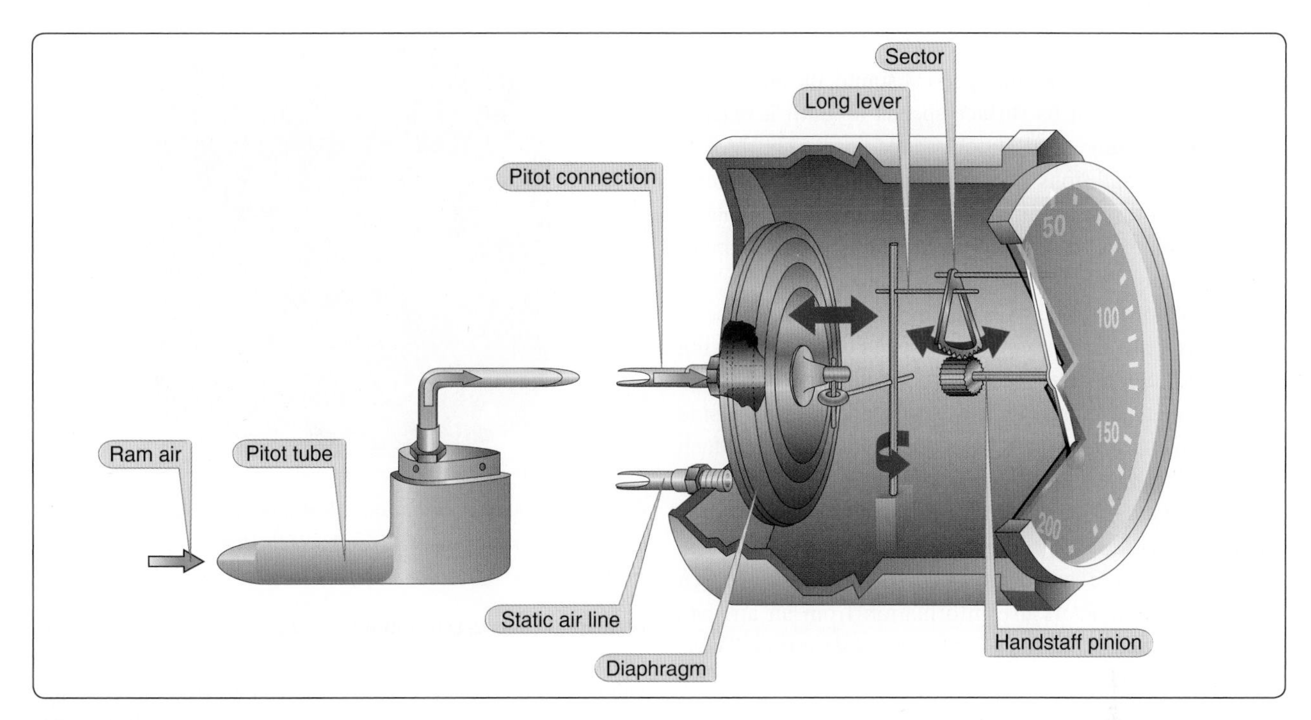

Figure 5-11. *Mechanism of an ASI.*

Types of Airspeed

Just as there are several types of altitude, there are multiple types of airspeed: indicated airspeed (IAS), calibrated airspeed (CAS), equivalent airspeed (EAS), and true airspeed (TAS).

Indicated Airspeed (IAS)

IAS is shown on the dial of the instrument, uncorrected for instrument or system errors.

Calibrated Airspeed (CAS)

CAS is the speed at which the aircraft is moving through the air, which is found by correcting IAS for instrument and position errors. The POH/AFM has a chart or graph to correct IAS for these errors and provide the correct CAS for the various flap and landing gear configurations.

Equivalent Airspeed (EAS)

EAS is CAS corrected for compression of the air inside the pitot tube. EAS is the same as CAS in standard atmosphere at sea level. As the airspeed and pressure altitude increase, the CAS becomes higher than it should be, and a correction for compression must be subtracted from the CAS.

True Airspeed (TAS)

TAS is CAS corrected for nonstandard pressure and temperature. TAS and CAS are the same in standard atmosphere at sea level. Under nonstandard conditions, TAS is found by applying a correction for pressure altitude and temperature to the CAS.

Some aircraft are equipped with true ASIs that have a temperature-compensated aneroid bellows inside the instrument case. This bellows modifies the movement of the rocking shaft inside the instrument case so the pointer shows the actual TAS.

The TAS indicator provides both true and IAS. These instruments have the conventional airspeed mechanism, with an added subdial visible through cutouts in the regular dial. A knob on the instrument allows the pilot to rotate the subdial and align an indication of the outside air temperature with the pressure altitude being flown. This alignment causes the instrument pointer to indicate the TAS on the subdial. *[Figure 5-12]*

Figure 5-12. *A true ASI allows the pilot to correct IAS for nonstandard temperature and pressure.*

Mach Number

As an aircraft approaches the speed of sound, the air flowing over certain areas of its surface speeds up until it reaches the speed of sound, and shock waves form. The IAS at which these conditions occur changes with temperature. Therefore, in this case, airspeed is not entirely adequate to warn the pilot of the impending problems. Mach number is more useful. Mach number is the ratio of the TAS of the aircraft to the speed of sound in the same atmospheric conditions. An aircraft flying at the speed of sound is flying at Mach 1.0. Some older mechanical Machmeters not driven from an air data computer use an altitude aneroid inside the instrument that converts pitot-static pressure into Mach number. These systems assume that the temperature at any altitude is standard; therefore, the indicated Mach number is inaccurate whenever the temperature deviates from standard. These systems are called indicated Machmeters. Modern electronic Machmeters use information from an air data computer system to correct for temperature errors. These systems display true Mach number.

Most high-speed aircraft are limited to a maximum Mach number at which they can fly. This is shown on a Machmeter as a decimal fraction. *[Figure 5-13]* For example, if the Machmeter indicates .83 and the aircraft is flying at 30,000 feet where the speed of sound under standard conditions is 589.5 knots, the airspeed is 489.3 knots. The speed of sound varies with the air temperature. If the aircraft were flying at Mach .83 at 10,000 feet where the air is much warmer, its airspeed would be 530 knots.

Figure 5-13. *A Machmeter shows the ratio of the speed of sound to the TAS the aircraft is flying.*

Maximum Allowable Airspeed

Some aircraft that fly at high subsonic speeds are equipped with maximum allowable ASIs like the one in *Figure 5-14.* This instrument looks much like a standard ASI, calibrated in knots, but has an additional pointer colored red, checkered, or striped. The maximum airspeed pointer is actuated by an aneroid, or altimeter mechanism, that moves it to a lower value as air density decreases. By keeping the airspeed pointer at a lower value than the maximum pointer, the pilot avoids the onset of transonic shock waves.

Figure 5-14. *A maximum allowable ASI has a movable pointer that indicates the never-exceed speed, which changes with altitude to avoid the onset of transonic shock waves.*

Airspeed Color Codes

The dial of an ASI is color coded to alert the pilot, at a glance, of the significance of the speed at which the aircraft is flying. These colors and their associated airspeeds are shown in *Figure 5-15.*

Magnetism

The Earth is a huge magnet, spinning in space, surrounded by a magnetic field made up of invisible lines of flux. These lines leave the surface at the magnetic North Pole and reenter at the magnetic South Pole.

Lines of magnetic flux have two important characteristics: any magnet that is free to rotate aligns with them, and an electrical current is induced into any conductor that cuts across them. Most direction indicators installed in aircraft make use of one of these two characteristics.

The Basic Aviation Magnetic Compass

One of the oldest and simplest instruments for indicating direction is the magnetic compass. It is also one of the basic instruments required by 14 CFR part 91 for both VFR and IFR flight.

Magnetic Compass Overview

A magnet is a piece of material, usually a metal containing iron, which attracts and holds lines of magnetic flux. Regardless of size, every magnet has two poles: a north pole and a south pole. When one magnet is placed in the

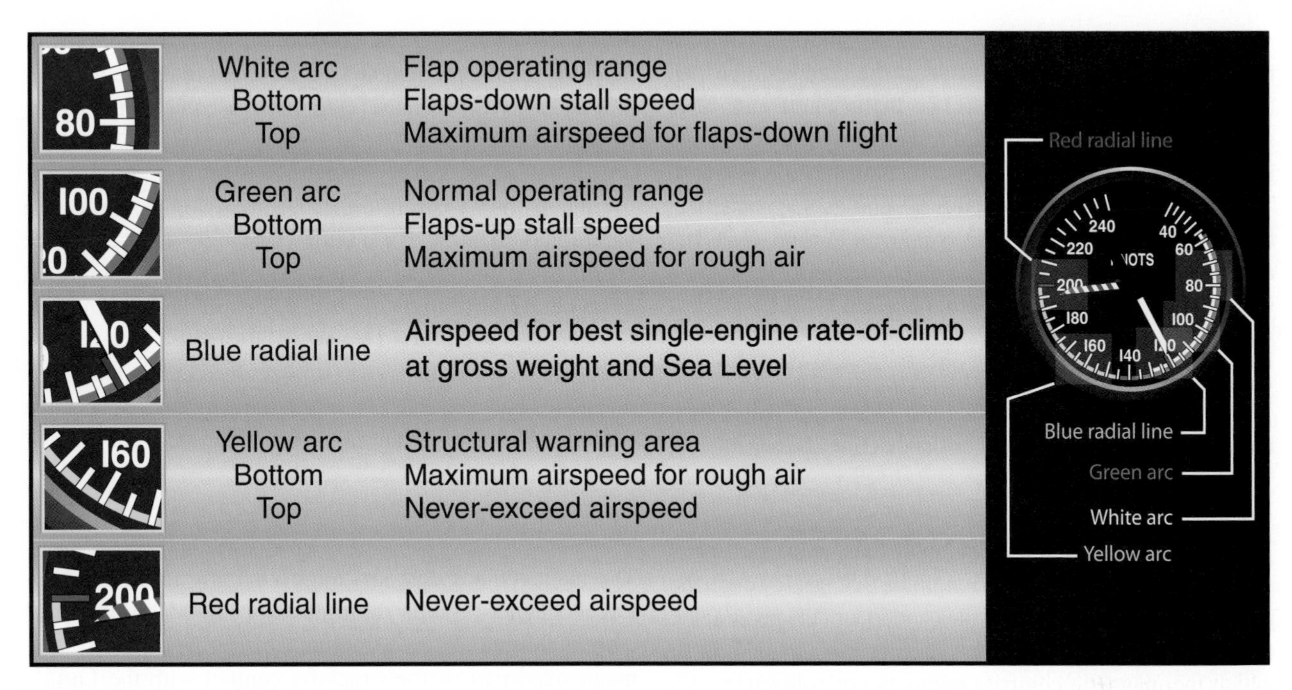

White arc
Bottom
Top
Flap operating range
Flaps-down stall speed
Maximum airspeed for flaps-down flight
Green arc
Bottom
Top
Normal operating range
Flaps-up stall speed
Maximum airspeed for rough air
Blue radial line
Airspeed for best single-engine rate-of-climb at gross weight and Sea Level
Yellow arc
Bottom
Top
Structural warning area
Maximum airspeed for rough air
Never-exceed airspeed
Red radial line
Never-exceed airspeed
Red radial line
Blue radial line
Green arc
White arc
Yellow arc

Figure 5-15. *Color codes for an ASI.*

field of another, the unlike poles attract each other and like poles repel.

An aircraft magnetic compass, such as the one in *Figure 5-16*, has two small magnets attached to a metal float sealed inside a bowl of clear compass fluid similar to kerosene. A graduated scale, called a card, is wrapped around the float and viewed through a glass window with a lubber line across it. The card is marked with letters representing the cardinal directions, north, east, south, and west, and a number for each 30° between these letters. The final "0" is omitted from these directions; for example, 3 = 30°, 6 = 60°, and 33 = 330°. There are long and short graduation marks between the letters and numbers, with each long mark representing 10° and each short mark representing 5°.

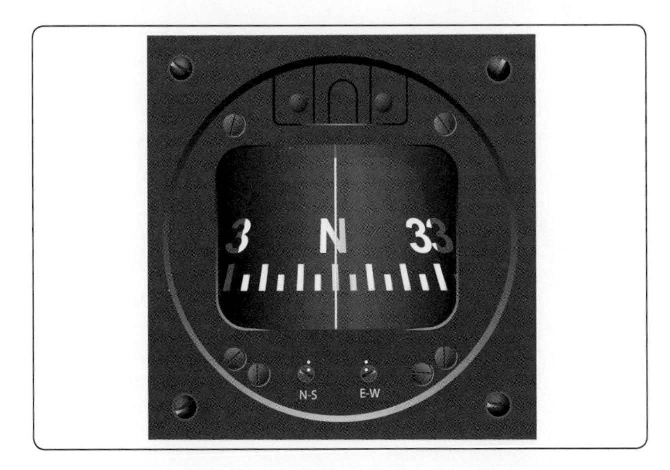

Figure 5-16. *A magnetic compass. The vertical line is called the lubber line.*

Magnetic Compass Construction

The float and card assembly has a hardened steel pivot in its center that rides inside a special, spring-loaded, hard-glass jewel cup. The buoyancy of the float takes most of the weight off the pivot, and the fluid damps the oscillation of the float and card. This jewel-and-pivot type mounting allows the float freedom to rotate and tilt up to approximately 18° angle of bank. At steeper bank angles, the compass indications are erratic and unpredictable.

The compass housing is entirely full of compass fluid. To prevent damage or leakage when the fluid expands and contracts with temperature changes, the rear of the compass case is sealed with a flexible diaphragm, or with a metal bellows in some compasses.

Magnetic Compass Theory of Operations

The magnets align with the Earth's magnetic field and the pilot reads the direction on the scale opposite the lubber line. Note that in *Figure 5-16*, the pilot sees the compass card from its backside. When the pilot is flying north as the compass shows, east is to the pilot's right, but on the card "33", which represents 330° (west of north), is to the right of north. The reason for this apparent backward graduation is that the card remains stationary, and the compass housing and the pilot turn around it, always viewing the card from its backside.

Magnetic fields caused by aircraft electronics and wiring can effect the accuracy of the magnetic compass. This induced error is called compass deviation. Compensator assemblies mounted on the compass allow aviation maintenance technicians (AMTs) to calibrate the compass by creating magnetic fields inside of the compass housing. The compensator assembly has two shafts whose ends have screwdriver slots accessible from the front of the compass. Each shaft rotates one or two small compensating magnets. The end of one shaft is marked E-W, and its magnets affect the compass when the aircraft is pointed east or west. The other shaft is marked N-S and its magnets affect the compass when the aircraft is pointed north or south.

Magnetic Compass Errors

The magnetic compass is the simplest instrument in the panel, but it is subject to a number of errors that must be considered.

Variation

The Earth rotates about its geographic axis; maps and charts are drawn using meridians of longitude that pass through the geographic poles. Directions measured from the geographic poles are called true directions. The north magnetic pole to which the magnetic compass points is not collocated with the geographic north pole, but is some 1,300 miles away; directions measured from the magnetic poles are called magnetic directions. In aerial navigation, the difference between true and magnetic directions is called variation. This same angular difference in surveying and land navigation is called declination.

Figure 5-17 shows the isogonic lines that identify the number of degrees of variation in their area. The line that passes near Chicago is called the agonic line. Anywhere along this line the two poles are aligned, and there is no variation. East of this line, the magnetic pole is to the west of the geographic pole and a correction must be applied to a compass indication to get a true direction.

Flying in the Washington, D.C. area, for example, the variation is 10° west. If the pilot wants to fly a true course of south (180°), the variation must be added to this resulting in a magnetic course to fly of 190°. Flying in the Los Angeles, CA area, the variation is 14° east. To fly a true course of 180° there, the pilot would have to subtract the variation and fly a magnetic course of 166°. The variation error does not change with the heading of the aircraft; it is the same anywhere along the isogonic line.

Deviation

The magnets in a compass align with any magnetic field. Local magnetic fields in an aircraft caused by electrical current flowing in the structure, in nearby wiring or any magnetized part of the structure, conflict with the Earth's magnetic field and cause a compass error called deviation.

Deviation, unlike variation, is different on each heading, but it is not affected by the geographic location. Variation error cannot be reduced or changed, but deviation error can be minimized when a pilot or AMT performs the maintenance task known as "swinging the compass."

Some airports have a compass rose, which is a series of lines marked out on a taxiway or ramp at some location where there

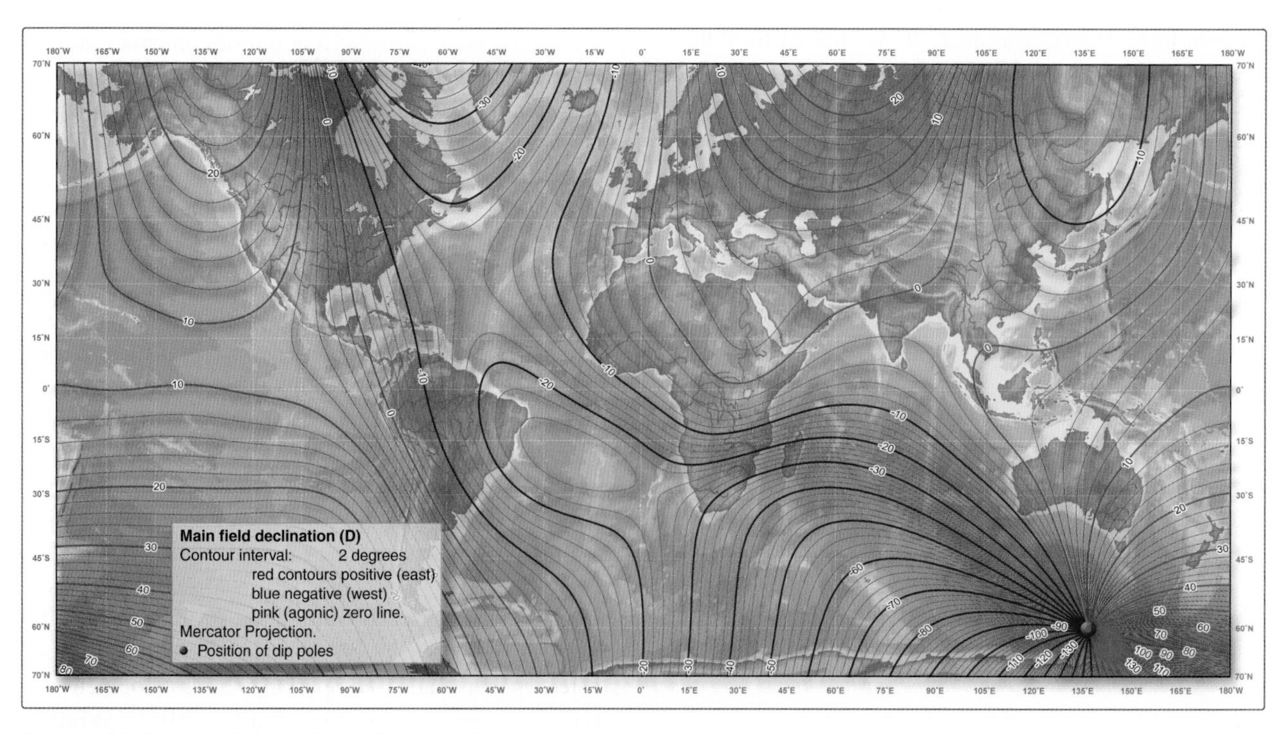

Figure 5-17. *Isogonic lines are lines of equal variation.*

is no magnetic interference. Lines, oriented to magnetic north, are painted every 30°, as shown in *Figure 5-18.*

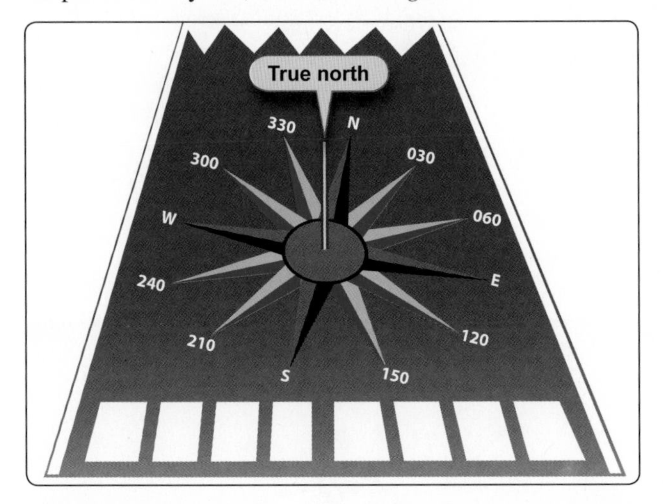

Figure 5-18. *Utilization of a compass rose aids compensation for deviation errors.*

The pilot or AMT aligns the aircraft on each magnetic heading and adjusts the compensating magnets to minimize the difference between the compass indication and the actual magnetic heading of the aircraft. Any error that cannot be removed is recorded on a compass correction card, like the one in *Figure 5-19*, and placed in a cardholder near the compass. If the pilot wants to fly a magnetic heading of 120° and the aircraft is operating with the radios on, the pilot should fly a compass heading of 123°.

FOR	000	030	060	090	120	150
STEER						
RDO. ON	001	032	062	095	123	155
RDO. OFF	002	031	064	094	125	157

FOR	180	210	240	270	300	330
STEER						
RDO. ON	176	210	243	271	296	325
RDO. OFF	174	210	240	273	298	327

Figure 5-19. *A compass correction card shows the deviation correction for any heading.*

The corrections for variation and deviation must be applied in the correct sequence as shown below starting from the true course desired.

Step 1: Determine the Magnetic Course
True Course (180°) ± Variation (+10°) = Magnetic Course (190°)

The Magnetic Course (190°) is steered if there is no deviation error to be applied. The compass card must now be considered for the compass course of 190°.

Step 2: Determine the Compass Course
Magnetic Course (190°, from step 1) ± Deviation (–2°, from correction card) = Compass Course (188°)

NOTE: Intermediate magnetic courses between those listed on the compass card need to be interpreted. Therefore, to steer a true course of 180°, the pilot would follow a compass course of 188°.

To find true course when the compass course is known, remove the variation and deviation corrections previously applied:

Compass Course ± Deviation = Magnetic Course ± Variation = True Course

Northerly Turning Errors

The center of gravity of the float assembly is located lower than the pivotal point. As the airplane turns, the force that results from the magnetic dip causes the float assembly to swing in the same direction that the float turns. The result is a false northerly turn indication. Because of this lead of the compass card, or float assembly, a northerly turn should be stopped prior to arrival at the desired heading. This compass error is amplified with the proximity to either pole. One rule of thumb to correct for this leading error is to stop the turn 15° plus half of the latitude (i.e., if the airplane is being operated in a position around the 40° of latitude, the turn should be stopped 15° + 20° = 35° prior to the desired heading). *[Figure 5-20A]*

Southerly Turning Errors

When turning in a southerly direction, the forces are such that the compass float assembly lags rather than leads. The result is a false southerly turn indication. The compass card, or float assembly, should be allowed to pass the desired heading prior to stopping the turn. As with the northerly error, this error is amplified with the proximity to either pole. To correct this lagging error, the aircraft should be allowed to pass the desired heading prior to stopping the turn. The same rule of 15° plus half of the latitude applies here (i.e., if the airplane is being operated in a position around the 30° of latitude, the turn should be stopped 15° + 15° + 30° after passing the desired heading). *[Figure 5-20B]*

Acceleration Error

The magnetic dip and the forces of inertia cause magnetic compass errors when accelerating and decelerating on Easterly and westerly headings. Because of the pendulous-type mounting, the aft end of the compass card is tilted upward when accelerating, and downward when decelerating during

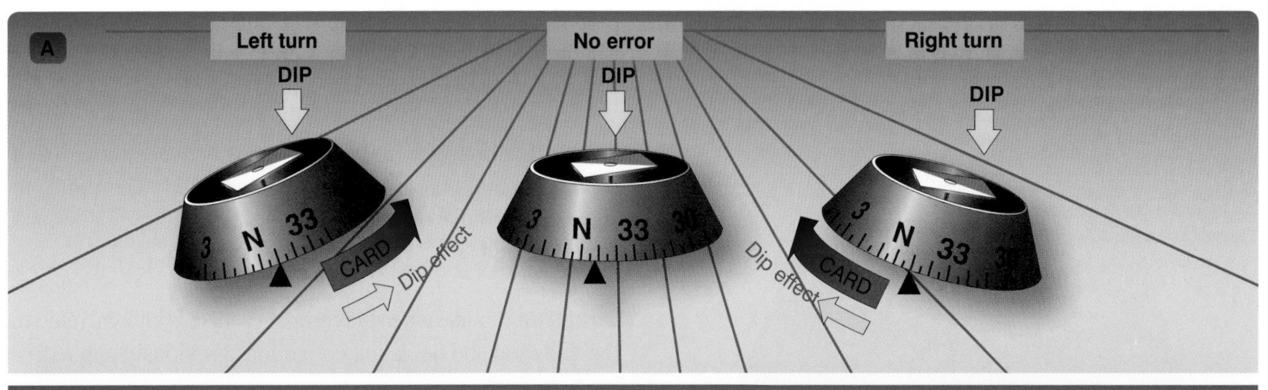

Figure 5-20. *Northerly turning error.*

changes of airspeed. When accelerating on either an easterly or westerly heading , the error appears as a turn indication toward north. When decelerating on either of these headings, the compass indicates a turn toward south. The word "ANDS" (Acceleration-North/Deceleration-South) may help you to remember the acceleration error. *[Figure 5-21]*

Oscillation Error

Oscillation is a combination of all of the other errors, and it results in the compass card swinging back and forth around the heading being flown. When setting the gyroscopic heading indicator to agree with the magnetic compass, use the average indication between the swings.

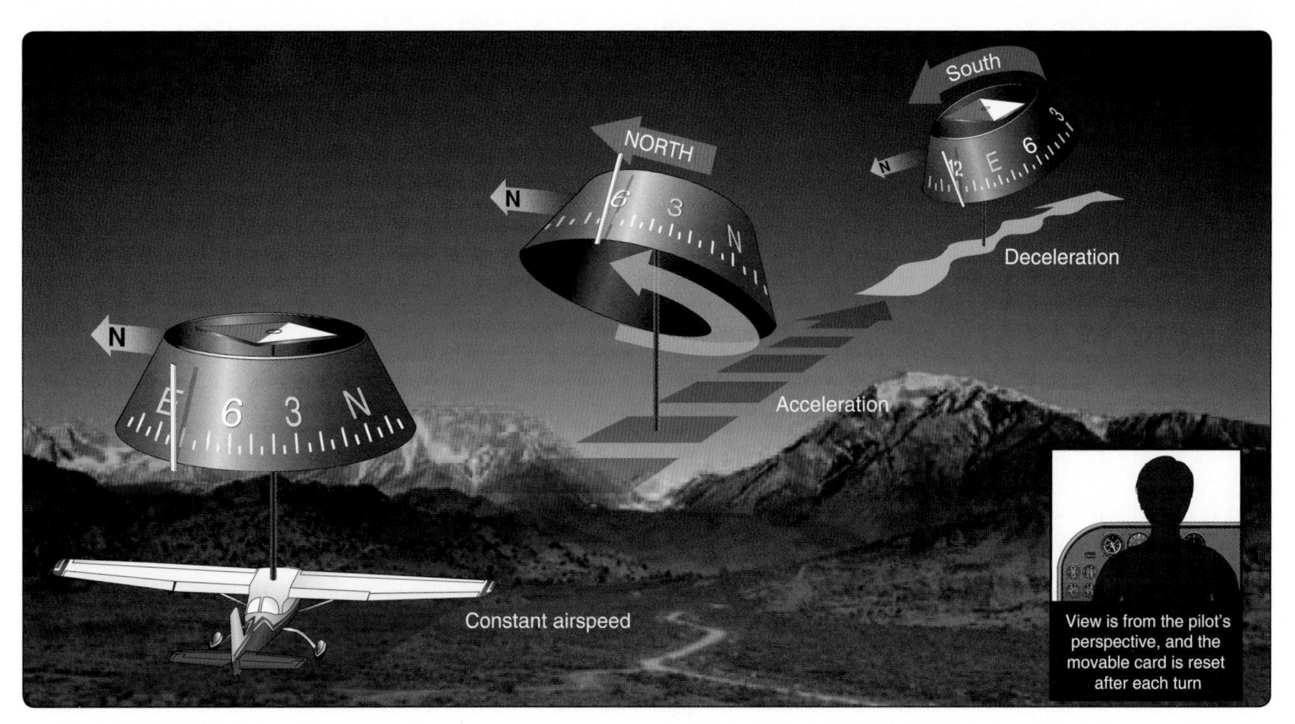

Figure 5-21. *The effects of acceleration error.*

The Vertical Card Magnetic Compass

The floating magnet type of compass not only has all the errors just described, but also lends itself to confused reading. It is easy to begin a turn in the wrong direction because its card appears backward. East is on what the pilot would expect to be the west side. The vertical card magnetic compass eliminates some of the errors and confusion. The dial of this compass is graduated with letters representing the cardinal directions, numbers every 30°, and marks every 5°. The dial is rotated by a set of gears from the shaft-mounted magnet, and the nose of the symbolic airplane on the instrument glass represents the lubber line for reading the heading of the aircraft from the dial. Eddy currents induced into an aluminum-damping cup damp oscillation of the magnet. *[Figure 5-22]*

Figure 5-22. *Vertical card magnetic compass.*

The Flux Gate Compass System

As mentioned earlier, the lines of flux in the Earth's magnetic field have two basic characteristics: a magnet aligns with these lines, and an electrical current is induced, or generated, in any wire crossed by them.

The flux gate compass that drives slaved gyros uses the characteristic of current induction. The flux valve is a small, segmented ring, like the one in *Figure 5-23*, made of soft iron that readily accepts lines of magnetic flux. An electrical coil is wound around each of the three legs to accept the current induced in this ring by the Earth's magnetic field. A coil wound around the iron spacer in the center of the frame has 400-Hz alternating current (A.C.) flowing through it. During the times when this current reaches its peak, twice during each cycle, there is so much magnetism produced by this coil that the frame cannot accept the lines of flux from the Earth's field.

But as the current reverses between the peaks, it demagnetizes the frame so it can accept the flux from the Earth's field. As this flux cuts across the windings in the three coils, it causes current to flow in them. These three coils are connected in

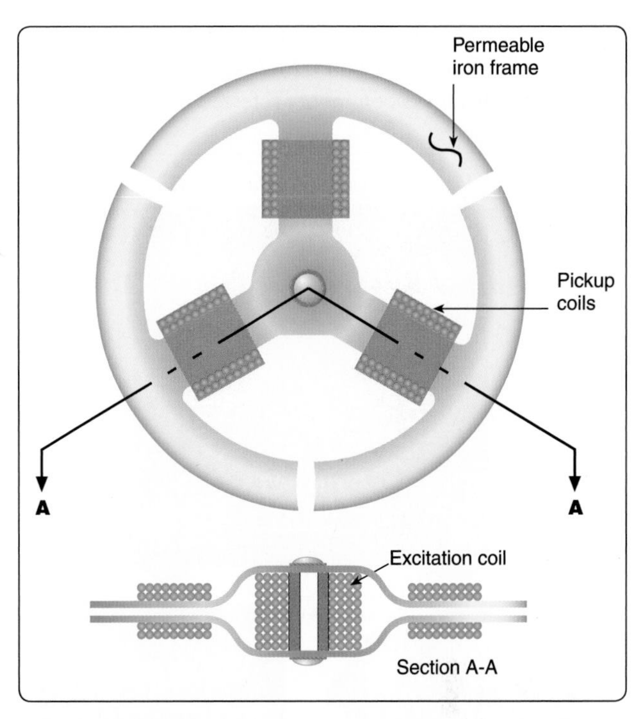

Figure 5-23. *The soft iron frame of the flux valve accepts the flux from the Earth's magnetic field each time the current in the center coil reverses. This flux causes current to flow in the three pickup coils.*

such a way that the current flowing in them changes as the heading of the aircraft changes. *[Figure 5-24]*

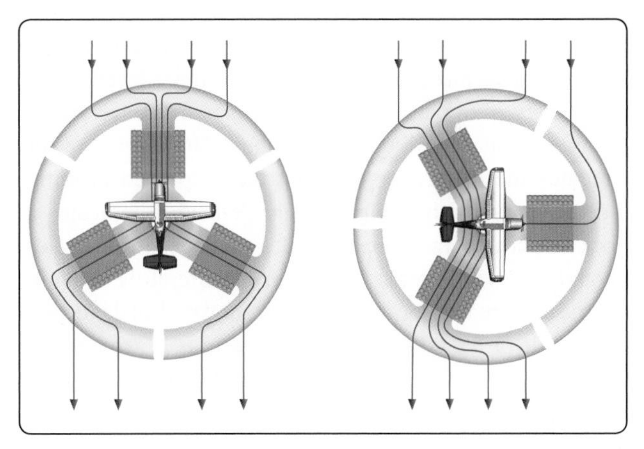

Figure 5-24. *The current in each of the three pickup coils changes with the heading of the aircraft.*

The three coils are connected to three similar but smaller coils in a synchro inside the instrument case. The synchro rotates the dial of a radio magnetic indicator (RMI) or a horizontal situation indicator (HSI).

Remote Indicating Compass

Remote indicating compasses were developed to compensate for the errors and limitations of the older type of heading indicators. The two panel-mounted components of a typical system are the pictorial navigation indicator and the slaving control and compensator unit. *[Figure 5-25]* The pictorial navigation indicator is commonly referred to as an HSI.

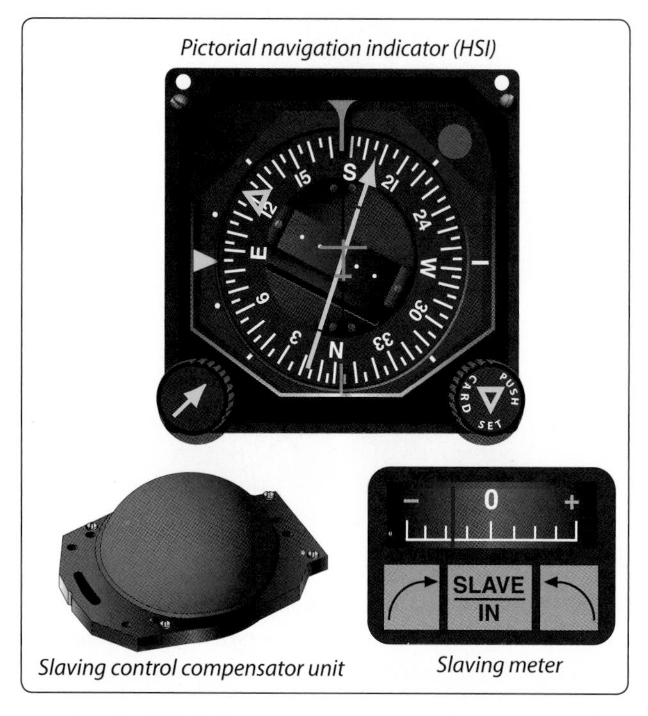

Figure 5-25. *The pictorial navigation indicator is commonly referred to as an HSI.*

The slaving control and compensator unit has a pushbutton that provides a means of selecting either the "slaved gyro" or "free gyro" mode. This unit also has a slaving meter and two manual heading-drive buttons. The slaving meter indicates the difference between the displayed heading and the magnetic heading. A right deflection indicates a clockwise error of the compass card; a left deflection indicates a counterclockwise error. Whenever the aircraft is in a turn and the card rotates, the slaving meter shows a full deflection to one side or the other. When the system is in "free gyro" mode, the compass card may be adjusted by depressing the appropriate heading-drive button.

A separate unit, the magnetic slaving transmitter is mounted remotely; usually in a wingtip to eliminate the possibility of magnetic interference. It contains the flux valve, which is the direction-sensing device of the system. A concentration of lines of magnetic force, after being amplified, becomes a signal relayed to the heading indicator unit, which is also remotely mounted. This signal operates a torque motor in the heading indicator unit that processes the gyro unit until it is aligned with the transmitter signal. The magnetic slaving transmitter is connected electrically to the HSI.

There are a number of designs of the remote indicating compass; therefore, only the basic features of the system are covered here. Instrument pilots must become familiar with the characteristics of the equipment in their aircraft.

As instrument panels become more crowded and the pilot's available scan time is reduced by a heavier flight deck workload, instrument manufacturers have worked toward combining instruments. One good example of this is the RMI in *Figure 5-26*. The compass card is driven by signals from the flux valve, and the two pointers are driven by an automatic direction finder (ADF) and a very high frequency omnidirectional range (VOR).

Figure 5-26. *Driven by signals from a flux valve, the compass card in this RMI indicates the heading of the aircraft opposite the upper center index mark. The green pointer is driven by the ADF. The yellow pointer is driven by the VOR receiver.*

Gyroscopic Systems

Flight without reference to a visible horizon can be safely accomplished by the use of gyroscopic instrument systems and the two characteristics of gyroscopes, which are rigidity and precession. These systems include attitude, heading, and rate instruments, along with their power sources. These instruments include a gyroscope (or gyro) that is a small wheel with its weight concentrated around its periphery. When this wheel is spun at high speed, it becomes rigid and resists tilting or turning in any direction other than around its spin axis.

Attitude and heading instruments operate on the principle of rigidity. For these instruments, the gyro remains rigid in its case and the aircraft rotates about it. Rate indicators, such as turn indicators and turn coordinators, operate on the principle of precession. In this case, the gyro precesses (or rolls over) proportionate to the rate the aircraft rotates about one or more of its axes.

Power Sources

Aircraft and instrument manufacturers have designed redundancy in the flight instruments so that any single failure does not deprive the pilot of the ability to safely conclude the flight. Gyroscopic instruments are crucial for instrument flight; therefore, they are powered by separate electrical or pneumatic sources.

Pneumatic Systems

Pneumatic gyros are driven by a jet of air impinging on buckets cut into the periphery of the wheel. On many aircraft this stream of air is obtained by evacuating the instrument case with a vacuum source and allowing filtered air to flow into the case through a nozzle to spin the wheel.

Venturi Tube Systems

Aircraft that do not have a pneumatic pump to evacuate the instrument case can use venturi tubes mounted on the outside of the aircraft, similar to the system shown in *Figure 5-27*. Air flowing through the venturi tube speeds up in the narrowest part and, according to Bernoulli's principle, the pressure drops. This location is connected to the instrument case by a piece of tubing. The two attitude instruments operate on approximately 4 "Hg of suction; the turn-and-slip indicator needs only 2 "Hg, so a pressure-reducing needle valve is used to decrease the suction. Air flows into the instruments through filters built into the instrument cases. In this system, ice can clog the venturi tube and stop the instruments when they are most needed.

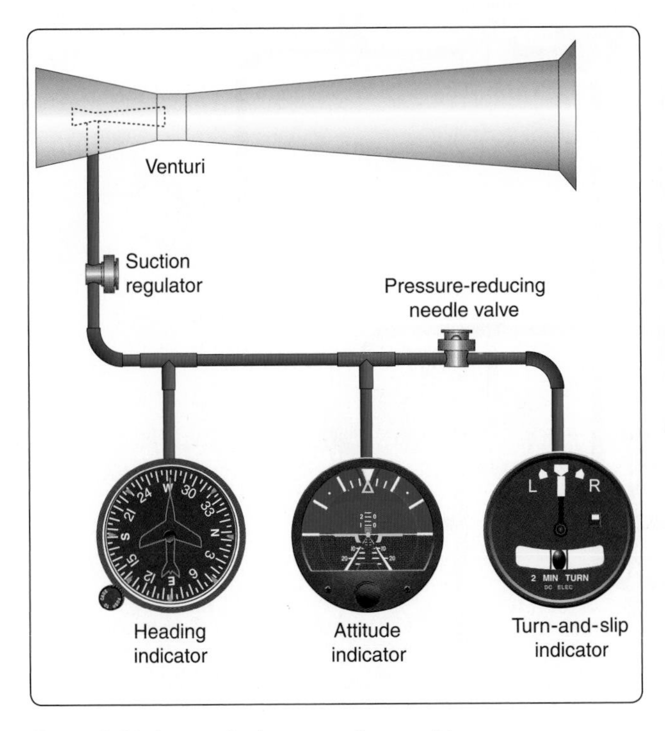

Figure 5-27. *A venturi tube system that provides necessary vacuum to operate key instruments.*

Vacuum Pump Systems

Wet-Type Vacuum Pump

Steel-vane air pumps have been used for many years to evacuate the instrument cases. The vanes in these pumps are lubricated by a small amount of engine oil metered into the pump and discharged with the air. In some aircraft the discharge air is used to inflate rubber deicer boots on the wing and empennage leading edges. To keep the oil from deteriorating the rubber boots, it must be removed with an oil separator like the one in *Figure 5-28*.

The vacuum pump moves a greater volume of air than is needed to supply the instruments with the suction needed, so a suction-relief valve is installed in the inlet side of the pump. This spring-loaded valve draws in just enough air to maintain the required low pressure inside the instruments, as is shown on the suction gauge in the instrument panel. Filtered air enters the instrument cases from a central air filter. As long as aircraft fly at relatively low altitudes, enough air is drawn into the instrument cases to spin the gyros at a sufficiently high speed.

Dry Air Vacuum Pump

As flight altitudes increase, the air is less dense and more air must be forced through the instruments. Air pumps that do not mix oil with the discharge air are used in high flying aircraft. Steel vanes sliding in a steel housing need to be lubricated, but vanes made of a special formulation of carbon sliding inside carbon housing provide their own lubrication in a microscopic amount as they wear.

Pressure Indicating Systems

Figure 5-29 is a diagram of the instrument pneumatic system of a twin-engine general aviation airplane. Two dry air pumps are used with filters in their inlets to filter out any contaminants that could damage the fragile carbon vanes in the pump. The discharge air from the pump flows through a regulator, where excess air is bled off to maintain the pressure in the system at the desired level. The regulated air then flows through inline filters to remove any contamination that could have been picked up from the pump, and from there into a manifold check valve. If either engine should become inoperative or either pump should fail, the check valve isolates the inoperative system and the instruments are driven by air from the operating system. After the air passes through the instruments and drives the gyros, it is exhausted from the case. The gyro pressure gauge measures the pressure drop across the instruments.

Electrical Systems

Many general aviation aircraft that use pneumatic attitude indicators use electric rate indicators and/or the reverse. Some

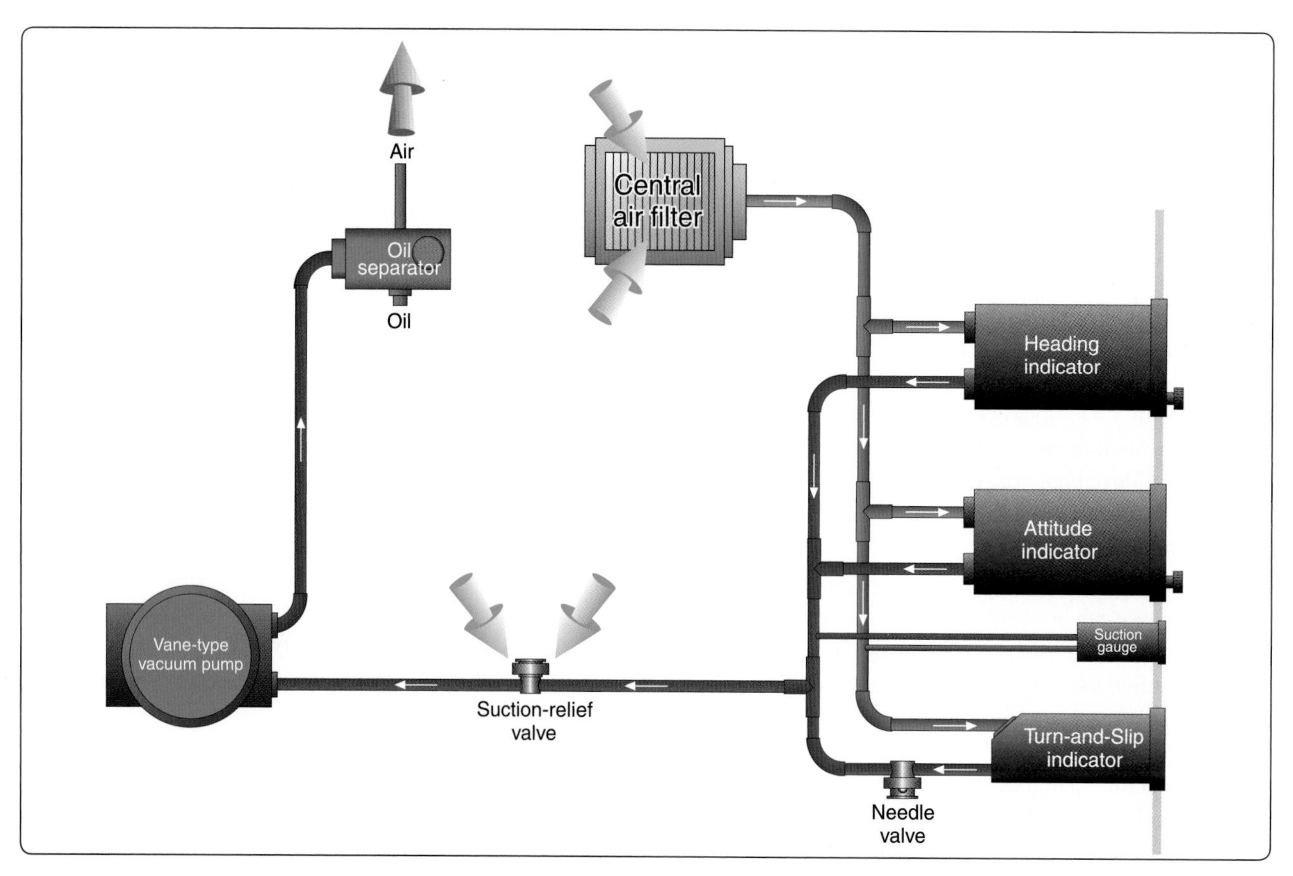

Figure 5-28. *Single-engine instrument vacuum system using a steel-vane, wet-type vacuum pump.*

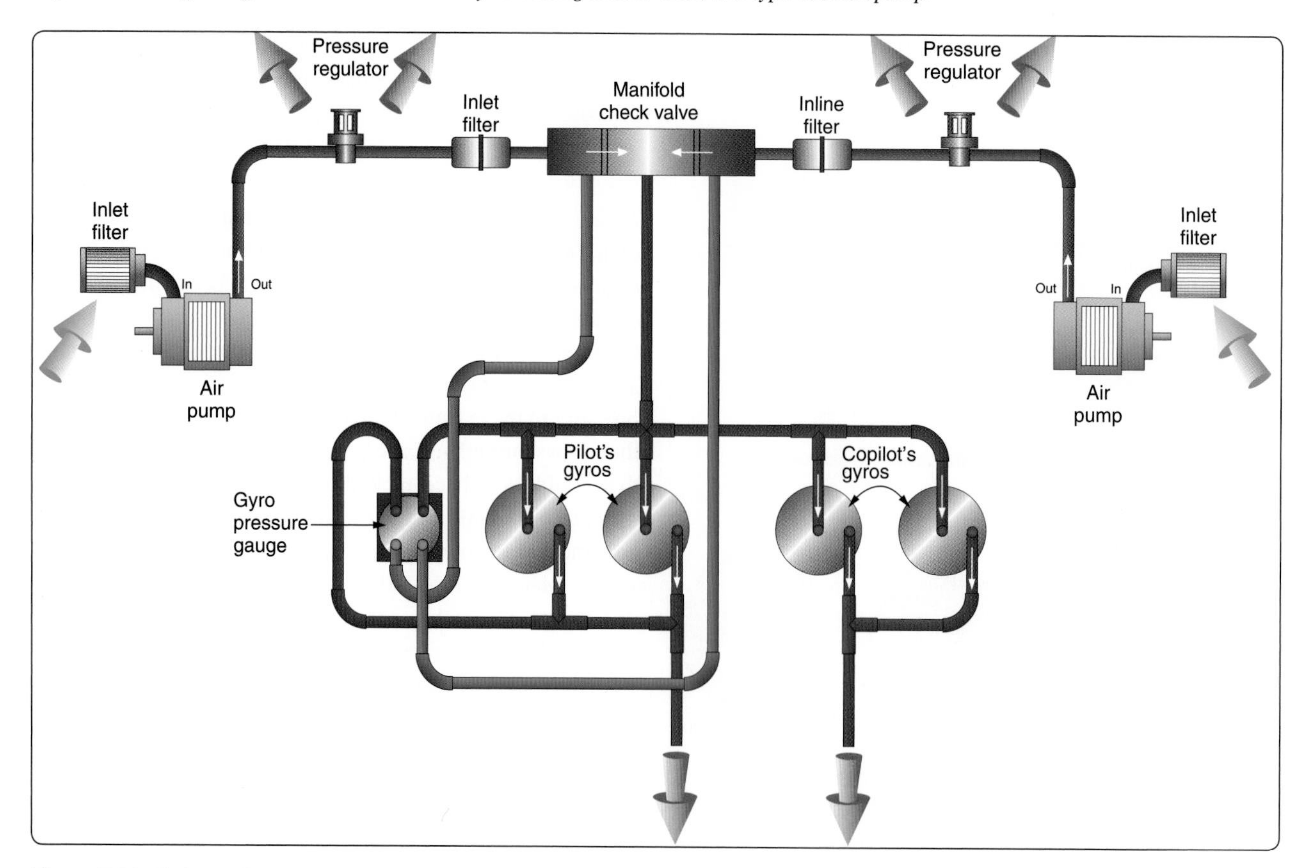

Figure 5-29. *Twin-engine instrument pressure system using a carbon-vane, dry-type air pump.*

instruments identify their power source on their dial, but it is extremely important that pilots consult the POH/AFM to determine the power source of all instruments to know what action to take in the event of an instrument failure. Direct current (D.C.) electrical instruments are available in 14- or 28-volt models, depending upon the electrical system in the aircraft. A.C. is used to operate some attitude gyros and autopilots. Aircraft with only D.C. electrical systems can use A.C. instruments via installation of a solid-state D.C. to A.C. inverter, which changes 14 or 28 volts D.C. into three-phase 115-volt, 400-Hz A.C.

Gyroscopic Instruments

Attitude Indicators

The first attitude instrument (AI) was originally referred to as an artificial horizon, later as a gyro horizon; now it is more properly called an attitude indicator. Its operating mechanism is a small brass wheel with a vertical spin axis, spun at a high speed by either a stream of air impinging on buckets cut into its periphery, or by an electric motor. The gyro is mounted in a double gimbal, which allows the aircraft to pitch and roll about the gyro as it remains fixed in space.

A horizon disk is attached to the gimbals so it remains in the same plane as the gyro, and the aircraft pitches and rolls about it. On early instruments, this was just a bar that represented the horizon, but now it is a disc with a line representing the horizon and both pitch marks and bank-angle lines. The top half of the instrument dial and horizon disc is blue, representing the sky; and the bottom half is brown, representing the ground. A bank index at the top of the instrument shows the angle of bank marked on the banking scale with lines that represent 10°, 20°, 30°, 45°, and 60°. *[Figure 5-30]*

Figure 5-30. *The dial of this attitude indicator has reference lines to show pitch and roll.*

A small symbolic aircraft is mounted in the instrument case so it appears to be flying relative to the horizon. A knob at the bottom center of the instrument case raises or lowers the aircraft to compensate for pitch trim changes as the airspeed changes. The width of the wings of the symbolic aircraft and the dot in the center of the wings represent a pitch change of approximately 2°.

For an AI to function properly, the gyro must remain vertically upright while the aircraft rolls and pitches around it. The bearings in these instruments have a minimum of friction; however, even this small amount places a restraint on the gyro producing precession and causing the gyro to tilt. To minimize this tilting, an erection mechanism inside the instrument case applies a force any time the gyro tilts from its vertical position. This force acts in such a way to return the spinning wheel to its upright position.

The older artificial horizons were limited in the amount of pitch or roll they could tolerate, normally about 60° in pitch and 100° in roll. After either of these limits was exceeded, the gyro housing contacted the gimbals, applying such a precessing force that the gyro tumbled. Because of this limitation, these instruments had a caging mechanism that locked the gyro in its vertical position during any maneuvers that exceeded the instrument limits. Newer instruments do not have these restrictive tumble limits; therefore, they do not have a caging mechanism.

When an aircraft engine is first started and pneumatic or electric power is supplied to the instruments, the gyro is not erect. A self-erecting mechanism inside the instrument actuated by the force of gravity applies a precessing force, causing the gyro to rise to its vertical position. This erection can take as long as 5 minutes, but is normally done within 2 to 3 minutes.

Attitude indicators are free from most errors, but depending upon the speed with which the erection system functions, there may be a slight nose-up indication during a rapid acceleration and a nose-down indication during a rapid deceleration. There is also a possibility of a small bank angle and pitch error after a 180° turn. These inherent errors are small and correct themselves within a minute or so after returning to straight-and-level flight.

Heading Indicators

A magnetic compass is a dependable instrument used as a backup instrument. Although very reliable, it has so many inherent errors that it has been supplemented with gyroscopic heading indicators.

The gyro in a heading indicator is mounted in a double gimbal, as in an attitude indicator, but its spin axis is horizontal permitting sensing of rotation about the vertical axis of the aircraft. Gyro heading indicators, with the exception of slaved gyro indicators, are not north seeking, therefore they must be manually set to the appropriate heading by referring to a magnetic compass. Rigidity causes them to maintain this heading indication, without the oscillation and other errors inherent in a magnetic compass.

Older directional gyros use a drum-like card marked in the same way as the magnetic compass card. The gyro and the card remain rigid inside the case with the pilot viewing the card from the back. This creates the possibility the pilot might start a turn in the wrong direction similar to using a magnetic compass. A knob on the front of the instrument, below the dial, can be pushed in to engage the gimbals. This locks the gimbals allowing the pilot to rotate the gyro and card until the number opposite the lubber line agrees with the magnetic compass. When the knob is pulled out, the gyro remains rigid and the aircraft is free to turn around the card.

Directional gyros are almost all air-driven by evacuating the case and allowing filtered air to flow into the case and out through a nozzle, blowing against buckets cut in the periphery of the wheel. The Earth constantly rotates at 15° per hour while the gyro is maintaining a position relative to space, thus causing an apparent drift in the displayed heading of 15° per hour. When using these instruments, it is standard practice to compare the heading indicated on the directional gyro with the magnetic compass at least every 15 minutes and to reset the heading as necessary to agree with the magnetic compass.

Heading indicators like the one in *Figure 5-31* work on the same principle as the older horizontal card indicators, except that the gyro drives a vertical dial that looks much like the dial of a vertical card magnetic compass. The heading of the aircraft is shown against the nose of the symbolic aircraft on

Figure 5-31. *The heading indicator is not north seeking, but must be set periodically (about every 15 minutes) to agree with the magnetic compass.*

the instrument glass, which serves as the lubber line. A knob in the front of the instrument may be pushed in and turned to rotate the gyro and dial. The knob is spring loaded so it disengages from the gimbals as soon as it is released. This instrument should be checked about every 15 minutes to see if it agrees with the magnetic compass.

Turn Indicators

Attitude and heading indicators function on the principle of rigidity, but rate instruments such as the turn-and-slip indicator operate on precession. Precession is the characteristic of a gyroscope that causes an applied force to produce a movement, not at the point of application, but at a point 90° from the point of application in the direction of rotation. *[Figure 5-32]*

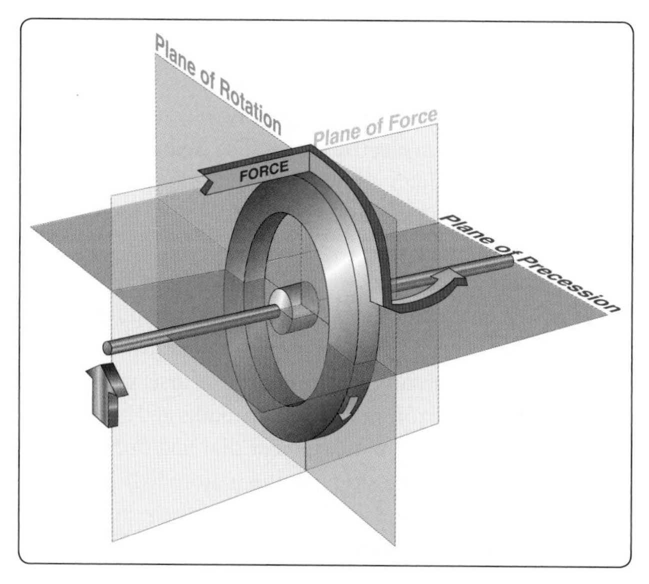

Figure 5-32. *Precession causes a force applied to a spinning wheel to be felt 90° from the point of application in the direction of rotation.*

Turn-and-Slip Indicator

The first gyroscopic aircraft instrument was the turn indicator in the needle and ball, or turn-and-bank indicator, which has more recently been called a turn-and-slip indicator. *[Figure 5-33]*

The inclinometer in the instrument is a black glass ball sealed inside a curved glass tube that is partially filled with a liquid for damping. This ball measures the relative strength of the force of gravity and the force of inertia caused by a turn. When the aircraft is flying straight-and-level, there is no inertia acting on the ball, and it remains in the center of the tube between two wires. In a turn made with a bank angle that is too steep, the force of gravity is greater than the inertia and the ball rolls down to the inside of the turn. If the turn is

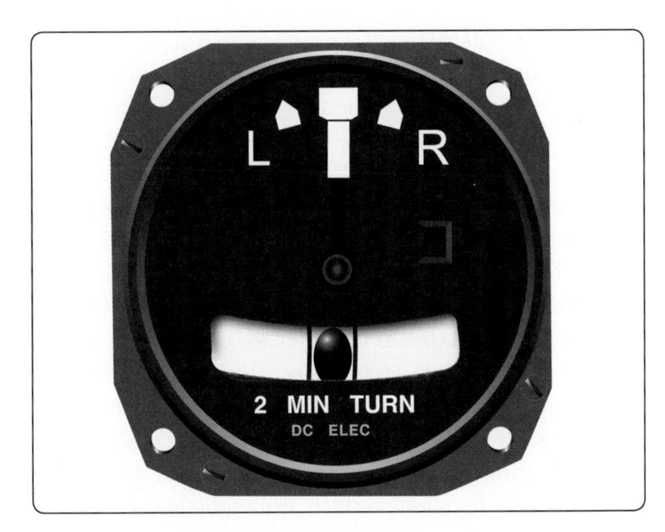

Figure 5-33. *Turn-and-slip indicator.*

made with too shallow a bank angle, the inertia is greater than gravity and the ball rolls upward to the outside of the turn.

The inclinometer does not indicate the amount of bank, nor does it indicate slip; it only indicates the relationship between the angle of bank and the rate of yaw.

The turn indicator is a small gyro spun either by air or by an electric motor. The gyro is mounted in a single gimbal with its spin axis parallel to the lateral axis of the aircraft and the axis of the gimbal parallel with the longitudinal axis. *[Figure 5-34]* When the aircraft yaws, or rotates about its vertical axis, it produces a force in the horizontal plane that, due to precession, causes the gyro and its gimbal to rotate about the gimbal's axis. It is restrained in this rotation plane by a calibration spring; it rolls over just enough to cause the pointer to deflect until it aligns with one of the doghouse-shaped marks on the dial, when the aircraft is making a standard rate turn.

The dial of these instruments is marked "2 MIN TURN." Some turn-and-slip indicators used in faster aircraft are marked "4 MIN TURN." In either instrument, a standard rate turn is being made whenever the needle aligns with a doghouse. A standard rate turn is 3° per second. In a 2 minute instrument, if the needle is one needle width either side of the center alignment mark, the turn is 3° per second and the turn takes 2 minutes to execute a 360° turn. In a 4 minute instrument, the same turn takes two widths deflection of the needle to achieve 3° per second.

Turn Coordinator

The major limitation of the older turn-and-slip indicator is that it senses rotation only about the vertical axis of the aircraft. It tells nothing of the rotation around the longitudinal axis, which in normal flight occurs before the aircraft begins to turn.

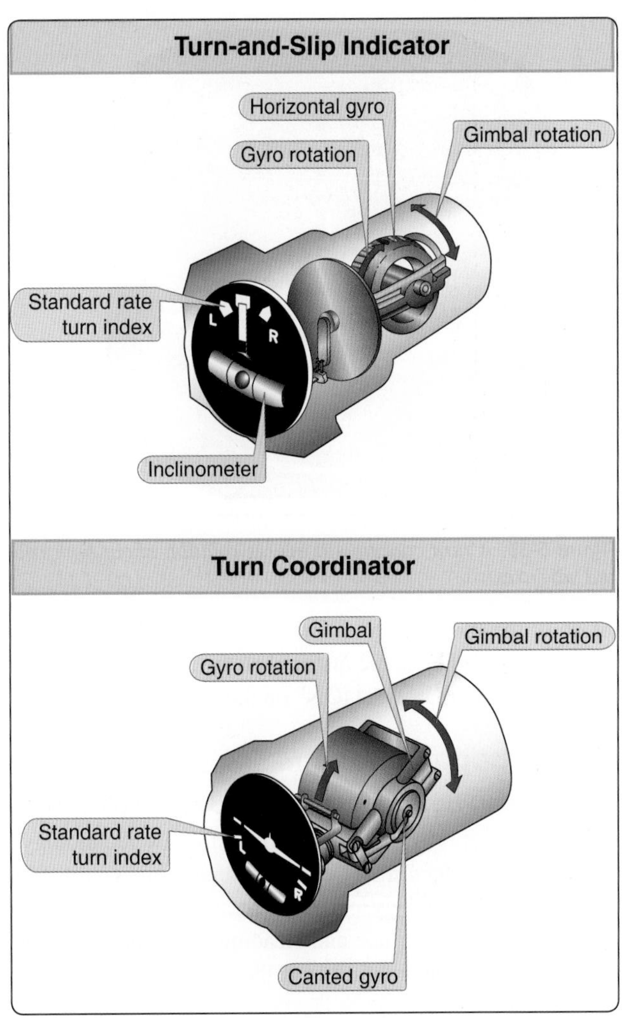

Figure 5-34. *The rate gyro in both turn-and-slip indicator and turn coordinator.*

A turn coordinator operates on precession, the same as the turn indicator, but its gimbals frame is angled upward about 30° from the longitudinal axis of the aircraft. *[Figure 5-34]* This allows it to sense both roll and yaw. Therefore during a turn, the indicator first shows the rate of banking and once stabilized, the turn rate. Some turn coordinator gyros are dual-powered and can be driven by either air or electricity.

Rather than using a needle as an indicator, the gimbal moves a dial that is the rear view of a symbolic aircraft. The bezel of the instrument is marked to show wings-level flight and bank angles for a standard rate turn. *[Figure 5-35]*

The inclinometer, similar to the one in a turn-and-slip indicator, is called a coordination ball, which shows the relationship between the bank angle and the rate of yaw. The turn is coordinated when the ball is in the center, between the marks. The aircraft is skidding when the ball rolls toward the outside of the turn and is slipping when it moves toward the

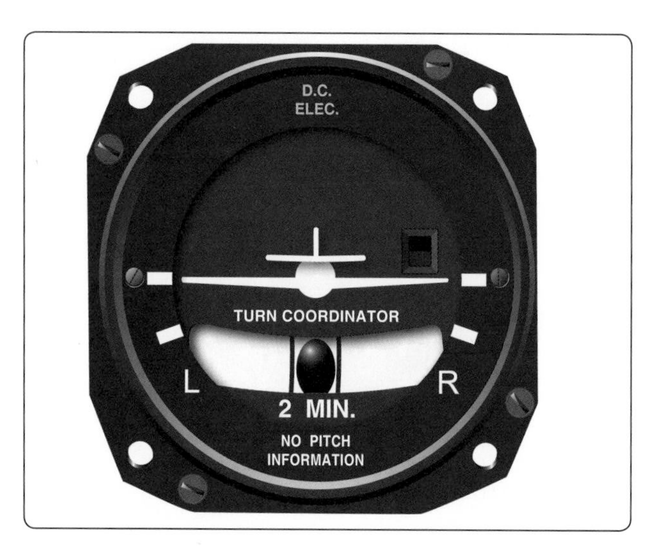

Figure 5-35. *A turn coordinator senses rotation about both roll and yaw axes.*

inside of the turn. A turn coordinator does not sense pitch. This is indicated on some instruments by placing the words "NO PITCH INFORMATION" on the dial.

Flight Support Systems

Attitude and Heading Reference System (AHRS)

As aircraft displays have transitioned to new technology, the sensors that feed them have also undergone significant change. Traditional gyroscopic flight instruments have been replaced by Attitude and Heading Reference Systems (AHRS) improving reliability and thereby reducing cost and maintenance.

The function of an AHRS is the same as gyroscopic systems; that is, to determine which way is level and which way is north. By knowing the initial heading the AHRS can determine both the attitude and magnetic heading of the aircraft.

The genesis of this system was initiated by the development of the ring-LASAR gyroscope developed by Kearfott located in Little Falls, New Jersey. *[Figure 5-36]* Their development of the Ring-LASAR gyroscope in the 1960s/1970s was in support of Department of Defense (DOD) programs to include cruise missile technology. With the precision of these gyroscopes, it became readily apparent that they could be leveraged for multiple tasks and functions. Gyroscopic miniaturization has become so common that solid-state gyroscopes are found in products from robotics to toys.

Because the AHRS system replaces separate gyroscopes, such as those associated with an attitude indicator, magnetic heading indicator and turn indicator these individual systems are no longer needed. As with many systems today, AHRS itself had matured with time. Early AHRS systems used expensive inertial sensors and flux valves. However, today the AHRS for aviation and general aviation in particular are small solid-state systems integrating a variety of technology such as low cost inertial sensors, rate gyros, and magnetometers, and have capability for satellite signal reception.

Air Data Computer (ADC)

An Air Data Computer (ADC) *[Figure 5-37]* is an aircraft computer that receives and processes pitot pressure, static

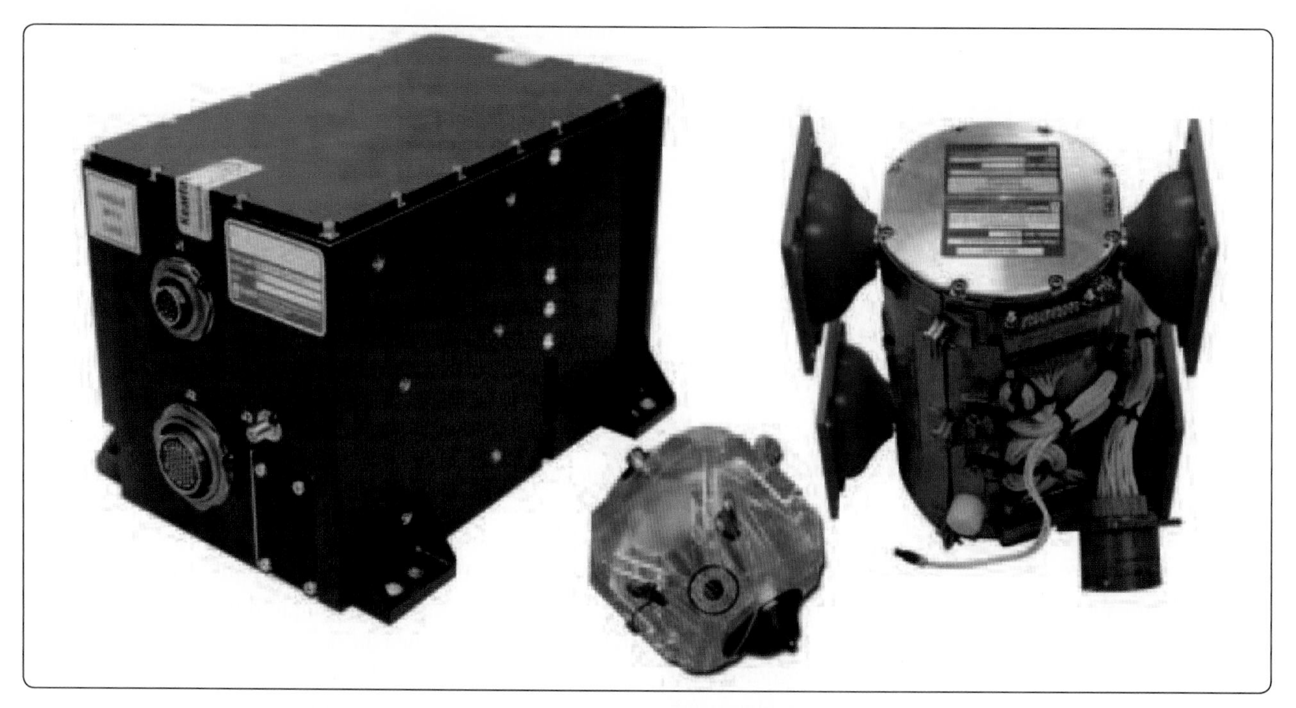

Figure 5-36. *The Kearfott Attitude Heading Reference System (AHRS) on the left incorporates a Monolithic Ring Laser Gyro (MRLG) (center), which is housed in an Inertial Sensor Assembly (ISA) on the right.*

Figure 5-37. *Air data computer (Collins).*

pressure, and temperature to calculate very precise altitude, IAS, TAS, and air temperature. The ADC outputs this information in a digital format that can be used by a variety of aircraft systems including an EFIS. Modern ADCs are small solid-state units. Increasingly, aircraft systems such as autopilots, pressurization, and FMS utilize ADC information for normal operations.

NOTE: In most modern general aviation systems, both the AHRS and ADC are integrated within the electronic displays themselves thereby reducing the number of units, reducing weight, and providing simplification for installation resulting in reduced costs.

Analog Pictorial Displays

Horizontal Situation Indicator (HSI)

The HSI is a direction indicator that uses the output from a flux valve to drive the dial, which acts as the compass card. This instrument, shown in *Figure 5-38*, combines the magnetic compass with navigation signals and a glideslope. This gives the pilot an indication of the location of the aircraft with relationship to the chosen course.

In *Figure 5-38*, the aircraft heading displayed on the rotating azimuth card under the upper lubber line is North or 360°. The course-indicating arrowhead shown is set to 020; the tail indicates the reciprocal, 200°. The course deviation bar operates with a VOR/Localizer (VOR/LOC) navigation receiver to indicate left or right deviations from the course selected with the course-indicating arrow, operating in the same manner that the angular movement of a conventional VOR/LOC needle indicates deviation from course.

The desired course is selected by rotating the course-indicating arrow in relation to the azimuth card by means of the course select knob. This gives the pilot a pictorial presentation: the fixed aircraft symbol and course deviation bar display the

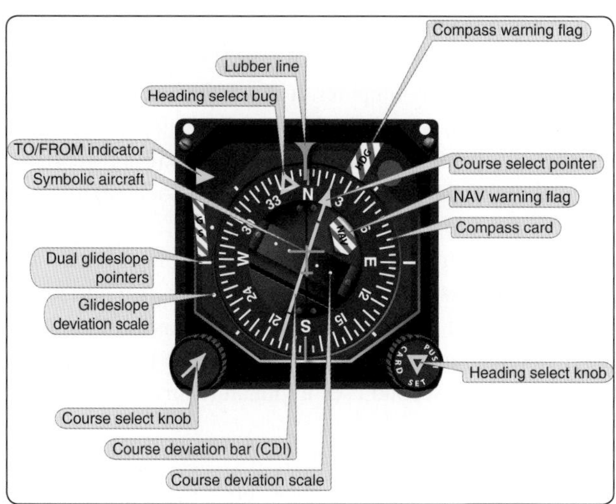

Figure 5-38. *Horizontal situation indicator (HSI).*

aircraft relative to the selected course, as though the pilot were above the aircraft looking down. The TO/FROM indicator is a triangular pointer. When the indicator points to the head of the course arrow, it shows that the course selected, if properly intercepted and flown, takes the aircraft to the selected facility. When the indicator points to the tail of the course arrow, it shows that the course selected, if properly intercepted and flown, takes the aircraft directly away from the selected facility.

The glideslope deviation pointer indicates the relation of the aircraft to the glideslope. When the pointer is below the center position, the aircraft is above the glideslope, and an increased rate of descent is required. In most installations, the azimuth card is a remote indicating compass driven by a fluxgate; however, in few installations where a fluxgate is not installed, or in emergency operation, the heading must be checked against the magnetic compass occasionally and reset with the course select knob.

Attitude Direction Indicator (ADI)

Advances in attitude instrumentation combine the gyro horizon with other instruments such as the HSI, thereby reducing the number of separate instruments to which the pilot must devote attention. The attitude direction indicator (ADI) is an example of such technological advancement. A flight director incorporates the ADI within its system, which is further explained below (Flight Director System). However, an ADI need not have command cues; however, it is normally equipped with this feature.

Flight Director System (FDS)

A Flight Director System (FDS) combines many instruments into one display that provides an easily interpreted understanding of the aircraft's flightpath. The computed solution furnishes the steering commands necessary to obtain and hold a desired path.

Major components of an FDS include an ADI, also called a Flight Director Indicator (FDI), an HSI, a mode selector, and a flight director computer. It should be noted that a flight director in use does not infer the aircraft is being manipulated by the autopilot (coupled), but is providing steering commands that the pilot (or the autopilot, if coupled) follows.

Typical flight directors use one of two display systems for steerage. The first is a set of command bars, one horizontal and one vertical. The command bars in this configuration are maintained in a centered position (much like a centered glideslope). The second uses a miniature aircraft aligned to a command cue.

A flight director displays steerage commands to the pilot on the ADI. As previously mentioned, the flight director receives its signals from one of various sources and provides that to the ADI for steerage commands. The mode controller provides signals through the ADI to drive the steering bars, e.g., the pilot flies the aircraft to place the delta symbol in the V of the steering bars. "Command" indicators tell the pilot in which direction and how much to change aircraft attitude to achieve the desired result.

The computed command indications relieve the pilot of many of the mental calculations required for instrument flight. The yellow cue in the ADI *[Figure 5-39]* provides all steering commands to the pilot. It is driven by a computer that receives information from the navigation systems, the ADC, AHRS, and other sources of data. The computer processes this information, providing the pilot with a single cue to follow. Following the cue provides the pilot with the necessary three-dimensional flight trajectory to maintain the desired path.

One of the first widely used flight directors was developed by Sperry and was called the Sperry Three Axis Attitude Reference System (STARS). Developed in the 1960s, it was commonly found on both commercial and business aircraft alike. STARS (with a modification) and successive flight directors were integrated with the autopilots and aircraft providing a fully integrated flight system.

The flight director/autopilot system described below is typical of installations in many general aviation aircraft. The components of a typical flight director include the mode controller, ADI, HSI, and annunciator panel. These units are illustrated in *Figure 5-40*.

The pilot may choose from among many modes including the HDG (heading) mode, the VOR/LOC (localizer tracking) mode, or the AUTO Approach (APP) or G/S (automatic capture and tracking of instrument landing system (ILS) localizers and glidepath) mode. The auto mode has a fully automatic pitch selection computer that takes into account aircraft performance and wind conditions, and operates once the pilot has reached the ILS glideslope. More sophisticated systems allow more flight director modes.

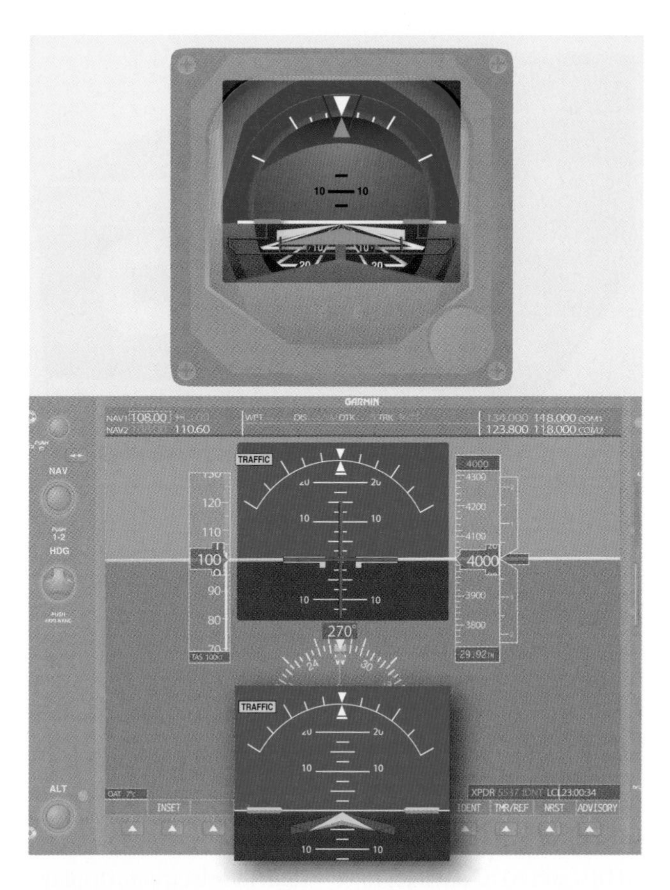

Figure 5-39. *A typical cue that a pilot would follow.*

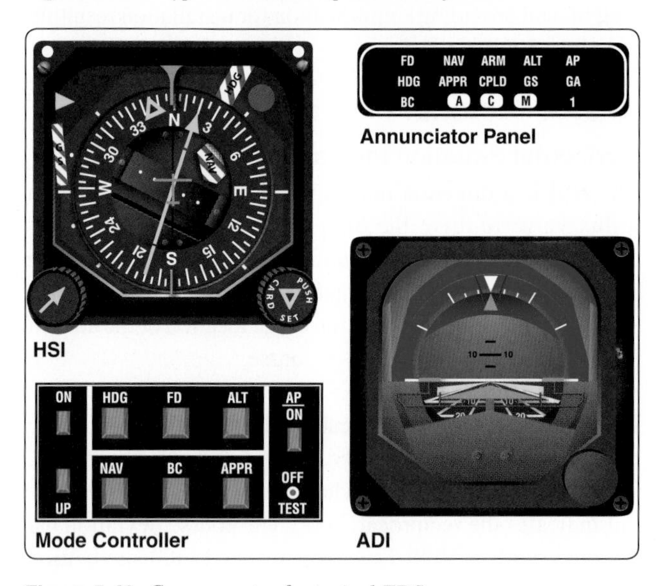

Figure 5-40. *Components of a typical FDS.*

Integrated Flight Control System

The integrated flight control system integrates and merges various systems into a system operated and controlled by one principal component. *Figure 5-41* illustrates key components of the flight control system that was developed from the onset as a fully integrated system comprised of the airframe, autopilot, and FDS. This trend of complete integration, once seen only in large commercial aircraft, is now becoming common in general aviation.

Autopilot Systems

An autopilot is a mechanical means to control an aircraft using electrical, hydraulic, or digital systems. Autopilots can control three axes of the aircraft: roll, pitch, and yaw. Most autopilots in general aviation control roll and pitch.

Autopilots also function using different methods. The first is position based. That is, the attitude gyro senses the degree of difference from a position such as wings level, a change in pitch, or a heading change.

Determining whether a design is position based and/or rate based lies primarily within the type of sensors used. In order for an autopilot to possess the capability of controlling an aircraft's attitude (i.e., roll and pitch), that system must be provided with constant information on the actual attitude of that aircraft. This is accomplished by the use of several different types of gyroscopic sensors. Some sensors are designed to indicate the aircraft's attitude in the form of position in relation to the horizon, while others indicate rate (position change over time).

Rate-based systems use the turn-and-bank sensor for the autopilot system. The autopilot uses rate information on two of the aircraft's three axes: movement about the vertical axis (heading change or yaw) and about the longitudinal axis (roll). This combined information from a single sensor is made possible by the 30° offset in the gyro's axis to the longitudinal axis.

Other systems use a combination of both position and rate-based information to benefit from the attributes of both systems while newer autopilots are digital. *Figure 5-42* illustrates an autopilot by Century.

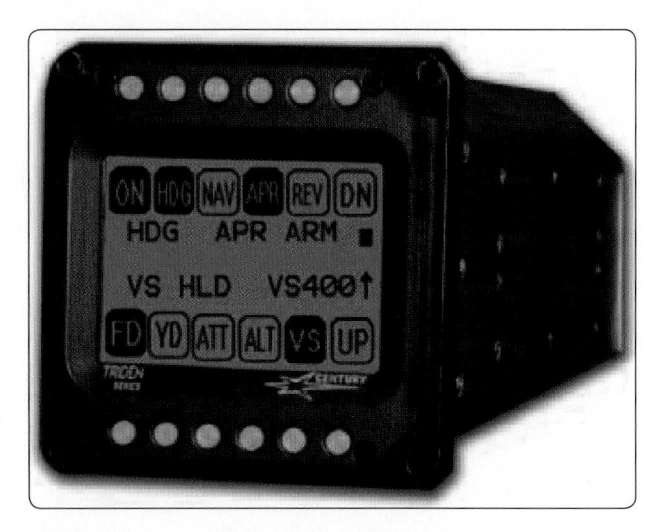

Figure 5-42. *An Autopilot by Century.*

Figure 5-43 is a diagram layout of a rate-based autopilot by S-Tec, which permits the purchaser to add modular capability form basic wing leveling to increased capability.

Figure 5-41. *The S-TEC/Meggit Corporation Integrated Autopilot installed in the Cirrus.*

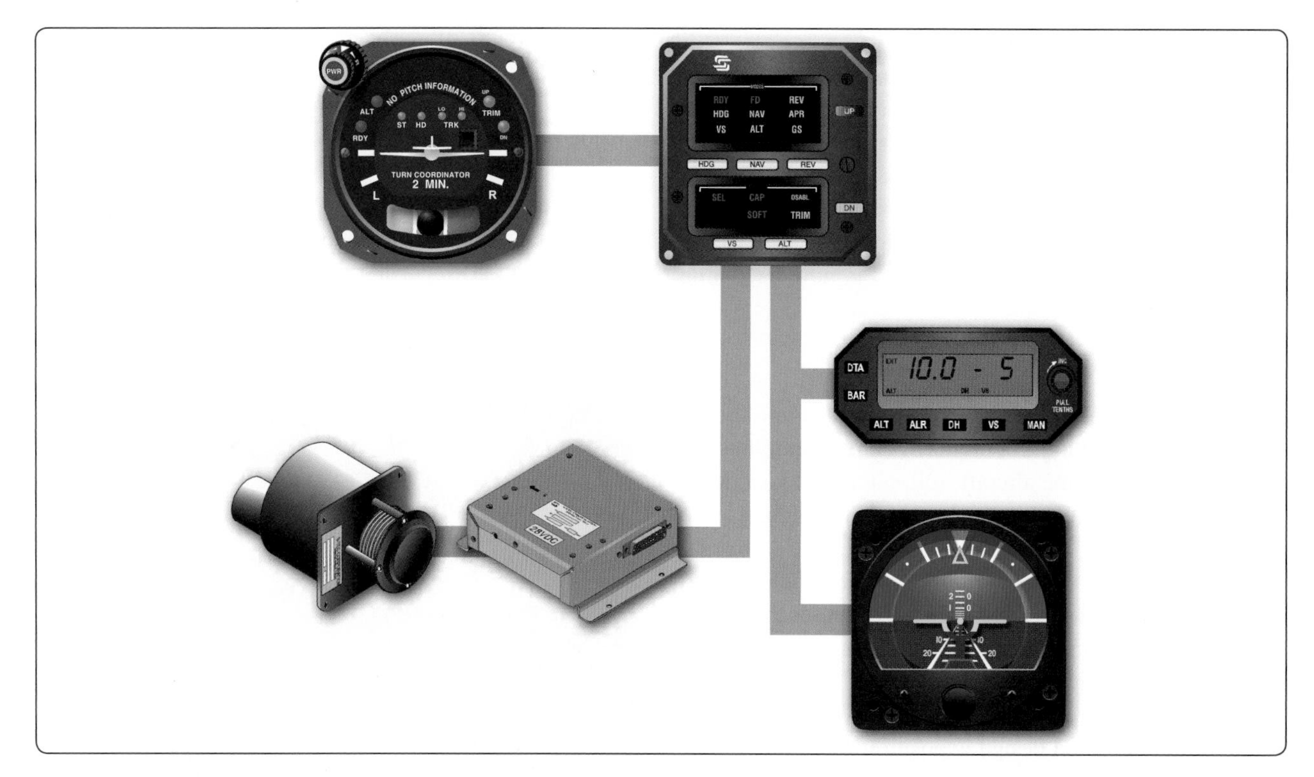

Figure 5-43. *A diagram layout of an autopilot by S-Tec.*

Flight Management Systems (FMS)

In the mid-1970s, visionaries in the avionics industry such as Hubert Naimer of Universal, and followed by others such as Ed King, Jr., were looking to advance the technology of aircraft navigation. As early as 1976, Naimer had a vision of a "Master Navigation System" that would accept inputs from a variety of different types of sensors on an aircraft and automatically provide guidance throughout all phases of flight.

At that time aircraft navigated over relatively short distances with radio systems, principally VOR or ADF. For long-range flight inertial navigation systems (INS), Omega, Doppler, and Loran were in common use. Short-range radio systems usually did not provide area navigation (RNAV) capability. Long-range systems were only capable of en route point-to-point navigation between manually entered waypoints described as longitude and latitude coordinates, with typical systems containing a limited number of waypoints.

The laborious process of manually entering cryptic latitude and longitude data for each flight waypoint created high crew workloads and frequently resulted in incorrect data entry. The requirement of a separate control panel for each long-range system consumed precious flight deck space and increased the complexity of interfacing the systems with display instruments, flight directors, and autopilots.

The concept employed a master computer interfaced with all of the navigation sensors on the aircraft. A common control display unit (CDU) interfaced with the master computer would provide the pilot with a single control point for all navigation systems, thereby reducing the number of required flight deck panels. Management of the various individual sensors would be transferred from the pilot to the new computer.

Since navigation sensors rarely agree exactly about position, Naimer believed that blending all available sensor position data through a highly sophisticated, mathematical filtering system would produce a more accurate aircraft position. He called the process output the "Best Computed Position." By using all available sensors to keep track of position, the system could readily provide area navigation capability. The master computer, not the individual sensors, would be integrated into the airplane, greatly reducing wiring complexity.

To solve the problems of manual waypoint entry, a pre-loaded database of global navigation information would be readily accessible by the pilot through the CDU. Using such a system a pilot could quickly and accurately construct a flight plan consisting of dozens of waypoints, avoiding the tedious typing of data and the error potential of latitude/longitude coordinates. Rather than simply navigating point-to-point, the master system would be able to maneuver the aircraft, permitting use of the system for terminal procedures including departures, arrivals, and approaches. The system would be able to automate any aspect of manual pilot navigation of the aircraft. When the first system, called the UNS-1, was released by Universal in 1982, it was called a flight management system (FMS). *[Figure 5-44]*

Figure 5-44. *A Control Display Unit (CDU) used to control the flight management system (FMS).*

An FMS uses an electronic database of worldwide navigational data including navigation aids, airways and intersections, Standard Instrument Departures (SIDs), STARs, and Instrument Approach Procedures (IAPs) together with pilot input through a CDU to create a flight plan. The FMS provides outputs to several aircraft systems including desired track, bearing and distance to the active waypoint, lateral course deviation and related data to the flight guidance system for the HSI displays, and roll steering command for the autopilot/flight director system. This allows outputs from the FMS to command the airplane where to go and when and how to turn. To support adaptation to numerous aircraft types, an FMS is usually capable of receiving and outputting both analog and digital data and discrete information. Currently, electronic navigation databases are updated every 28 days.

The introduction of the Global Positioning System (GPS) has provided extremely precise position at low cost, making GPS the dominant FMS navigation sensor today. Currently, typical FMS installations require that air data and heading information be available electronically from the aircraft. This limits FMS usage in smaller aircraft, but emerging technologies allow this data from increasingly smaller and less costly systems. Some systems interface with a dedicated Distance Measuring Equipment (DME) receiver channel under the control of the FMS to provide an additional sensor. In these systems, the FMS determines which DME sites should be interrogated for distance information using aircraft position and the navigation database to locate appropriate DME sites. The FMS then compensates aircraft altitude and station altitude with the aid of the database to determine the precise distance to the station. With the distances from a number of sites the FMS can compute a position nearly as accurately as GPS.

Aimer visualized three-dimensional aircraft control with an FMS. Modern systems provide Vertical Navigation (VNAV) as well as Lateral Navigation (LNAV) allowing the pilot to create a vertical flight profile synchronous with the lateral flight plan. Unlike early systems, such as Inertial Reference Systems (IRS) that were only suitable for en route navigation, the modern FMS can guide an aircraft during instrument approaches.

Today, an FMS provides not only real-time navigation capability but typically interfaces with other aircraft systems providing fuel management, control of cabin briefing and display systems, display of uplinked text and graphic weather data and air/ground data link communications.

Electronic Flight Instrument Systems

Modern technology has introduced into aviation a new method of displaying flight instruments, such as electronic flight instrument systems, integrated flight deck displays, and others. For the purpose of the practical test standards, any flight instrument display that utilizes LCD or picture tube like displays is referred to as "electronic flight instrument display" and/or a glass flight deck. In general aviation there is typically a primary flight display (PFD) and a multi-function display (MFD). Although both displays are in many cases identical, the PFD provides the pilot instrumentation necessary for flight to include altitude, airspeed, vertical velocity, attitude, heading and trim and trend information.

Glass flight decks (a term coined to describe electronic flight instrument systems) are becoming more widespread as cost falls and dependability continually increases. These systems provide many advantages such as being lighter, more reliable, no moving parts to wear out, consuming less power, and replacing numerous mechanical indicators with a single glass display. Because the versatility offered by glass displays is much greater than that offered by analog displays, the use of such systems only increases with time until analog systems are eclipsed.

Primary Flight Display (PFD)

PFDs provide increased situational awareness (SA) to the pilot by replacing the traditional six instruments used for instrument flight with an easy-to-scan display that provides the horizon, airspeed, altitude, vertical speed, trend, trim, rate of turn among other key relevant indications. Examples of PFDs are illustrated in *Figure 5-45*.

Synthetic Vision

Synthetic vision provides a realistic depiction of the aircraft in relation to terrain and flightpath. Systems such as those produced by Chelton Flight Systems, Universal Flight Systems, and others provide for depictions of terrain and course. *Figure 5-46* is an example of the Chelton Flight System providing both 5-dimensional situational awareness and a synthetic highway in the sky, representing the desired flightpath. Synthetic vision is used as a PFD, but provides guidance in a more normal, outside reference format.

Multi-Function Display (MFD)

In addition to a PFD directly in front of the pilot, an MFD that provides the display of information in addition to primary flight information is used within the flight deck. *[Figure 5-47]* Information such as a moving map, approach charts, Terrain Awareness Warning System, and weather depiction can all be illustrated on the MFD. For additional redundancy both the PFD and MFD can display all critical information that the other normally presents thereby providing redundancy (using a reversionary mode) not normally found in general aviation flight decks.

Advanced Technology Systems

Automatic Dependent Surveillance—Broadcast (ADS-B)

Although standards for Automatic Dependent Surveillance (Broadcast) (ADS-B) are still under continuing development, the concept is simple: aircraft broadcast a message on a regular basis, which includes their position (such as latitude, longitude and altitude), velocity, and possibly other information. Other aircraft or systems can receive this information for use in a wide variety of applications. The key to ADS-B is GPS, which provides three-dimensional position of the aircraft.

Figure 5-46. *The benefits of realistic visualization imagery, as illustrated by Synthetic Vision manufactured by Chelton Flight Systems. The system provides the pilot a realistic, real-time, three-dimensional depiction of the aircraft and its relation to terrain around it.*

As an simplified example, consider air-traffic radar. The radar measures the range and bearing of an aircraft. The bearing is measured by the position of the rotating radar antenna when it receives a reply to its interrogation from the aircraft, and the range by the time it takes for the radar to receive the reply.

An ADS-B based system, on the other hand, would listen for position reports broadcast by the aircraft. *[Figure 5-48]* These position reports are based on satellite navigation systems. These transmissions include the transmitting aircraft's position, which the receiving aircraft processes into

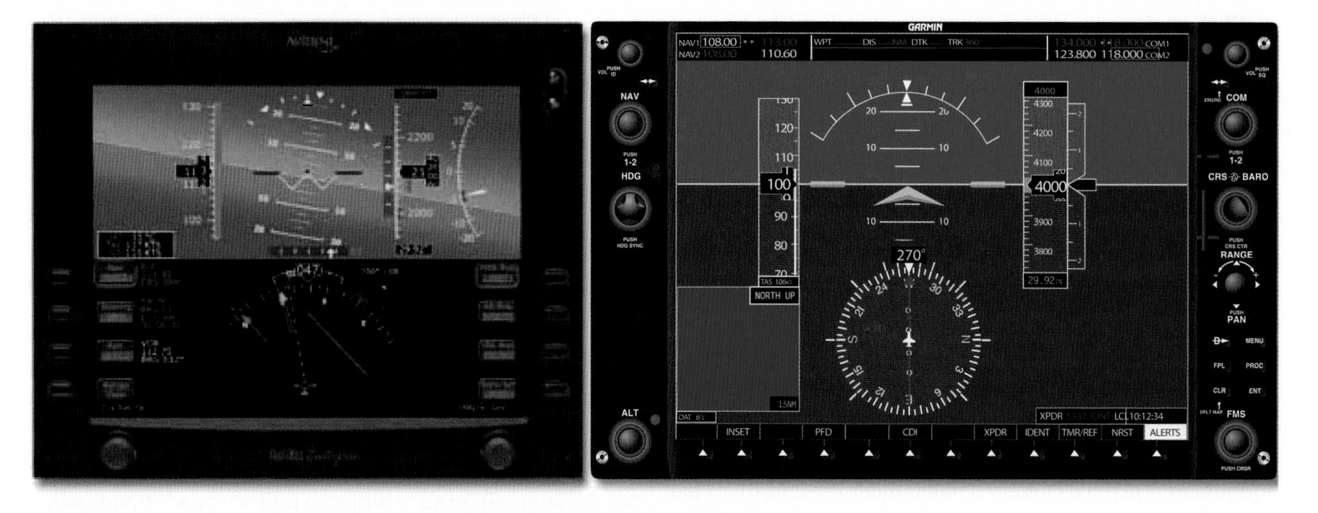

Figure 5-45. *Two primary flight displays (Avidyne on the left and Garmin on the right).*

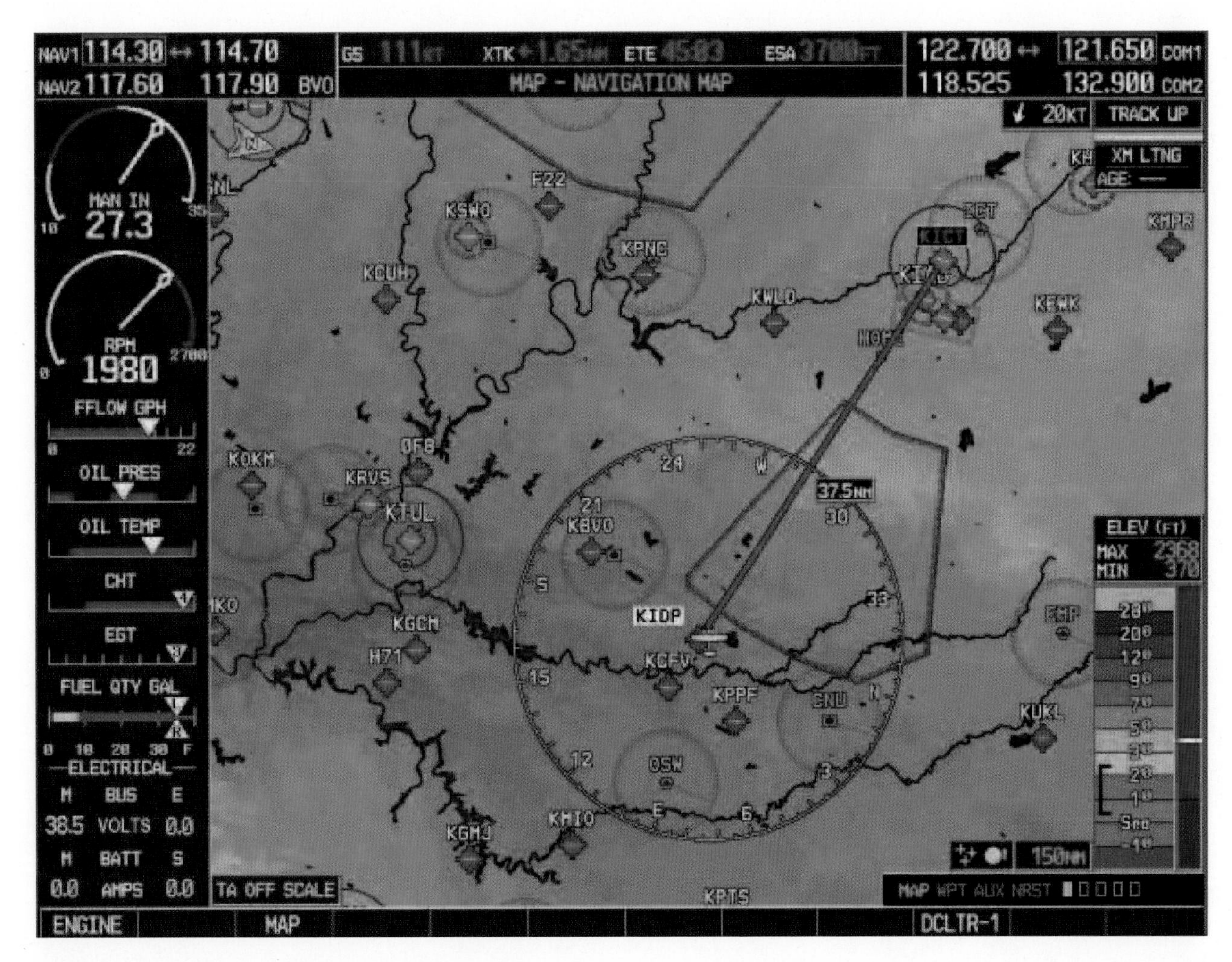

Figure 5-47. *Example of a multi-function display (MFD).*

usable pilot information. The accuracy of the system is now determined by the accuracy of the navigation system, not measurement errors. Furthermore the accuracy is unaffected by the range to the aircraft as in the case of radar. With radar, detecting aircraft speed changes require tracking the data and changes can only be detected over a period of several position updates. With ADS-B, speed changes are broadcast almost instantaneously and received by properly equipped aircraft. Additionally, other information can be obtained by properly equipped aircraft to include notices to airmen (NOTAM), weather, etc. *[Figures 5-49* and *5-50]* At the present time, ADS-B is predominantly available along the east coast of the United States where it is matured.

Safety Systems

Radio Altimeters

A radio altimeter, commonly referred to as a radar altimeter, is a system used for accurately measuring and displaying the height above the terrain directly beneath the aircraft. It sends a signal to the ground and processes the timed information. Its primary application is to provide accurate absolute altitude information to the pilot during approach and landing. In advanced aircraft today, the radar altimeter also provides its information to other onboard systems such as the autopilot and flight directors while they are in the glideslope capture mode below 200-300 feet above ground level (AGL).

A typical system consists of a receiver-transmitter (RT) unit, antenna(s) for receiving and transmitting the signal, and an indicator. *[Figure 5-51]* Category II and III precision approach procedures require the use of a radar altimeter and specify the exact minimum height above the terrain as a decision height (DH) or radio altitude (RA).

Traffic Advisory Systems

Traffic Information System

The Traffic Information Service (TIS) is a ground-based service providing information to the flight deck via data link using the S-mode transponder and altitude encoder. TIS improves the safety and efficiency of "see and avoid" flight through an automatic display that informs the pilot of nearby traffic. The display can show location, direction, altitude and the climb/descent trend of other transponder-equipped aircraft. TIS provides estimated position, altitude, altitude

Figure 5-48. *Aircraft equipped with Automatic Dependent Surveillance—Broadcast (ADS-B) continuously broadcast their identification, altitude, direction, and vertical trend. The transmitted signal carries significant information for other aircraft and ground stations alike. Other ADS-equipped aircraft receive this information and process it in a variety of ways. It is possible that in a saturated environment (assuming all aircraft are ADS equipped), the systems can project tracks for their respective aircraft and retransmit to other aircraft their projected tracks, thereby enhancing collision avoidance. At one time, there was an Automatic Dependent Surveillance—Addressed (ADS-A) and that is explained in the Pilot's Handbook of Aeronautical Knowledge.*

trend, and ground track information for up to several aircraft simultaneously within about 7 NM horizontally, 3,500 feet above and 3,500 feet below the aircraft. *[Figure 5-52]* This data can be displayed on a variety of MFDs. *[Figure 5-53]*

Figure 5-54 displays the pictorial concept of the traffic information system. Noteworthy is the requirement to have Mode S and that the ground air traffic station processes the Mode S signal.

Traffic Alert Systems

Traffic alert systems receive transponder information from nearby aircraft to help determine their relative position to the equipped aircraft. They provide three-dimensional location of other aircraft *[Figures 5-55, 5-56,* and *5-57]* and are cost effective alternatives to TCAS equipage for smaller aircraft.

Traffic Avoidance Systems

Traffic Alert and Collision Avoidance System (TCAS)

The TCAS is an airborne system developed by the FAA that operates independently from the ground-based ATC system. TCAS was designed to increase flight deck awareness of proximate aircraft and to serve as a "last line of defense" for the prevention of mid-air collisions.

There are two levels of TCAS systems. TCAS I was developed to accommodate the general aviation (GA) community and the regional airlines. This system issues traffic advisories (TAs) to assist pilots in visual acquisition of intruder aircraft. TCAS I provides approximate bearing and relative altitude of aircraft with a selectable range. It provides the pilot with TA alerting him or her to potentially conflicting traffic. The pilot then visually acquires the traffic and takes appropriate action for collision avoidance.

Figure 5-49. *An aircraft equipped with ADS will receive identification, altitude in hundreds of feet (above or below using + or–), direction of the traffic, and aircraft descent or climb using an up or down arrow. The yellow target is an illustration of how a non-ADS equipped aircraft would appear on an ADS-equipped aircraft's display.*

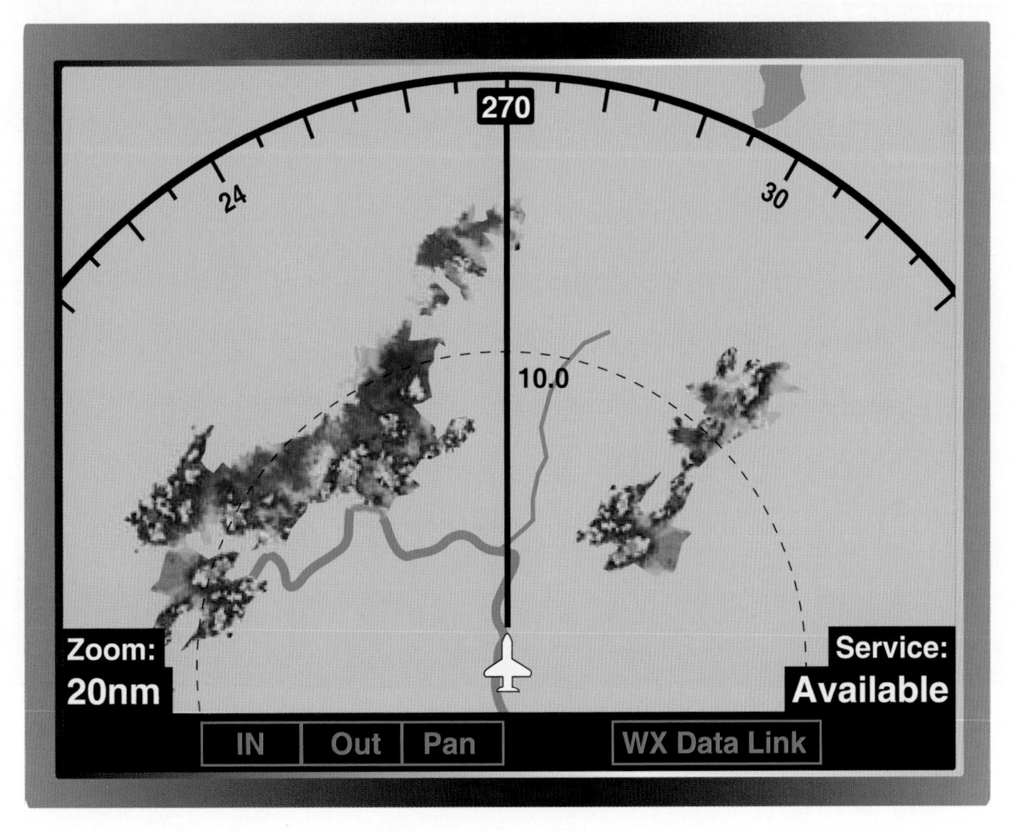

Figure 5-50. *An aircraft equipped with ADS has the ability to upload and display weather.*

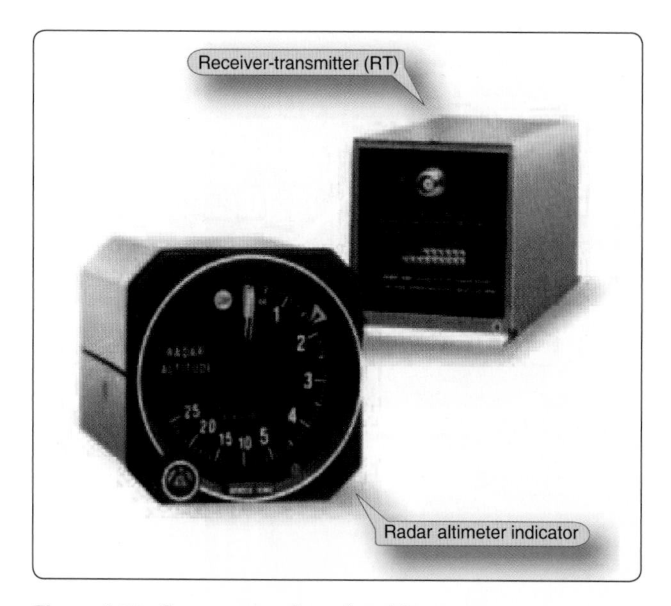

Figure 5-51. *Components of a radar altimeter.*

Figure 5-52. *Coverage provided by a traffic information system.*

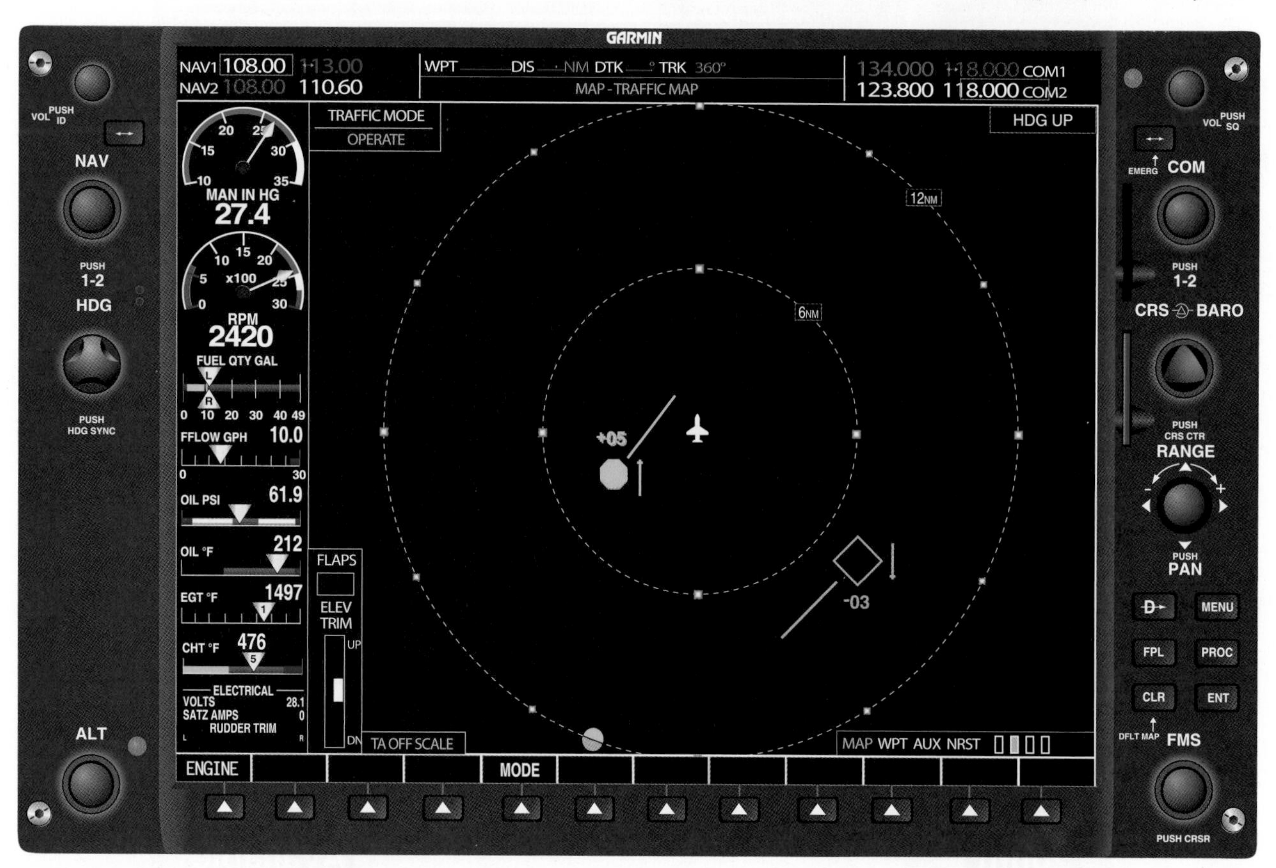

Figure 5-53. *Multi-function display (MFD).*

Figure 5-54. *Concept of the traffic information system.*

Figure 5-55. *Theory of a typical alert system.*

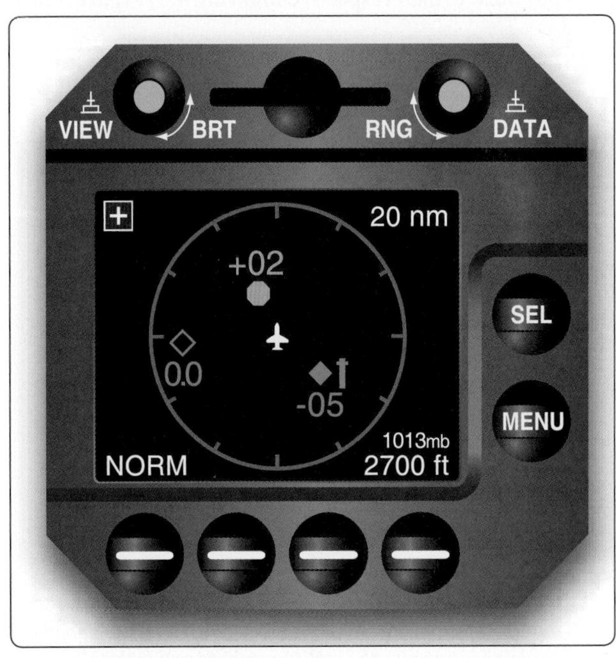

Figure 5-56. *A Skywatch System.*

Figure 5-57. *Alert System by Avidyne (Ryan).*

TCAS II is a more sophisticated system which provides the same information of TCAS I. It also analyzes the projected flightpath of approaching aircraft and issues resolution advisories to the pilot to resolve potential mid-air collisions. Additionally, if communicating with another TCAS II equipped aircraft, the two systems coordinate the resolution alerts provided to their respective flight crews. *[Figure 5-58]*

Figure 5-58. *An example of a resolution advisory being provided to the pilot. In this case, the pilot is requested to climb, with 1,750 feet being the appropriate rate of ascent to avoid traffic conflict. This visual indication plus the audio warning provide the pilot with excellent traffic awareness that augments see-and-avoid practices.*

Terrain Alerting Systems

Ground Proximity Warning System (GPWS)

An early application of technology to reduce controlled flight into terrain (CFIT) was the GPWS. In airline use since the early 1970s, GPWS uses the radio altimeter, speed, and barometric altitude to determine the aircraft's position relative to the ground. The system uses this information in determining aircraft clearance above the Earth and provides limited predictability about aircraft position relative to rising terrain. It does this based upon algorithms within the system and developed by the manufacturer for different airplanes or helicopters. However, in mountainous areas the system is unable to provide predictive information due to the unusual slope encountered.

This inability to provide predictive information was evidenced in 1999 when a DH-7 crashed in South America. The crew had a GPWS onboard, but the sudden rise of the terrain rendered it ineffective; the crew continued unintentionally into a mountain with steep terrain. Another incident involved Secretary of Commerce Brown who, along with all on board, was lost when the crew flew over rapidly rising terrain where the GPWS capability is offset by terrain gradient. However, the GPWS is tied into and considers landing gear status, flap position, and ILS glideslope deviation to detect unsafe aircraft operation with respect to terrain, excessive descent rate, excessive closure rate to terrain, unsafe terrain clearance while not in a landing configuration, excessive deviation below an ILS glideslope. It also provides advisory callouts.

Generally, the GPWS is tied into the hot bus bar of the electrical system to prevent inadvertent switch off. This was demonstrated in an accident involving a large four-engine turboprop airplane. While on final for landing with the landing gear inadvertently up, the crew failed to heed the GPWS warning as the aircraft crossed a large berm close to the threshold. In fact, the crew attempted without success to shut the system down and attributed the signal to a malfunction. Only after the mishap did the crew realize the importance of the GPWS warning.

Terrain Awareness and Warning System (TAWS)

A TAWS uses GPS positioning and a database of terrain and obstructions to provide true predictability of the upcoming terrain and obstacles. The warnings it provides pilots are both aural and visual, instructing the pilot to take specific action. Because TAWS relies on GPS and a database of terrain/obstacle information, predictability is based upon aircraft location and projected location. The system is time based and therefore compensates for the performance of the aircraft and its speed. *[Figure 5-59]*

Head-Up Display (HUD)

The HUD is a display system that provides a projection of navigation and air data (airspeed in relation to approach reference speed, altitude, left/right and up/down glideslope) on a transparent screen between the pilot and the windshield. The concept of a HUD is to diminish the shift between looking at the instrument panel and outside. Virtually any

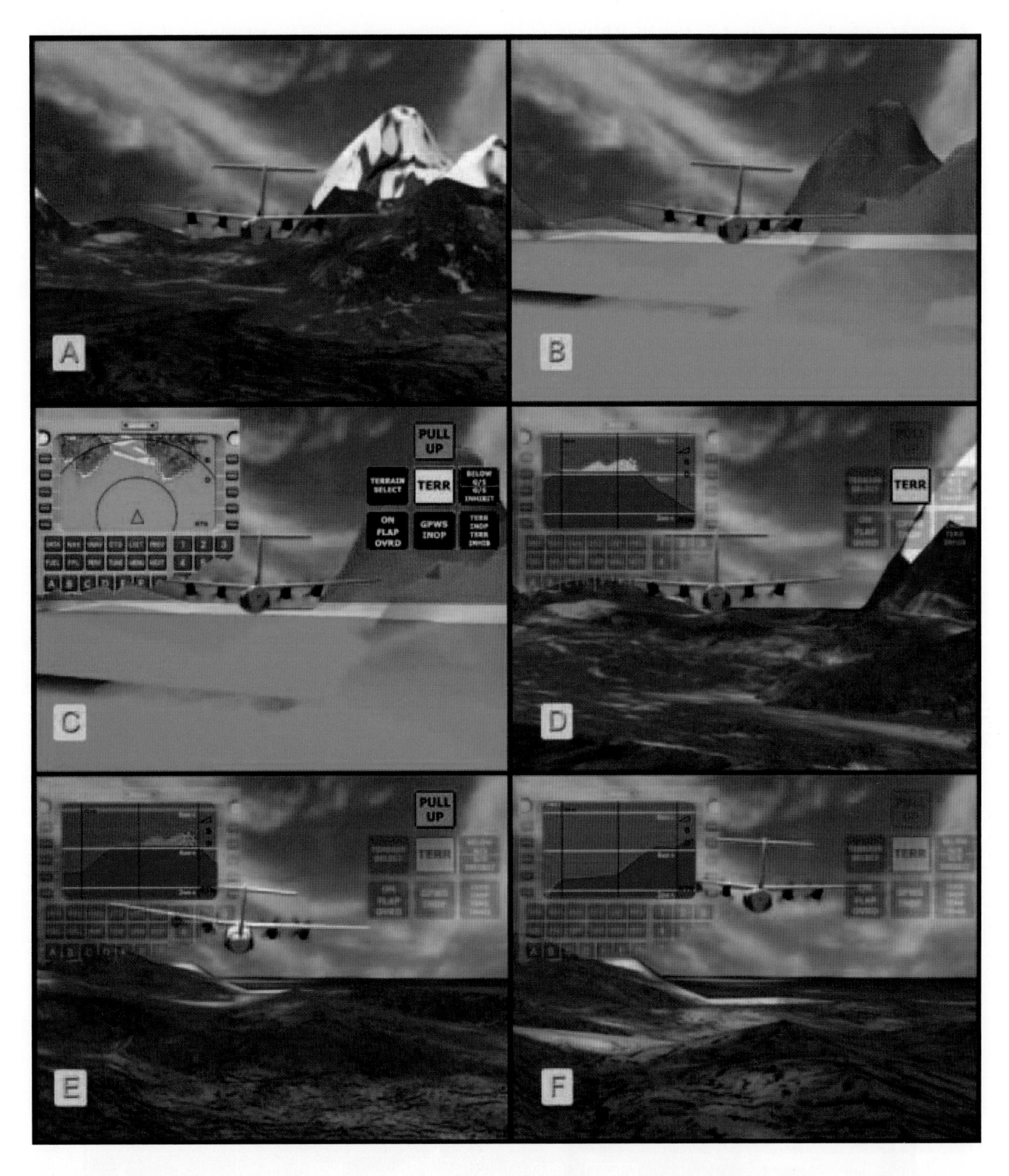

Figure 5-59. *A six-frame sequence illustrating the manner in which TAWS operates. A TAWS installation is aircraft specific and provides warnings and cautions based upon time to potential impact with terrain rather than distance. The TAWS is illustrated in an upper left window while aircrew view is provided out of the windscreen.* A *illustrates the aircraft in relation to the outside terrain while* B *and* C *illustrate the manner in which the TAWS system displays the terrain.* D *is providing a caution of terrain to be traversed, while* E *provides an illustration of a warning with an aural and textural advisory (red) to pull up.* E *also illustrates a pilot taking appropriate action (climb in this case) while* F *illustrates that a hazard is no longer a factor.*

information desired can be displayed on the HUD if it is available in the aircraft's flight computer. The display for the HUD can be projected on a separate panel near the windscreen or as shown in *Figure 5-60* on an eye piece. Other information may be displayed, including a runway target in relation to the nose of the aircraft, which allows the pilot to see the information necessary to make the approach while also being able to see out the windshield.

Required Navigation Instrument System Inspection

Systems Preflight Procedures

Inspecting the instrument system requires a relatively small part of the total time required for preflight activities, but its importance cannot be overemphasized. Before any flight involving aircraft control by instrument reference, the pilot should check all instruments and their sources of power for proper operation.

NOTE: The following procedures are appropriate for conventional aircraft instrument systems. Aircraft equipped with electronic instrument systems utilize different procedures.

Before Engine Start

1. Walk-around inspection: Check the condition of all antennas and check the pitot tube for the presence of any obstructions and remove the cover. Check the static ports to be sure they are free from dirt and obstructions, and ensure there is nothing on the structure near the ports that would disturb the air flowing over them.
2. Aircraft records: Confirm that the altimeter and static system have been checked and found within approved limits within the past 24 calendar months. Check the replacement date for the emergency locator transmitter (ELT) batteries noted in the maintenance record, and be sure they have been replaced within this time interval.
3. Preflight paperwork: Check the Airport/Facility Directory (A/FD) and all NOTAMs for the condition and frequencies of all the navigation aid (NAVAIDs) that are used on the flight. Handbooks, en route charts, approach charts, computer and flight log should be appropriate for the departure, en route, destination, and alternate airports.
4. Radio equipment: Switches OFF.
5. Suction gauge: Proper markings as applicable if electronic flight instrumentation is installed.
6. ASI: Proper reading, as applicable. If electronic flight instrumentation is installed, check emergency instrument.

Figure 5-60. *A head-up display (HUD).*

7. Attitude indicator: Uncaged, if applicable. If electronic flight instrumentation is installed, check emergency system to include its battery as appropriate.
8. Altimeter: Set the current altimeter setting and ensure that the pointers indicate the elevation of the airport.
9. VSI: Zero indication, as applicable (if electronic flight instrumentation is installed).
10. Heading indicator: Uncaged, if applicable.
11. Turn coordinator: If applicable, miniature aircraft level, ball approximately centered (level terrain).
12. Magnetic compass: Full of fluid and the correction card is in place and current.
13. Clock: Set to the correct time and running.
14. Engine instruments: Proper markings and readings, as applicable if electronic flight instrumentation is installed.
15. Deicing and anti-icing equipment: Check availability and fluid quantity.
16. Alternate static-source valve: Be sure it can be opened if needed, and that it is fully closed.
17. Pitot tube heater: Check by watching the ammeter when it is turned on, or by using the method specified in the POH/AFM.

After Engine Start

1. When the master switch is turned on, listen to the gyros as they spin up. Any hesitation or unusual noises should be investigated before flight.
2. Suction gauge or electrical indicators: Check the source of power for the gyro instruments. The suction developed should be appropriate for the instruments in that particular aircraft. If the gyros are electrically driven, check the generators and inverters for proper operation.
3. Magnetic compass: Check the card for freedom of movement and confirm the bowl is full of fluid. Determine compass accuracy by comparing the indicated heading against a known heading (runway heading) while the airplane is stopped or taxiing straight. Remote indicating compasses should also be checked against known headings. Note the compass card correction for the takeoff runway heading.
4. Heading indicator: Allow 5 minutes after starting engines for the gyro to spin up. Before taxiing, or while taxiing straight, set the heading indicator to correspond with the magnetic compass heading. A slaved gyrocompass should be checked for slaving action and its indications compared with those of the magnetic compass. If an electronic flight instrument system is installed, consult the flight manual for proper procedures.
5. Attitude indicator: Allow the same time as noted above for gyros to spin up. If the horizon bar erects to the horizontal position and remains at the correct position for the attitude of the airplane, or if it begins to vibrate after this attitude is reached and then slowly stops vibrating altogether, the instrument is operating properly. If an electronic flight instrument system is installed, consult the flight manual for proper procedures.
6. Altimeter: With the altimeter set to the current reported altimeter setting, note any variation between the known field elevation and the altimeter indication. If the indication is not within 75 feet of field elevation, the accuracy of the altimeter is questionable and the problem should be referred to a repair station for evaluation and possible correction. Because the elevation of the ramp or hangar area might differ significantly from field elevation, recheck when in the run-up area if the error exceeds 75 feet. When no altimeter setting is available, set the altimeter to the published field elevation during the preflight instrument check.
7. VSI: The instrument should read zero. If it does not, tap the panel gently. If an electronic flight instrument system is installed, consult the flight manual for proper procedures.
8. Engine instruments: Check for proper readings.
9. Radio equipment: Check for proper operation and set as desired.
10. Deicing and anti-icing equipment: Check operation.

Taxiing and Takeoff

Ensuring the functionality of the turn coordinator, heading indicator, magnetic compass, and attitude indicator prior to taxiing and takeoff is essential to flight safety. Runway incursion is an incident at an airport that adversely affects runway safety and pilots must mitigate this risk by ensuring that all of the directional flight instruments are checked properly before taxiing or taking off so that the position of the aircraft in relation to the runway and other traffic is always known.

1. Turn coordinator: During taxi turns, check the miniature aircraft for proper turn indications. The ball or slip/skid should move freely. The ball or slip/skid indicator should move opposite to the direction of turns. The turn instrument should indicate the direction of the turn. While taxiing straight, the miniature aircraft (as appropriate) should be level.

2. Heading indicator: Before takeoff, recheck the heading indicator. If the magnetic compass and deviation card are accurate, the heading indicator should show the known taxiway or runway direction when the airplane is aligned with them (within 5°).
3. Attitude indicator: If the horizon bar fails to remain in the horizontal position during straight taxiing, or tips in excess of 5° during taxi turns, the instrument is unreliable. Adjust the miniature aircraft with reference to the horizon bar for the particular airplane while on the ground. For some tricycle-gear airplanes, a slightly nose-low attitude on the ground gives a level flight attitude at normal cruising speed.

Engine Shut Down

When shutting down the engine, note any abnormal instrument indications.

Chapter 6, Section I

Airplane Attitude Instrument Flying

Using Analog Instrumentation

Introduction

Attitude instrument flying is defined as the control of an aircraft's spatial position by using instruments rather than outside visual references. Today's aircraft come equipped with analog and/or digital instruments. Analog instrument systems are mechanical and operate with numbers representing directly measurable quantities, such as a watch with a sweep second hand. In contrast, digital instrument systems are electronic and operate with numbers expressed in digits. Although more manufacturers are providing aircraft with digital instrumentation, analog instruments remain more prevalent. This section acquaints the pilot with the use of analog flight instruments.

Any flight, regardless of the aircraft used or route flown, consists of basic maneuvers. In visual flight, aircraft attitude is controlled by using certain reference points on the aircraft with relation to the natural horizon. In instrument flight, the aircraft attitude is controlled by reference to the flight instruments. Proper interpretation of the flight instruments provides essentially the same information that outside references do in visual flight. Once the role of each instrument in establishing and maintaining a desired aircraft attitude is learned, a pilot is better equipped to control the aircraft in emergency situations involving failure of one or more key instruments.

Learning Methods

The two basic methods used for learning attitude instrument flying are "control and performance" and "primary and supporting." Both methods utilize the same instruments and responses for attitude control. They differ in their reliance on the attitude indicator and interpretation of other instruments.

Attitude Instrument Flying Using the Control and Performance Method

Aircraft performance is achieved by controlling the aircraft attitude and power. Aircraft attitude is the relationship of both the aircraft's pitch and roll axes in relation to the Earth's horizon. An aircraft is flown in instrument flight by controlling the attitude and power, as necessary, to produce both controlled and stabilized flight without reference to a visible horizon. This overall process is known as the control and performance method of attitude instrument flying. Starting with basic instrument maneuvers, this process can be applied through the use of control, performance, and navigation instruments resulting in a smooth flight from takeoff to landing.

Control Instruments

The control instruments display immediate attitude and power indications and are calibrated to permit those respective adjustments in precise increments. In this discussion, the term "power" is used in place of the more technically correct term "thrust or drag relationship." Control is determined by reference to the attitude and power indicators. Power indicators vary with aircraft and may include manifold pressure, tachometers, fuel flow, etc. *[Figure 6-1]*

Performance Instruments

The performance instruments indicate the aircraft's actual performance. Performance is determined by reference to the altimeter, airspeed, or vertical speed indicator (VSI). *[Figure 6-2]*

Navigation Instruments

The navigation instruments indicate the position of the aircraft in relation to a selected navigation facility or fix. This group of instruments includes various types of course indicators, range indicators, glideslope indicators, and bearing pointers. *[Figure 6-3]* Newer aircraft with more technologically advanced instrumentation provide blended information, giving the pilot more accurate positional information.

Procedural Steps in Using Control and Performance

1. Establish an attitude and power setting on the control instruments that results in the desired performance. Known or computed attitude changes and approximated power settings helps to reduce the pilot's workload.
2. Trim (fine tune the control forces) until control pressures are neutralized. Trimming for hands-off flight is essential for smooth, precise aircraft control.

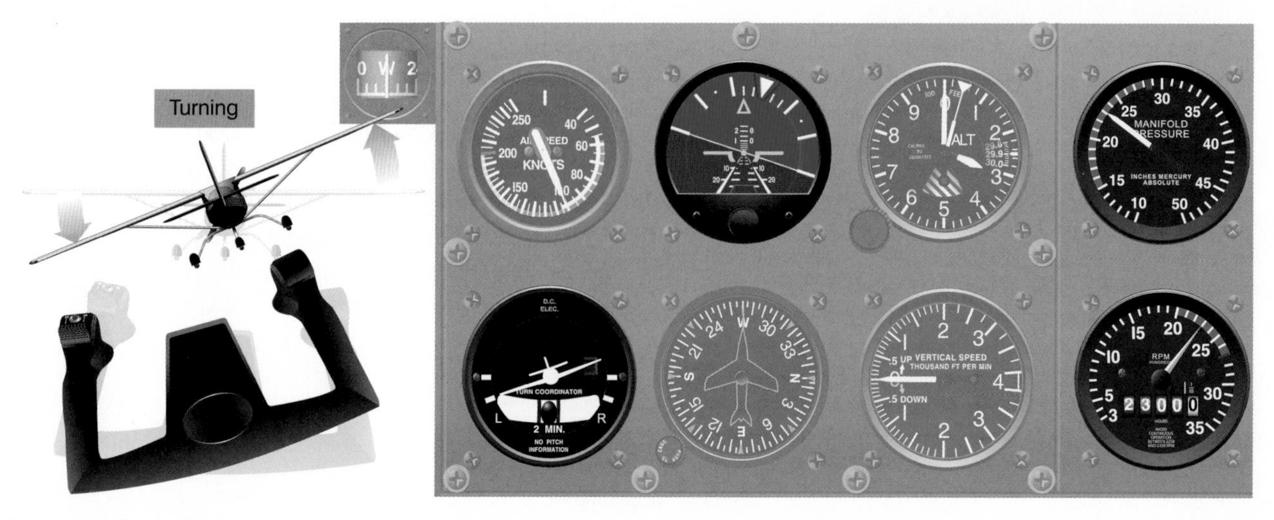

Figure 6-1. *Control instruments.*

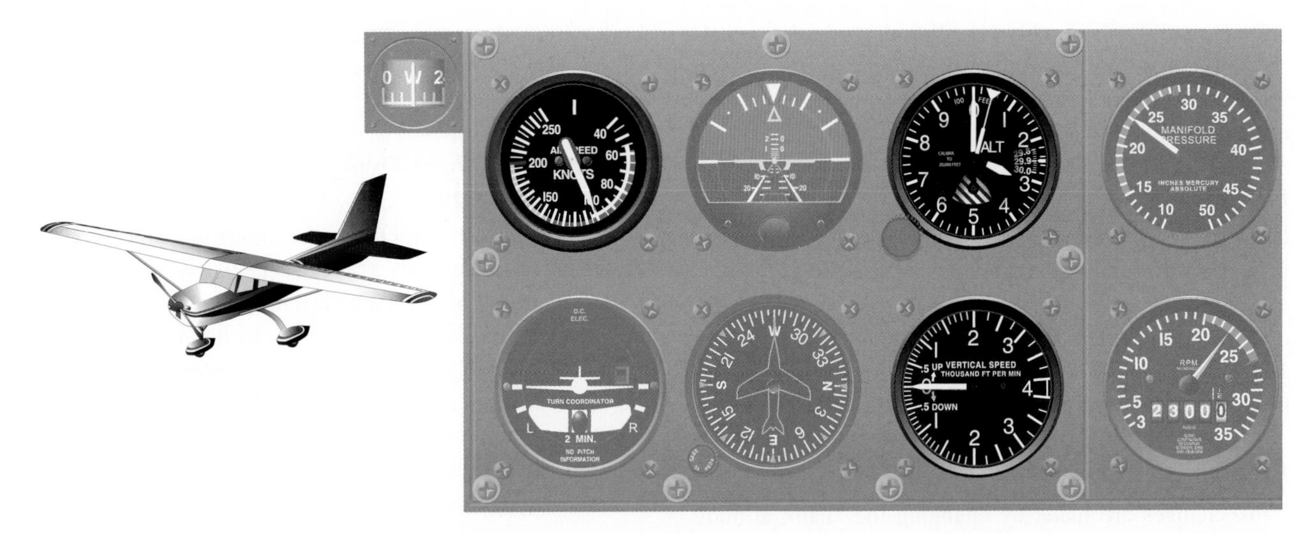

Figure 6-2. *Performance instruments.*

Figure 6-3. *Navigation instruments.*

It allows a pilot to attend to other flight deck duties with minimum deviation from the desired attitude.

3. Cross-check the performance instruments to determine if the established attitude or power setting is providing the desired performance. The cross-check involves both seeing and interpreting. If a deviation is noted, determine the magnitude and direction of adjustment required to achieve the desired performance.

4. Adjust the attitude and/or power setting on the control instruments as necessary.

Aircraft Control During Instrument Flight

Attitude Control

Proper control of aircraft attitude is the result of proper use of the attitude indicator, knowledge of when to change the attitude, and then smoothly changing the attitude a precise amount. The attitude reference provides an immediate, direct, and corresponding indication of any change in aircraft pitch or bank attitude.

Pitch Control

Changing the "pitch attitude" of the miniature aircraft or fuselage dot by precise amounts in relation to the horizon makes pitch changes. These changes are measured in degrees, or fractions thereof, or bar widths depending upon the type of attitude reference. The amount of deviation from the desired performance determines the magnitude of the correction.

Bank Control

Bank changes are made by changing the "bank attitude" or bank pointers by precise amounts in relation to the bank scale. The bank scale is normally graduated at 0°, 10°, 20°, 30°, 60°, and 90° and is located at the top or bottom of the attitude reference. Bank angle use normally approximates the degrees to turn, not to exceed 30°.

Power Control

Proper power control results from the ability to smoothly establish or maintain desired airspeeds in coordination with attitude changes. Power changes are made by throttle adjustments and reference to the power indicators. Power indicators are not affected by such factors as turbulence, improper trim, or inadvertent control pressures. Therefore, in most aircraft little attention is required to ensure the power setting remains constant.

Experience in an aircraft teaches a pilot approximately how far to move the throttle to change the power a given amount. Power changes are made primarily by throttle movement, followed by an indicator cross-check to establish a more precise setting. The key is to avoid fixating on the indicators while setting the power. Knowledge of approximate power settings for various flight configurations helps the pilot avoid overcontrolling power.

Attitude Instrument Flying Using the Primary and Supporting Method

Another basic method for teaching attitude instrument flying classifies the instruments as they relate to control function, as well as aircraft performance. All maneuvers involve some degree of motion about the lateral (pitch), longitudinal (bank/roll), and vertical (yaw) axes. Attitude control is stressed in this handbook in terms of pitch control, bank control, power control, and trim control. Instruments are grouped as they relate to control function and aircraft performance as pitch control, bank control, power control, and trim.

Pitch Control

Pitch control is controlling the rotation of the aircraft about the lateral axis by movement of the elevators. After interpreting the pitch attitude from the proper flight instruments, exert control pressures to effect the desired pitch attitude with reference to the horizon. These instruments include the attitude indicator, altimeter, VSI, and airspeed indicator. *[Figure 6-4]* The attitude indicator displays a direct indication of the aircraft's pitch attitude while the other pitch attitude control instruments indirectly indicate the pitch attitude of the aircraft.

Attitude Indicator

The pitch attitude control of an aircraft controls the angular relationship between the longitudinal axis of the aircraft and the actual horizon. The attitude indicator gives a direct and immediate indication of the pitch attitude of the aircraft. The aircraft controls are used to position the miniature aircraft in relation to the horizon bar or horizon line for any pitch attitude required. *[Figure 6-5]*

Figure 6-4. *Pitch instruments.*

Figure 6-5. *Attitude indicator.*

The miniature aircraft should be placed in the proper position in relation to the horizon bar or horizon line before takeoff. The aircraft operator's manual explains this position. As soon as practicable in level flight and at desired cruise airspeed, the miniature aircraft should be moved to a position that aligns its wings in front of the horizon bar or horizon line. This adjustment can be made any time varying loads or other conditions indicate a need. Otherwise, the position of the miniature aircraft should not be changed for flight at other than cruise speed. This is to make sure that the attitude indicator displays a true picture of pitch attitude in all maneuvers.

When using the attitude indicator in applying pitch attitude corrections, control pressure should be extremely light. Movement of the horizon bar above or below the miniature aircraft of the attitude indicator in an airplane should not exceed one-half the bar width. *[Figure 6-6]* If further change is required, an additional correction of not more than one-half horizon bar wide normally counteracts any deviation from normal flight.

Figure 6-6. *Pitch correction using the attitude indicator.*

Altimeter

If the aircraft is maintaining level flight, the altimeter needles maintain a constant indication of altitude. If the altimeter indicates a loss of altitude, the pitch attitude must be adjusted upward to stop the descent. If the altimeter indicates a gain in altitude, the pitch attitude must be adjusted downward to stop the climb. *[Figure 6-7]* The altimeter can also indicate the pitch attitude in a climb or descent by how rapidly the needles move. A minor adjustment in pitch attitude may be made to control the rate at which altitude is gained or lost. Pitch attitude is used only to correct small altitude changes caused by external forces, such as turbulence or up and down drafts.

Figure 6-7. *Pitch correction using the altimeter.*

Vertical Speed Indicator (VSI)

In flight at a constant altitude, the VSI (sometimes referred to as vertical velocity indicator or rate-of-climb indicator) remains at zero. If the needle moves above zero, the pitch attitude must be adjusted downward to stop the climb and return to level flight. Prompt adjustments to the changes in the indications of the VSI can prevent any significant change in altitude. *[Figure 6-8]* Turbulent air causes the needle to fluctuate near zero. In such conditions, the average of the

Figure 6-8. *Vertical speed indicator.*

fluctuations should be considered as the correct reading. Reference to the altimeter helps in turbulent air because it is not as sensitive as the VSI.

Vertical speed is represented in feet per minute (fpm). *[Figure 6-8]* The face of the instrument is graduated with numbers such as 1, 2, 3, etc. These represent thousands of feet up or down in a minute. For instance, if the pointer is aligned with .5 (½ of a thousand or 500 fpm), the aircraft climbs 500 feet in one minute. The instrument is divided into two regions: one for climbing (up) and one for descending (down).

During turbulence, it is not uncommon to see large fluctuations on the VSI. It is important to remember that small corrections should be employed to avoid further exacerbating a potentially divergent situation.

Overcorrecting causes the aircraft to overshoot the desired altitude; however, corrections should not be so small that the return to altitude is unnecessarily prolonged. As a guide, the pitch attitude should produce a rate of change on the VSI about twice the size of the altitude deviation. For example, if the aircraft is 100 feet off the desired altitude, a 200 fpm rate of correction would be used.

During climbs or descents, the VSI is used to change the altitude at a desired rate. Pitch attitude and power adjustments are made to maintain the desired rate of climb or descent on the VSI.

When pressure is applied to the controls and the VSI shows an excess of 200 fpm from that desired, overcontrolling is indicated. For example, if attempting to regain lost altitude at the rate of 500 fpm, a reading of more than 700 fpm would indicate overcontrolling. Initial movement of the needle indicates the trend of vertical movement. The time for the VSI to reach its maximum point of deflection after a correction is called lag. The lag is proportional to speed and magnitude of pitch change. In an airplane, overcontrolling may be reduced by relaxing pressure on the controls, allowing the pitch attitude to neutralize. In some helicopters with servo-assisted controls, no control pressures are apparent. In this case, overcontrolling can be reduced by reference to the attitude indicator.

Some aircraft are equipped with an instantaneous vertical speed indicator (IVSI). The letters "IVSI" appear on the face of the indicator. This instrument assists in interpretation by instantaneously indicating the rate of climb or descent at a given moment with little or no lag as displayed in a VSI.

Occasionally, the VSI is slightly out of calibration and indicates a gradual climb or descent when the aircraft is in level flight. If readjustments cannot be accomplished, the error in the indicator should be considered when the instrument is used for pitch control. For example, an improperly set VSI may indicate a descent of 100 fpm when the aircraft is in level flight. Any deviation from this reading would indicate a change in pitch attitude.

Airspeed Indicator

The airspeed indicator gives an indirect reading of the pitch attitude. With a constant power setting and a constant altitude, the aircraft is in level flight and airspeed remains constant. If the airspeed increases, the pitch attitude has lowered and should be raised. *[Figure 6-9]* If the airspeed decreases, the pitch attitude has moved higher and should be lowered. *[Figure 6-10]* A rapid change in airspeed indicates a large change in pitch; a slow change in airspeed indicates a small change in pitch. Although the airspeed indicator is used as a pitch instrument, it may be used in level flight for power control. Changes in pitch are reflected immediately by a change in airspeed. There is very little lag in the airspeed indicator.

Figure 6-9. *Pitch attitude has lowered.*

Figure 6-10. *Pitch attitude has moved higher.*

Pitch Attitude Instrument Cross-Check

The altimeter is an important instrument for indicating pitch attitude in level flight except when used in conditions of exceptionally strong vertical currents, such as thunderstorms. With proper power settings, any of the pitch attitude instruments can be used to hold reasonably level flight attitude. However, only the altimeter gives the exact altitude information. Regardless of which pitch attitude control instrument indicates a need for a pitch attitude adjustment, the attitude indicator, if available, should be used to make the adjustment. Common errors in pitch attitude control are:

- Overcontrolling;
- Improperly using power; and
- Failing to adequately cross-check the pitch attitude instruments and take corrective action when pitch attitude change is needed.

Bank Control

Bank control is controlling the angle made by the wing and the horizon. After interpreting the bank attitude from the appropriate instruments, exert the necessary pressures to move the ailerons and roll the aircraft about the longitudinal axis. As illustrated in *Figure 6-11*, these instruments include:

- Attitude indicator
- Heading indicator
- Magnetic compass
- Turn coordinator/turn-and-slip indicator

Attitude Indicator

As previously discussed, the attitude indicator is the only instrument that portrays both instantly and directly the actual flight attitude and is the basic attitude reference.

Heading Indicator

The heading indicator supplies the pertinent bank and heading information and is considered a primary instrument for bank.

Magnetic Compass

The magnetic compass provides heading information and is considered a bank instrument when used with the heading indicator. Care should be exercised when using the magnetic compass as it is affected by acceleration, deceleration in flight caused by turbulence, climbing, descending, power changes, and airspeed adjustments. Additionally, the magnetic compass indication will lead and lag in its reading depending upon the direction of turn. As a result, acceptance of its indication should be considered with other instruments that indicate turn information. These include the already mentioned attitude and heading indicators, as well as the turn-and-slip indicator and turn coordinator.

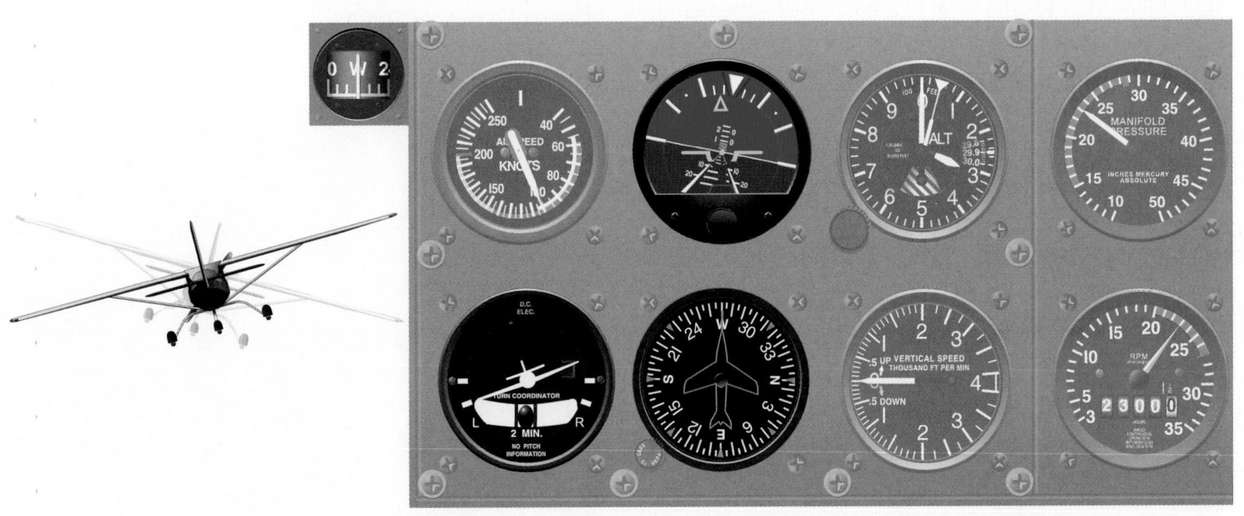

Figure 6-11. *Bank instruments.*

Turn Coordinator/Turn-and-Slip Indicator

Both of these instruments provide turn information. *[Figure 6-12]* The turn coordinator provides both bank rate and then turn rate once stabilized. The turn-and-slip indicator provides only turn rate.

Figure 6-12. *Turn coordinator and turn-and-slip indicator.*

Power Control

A power change to adjust airspeed may cause movement around some or all of the aircraft axes. The amount and direction of movement depends on how much or how rapidly the power is changed, whether single-engine or multiengine airplane or helicopter. The effect on pitch attitude and airspeed caused by power changes during level flight is illustrated in *Figures 6-13* and *6-14*. During or immediately after adjusting the power control(s), the power instruments should be cross-checked to see if the power adjustment is as desired. Whether or not the need for a power adjustment is indicated by another instrument(s), adjustment is made by cross-checking the power instruments. Aircraft are powered by a variety of powerplants, each powerplant having certain instruments that indicate the amount of power being applied to operate the aircraft. During instrument flight, these instruments must be used to make the required power adjustments.

As illustrated in *Figure 6-15,* power indicator instruments include:

- Airspeed indicator
- Engine instruments

Airspeed Indicator

The airspeed indicator provides an indication of power best observed initially in level flight where the aircraft is in balance and trim. If in level flight the airspeed is increasing, it can generally be assumed that the power has increased, necessitating the need to adjust power or re-trim the aircraft.

Engine Instruments

Engine instruments, such as the manifold pressure (MP) indicator, provide an indication of aircraft performance for a given setting under stable conditions. If the power conditions are changed, as reflected in the respective engine instrument readings, there is an affect upon the aircraft performance, either an increase or decrease of airspeed. When the propeller rotational speed (revolutions per minute (RPM) as viewed on a tachometer) is increased or decreased on fixed-pitch propellers, the performance of the aircraft reflects a gain or loss of airspeed as well.

Trim Control

Proper trim technique is essential for smooth and accurate instrument flying and utilizes instrumentation illustrated in *Figure 6-16.* The aircraft should be properly trimmed while executing a maneuver. The degree of flying skill, which ultimately develops, depends largely upon how well the aviator learns to keep the aircraft trimmed.

Airplane Trim

An airplane is correctly trimmed when it is maintaining a desired attitude with all control pressures neutralized. By relieving all control pressures, it is much easier to maintain the

Figure 6-13. *An increase in power—increasing airspeed accordingly in level flight.*

Figure 6-14. *Pitch control and power adjustment required to bring aircraft to level flight.*

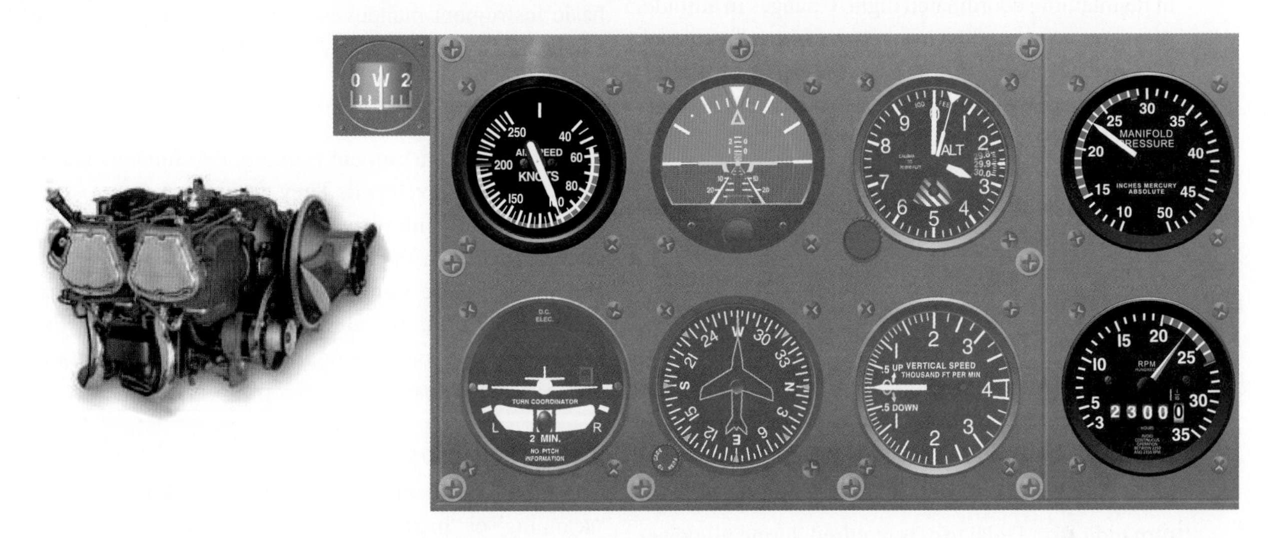

Figure 6-15. *Power instruments.*

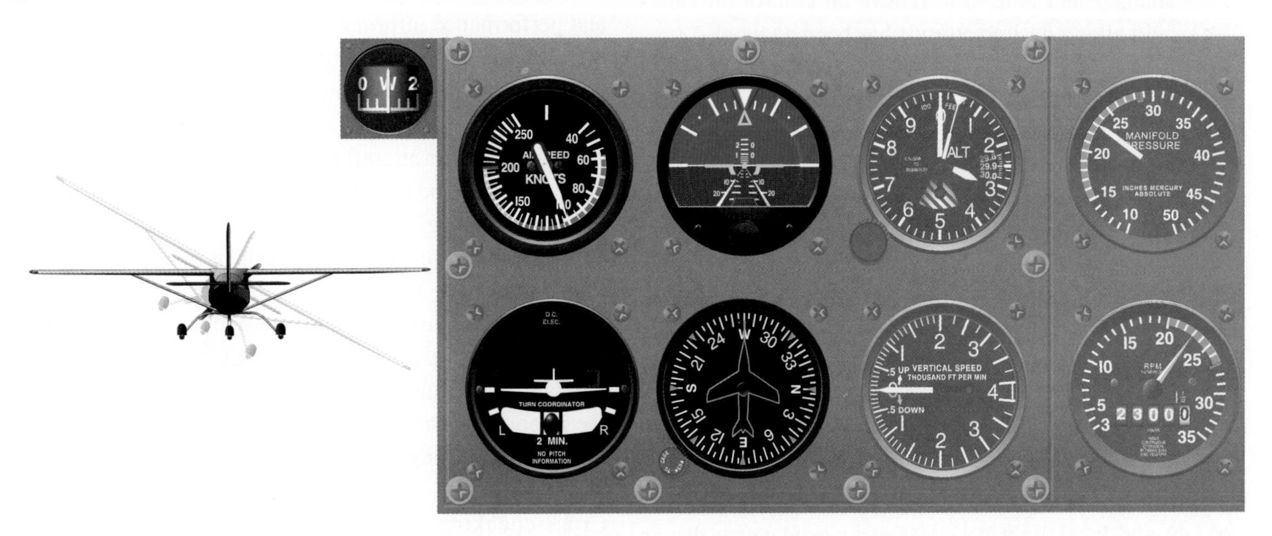

Figure 6-16. *Trim instruments.*

aircraft at a certain attitude. This allows more time to devote to the navigation instruments and additional flight deck duties.

An aircraft is placed in trim by:

- Applying control pressure(s) to establish a desired attitude. Then, the trim is adjusted so that the aircraft maintains that attitude when flight controls are released. The aircraft is trimmed for coordinated flight by centering the ball of the turn-and-slip indicator.
- Moving the rudder trim in the direction where the ball is displaced from center. Aileron trim may then be adjusted to maintain a wings-level attitude.
- Using balanced power or thrust when possible to aid in maintaining coordinated flight. Changes in attitude, power, or configuration may require trim adjustments. Use of trim alone to establish a change in aircraft attitude usually results in erratic aircraft control. Smooth and precise attitude changes are best attained by a combination of control pressures and subsequent trim adjustments. The trim controls are aids to smooth aircraft control.

Helicopter Trim

A helicopter is placed in trim by continually cross-checking the instruments and performing the following:

- Using the cyclic-centering button. If the helicopter is so equipped, this relieves all possible cyclic pressures.
- Using the pedal adjustment to center the ball of the turn indicator. Pedal trim is required during all power changes and is used to relieve all control pressures held after a desired attitude has been attained.

An improperly trimmed helicopter requires constant control pressures, produces tension, distracts attention from cross-checking, and contributes to abrupt and erratic attitude control. The pressures felt on the controls should be only those applied while controlling the helicopter.

Adjust the pitch attitude, as airspeed changes, to maintain desired attitude for the maneuver being executed. The bank must be adjusted to maintain a desired rate of turn, and the pedals must be used to maintain coordinated flight. Trim must be adjusted as control pressures indicate a change is needed.

Example of Primary and Support Instruments

Straight-and-level flight at a constant airspeed means that an exact altitude is to be maintained with zero bank (constant heading). The primary pitch, bank, and power instruments used to maintain this flight condition are:

- Altimeter—supplies the most pertinent altitude information and is primary for pitch.
- Heading Indicator—supplies the most pertinent bank or heading information and is primary for bank.
- Airspeed Indicator—supplies the most pertinent information concerning performance in level flight in terms of power output and is primary for power.

Although the attitude indicator is the basic attitude reference, the concept of primary and supporting instruments does not devalue any particular flight instrument, when available, in establishing and maintaining pitch-and-bank attitudes. It is the only instrument that instantly and directly portrays the actual flight attitude. It should always be used, when available, in establishing and maintaining pitch-and-bank attitudes. The specific use of primary and supporting instruments during basic instrument maneuvers is presented in more detail in Chapter 7, Airplane Basic Flight Maneuvers.

Fundamental Skills

During attitude instrument training, two fundamental flight skills must be developed. They are instrument cross-check and instrument interpretation, both resulting in positive aircraft control. Although these skills are learned separately and in deliberate sequence, a measure of proficiency in precision flying is the ability to integrate these skills into unified, smooth, positive control responses to maintain any prescribed flightpath.

Instrument Cross-Check

The first fundamental skill is cross-checking (also called "scanning" or "instrument coverage"). Cross-checking is the continuous and logical observation of instruments for attitude and performance information. In attitude instrument flying, the pilot maintains an attitude by reference to instruments, producing the desired result in performance. Observing and interpreting two or more instruments to determine attitude and performance of an aircraft is called cross-checking. Although no specific method of cross-checking is recommended, those instruments that give the best information for controlling the aircraft in any given maneuver should be used. The important instruments are the ones that give the most pertinent information for any particular phase of the maneuver. These are usually the instruments that should be held at a constant indication. The remaining instruments should help maintain the important instruments at the desired indications, which is also true in using the emergency panel.

Cross-checking is mandatory in instrument flying. In visual flight, a level attitude can be maintained by outside references. However, even then the altimeter must be checked to determine if altitude is being maintained. Due to human error, instrument error, and airplane performance differences in various atmospheric and loading conditions, it is impossible to establish an attitude and have performance remain constant for a long period of time. These variables make it necessary for the pilot to constantly check the instruments and make appropriate changes in airplane attitude using cross-checking of instruments. Examples of cross-checking are explained in the following paragraphs.

Selected Radial Cross-Check

When the selected radial cross-check is used, a pilot spends 80 to 90 percent of flight time looking at the attitude indicator, taking only quick glances at the other flight instruments (for this discussion, the five instruments surrounding the attitude indicator are called the flight instruments). With this method, the pilot's eyes never travel directly between the flight instruments but move by way of the attitude indicator. The maneuver being performed determines which instruments to look at in the pattern. *[Figure 6-17]*

Inverted-V Cross-Check

In the inverted-V cross-check, the pilot scans from the attitude indicator down to the turn coordinator, up to the attitude indicator, down to the VSI, and back up to the attitude indicator. *[Figure 6-18]*

Rectangular Cross-Check

In the rectangular cross-check, the pilot scans across the top three instruments (airspeed indicator, attitude indicator, and altimeter), and then drops down to scan the bottom three instruments (VSI, heading indicator, and turn instrument). This scan follows a rectangular path (clockwise or counterclockwise rotation is a personal choice). *[Figure 6-19]*

This cross-checking method gives equal weight to the information from each instrument, regardless of its importance to the maneuver being performed. However, this method lengthens the time it takes to return to an instrument critical to the successful completion of the maneuver.

Common Cross-Check Errors

A beginner might cross-check rapidly, looking at the instruments without knowing exactly what to look for. With increasing experience in basic instrument maneuvers and familiarity with the instrument indications associated with them, a pilot learns what to look for, when to look for it, and what response to make. As proficiency increases, a pilot cross-checks primarily from habit, suiting scanning rate and sequence to the demands of the flight situation. Failure to maintain basic instrument proficiency through practice can result in many of the following common scanning errors, both during training and at any subsequent time.

Fixation, or staring at a single instrument, usually occurs for a reason, but has poor results. For example, a pilot may stare

Figure 6-17. *Radial cross-check.*

Figure 6-18. *Inverted-V cross-check.*

Figure 6-19. *Rectangular cross-check.*

at the altimeter reading 200 feet below the assigned altitude, and wonder how the needle got there. While fixated on the instrument, increasing tension may be unconsciously exerted on the controls, which leads to an unnoticed heading change that leads to more errors. Another common fixation is likely when initiating an attitude change. For example, a shallow bank is established for a 90° turn and, instead of maintaining a cross-check of other pertinent instruments, the pilot stares at the heading indicator throughout the turn. Since the aircraft is turning, there is no need to recheck the heading indicator for approximately 25 seconds after turn entry. The problem here may not be entirely due to cross-check error. It may be related to difficulties with instrument interpretation. Uncertainty about reading the heading indicator (interpretation) or uncertainty because of inconsistency in rolling out of turns (control) may cause the fixation.

Omission of an instrument from a cross-check is another likely fault. It may be caused by failure to anticipate significant instrument indications following attitude changes. For example, in a roll-out from a 180° steep turn, straight-and-level flight is established with reference only to the attitude indicator, and the pilot neglects to check the heading indicator for constant heading information. Because of precession error, the attitude indicator temporarily shows a slight error, correctable by quick reference to the other flight instruments.

Emphasis on a single instrument, instead of on the combination of instruments necessary for attitude information, is an understandable fault during the initial stages of training. It is a natural tendency to rely on the instrument that is most readily understood, even when it provides erroneous or inadequate information. Reliance on a single instrument is poor technique. For example, a pilot can maintain reasonably close altitude control with the attitude indicator, but cannot hold altitude with precision without including the altimeter in the cross-check.

Instrument Interpretation

The second fundamental skill, instrument interpretation, requires more thorough study and analysis. It begins by understanding each instrument's construction and operating principles. Then, this knowledge must be applied to the performance of the aircraft being flown, the particular maneuvers to be executed, the cross-check and control techniques applicable to that aircraft, and the flight conditions.

For example, a pilot uses full power in a small airplane for a 5-minute climb from near sea level, and the attitude indicator shows the miniature aircraft two bar widths (twice the thickness of the miniature aircraft wings) above the artificial horizon. *[Figure 6-20]* The airplane is climbing at 500 fpm as shown on the VSI, and at airspeed of 90 knots, as shown on the airspeed indicator. With the power available in this particular airplane and the attitude selected by the pilot, the performance is shown on the instruments. Now, set up the identical picture on the attitude indicator in a jet airplane. With the same airplane attitude as shown in the first example, the VSI in the jet reads 2,000 fpm and the airspeed indicator reads 250 knots.

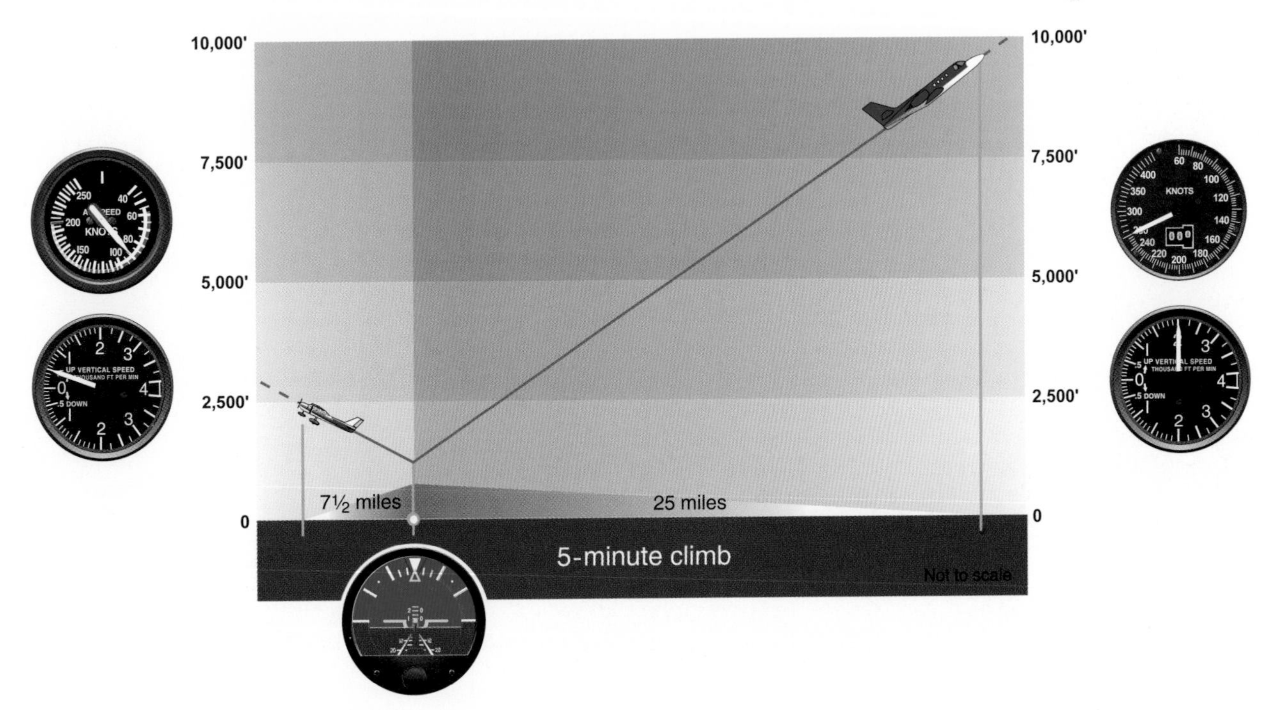

Figure 6-20. *Power and attitude equal performance.*

As the performance capabilities of the aircraft are learned, a pilot interprets the instrument indications appropriately in terms of the attitude of the aircraft. If the pitch attitude is to be determined, the airspeed indicator, altimeter, VSI, and attitude indicator provide the necessary information. If the bank attitude is to be determined, the heading indicator, turn coordinator, and attitude indicator must be interpreted. For each maneuver, learn what performance to expect and the combination of instruments to be interpreted in order to control aircraft attitude during the maneuver. It is the two fundamental flight skills, instrument cross-check and instrument interpretation, that provide the smooth and seamless control necessary for basic instrument flight as discussed at the beginning of the chapter.

Chapter 6, Section II

Airplane Attitude Instrument Flying

Using an Electronic Flight Display

Introduction

Attitude instrument flying is defined as the control of an aircraft's spatial position by using instruments rather than outside visual references. As noted in Section I, today's aircraft come equipped with analog and/or digital instruments. Section II acquaints the pilot with the use of digital instruments known as an electronic flight display (EFD).

The improvements in avionics coupled with the introduction of EFDs to general aviation aircraft offer today's pilot an unprecedented array of accurate instrumentation to use in the support of instrument flying.

Until recently, most general aviation aircraft were equipped with individual instruments utilized collectively to safely maneuver the aircraft by instrument reference alone. With the release of the EFD system, the conventional instruments have been replaced by multiple liquid crystal display (LCD) screens. The first screen is installed in front of the left seat pilot position and is referred to as the primary flight display (PFD). *[Figure 6-21]* The second screen is positioned in approximately the center of the instrument panel and is referred to as the multifunction display (MFD). *[Figure 6-22]* The pilot can use the MFD to display navigation information (moving maps), aircraft systems information (engine monitoring), or should the need arise, a PFD. *[Figure 6-23]* With just these two screens, aircraft designers have been able to declutter instrument panels while increasing safety. This has been accomplished through the utilization of solid-state instruments that have a failure rate far lower than those of conventional analog instrumentation.

However, in the event of electrical failure, the pilot still has emergency instruments as a backup. These instruments either do not require electrical power, or as in the case of many attitude indicators, they are battery equipped. *[Figure 6-24]*

Pilots flying under visual flight rules (VFR) maneuver their aircraft by reference to the natural horizon, utilizing specific reference points on the aircraft. In order to operate the aircraft in other than VFR weather, with no visual reference to the natural horizon, pilots need to develop additional skills. These skills come from the ability to maneuver the aircraft by reference to flight instruments alone. These flight instruments replicate all the same key elements that a VFR pilot utilizes during a normal flight. The natural horizon is replicated on the attitude indicator by the artificial horizon.

Understanding how each flight instrument operates and what role it plays in controlling the attitude of the aircraft is fundamental in learning attitude instrument flying. When the pilot understands how all the instruments are used in establishing and maintaining a desired aircraft attitude, the pilot is better prepared to control the aircraft should one or more key instruments fail or if the pilot should enter instrument flight conditions.

Learning Methods

There are two basic methods utilized for learning attitude instrument flying. They are "control and performance" and "primary and supporting." These methods rely on the same flight instruments and require the pilot to make the same adjustments to the flight and power controls to control aircraft attitude. The main difference between the two methods is the importance that is placed on the attitude indicator and the interpretation of the other flight instruments.

Figure 6-21. *Primary flight display (PFD) and analog counterparts.*

Figure 6-22. *Multifunction display (MFD).*

Figure 6-23. *Reversionary displays.*

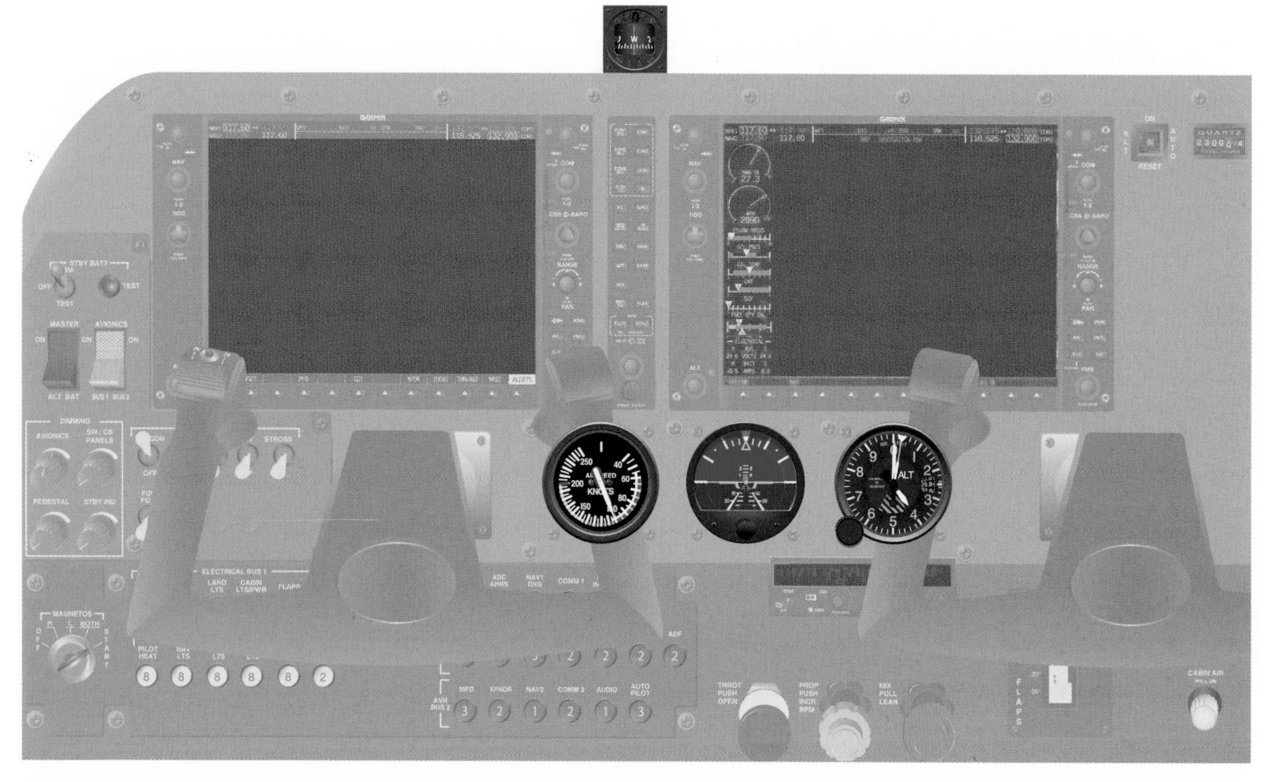

Figure 6-24. *Emergency back-up of the airspeed indicator, attitude indicator, and altitude indicator.*

Control and Performance Method

Aircraft performance is accomplished by controlling the aircraft attitude and power output. Aircraft attitude is the relationship of its longitudinal and lateral axes to the Earth's horizon. When flying in instrument flight conditions, the pilot controls the attitude of the aircraft by referencing the flight instruments and manipulating the power output of the engine to achieve the performance desired. This method can be used to achieve a specific performance level enabling a pilot to perform any basic instrument maneuver.

The instrumentation can be broken up into three different categories: control, performance, and navigation.

Control Instruments

The control instruments depict immediate attitude and power changes. The instrument for attitude display is the attitude indicator. Power changes are directly reflected on the manifold pressure gauge and the tachometer. *[Figure 6-25]* All three of these instruments can reflect small adjustments, allowing for precise control of aircraft attitude.

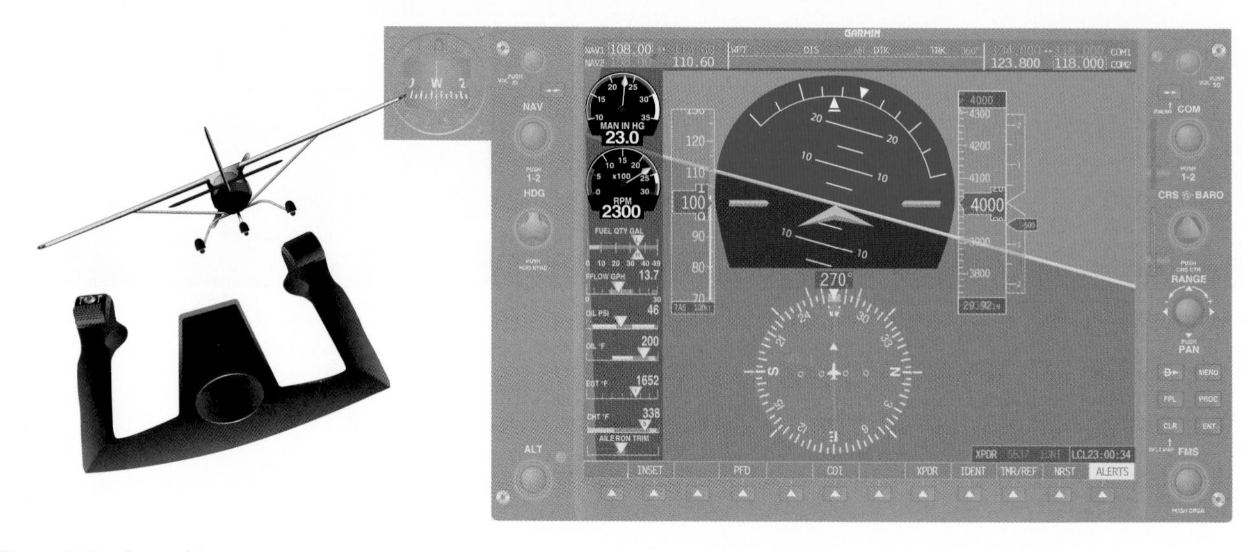

Figure 6-25. *Control instruments.*

In addition, the configuration of the power indicators installed in each aircraft may vary to include the following types of power indicators: tachometers, manifold pressure indicator, engine pressure ratio indicator, fuel flow gauges, etc.

The control instruments do not indicate how fast the aircraft is flying or at what altitude it is flying. In order to determine these variables and others, a pilot needs to refer to the performance instruments.

Performance Instruments

The performance instruments directly reflect the performance the aircraft is achieving. The speed of the aircraft can be referenced on the airspeed indicator. The altitude can be referenced on the altimeter. The aircraft's climb performance can be determined by referencing the vertical speed indicator (VSI). *[Figure 6-26]* Other performance instruments available are the heading indicator, pitch attitude indicator, and the slip/skid indicator.

The performance instruments most directly reflect a change in acceleration, which is defined as change in velocity or direction. Therefore, these instruments indicate if the aircraft is changing airspeed, altitude, or heading, which are horizontal, vertical, or lateral vectors.

Navigation Instruments

The navigation instruments are comprised of global positioning system (GPS) displays and indicators, very high frequency omnidirectional range/nondirectional radio beacon (VOR/NDB) indicators, moving map displays, localizer, and glideslope (GS) indicators. *[Figure 6-27]* The instruments indicate the position of the aircraft relative to a selected navigation facility or fix. Navigation instruments allow the pilot to maneuver the aircraft along a predetermined path of ground-based or spaced-based navigation signals without reference to any external visual cues. The navigation instruments can support both lateral and visual inputs.

The Four-Step Process Used to Change Attitude

In order to change the attitude of the aircraft, the pilot must make the proper changes to the pitch, bank, or power settings of the aircraft. Four steps (establish, trim, cross-check, and adjust) have been developed in order to aid in the process.

Establish

Any time the attitude of the aircraft requires changing, the pilot must adjust the pitch and/or bank in conjunction with power to establish the desired performance. The changes in pitch and bank require the pilot to reference the attitude indicator in order to make precise changes. Power changes should be verified on the tachometer, manifold pressure gauge, etc. To ease the workload, the pilot should become

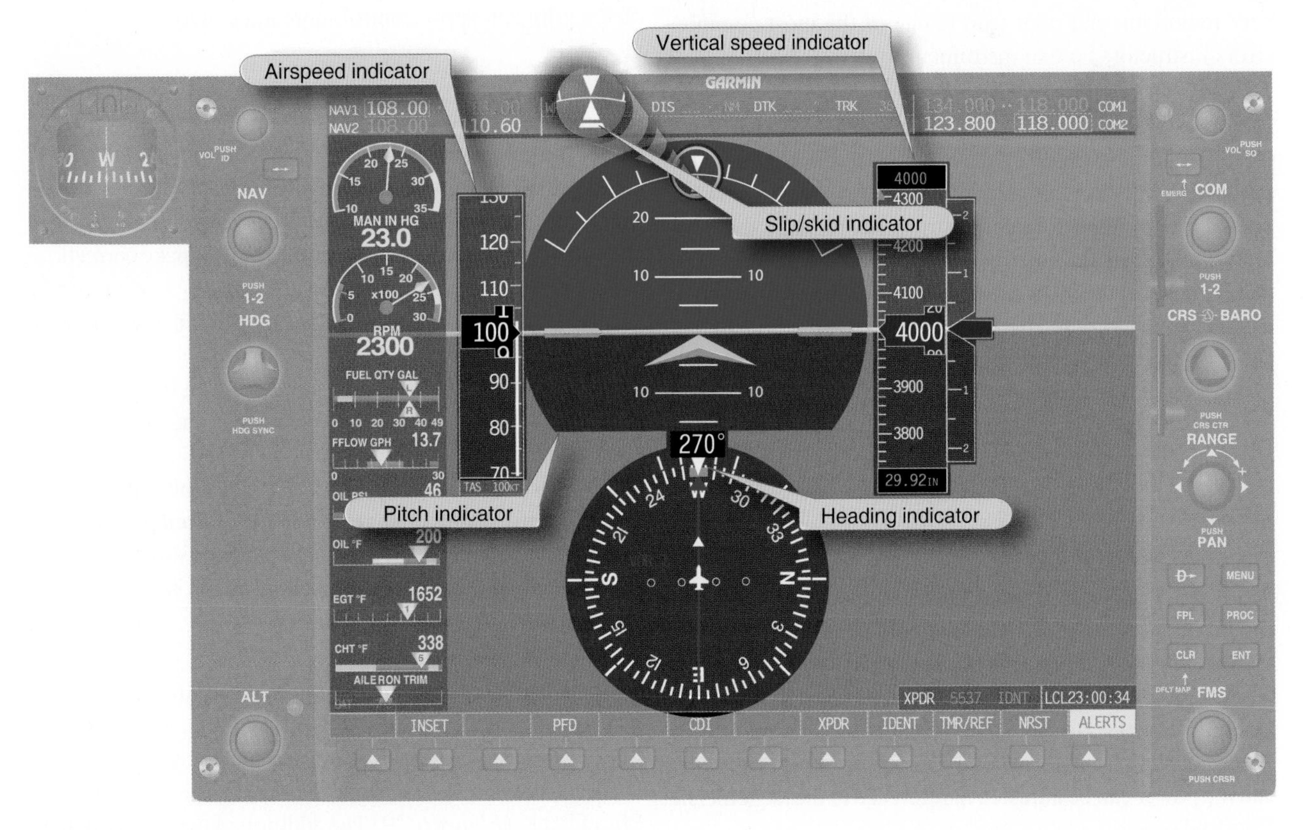

Figure 6-26. *Performance instruments.*

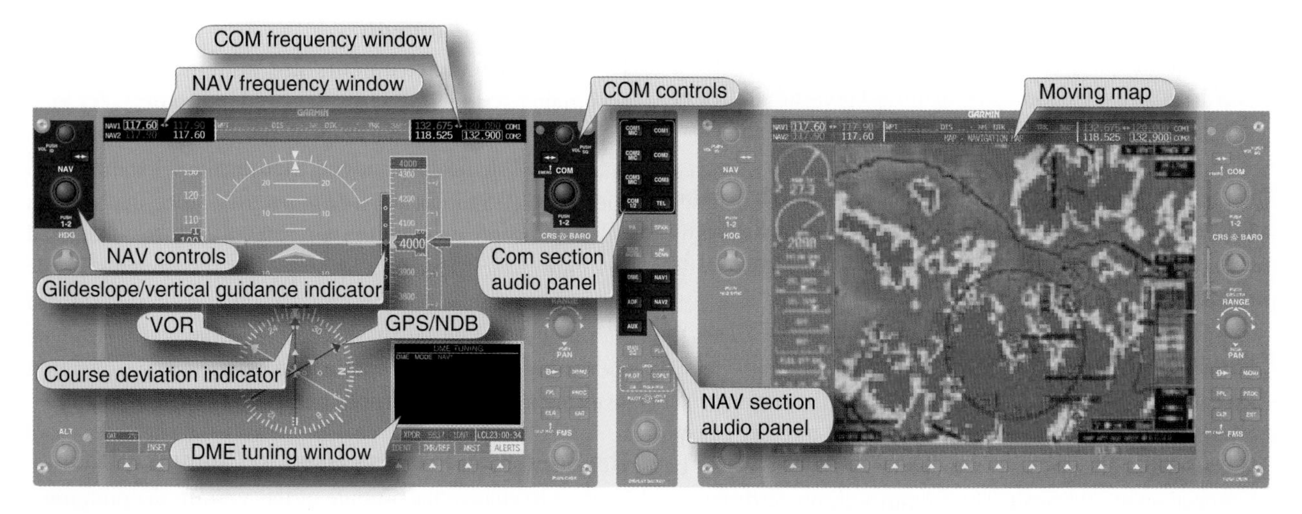

Figure 6-27. *Navigation instruments.*

familiar with the approximate pitch and power changes necessary to establish a specified attitude.

Trim

Another important step in attitude instrument flying is trimming the aircraft. Trim is utilized to eliminate the need to apply force to the control yoke in order to maintain the desired attitude. When the aircraft is trimmed appropriately, the pilot is able to relax pressure on the control yoke and momentarily divert attention to another task at hand without deviating from the desired attitude. Trimming the aircraft is very important, and poor trim is one of the most common errors instructors note in instrument students.

Cross-Check

Once the initial attitude changes have been made, the pilot should verify the performance of the aircraft. Cross-checking the control and performance instruments requires the pilot to visually scan the instruments, as well as interpret the indications. All the instruments must be utilized collectively in order to develop a full understanding of the aircraft attitude. During the cross-check, the pilot needs to determine the magnitude of any deviations and determine how much of a change is required. All changes are then made based on the control instrument indications.

Adjust

The final step in the process is adjusting for any deviations that have been noted during the cross-check. Adjustments should be made in small increments. The attitude indicator and the power instruments are graduated in small increments to allow for precise changes to be made. The pitch should be made in reference to bar widths on the miniature airplane. The bank angle can be changed in reference to the roll scale and the power can be adjusted in reference to the tachometer, manifold pressure gauge, etc.

By utilizing these four steps, pilots can better manage the attitude of their aircraft. One common error associated with this process is making a larger than necessary change when a deviation is noted. Pilots need to become familiar with the aircraft and learn how great a change in attitude is needed to produce the desired performance.

Applying the Four-Step Process

In attitude instrument flight, the four-step process is used to control pitch attitude, bank attitude, and power application of the aircraft. The EFD displays indications precisely enough that a pilot can apply control more accurately.

Pitch Control

The pitch control is indicated on the attitude indicator,which spans the full width of the PFD. Due to the increased size of the display, minute changes in pitch can be made and corrected. The pitch scale on the attitude indicator is graduated in 5-degree increments that allow the pilot to make corrections with precision to approximately ½ degree. The miniature airplane utilized to represent the aircraft in conventional attitude indicators is replaced in glass panel displays by a yellow chevron. *[Figure 6-28]* Representing the nose of the aircraft, the point of the chevron affords the pilot a much more precise indication of the degree of pitch and allows the pilot to make small, precise changes should the desired aircraft performance change. When the desired performance is not being achieved, precise pitch changes should be made by referencing the point of the yellow chevron.

Bank Control

Precise bank control can be developed utilizing the roll pointer in conjunction with the roll index displayed on the attitude indicator. The roll index is sectioned by hash marks at 0°, 10°, 20°, 30°, 45°, 60° and the horizon line, which depicts 90° of bank. *[Figure 6-29]* The addition of the 45° hash mark is an improvement over conventional attitude indicators.

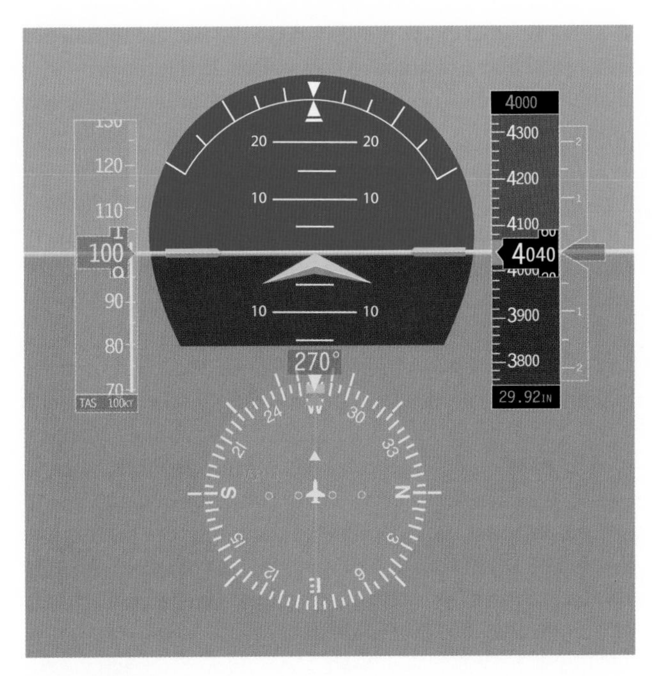

Figure 6-28. *The chevron's relationship to the horizon line indicates the pitch of the aircraft.*

Figure 6-29. *Bank indicator index lines.*

In addition to the roll index, the instrument pilot utilizes the turn rate indicator to maintain the aircraft in a standard rate turn (3° per second). Most instrument maneuvers can be done comfortably, safely, and efficiently by utilizing a standard rate turn.

Power Control

The power instruments indicate how much power is being generated by the engine. They are not affected by turbulence, improper trim, or control pressures. All changes in power should be made with reference to power instruments and cross-checked on performance instruments.

Power control needs to be learned from the beginning of flight training. Attitude instrument flying demands increased precision when it comes to power control. As experience increases, pilots begin to know approximately how much change in throttle position is required to produce the desired change in airspeed. Different aircraft demand differing amounts of throttle change to produce specific performance. It is imperative that the pilot make the specific changes on the power instruments and allow the performance to stabilize. Avoid the tendency to overcontrol.

One common error encountered with glass panel displays is associated with the precision of the digital readouts. This precision causes pilots to focus too much attention on establishing the exact power setting.

Control and power instruments are the foundation for precise attitude instrument flying. The keys to attitude instrument flying are establishing the desired aircraft attitude on the attitude indicator and selecting the desired engine output on the power instruments. Cross-checking is the vital ingredient in maintaining precise attitude instrument flight.

Attitude Instrument Flying—Primary and Supporting Method

The second method for performing attitude instrument flight is a direct extension of the control/power method. By utilizing the primary and supporting flight instruments in conjunction with the control and power instruments, the pilot can precisely maintain aircraft attitude. This method utilizes the same instruments as the control/power method; however, it focuses more on the instruments that depict the most accurate indication for the aspect of the aircraft attitude being controlled. The four key elements (pitch, bank, roll, and trim) are discussed in detail.

Similar to the control/power method, all changes to aircraft attitude need to be made using the attitude indicator and the power instruments (tachometer, manifold pressure gauge, etc.). The following explains how each component of the aircraft attitude is monitored for performance.

Pitch Control

The pitch of the aircraft refers to the angle between the longitudinal axis of the aircraft and the natural horizon. When flying in instrument meteorological conditions (IMC), the natural horizon is unavailable for reference and an artificial horizon is utilized in its place. *[Figure 6-30]* The only instrument capable of depicting the aircraft attitude is the attitude indicator displayed on the PFD. The attitude and heading reference system (AHRS) is the engine that drives the attitude display. The AHRS unit is capable of precisely tracking minute changes in the pitch, bank, and yaw axes, thereby making the PFD very accurate and reliable. The AHRS unit determines the angle between the aircraft's longitudinal axis and the horizon line on initialization. There is no need or means for the pilot to adjust the position of the yellow chevron, which represents the nose of the aircraft.

Straight-and-Level Flight

In straight-and-level flight, the pilot maintains a constant altitude, airspeed and, for the most part, heading for extended periods of time. To achieve this, the three primary instruments that need to be referenced in order to maintain these three variables are the altitude, airspeed, and heading indicators.

Primary Pitch

When the pilot is maintaining a constant altitude, the primary instrument for pitch is the altimeter. As long as the aircraft maintains a constant airspeed and pitch attitude, the altitude should remain constant.

Two factors that cause the altitude to deviate are turbulence and momentary distractions. When a deviation occurs, a change in the pitch needs to be made on the attitude indicator. Small deviations require small corrections, while large deviations require larger corrections. Pilots should avoid making large corrections that result in rapid attitude changes, for this may lead to spatial disorientation. Smooth, timely corrections should be made to bring the aircraft back to the desired attitude.

Pay close attention to indications on the PFD. An increase in pitch of 2.5° produces a climb rate of 450 feet per minute (fpm). Small deviations do not require large attitude changes. A rule of thumb for correcting altitude deviations is to establish a change rate of twice the altitude deviation, not to exceed 500 fpm. For example, if the aircraft is off altitude by 40 feet, $2 \times 40 = 80$ feet, so a descent of approximately 100 fpm allows the aircraft to return to the desired altitude in a controlled, timely fashion.

In addition to the primary instrument, there are also supporting instruments that assist the pilot in cross-checking the pitch attitude. The supporting instruments indicate trend, but they do not indicate precise attitude indications. Three instruments (vertical speed, airspeed, and altitude trend tape) indicate when the pitch attitude has changed and that the altitude is changing. *[Figure 6-31]* When the altitude is constant, the VSI and altitude trend tape are not shown on the PFD. When these two trend indicators are displayed, the

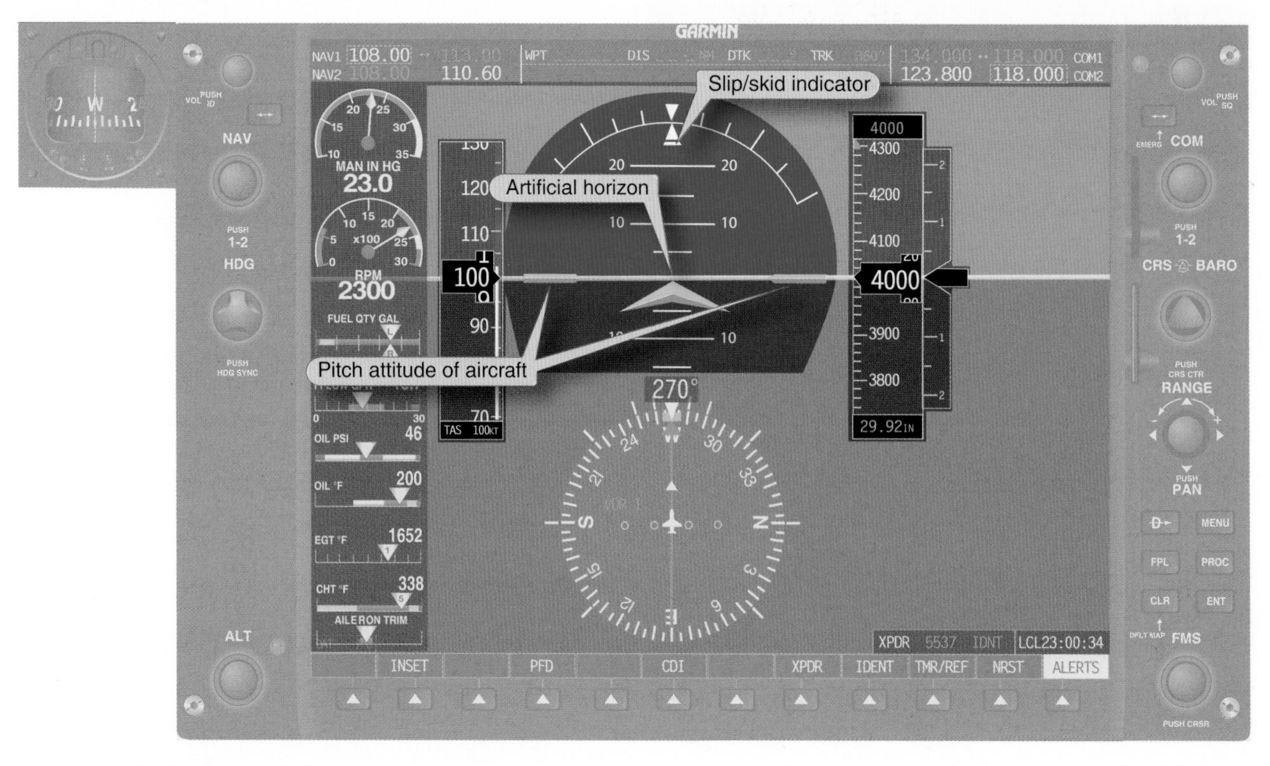

Figure 6-30. *Pitch of the aircraft.*

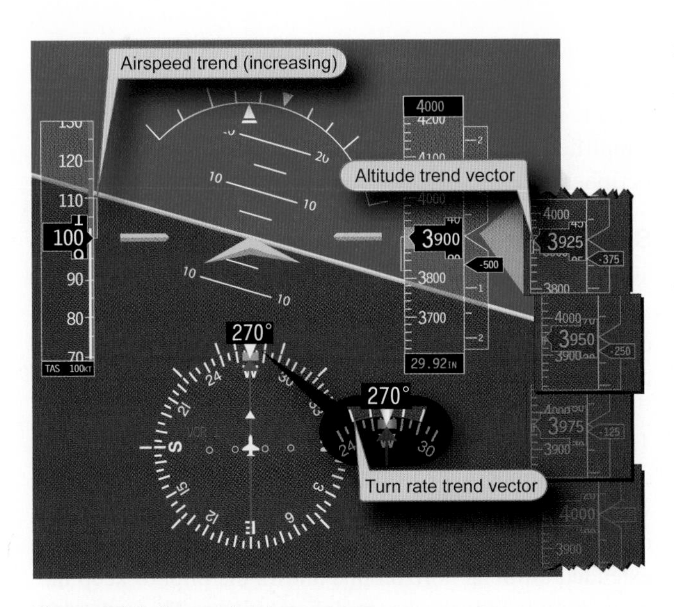

Figure 6-31. *Supporting instruments.*

pilot is made aware that the pitch attitude of the aircraft has changed and may need adjustment. Notice in *Figure 6-31* that the aircraft is descending at a rate of 500 fpm.

The instrument cross-check necessitates utilizing these supporting instruments to better manage altitude control. The VSI and trend tape provide the pilot with information regarding the direction and rate of altitude deviations. The pilot is thus able to make corrections to the pitch attitude before a large deviation in altitude occurs. The airspeed indicator depicts an increase if the pitch attitude is lowered. Conversely, when the pitch attitude increases, the pilot should note a decrease in the airspeed.

Primary Bank

When flying in IMC, pilots maintain preplanned or assigned headings. With this in mind, the primary instrument for bank angle is the heading indicator. Heading changes are displayed instantaneously. The heading indicator is the only instrument that displays the current magnetic heading, provided that it is matched to the magnetic compass with all deviation adjustments accounted for. *[Figure 6-32]*

There are supporting instruments associated with bank as well. The turn rate trend indicator shows the pilot when the aircraft is changing heading. The magnetic compass is also useful for maintaining a heading; however, it is influenced by several errors in various phases of flight.

Primary Yaw

The slip/skid indicator is the primary instrument for yaw. It is the only instrument that can indicate if the aircraft is

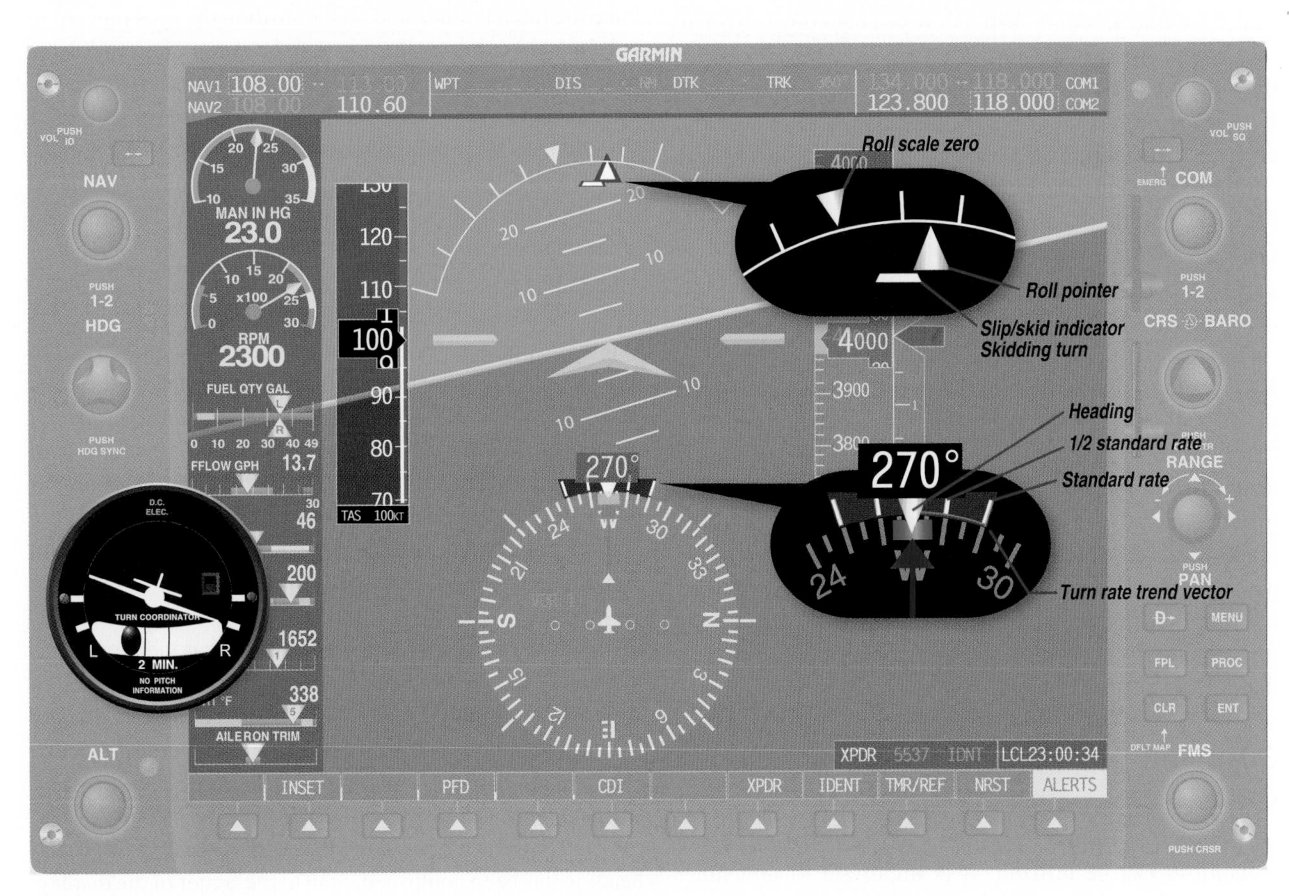

Figure 6-32. *Primary bank.*

moving through the air with the longitudinal axis of the aircraft aligned with the relative wind.

Primary Power

The primary power instrument for straight-and-level flight is the airspeed indicator. The main focus of power is to maintain a desired airspeed during level flight. No other instrument delivers instantaneous indication.

Learning the primary and supporting instruments for each variable is the key to successfully mastering attitude instrument flying. At no point does the primary and supporting method devalue the importance of the attitude indicator or the power instruments. All instruments (control, performance, primary, and supporting) must be utilized collectively.

Fundamental Skills of Attitude Instrument Flying

When first learning attitude instrument flying, it is very important that two major skills be mastered. Instrument cross-check and instrument interpretation comprise the foundation for safely maneuvering the aircraft by reference to instruments alone. Without mastering both skills, the pilot is not able to maintain precise control of aircraft attitude.

Instrument Cross-Check

The first fundamental skill is cross-checking (also call "scanning"). Cross-checking is the continuous observation of the indications on the control and performance instruments. It is imperative that the new instrument pilot learn to observe and interpret the various indications in order to control the attitude and performance of the aircraft. Due to the configuration of some glass panel displays, such as the Garmin G1000, one or more of the performance instruments may be located on an MFD installed to the right of the pilot's direct forward line of sight.

How a pilot gathers the necessary information to control the aircraft varies by individual pilot. No specific method of cross-checking (scanning) is recommended; the pilot must learn to determine which instruments give the most pertinent information for any particular phase of a maneuver. With practice, the pilot is able to observe the primary instruments quickly and cross-check with the supporting instruments in order to maintain the desired attitude. At no time during instrument flying should the pilot stop cross-checking the instrumentation.

Scanning Techniques

Since most of the primary and supporting aircraft attitude information is displayed on the PFD, standard scanning techniques can be utilized. It is important to remember to include the stand-by flight instruments as well as the engine indications in the scan. Due to the size of the attitude instrument display, scanning techniques have been simplified because the attitude indicator is never out of peripheral view.

Selected Radial Cross-Check

The radial scan is designed so that your eyes remain on the attitude indicator 80–90 percent of the time. The remainder of the time is spent transitioning from the attitude indicator to the various other flight instruments. *[Figure 6-33]*

The radial scan pattern works well for scanning the PFD. The close proximity of the instrument tape displays necessitates very little eye movement in order to focus in on the desired instrument. While the eyes move in any direction, the extended artificial horizon line allows the pilot to keep the pitch attitude in his or her peripheral vision. This extended horizon line greatly reduces the tendency to fixate on one instrument and completely ignore all others. Because of the size of the attitude display, some portion of the attitude indicator is always visible while viewing another instrument display on the PFD.

Starting the Scan

Start the scan in the center of the PFD on the yellow chevron. Note the pitch attitude and then transition the eyes upward to the slip/skid indicator. Ensure that the aircraft is coordinated by aligning the split triangle symbol. The top of the split triangle is referred to as the roll pointer. The lower portion of the split triangle is the slip/skid indicator. If the lower portion of the triangle is off to one side, step on the rudder pedal on the same side to offset it. *[Figure 6-34 NOTE: The aircraft is not changing heading. There is no trend vector on the turn rate indicator.]*

While scanning that region, check the roll pointer and assure that the desired degree of roll is being indicated on the bank scale. The roll index and the bank scale remain stationary at the top of the attitude indicator. The index is marked with angles of 10°, 20°, 30°, 45°, and 60° in both directions. If the desired bank angle is not indicated, make the appropriate aileron corrections. Verify the bank angle is correct and continue scanning back to the yellow chevron.

Scan left to the airspeed tape and verify that the airspeed is as desired, then return back to the center of the display. Scan right to the altimeter tape. Verify that the desired altitude is being maintained. If it is not, make the appropriate pitch change and verify the result. Once the desired altitude has been verified, return to the center of the display. Transition down to the heading indicator to verify the desired heading. When the heading has been confirmed, scan to the center of the display.

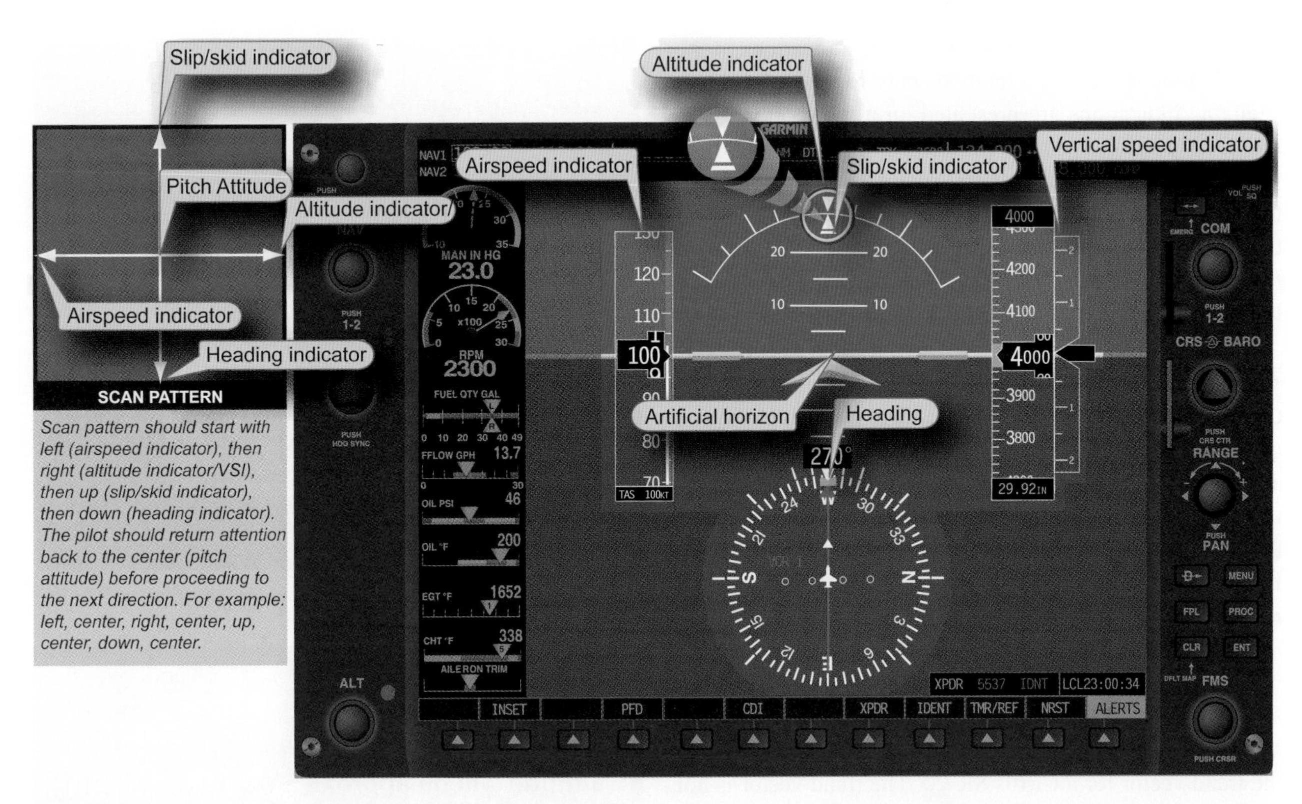

Figure 6-33. *Selected radial cross-check.*

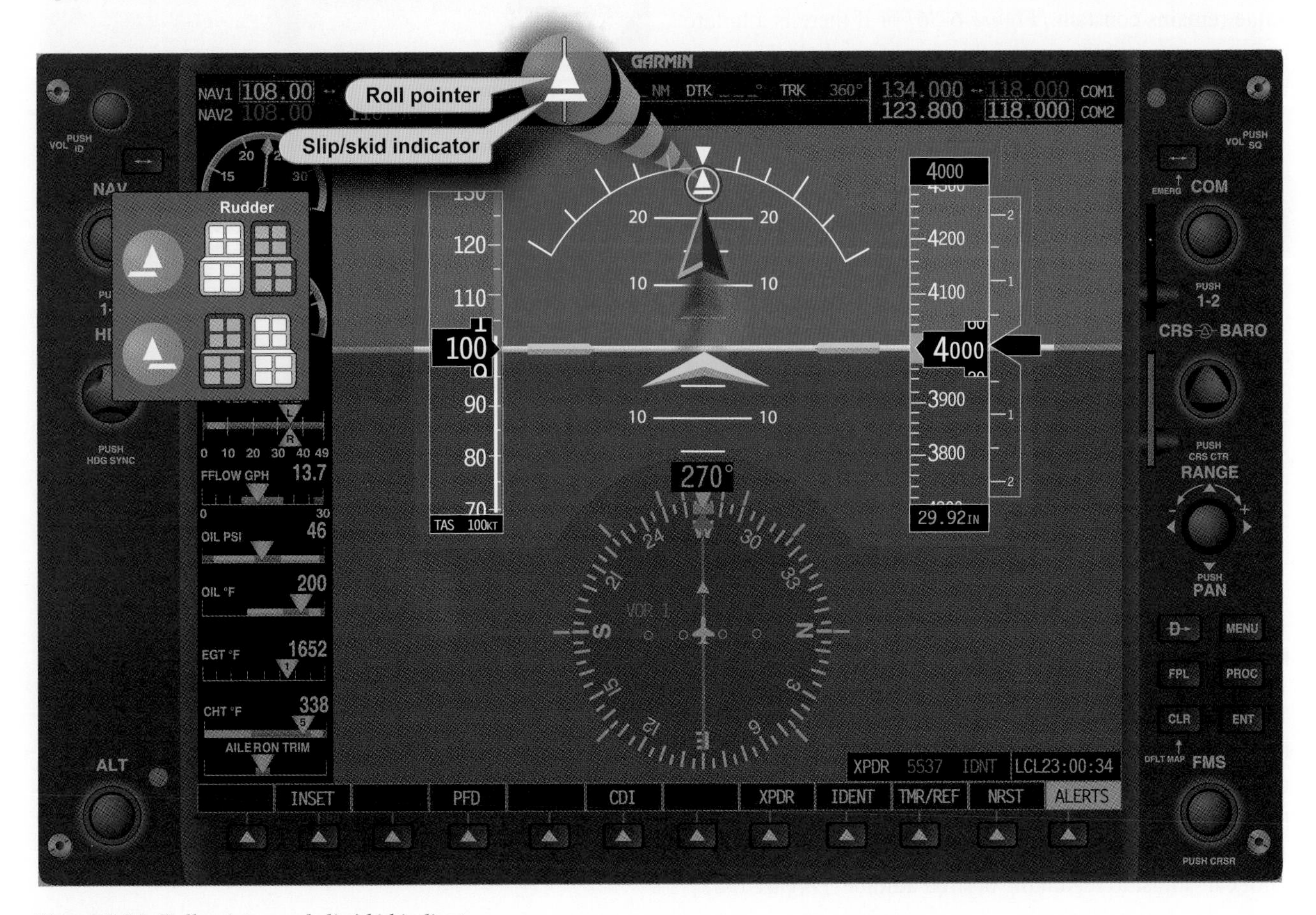

Figure 6-34. *Roll pointer and slip/skid indicator.*

It is also important to include the engine indications in the scan. Individualized scan methods may require adjustment if engine indications are presented on a separate MFD. A modified radial scan can be performed to incorporate these instruments into the scan pattern. Another critical component to include in the scan is the moving map display located on the MFD. To aid in situational awareness and facilitate a more centralized scan, a smaller inset map can be displayed in the lower left corner of the PFD screen.

Trend Indicators

One improvement the glass panel displays brought to the general aviation industry is the trend vector. Trend vectors are colored lines that appear on the airspeed and altitude tapes, as well as on the turn rate indicator. The color of the line may vary depending on the airplane manufacturer. For example, on a Cirrus SR-20, the trend vector lines are magenta and on the B-737 they are green. These colored lines indicate what the associated airspeed, altitude, or heading will be in 6 seconds for the Cirrus SR-20 and 10 seconds for the B-737 if the current rate is maintained. The example shown in *Figure 6-35* uses the color and data that represents the trend vector for a Cirrus SR-20. The trend vector is not displayed if there is no change to the associated tape and the value remains constant *[Figure 6-36]* or if there is a failure in some portion of the system that would preclude the vector from being determined.

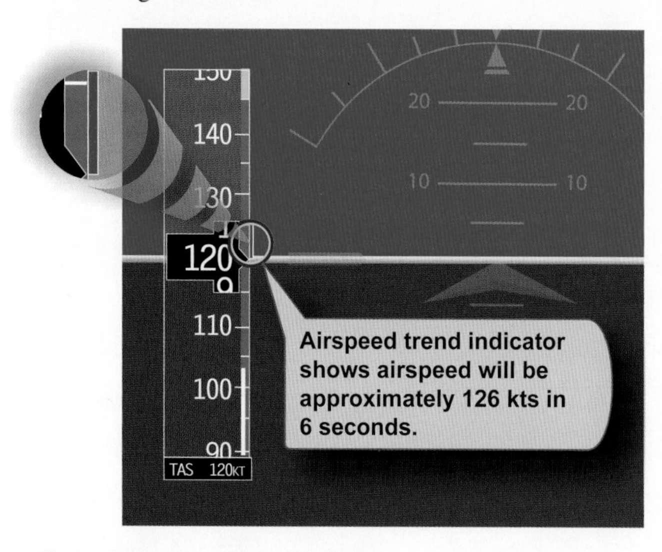

Figure 6-35. *Airspeed trend indicators.*

Trend vectors are a very good source of information for the new instrument rated pilot(s). Pilots who utilize good scanning techniques can pick up subtle deviations from desired parameters and make small correction to the desired attitude. As soon as a trend is indicated on the PFD, a conscientious pilot can adjust to regain the desired attitude. *[Figure 6-37]*

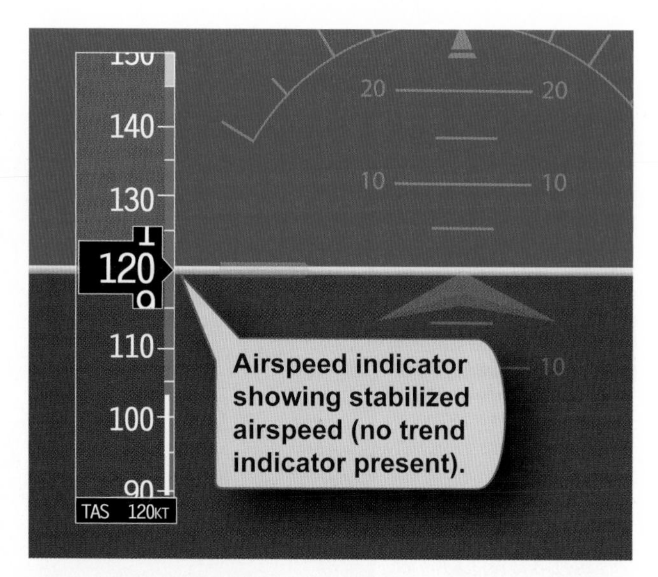

Figure 6-36. *Airspeed indicators with no trend present.*

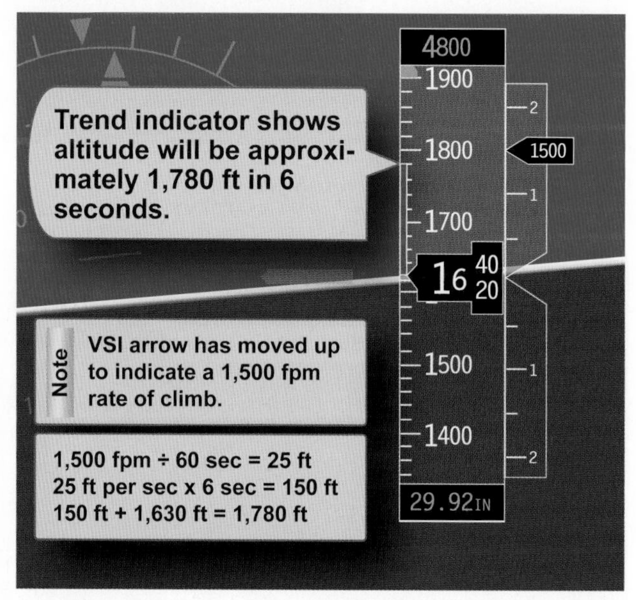

Figure 6-37. *Altimeter trend indicators.*

Another advancement in attitude instrument flying is the turn rate trend indicator. As in the cases of airspeed, altitude, and vertical speed trend indicators, the turn rate trend indicator depicts what the aircraft's heading will be in 6 seconds. While examining the top of the heading indicator, notice two white lines on the exterior of the compass rose. *[Figure 6-38]* These two tick marks located on both sides of the top of the heading indicator show half-standard rate turns as well as standard rate turns.

In *Figure 6-39*, when the aircraft begins its turn to the left, the magenta trend indicator elongates proportionally with the rate of turn. To initiate a half-standard rate turn, position the

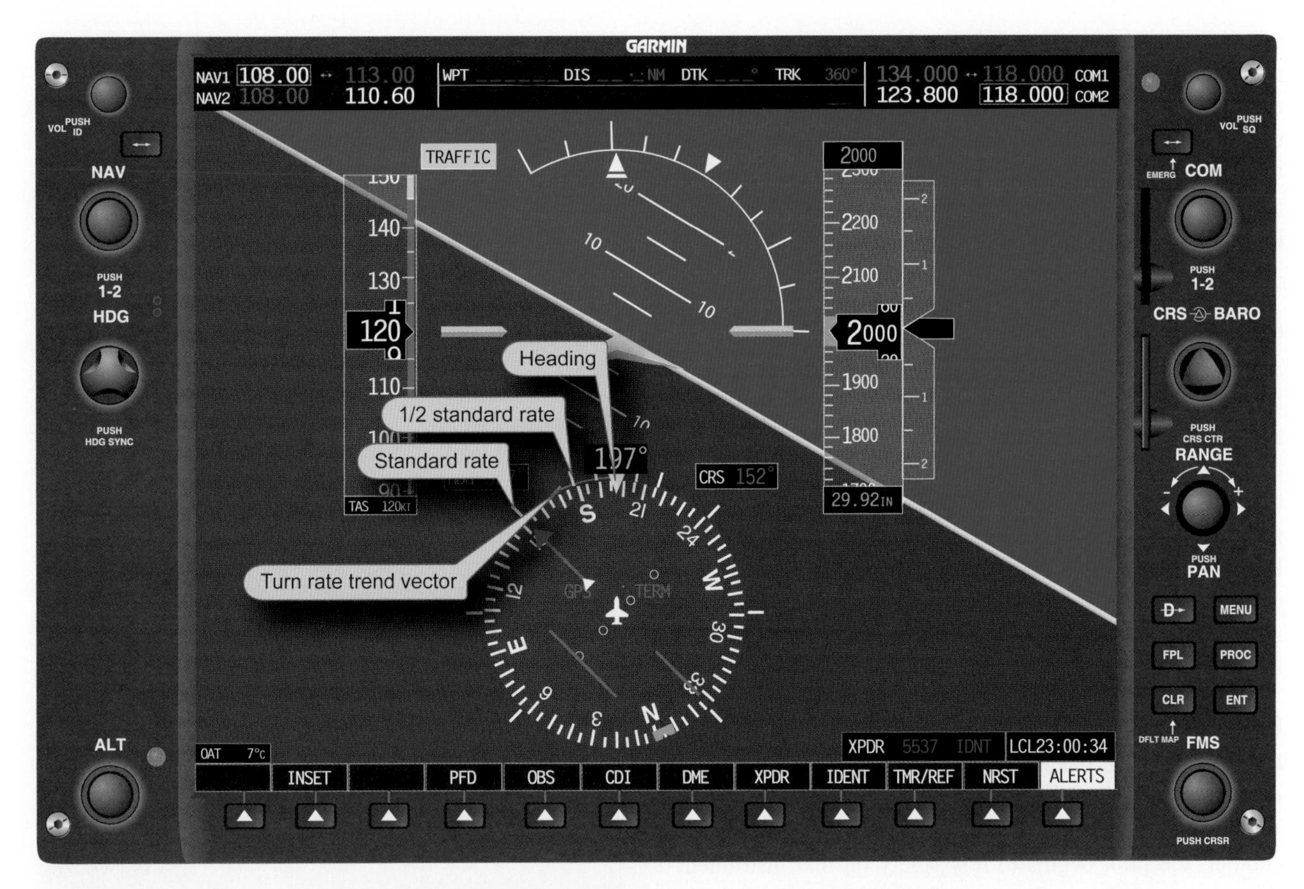

Figure 6-38. *Horizontal situation indicator (HSI) trend indicator elongates proportionally with the rate of turn.*

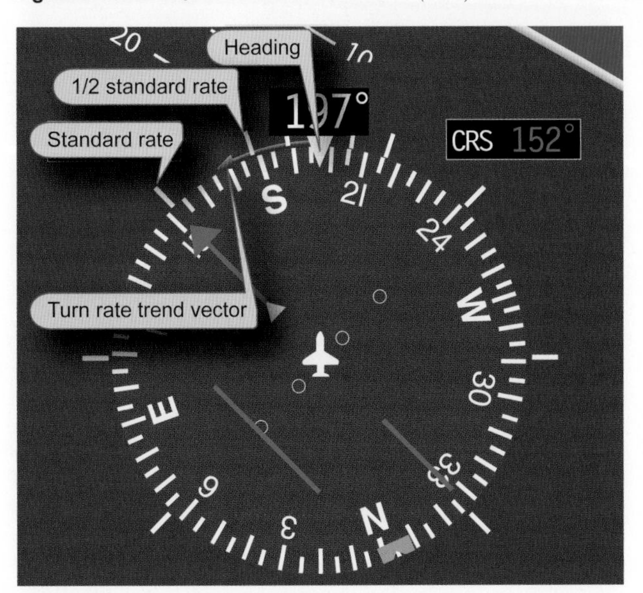

Figure 6-39. *HSI indicator (enlargement).*

indicator on the first tick mark. A standard rate turn would be indicated by the trend indicator extending to the second tick mark. A turn rate in excess of standard rate would be indicated by the trend indicator extending past the second tick mark. This trend indicator shows what the aircraft's heading will be in 6 seconds, but is limited to indicate no more than 24° in front of the aircraft or 4° per second. When the aircraft exceeds a turning rate of 25° in 6 seconds, the trend indicator has an arrowhead attached to it.

Trend indicators are very useful when leveling off at a specific altitude, when rolling out on a heading, or when stabilizing airspeed. One method of determining when to start to level off from a climb or descent is to start leveling at 10 percent of the vertical speed rate prior to the desired altitude.

As the aircraft approaches the desired altitude, adjust the pitch attitude to keep the trend indicator aligned with the target altitude. As the target approaches, the trend indicator gradually shrinks until altitude stabilizes. Trend indicators should be used as a supplement, not as a primary means of determining pitch change.

Common Errors

Fixation

Fixation, or staring at one instrument, is a common error observed in pilots first learning to utilize trend indicators. The pilot may initially fixate on the trend indicator and make adjustments with reference to that alone. Trend indicators are not the only tools to aid the pilot in maintaining the desired power or attitude; they should be used in conjunction with the primary and supporting instruments in order to better manage the flight. With the introduction of airspeed tapes, the pilot can monitor airspeed to within one knot. Fixation can lead to attempting to keep the airspeed to an unnecessarily tight tolerance. There is no need to hold airspeed to within one knot; the Instrument Rating Practical Test Standards (PTS) allows greater latitude.

Omission

Another common error associated with attitude instrument flying is omission of an instrument from the cross-check. Due to the high reliability of the PFD and associated components, pilots tend to omit the stand-by instruments as well as the magnetic compass from their scans. An additional reason for the omission is the position of the stand-by instruments. Pilots should continue to monitor the stand-by instruments in order to detect failures within those systems. One of the most commonly omitted instruments from the scan is the slip/skid indicator.

Emphasis

In initial training, placing emphasis on a single instrument is very common and can become a habit if not corrected. When the importance of a single instrument is elevated above another, the pilot begins to rely solely on that instrument for guidance. When rolling out of a 180° turn, the attitude indicator, heading indicator, slip/skid indicator, and altimeter need to be referenced. If a pilot omits the slip/skid indicator, coordination is sacrificed.

Chapter 7, Section I

Airplane Basic Flight Maneuvers

Using Analog Instrumentation

Introduction

Instrument flying techniques differ according to aircraft type, class, performance capability, and instrumentation. Therefore, the procedures and techniques that follow need to be modified to suit individual aircraft. Recommended procedures, performance data, operating limitations, and flight characteristics of a particular aircraft are available in the Pilot's Operating Handbook/Airplane Flight Manual (POH/AFM) for study before practicing the flight maneuvers.

The flight maneuvers discussed in Chapter 7-I assume the use of a single-engine, propeller-driven small airplane with retractable gear and flaps and a panel with instruments representative of those discussed earlier in Chapter 5, Flight Instruments. With the exception of the instrument takeoff, all of the maneuvers can be performed on "partial panel," with the attitude gyro and heading indicator covered or inoperative.

Straight-and-Level Flight

Pitch Control

The pitch attitude of an airplane is the angle between the longitudinal axis of the airplane and the actual horizon. In level flight, the pitch attitude varies with airspeed and load. For training purposes, the latter factor can normally be disregarded in small airplanes. At a constant airspeed, there is only one specific pitch attitude for level flight. At slow cruise speeds, the level flight attitude is nose high with indications as in *Figure 7-1*; at fast cruise speeds, the level-flight attitude is nose low. *[Figure 7-2] Figure 7-3* shows the indications for the attitude at normal cruise speeds. The instruments used to determine the pitch attitude of the aircraft are the attitude indicator, the altimeter, the vertical speed indicator (VSI), and the airspeed indicator (ASI).

Attitude Indicator

The attitude indicator gives the direct indication of pitch attitude. The desired pitch attitude is gained by using the elevator control to raise or lower the miniature aircraft in relation to the horizon bar. This corresponds to the way pitch attitude is adjusted in visual flight by raising or lowering the nose of the airplane in relation to the natural horizon. However, unless the airspeed is constant, and until the level flight attitude for that airspeed has been identified and established, there is no way to know whether level flight as

Figure 7-2. *Pitch attitude and airspeed in level flight, fast cruise speed.*

Figure 7-1. *Pitch attitude and airspeed in level flight, slow cruise speed.*

Figure 7-3. *Pitch attitude and airspeed in level flight, normal cruise speed.*

indicated on the attitude indicator is resulting in level flight as shown on the altimeter, VSI, and ASI. If the miniature aircraft of the attitude indicator is properly adjusted on the ground before takeoff, it shows approximately level flight at normal cruise speed when the pilot completes the level off from a climb. If further adjustment of the miniature aircraft is necessary, the other pitch instruments must be used to maintain level flight while the adjustment is made.

To practice pitch control for level flight using only the attitude indicator, use the following exercise. Restrict the displacement of the horizon bar to a one-half bar width, a bar width up or down, then a one-and-one-half bar width. One-half, one, and one-and-one-half bar width nose-high attitudes are shown in *Figures 7-4, 7-5,* and *7-6.*

Figure 7-4. *Pitch correction for level flight, one-half bar width.*

Figure 7-5. *Pitch correction for level flight, one bar width.*

An instructor pilot can demonstrate these normal pitch corrections and compare the indications on the attitude indicator with the airplane's position to the natural horizon.

Figure 7-6. *Pitch correction for level flight, one-and-one-half bar width.*

Pitch attitude changes for corrections to level flight by reference to instruments are much smaller than those commonly used for visual flight. With the airplane correctly trimmed for level flight, the elevator displacement and the control pressures necessary to effect these standard pitch changes are usually very slight. The following are a few helpful hints to help determine how much elevator control pressure is required.

First, a tight grip on the controls makes it difficult to feel control pressure changes. Relaxing and learning to control the aircraft usually takes considerable conscious effort during the early stages of instrument training.

Second, make smooth and small pitch changes with positive pressure. With practice, a pilot can make these small pitch corrections up or down, "freezing" (holding constant) the one-half, full, and one-and-one-half bar widths on the attitude indicator.

Third, with the airplane properly trimmed for level flight, momentarily release all pressure on the elevator control when becoming aware of tenseness. This is a reminder that the airplane is stable; except under turbulent conditions, it maintains level flight if left alone. Even when no control change is called for, it is difficult to resist the impulse to move the controls. This may be one of the most difficult initial training problems in instrument flight.

Altimeter

At constant power, any deviation from level flight (except in turbulent air) is the result of a pitch change. Therefore, the altimeter gives an indirect indication of the pitch attitude in level flight, assuming constant power. Since the altitude should remain constant when the airplane is in level flight, any deviation from the desired altitude signals the need for a pitch change. If the aircraft is gaining altitude, the nose must be lowered. *[Figures 7-7* and *7-8]*

Figure 7-7. *Using the altimeter for pitch interpretation, a high altitude means a nose-high pitch attitude.*

Figure 7-8. *Pitch correction following altitude increase—lower nose to correct altitude error.*

The rate of movement of the altimeter needle is as important as its direction of movement in maintaining level flight without the use of the attitude indicator. An excessive pitch deviation from level flight results in a relatively rapid change of altitude; a slight pitch deviation causes a slow change. Thus, if the altimeter needle moves rapidly clockwise, assume a considerable nose-high deviation from level flight attitude. Conversely, if the needle moves slowly counterclockwise to indicate a slightly nose-low attitude, assume that the pitch correction necessary to regain the desired altitude is small. As the altimeter is added to the attitude indicator in a cross-check, a pilot learns to recognize the rate of movement of the altimeter needle for a given pitch change as shown on the attitude indicator.

To practice precision control of pitch in an airplane without an attitude indicator, make small pitch changes by visual reference to the natural horizon and note the rate of movement of the altimeter. Note what amount of pitch change gives the slowest steady rate of change on the altimeter. Then practice small pitch corrections by accurately interpreting and controlling the rate of needle movement.

An instructor pilot can demonstrate an excessive nose-down deviation (indicated by rapid movement of the altimeter needle) and then, as an example, show the result of improper corrective technique. The normal impulse is to make a large pitch correction in a hurry, but this inevitably leads to overcontrolling. The needle slows down, then reverses direction, and finally indicates an excessive nose-high deviation. The result is tension on the controls, erratic control response, and increasingly extreme control movements. The correct technique, which is slower and smoother, returns the airplane to the desired attitude more quickly, with positive control and no confusion.

When a pitch error is detected, corrective action should be taken promptly, but with light control pressures and two distinct changes of attitude: (1) a change of attitude to stop the needle movement and (2) a change of attitude to return to the desired altitude.

When the altimeter indicates an altitude deviation, apply just enough elevator pressure to decrease the rate of needle movement. If it slows down abruptly, ease off some of the pressure until the needle continues to move, but ease off slowly. Slow needle movement means the airplane attitude is close to level flight. Add slightly more corrective pressure to stop the direction of needle movement. At this point, level flight is achieved; a reversal of needle movement means the aircraft has passed through it. Relax control pressures carefully, continuing to cross-check since changing airspeed causes changes in the effectiveness of a given control pressure. Next, adjust the pitch attitude with elevator pressure for the rate of change of altimeter needle movement that is correlated with normal pitch corrections and return to the desired altitude.

As a rule of thumb, for errors of less than 100 feet, use a half bar width correction. *[Figures 7-9* and *7-10]* For errors in excess of 100 feet, use an initial full bar width correction. *[Figures 7-11* and *7-12]* Practice predetermined altitude changes using the altimeter alone, then in combination with the attitude indicator.

Vertical Speed Indicator (VSI)

The VSI, like the altimeter, gives an indirect indication of pitch attitude and is both a trend and a rate instrument. As a trend instrument, it shows immediately the initial vertical movement of the airplane, which disregarding turbulence can be considered a reflection of pitch change. To maintain level flight, use the VSI in conjunction with the altimeter and attitude indicator. Note any positive or negative trend of the needle from zero and apply a very light corrective elevator pressure. As the needle returns to zero, relax the corrective pressure. If control pressures have been smooth and light, the needle reacts immediately and slowly, and the altimeter shows little or no change of altitude. As a rate instrument, the VSI requires consideration of lag characteristics.

Lag refers to the delay involved before the needle attains a stable indication following a pitch change. Lag is directly proportional to the speed and magnitude of a pitch change. If a slow, smooth pitch change is initiated, the needle moves with minimum lag to a point of deflection corresponding to the extent of the pitch change, and then stabilizes as the aerodynamic forces are balanced in the climb or descent. A large and abrupt pitch change produces erratic needle movement, a reverse indication, and introduces greater time delay (lag) before the needle stabilizes. Pilots are cautioned not to chase the needle when flight through turbulent conditions produces erratic needle movements. The apparent lag in airspeed indications with pitch changes varies greatly among different airplanes and is due to the time required for the airplane to accelerate or decelerate when the pitch attitude is changed. There is no appreciable lag due to the construction or operation of the instrument. Small pitch changes, smoothly executed, result in an immediate change of airspeed.

When using the VSI as a rate instrument and combining it with the altimeter and attitude indicator to maintain level flight, a pilot should know that the amount the altimeter needle moves from the desired altitude governs the rate that should be used to return to that altitude. A rule of thumb is to make an attitude change that results in a vertical-speed rate approximately double the error in altitude. For example, if altitude is off by 100 feet, the rate of return to the desired altitude should be approximately 200 feet per minute (fpm). If it is off by more than 100 feet, the correction should be correspondingly greater, but should never exceed the optimum rate of climb or descent for the airplane at a given airspeed and configuration.

A deviation of more than 200 fpm from the desired rate of return is considered overcontrolling. For example, if attempting to change altitude by 200 feet, a rate in excess of 400 fpm indicates overcontrolling.

When returning to an altitude, the VSI is the primary pitch instrument. Occasionally, the VSI is slightly out of calibration and may indicate a climb or descent when the airplane is in level flight. If the instrument cannot be adjusted, take the error into consideration when using it for pitch control. For

Figure 7-9. *Altitude error, less than 100 feet.*

Figure 7-10. *Pitch correction, less than 100 feet—one-half bar low to correct altitude error.*

Figure 7-11. *Altitude error, greater than 100 feet.*

Figure 7-12. *Pitch correction, greater than 100 feet—one bar correction initially.*

example, if the needle indicates a descent of 200 fpm while in level flight, use this indication as the zero position.

Airspeed Indicator (ASI)

The ASI presents an indirect indication of the pitch attitude. In non-turbulent conditions with a constant power setting and pitch attitude, airspeed remains constant. *[Figure 7-13]* As the pitch attitude lowers, airspeed increases, and the nose should be raised. *[Figure 7-14]* As the pitch attitude rises, airspeed decreases, and the nose should be lowered. *[Figure 7-15]* A rapid change in airspeed indicates a large pitch change, and a slow change of airspeed indicates a small pitch change.

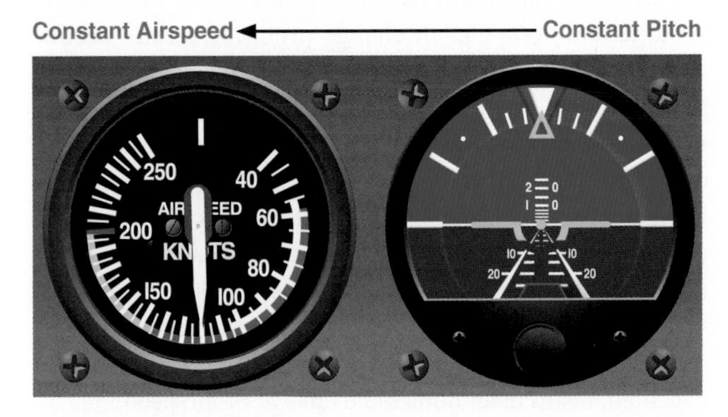

Figure 7-13. *Constant power plus constant pitch equals constant speed.*

Increased Airspeed ← Decreased Pitch

Figure 7-14. *Constant power plus decreased pitch equals increased airspeed.*

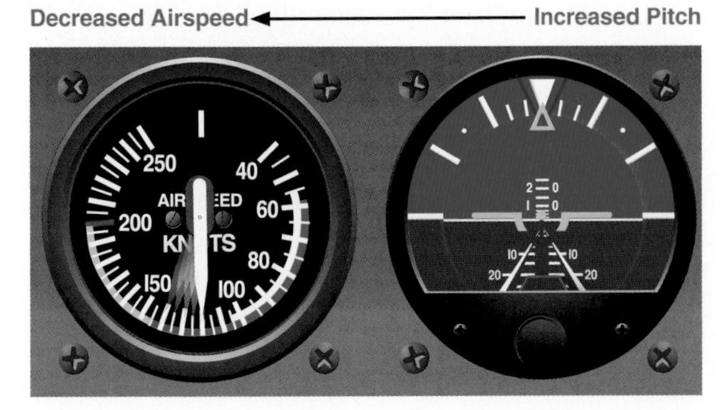

Figure 7-15. *Constant power plus increased pitch equals decreased airspeed.*

Pitch control in level flight is a question of cross-check and interpretation of the instrument panel for the instrument information that enables a pilot to visualize and control pitch attitude. Regardless of individual differences in cross-check technique, all pilots should use the instruments that give the best information for controlling the airplane in any given maneuver. Pilots should also check the other instruments to aid in maintaining the primary instruments at the desired indication.

As noted previously, the primary instrument is the one that gives the most pertinent information for a particular maneuver. It is usually the one that should be held at a constant indication. Which instrument is primary for pitch control in level flight, for example? This question should be considered in the context of specific airplane, weather conditions, pilot experience, operational conditions, and other factors. Attitude changes must be detected and interpreted instantly for immediate control action in high-performance airplanes. On the other hand, a reasonably proficient instrument pilot in a slower airplane may rely more on the altimeter for primary pitch information, especially if it is determined that too much reliance on the attitude indicator fails to provide the necessary precise attitude information. Whether the pilot decides to regard the altimeter or the attitude indicator as primary depends on which approach will best help control the attitude. In this handbook, the altimeter is normally considered as the primary pitch instrument during level flight.

Bank Control

The bank attitude of an airplane is the angle between the airplane's wings and the natural horizon. To maintain a straight-and-level flightpath, the wings of the airplane are kept level with the horizon (assuming the airplane is in coordinated flight). The instruments used for bank control are the attitude indicator, the heading indicator, and the turn coordinator. *Figure 7-16* illustrates coordinated flight. The aircraft is banked left with the attitude indicator and turn coordinator indicating the bank. The heading indicator indicates a left turn by apparent clockwise rotation of the compass card behind the airplane silhouette.

Attitude Indicator

The attitude indicator shows any change in bank attitude directly and instantly and is, therefore, a direct indicator. On the standard attitude indicator, the angle of bank is shown pictorially by the relationship of the miniature aircraft to the artificial horizon bar and by the alignment of the pointer with the banking scale at the top of the instrument. On the face of the standard three-inch instrument, small angles of bank can be difficult to detect by reference to the miniature aircraft, especially if leaning to one side or changing a seating position

Figure 7-16. *Instruments used for bank control.*

slightly. The position of the scale pointer is a good check against the apparent miniature aircraft position. Disregarding precession error, small deviations from straight coordinated flight can be readily detected on the scale pointer. The banking index may be graduated as shown in *Figure 7-17*, or it may be graduated in 30° increments.

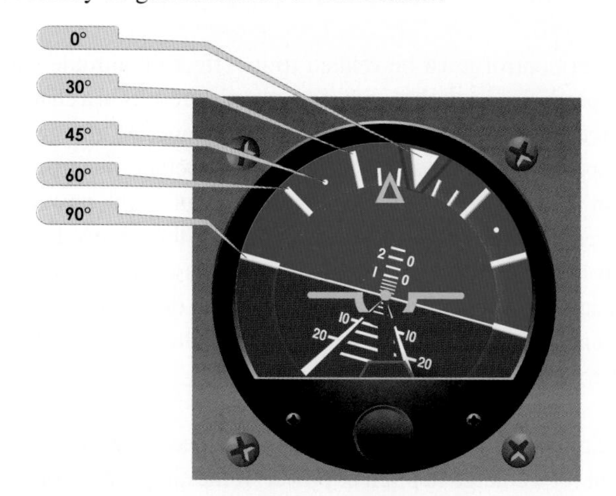

Figure 7-17. *Bank interpretation with the attitude indicator.*

The instrument depicted in *Figure 7-17* has a scale pointer that moves in the same direction of bank shown by the miniature aircraft. In this case, the aircraft is in a left 15° bank. Precession errors in this instrument are common and predictable, but the obvious advantage of the attitude indicator is an immediate indication of both pitch attitude and bank attitude in a single glance. Even with the precession errors associated with many attitude indicators, the quick attitude presentation requires less visual effort and time for positive control than other flight instruments.

Heading Indicator

The bank attitude of an aircraft in coordinated flight is shown indirectly on the heading indicator, since banking results in a turn and change in heading. Assuming the same airspeed in both instances, a rapid movement of the heading indicator (azimuth card in a directional gyro) indicates a large angle of bank, whereas slow movement reflects a small angle of bank. Note the rate of movement of the heading indicator and compare it to the attitude indicator's degrees of bank. The attitude indicator's precession error makes a precise check of heading information necessary in order to maintain straight flight.

When deviations from straight flight are noted on the heading indicator, correct to the desired heading using a bank angle no greater than the number of degrees to be turned. In any case, limit bank corrections to a bank angle no greater than that required for a standard rate turn. Use of larger bank angles requires a very high level of proficiency, and normally results in overcontrolling and erratic bank control.

Turn Coordinator

The miniature aircraft of the turn coordinator gives an indirect indication of the bank attitude of the airplane. When the miniature aircraft is level, the airplane is in straight flight. When the miniature airplane is aligned with one of the alignment marks and the aircraft is rolling to the left or right the indication represents the roll rate, with the alignment marks indicating a roll of 3 degrees per second in the direction of the miniature aircraft. This can be seen in level flight when a bank is introduced either to the left or the right. The turn coordinator's indicator will indicate the rolling motion although there is no turn being made. Conversely, a pedal input to the right or left causes the aircraft to turn momentarily about its vertical axis (with no rolling motion) with an indication of turn on the turn coordinator. After the turn becomes stabilized and the aircraft is no longer rolling, the turn coordinator displays the rate of turn with the alignment marks equaling a turn of 3 degrees per second. The turn coordinator is able to display both roll and turn parameters because its electrically-powered gyroscope is canted at an angle. As a result, the turn-and-slip indicator provides both roll and turn indications. Autopilots in general aviation today use this instrument in determining both roll and turn information. After the completion of a turn, return to straight flight is accomplished by coordinated aileron and rudder pressure to level the miniature aircraft. Include the miniature aircraft in the cross-check and correct for even the smallest deviations from the desired position. When this instrument is used to maintain straight flight, control pressures must be applied very lightly and smoothly.

The ball of the turn coordinator is actually a separate instrument, conveniently located under the miniature aircraft because the two instruments are used together. The ball instrument indicates the quality of the turn. If the ball is off-center, the airplane is slipping or skidding. That is, if the coordinator's miniature airplane is tilted left and the ball is displaced to the right, the aircraft is in a skid. *[Figure 7-18]* If however, the miniature airplane is tilted to the right with the ball off-center to the right, the aircraft is in a slip. *[Figure 7-19]* If the wings are level and the airplane is properly trimmed, the ball remains in the center, and the airplane is in straight flight. If the ball is not centered, the airplane is improperly trimmed.

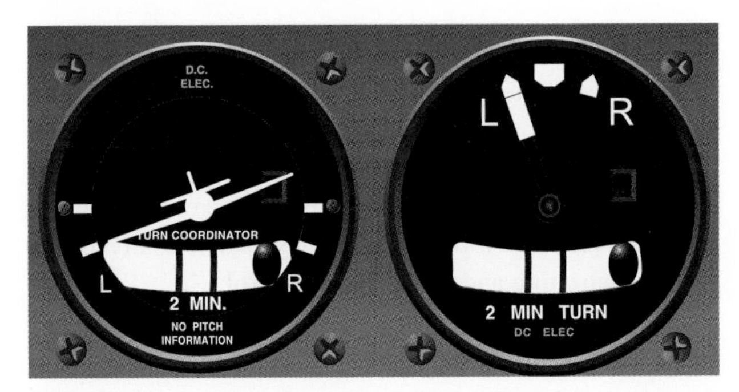

Figure 7-18. *Skid indication.*

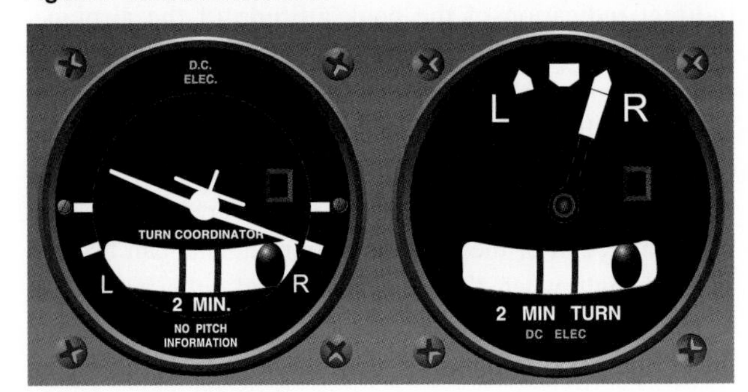

Figure 7-19. *Slip indication.*

To maintain straight-and-level flight with proper trim, note the direction of ball displacement. If the ball is to the left of center and the left wing is low, apply left rudder pressure to center the ball and correct the slip. At the same time, apply right aileron pressure as necessary to level the wings, cross-checking the heading indicator and attitude indicator while centering the ball. If the wings are level and the ball is displaced from the center, the airplane is skidding. Note the direction of ball displacement and use the same corrective technique as for an indicated slip. Center the ball (left ball/left rudder, right ball/right rudder), use aileron as necessary for bank control and retrim.

To trim the airplane using only the turn coordinator, use aileron pressure to level the miniature aircraft and rudder pressure to center the ball. Hold these indications with control pressures, gradually releasing them while applying rudder trim sufficient to relieve all rudder pressure. Apply aileron trim, if available, to relieve aileron pressure. With a full instrument panel, maintain a wings-level attitude by reference to all available instruments while trimming the airplane.

Turn-and-Slip Indicator (Needle and Ball)

Unlike the turn coordinator that provides three indications (roll, turn, and trim), the turn-and-slip indicator provides two: turn-rate and trim. Although the turn-and-slip indicator needle provides an indication of turn only, it provides an indirect indication of aircraft attitude when used with roll indicators, such as a heading indicator or magnetic compass. As with the turn coordinator (after stabilizing from a roll), when the turn-and-slip indicator's needle is aligned with the alignment marks, the aircraft is in a standard turn of 3 degrees per second or 360° in 2 minutes.

The ball of the turn-and-bank indicator provides important trim in the same manner that the ball in the turn coordinator does. *Figures 7-18* and *7-19* provide a comparison of the two instruments.

Power Control

Power produces thrust which, with the appropriate angle of attack of the wing, overcomes the forces of gravity, drag, and inertia to determine airplane performance.

Power control must be related to its effect on altitude and airspeed, since any change in power setting results in a change in the airspeed or the altitude of the airplane. At any given airspeed, the power setting determines whether the airplane is in level flight, in a climb, or in a descent. If the power is increased in straight-and-level flight and the airspeed held constant, the airplane climbs. If power is decreased while the airspeed is held constant, the airplane descends. On the other hand, if altitude is held constant, the power applied determines the airspeed.

The relationship between altitude and airspeed determines the need for a change in pitch or power. If the airspeed is not the desired value, always check the altimeter before deciding that a power change is necessary. Think of altitude and airspeed as interchangeable; altitude can be traded for airspeed by lowering the nose or convert airspeed to altitude by raising the nose. If altitude is higher than desired and airspeed is low, or vice versa, a change in pitch alone may return the airplane to the desired altitude and airspeed. *[Figure 7-20]* If both airspeed and altitude are high or if both are low, then a change in both pitch and power is necessary in order to return to the desired airspeed and altitude. *[Figure 7-21]*

For changes in airspeed in straight-and-level flight, pitch, bank, and power must be coordinated in order to maintain constant altitude and heading. When power is changed to vary airspeed in straight-and-level flight, a single-engine, propeller-driven airplane tends to change attitude around all axes of movement. Therefore, to maintain constant altitude and heading, apply various control pressures in proportion to the change in power. When power is added to increase airspeed, the pitch instruments indicate a climb unless forward elevator control pressure is applied as the airspeed changes. With an increase in power, the airplane tends to yaw and roll to the left unless counteracting aileron and rudder pressures are applied. Keeping ahead of these changes requires increasing cross-check speed, which varies with the type of airplane and its torque characteristics, the extent of power, and speed change involved.

Power Settings

Power control and airspeed changes are much easier when approximate power settings necessary to maintain various airspeeds in straight-and-level flight are known in advance. However, to change airspeed by any appreciable amount, the common procedure is to underpower or overpower on initial power changes to accelerate the rate of airspeed change. (For small speed changes, or in airplanes that decelerate or accelerate rapidly, overpowering or underpowering is not necessary.)

Consider the example of an airplane that requires 23 inches of mercury ("Hg) of manifold pressure to maintain a normal cruising airspeed of 120 knots, and 18 "Hg of manifold pressure to maintain an airspeed of 100 knots. The reduction in airspeed from 120 knots to 100 knots while maintaining straight-and-level flight is discussed below and illustrated in *Figures 7-22, 7-23,* and *7-24.*

Instrument indications, prior to the power reduction, are shown in *Figure 7-22*. The basic attitude is established and maintained on the attitude indicator. The specific pitch, bank, and power control requirements are detected on these primary instruments:

Altimeter—Primary Pitch
Heading Indicator—Primary Bank
Airspeed Indicator—Primary Power

Supporting pitch-and-bank instruments are shown in *Figure 7-23*. Note that the supporting power instrument is the manifold pressure gauge (or tachometer if the propeller is fixed pitch). However, when a smooth power reduction to approximately 15 "Hg (underpower) is made, the manifold pressure gauge becomes the primary power instrument. *[Figure 7-23]* With practice, power setting can be changed with only a brief glance at the power instrument, by sensing

Figure 7-20. *Airspeed low and altitude high—lower pitch.*

Figure 7-21. *Airspeed and altitude high—lower pitch and reduce power.*

Figure 7-22. *Straight-and-level flight (normal cruising speed).*

Figure 7-23. *Straight-and-level flight (airspeed decreasing).*

Figure 7-24. *Straight-and-level flight (reduced airspeed stabilized).*

the movement of the throttle, the change in sound, and the changes in the feel of control pressures.

As thrust decreases, increase the speed of the cross-check and be ready to apply left rudder, back-elevator, and aileron control pressure the instant the pitch-and-bank instruments show a deviation from altitude and heading. As proficiency is obtained, a pilot learns to cross-check, interpret, and control the changes with no deviation of heading and altitude. Assuming smooth air and ideal control technique as airspeed decreases, a proportionate increase in airplane pitch attitude is required to maintain altitude. Similarly, effective torque control means counteracting yaw with rudder pressure.

As the power is reduced, the altimeter is primary for pitch, the heading indicator is primary for bank, and the manifold pressure gauge is momentarily primary for power (at 15 "Hg in this example). Control pressures should be trimmed off as the airplane decelerates. As the airspeed approaches the desired airspeed of 100 knots, the manifold pressure is adjusted to approximately 18 "Hg and becomes the supporting power instrument. The ASI again becomes primary for power. *[Figure 7-24]*

Airspeed Changes in Straight-and-Level Flight

Practice of airspeed changes in straight-and-level flight provides an excellent means of developing increased proficiency in all three basic instrument skills and brings out some common errors to be expected during training in straight-and-level flight. Having learned to control the airplane in a clean configuration (minimum drag conditions), increase proficiency in cross-check and control by practicing speed changes while extending or retracting the flaps and landing gear. While practicing, be sure to comply with the airspeed limitations specified in the POH/AFM for gear and flap operation.

Sudden and exaggerated attitude changes may be necessary in order to maintain straight-and-level flight as the landing gear is extended and the flaps are lowered in some airplanes. The nose tends to pitch down with gear extension, and when flaps are lowered, lift increases momentarily (at partial flap settings) followed by a marked increase in drag as the flaps near maximum extension.

Control technique varies according to the lift and drag characteristics of each airplane. Accordingly, knowledge of the power settings and trim changes associated with different combinations of airspeed, gear, and flap configurations reduces instrument cross-check and interpretation problems.

For example, assume that in straight-and-level flight instruments indicate 120 knots with power at 23 "Hg/2,300 revolutions per minute (rpm), gear and flaps up. After reduction in airspeed, with gear and flaps fully extended, straight-and-level flight at the same altitude requires 25 "Hg manifold pressure/2,500 rpm. Maximum gear extension speed is 115 knots; maximum flap extension speed is 105 knots. Airspeed reduction to 95 knots, gear and flaps down, can be made in the following manner:

1. Maintain rpm at 2,500, since a high power setting is used in full drag configuration.
2. Reduce manifold pressure to 10 "Hg. As the airspeed decreases, increase cross-check speed.
3. Make trim adjustments for an increased angle of attack and decrease in torque.
4. Lower the gear at 115 knots. The nose may tend to pitch down and the rate of deceleration increases. Increase pitch attitude to maintain constant altitude, and trim off some of the back-elevator pressures. If full flaps are lowered at 105 knots, cross-check, interpretation, and control must be very rapid. A simpler technique is to stabilize attitude with gear down before lowering the flaps.
5. Since 18 "Hg manifold pressure will hold level flight at 100 knots with the gear down, increase power smoothly to that setting until the ASI shows approximately 105 knots. The attitude indicator now shows approximately two-and-a-half bar width nose-high in straight-and-level flight.
6. Actuate the flap control and simultaneously increase power to the predetermined setting (25 "Hg) for the desired airspeed, and trim off the pressures necessary to hold constant altitude and heading. The attitude indicator now shows a bar width nose-low in straight-and-level flight at 95 knots.

Proficiency in straight-and-level flight is attained when a pilot can consistently maintain constant altitude and heading with smooth pitch, bank, power, and trim control during the pronounced changes in aircraft attitude.

Trim Technique

Proper trim technique is essential for smooth and precise aircraft control during all phases of flight. By relieving all control pressures, it is much easier to hold a given attitude constant and devote more attention to other flight deck duties.

An aircraft is trimmed by applying control pressures to establish a desired attitude, then adjusting the trim so the aircraft maintains that attitude when the flight controls are released. Trim the aircraft for coordinated flight by centering the ball of the turn-and-slip indicator, by using rudder trim in the direction the ball is displaced from the center. Differential power control on multiengine aircraft is an additional factor affecting coordinated flight. Use balanced power or thrust, when possible, to aid in maintaining coordinated flight.

Changes in attitude, power, or configuration requires a trim adjustment in most cases. Using trim alone to establish a change in aircraft attitude invariably leads to erratic aircraft control. Smooth and precise attitude changes are best attained by a combination of control pressures and trim adjustments. Therefore, when used correctly, trim adjustment is an aid to smooth aircraft control.

Common Errors in Straight-and-Level Flight

Pitch

Pitch errors usually result from the following faults:

1. Improper adjustment of the attitude indicator's miniature aircraft to the wings-level attitude. Following the initial level off from a climb, check the attitude indicator and make any necessary adjustment in the miniature aircraft for level flight indication at normal cruise airspeed.
2. Insufficient cross-check and interpretation of pitch instruments. For example, the airspeed indication is low. The pilot, believing a nose-high attitude exists, applies forward pressure without noting that a low power setting is the cause of the airspeed discrepancy. Increase cross-check speed to include all relevant instrument indications before making a control input.
3. Uncaging the attitude indicator (if caging feature is present) when the airplane is not in level flight. The altimeter and heading indicator must be stabilized with airspeed indication at normal cruise before pulling out the caging knob to obtain correct indications in straight-and-level flight at normal cruise airspeed.
4. Failure to interpret the attitude indicator in terms of the existing airspeed.
5. Late pitch corrections. Pilots commonly like to leave well enough alone. When the altimeter indicates a 20 foot error, there is a reluctance to correct it, perhaps because of fear of overcontrolling. If overcontrolling is the anticipated error, practice small corrections and find the cause of overcontrolling. If any deviation is tolerated, errors increase.
6. Chasing the vertical speed indications. This tendency can be corrected by proper cross-check of other pitch instruments, as well as by increasing overall understanding of instrument characteristics.
7. Using excessive pitch corrections for the altimeter evaluation. Rushing a pitch correction by making a large pitch change usually aggravates the existing error, saving neither time nor effort.

8. Failure to maintain established pitch corrections, a common error associated with cross-check and trim errors. For example, having established a pitch change to correct an altitude error, there is a tendency to slow down the cross-check, waiting for the airplane to stabilize in the new pitch attitude. To maintain the attitude, continue to cross-check and trim off the pressures.
9. Fixations during cross-check. After initiating a heading correction, for example, there is a tendency to become preoccupied with bank control and miss errors in pitch attitude. Likewise, during an airspeed change, unnecessary gazing at the power instrument is common. A small error in power setting is of less consequence than large altitude and heading errors. The airplane will not decelerate any faster by staring at the manifold pressure gauge.

Heading

Heading errors usually result from the following faults:

1. Failure to cross-check the heading indicator, especially during changes in power or pitch attitude.
2. Misinterpretation of changes in heading, with resulting corrections in the wrong direction.
3. Failure to note and remember a preselected heading.
4. Failure to observe the rate of heading change and its relation to bank attitude.
5. Overcontrolling in response to heading changes, especially during changes in power settings.
6. Anticipating heading changes with premature application of rudder control.
7. Failure to correct small heading deviations. Unless zero error in heading is the goal, a pilot will tolerate larger and larger deviations. Correction of a 1 degree error takes a lot less time and concentration than correction of a 20° error.
8. Correcting with improper bank attitude. If correcting a 10° heading error with 20° of bank, the airplane rolls past the desired heading before the bank is established, requiring another correction in the opposite direction. Do not multiply existing errors with errors in corrective technique.
9. Failure to note the cause of a previous heading error and thus repeating the same error. For example, the airplane is out of trim, with a left wing low tendency. Repeated corrections for a slight left turn are made, yet trim is ignored.
10. Failure to set the heading indicator properly or failure to uncage it.

Power

Power errors usually result from the following faults:

1. Failure to know the power settings and pitch attitudes appropriate to various airspeeds and airplane configurations.
2. Abrupt use of throttle.
3. Failure to lead the airspeed when making power changes. For example, during airspeed reduction in level flight, especially with gear and flaps extended, adjust the throttle to maintain the slower speed before the airspeed actually reaches the desired speed. Otherwise, the airplane decelerates to a speed lower than that desired, resulting in additional power adjustments. The amount of lead depends upon how fast the airplane responds to power changes.
4. Fixation on airspeed or manifold pressure instruments during airspeed changes, resulting in erratic control of both airspeed and power.

Trim

Trim errors usually result from the following faults:

1. Improper adjustment of seat or rudder pedals for comfortable position of legs and feet. Tension in the ankles makes it difficult to relax rudder pressures.
2. Confusion about the operation of trim devices that differ among various airplane types. Some trim wheels are aligned appropriately with the airplane's axes; others are not. Some rotate in a direction contrary to what is expected.
3. Faulty sequence in trim technique. Trim should be used not as a substitute for control with the wheel (stick) and rudders, but to relieve pressures already held to stabilize attitude. As proficiency is gained, little conscious effort is required to trim off the pressures as they occur.
4. Excessive trim control. This induces control pressures that must be held until the airplane is trimmed properly. Use trim frequently and in small amounts.
5. Failure to understand the cause of trim changes. Lack of understanding the basic aerodynamics related to basic instrument skills causes a pilot to continually lag behind the airplane.

Straight Climbs and Descents

Climbs

For a given power setting and load condition, there is only one attitude that gives the most efficient rate of climb. The airspeed and climb power setting that determines this climb attitude are given in the performance data found in the POH/AFM. Details of the technique for entering a climb vary according to airspeed on entry and the type of climb (constant airspeed or constant rate) desired. (Heading and trim control are maintained as discussed in straight-and-level flight.)

Entry

To enter a constant-airspeed climb from cruising airspeed, raise the miniature aircraft to the approximate nose-high indication for the predetermined climb speed. The attitude varies according to the type of airplane. Apply light back-elevator pressure to initiate and maintain the climb attitude. The pressures vary as the airplane decelerates. Power may be advanced to the climb power setting simultaneously with the pitch change or after the pitch change is established and the airspeed approaches climb speed. If the transition from level flight to climb is smooth, the VSI shows an immediate trend upward, continues to move slowly, and then stops at a rate appropriate to the stabilized airspeed and attitude. (Primary and supporting instruments for the climb entry are shown in *Figure 7-25*.)

Once the airplane stabilizes at a constant airspeed and attitude, the ASI is primary for pitch and the heading indicator remains primary for bank. *[Figure 7-26]* Monitor the tachometer or manifold pressure gauge as the primary power instrument to ensure the proper climb power setting is being maintained. If the climb attitude is correct for the power setting selected, the airspeed will stabilize at the desired speed. If the airspeed is low or high, make an appropriately small pitch correction.

To enter a constant airspeed climb, first complete the airspeed reduction from cruise airspeed to climb speed in straight-and-level flight. The climb entry is then identical to entry from cruising airspeed, except that power must be increased simultaneously to the climb setting as the pitch attitude is increased. Climb entries on partial panel are more easily and accurately controlled if entering the maneuver from climbing speed.

The technique for entering a constant-rate climb is very similar to that used for entry to a constant-airspeed climb from climb airspeed. As the power is increased to the approximate setting for the desired rate, simultaneously raise the miniature aircraft to the climbing attitude for the desired airspeed and rate of climb. As the power is increased, the ASI is primary for pitch control until the vertical speed approaches the desired value. As the vertical speed needle stabilizes, it becomes primary for pitch control and the ASI becomes primary for power control. *[Figure 7-27]*

Figure 7-25. *Climb entry for constant airspeed climb.*

Figure 7-26. *Stabilized climb at constant airspeed.*

Figure 7-27. *Stabilized climb at constant rate.*

Pitch and power corrections must be promptly and closely coordinated. For example, if the vertical speed is correct, but the airspeed is low, add power. As the power is increased, the miniature aircraft must be lowered slightly to maintain constant vertical speed. If the vertical speed is high and the airspeed is low, lower the miniature aircraft slightly and note the increase in airspeed to determine whether or not a power change is also necessary. *[Figure 7-28]* Familiarity with the approximate power settings helps to keep pitch and power corrections at a minimum.

Leveling Off

To level off from a climb and maintain an altitude, it is necessary to start the level off before reaching the desired altitude. The amount of lead varies with rate of climb and pilot technique. If the airplane is climbing at 1,000 fpm, it continues to climb at a decreasing rate throughout the transition to level flight. An effective practice is to lead the altitude by 10 percent of the vertical speed shown (500 fpm/ 50-foot lead, 1,000 fpm/100-foot lead).

To level off at cruising airspeed, apply smooth, steady forward-elevator pressure toward level flight attitude for the speed desired. As the attitude indicator shows the pitch change, the vertical speed needle moves slowly toward zero, the altimeter needle moves more slowly, and the airspeed shows acceleration. *[Figure 7-29]* When the altimeter, attitude indicator, and VSI show level flight, constant changes in pitch and torque control have to be made as the airspeed increases. As the airspeed approaches cruising speed, reduce power to the cruise setting. The amount of lead depends upon the rate of acceleration of the airplane.

To level off at climbing airspeed, lower the nose to the pitch attitude appropriate to that airspeed in level flight. Power is simultaneously reduced to the setting for that airspeed as the pitch attitude is lowered. If power reduction is at a rate proportionate to the pitch change, airspeed will remain constant.

Descents

A descent can be made at a variety of airspeeds and attitudes by reducing power, adding drag, and lowering the nose to a predetermined attitude. The airspeed eventually stabilizes at a constant value. Meanwhile, the only flight instrument providing a positive attitude reference is the attitude indicator. Without the attitude indicator (such as during a partial panel descent), the ASI, altimeter, and VSI show varying rates of change until the airplane decelerates to a constant airspeed at a constant attitude. During the transition, changes in control pressure and trim, as well as cross-check and interpretation, must be accurate to maintain positive control.

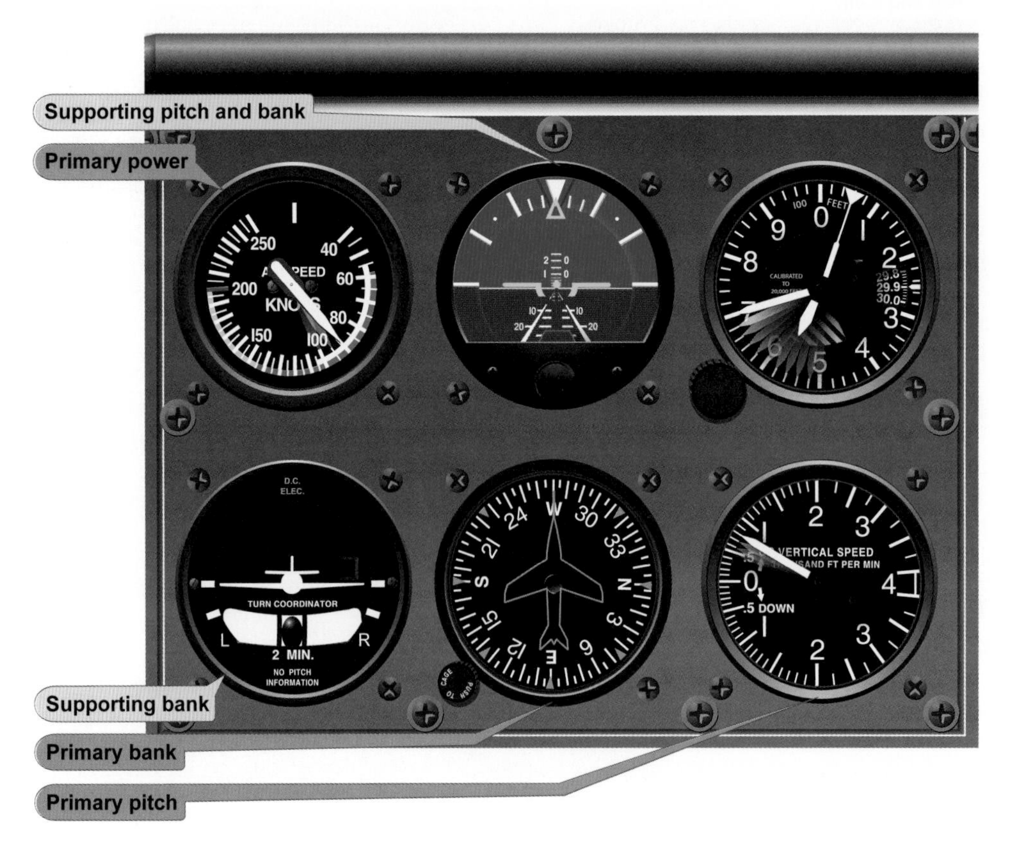

Figure 7-28. *Airspeed low and vertical speed high—reduce pitch.*

Figure 7-29. *Level off at cruising speed.*

Entry

The following method for entering descents is effective with or without an attitude indicator. First, reduce airspeed to a selected descent airspeed while maintaining straight-and-level flight, then make a further reduction in power (to a predetermined setting). As the power is adjusted, simultaneously lower the nose to maintain constant airspeed, and trim off control pressures.

During a constant airspeed descent, any deviation from the desired airspeed calls for a pitch adjustment. For a constant rate descent, the entry is the same, but the VSI is primary for pitch control (after it stabilizes near the desired rate), and the ASI is primary for power control. Pitch and power must be closely coordinated when corrections are made, as they are in climbs. *[Figure 7-30]*

Leveling Off

The level off from a descent must be started before reaching the desired altitude. The amount of lead depends upon the rate of descent and control technique. With too little lead, the airplane tends to overshoot the selected altitude unless technique is rapid. Assuming a 500 fpm rate of descent, lead the altitude by 100–150 feet for a level off at an airspeed higher than descending speed. At the lead point, add power to the appropriate level flight cruise setting. *[Figure 7-31]* Since the nose tends to rise as the airspeed increases, hold forward elevator pressure to maintain the vertical speed at the descending rate until approximately 50 feet above the altitude, and then smoothly adjust the pitch attitude to the level flight attitude for the airspeed selected.

To level off from a descent at descent airspeed, lead the desired altitude by approximately 50 feet, simultaneously adjusting the pitch attitude to level flight and adding power to a setting that holds the airspeed constant. *[Figure 7-32]* Trim off the control pressures and continue with the normal straight-and-level flight cross-check.

Common Errors in Straight Climbs and Descents

Common errors result from the following faults:

1. Overcontrolling pitch on climb entry. Until the pitch attitudes related to specific power settings used in climbs and descents are known, larger than necessary pitch adjustments are made. One of the most difficult habits to acquire during instrument training is to restrain the impulse to disturb a flight attitude until the result is known. Overcome the inclination to make a large control movement for a pitch change, and learn to apply small control pressures smoothly, cross-checking rapidly for the results of the change, and continuing with the pressures as instruments show the desired results. Small pitch changes can be easily controlled, stopped, and corrected; large changes are more difficult to control.

Figure 7-30. *Constant airspeed descent, airspeed high—reduce power.*

Figure 7-31. *Level off airspeed higher than descent airspeed.*

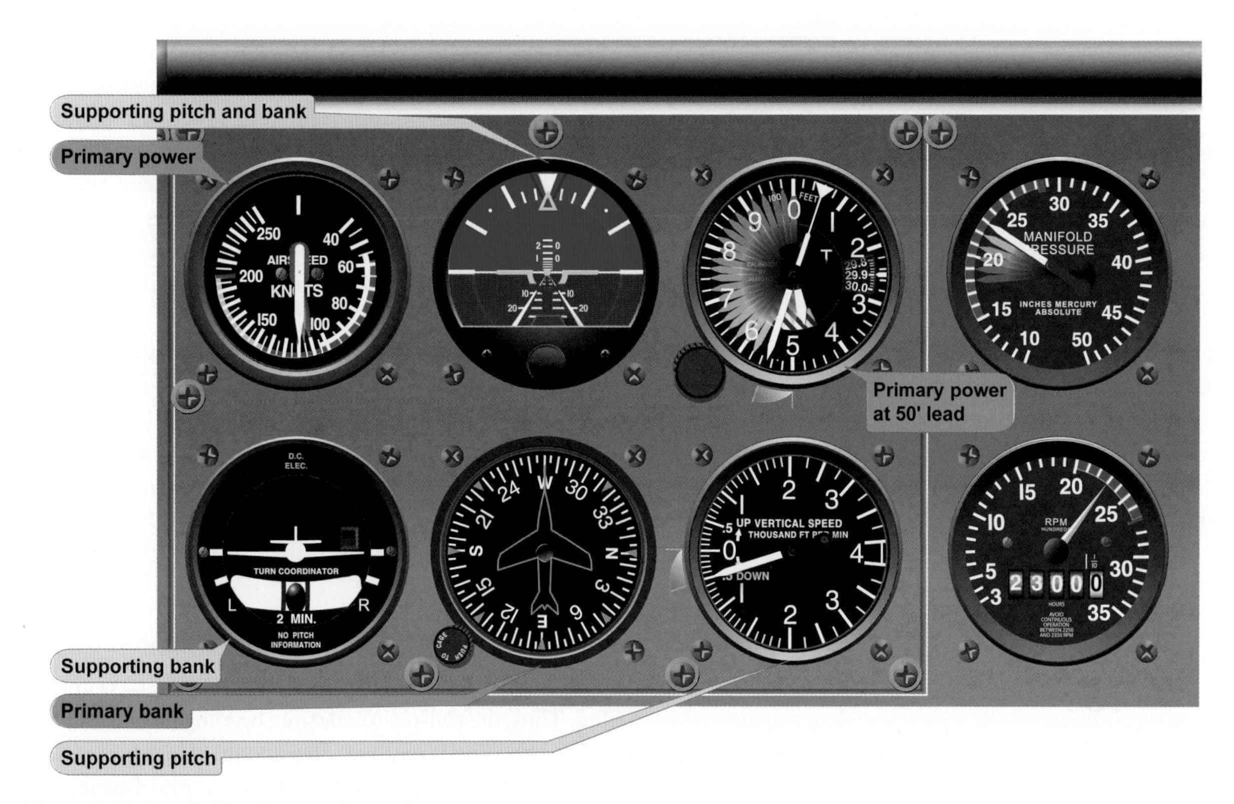

Figure 7-32. *Level off at descent airspeed.*

2. Failure to vary the rate of cross-check during speed, power, or attitude changes or climb or descent entries.

3. Failure to maintain a new pitch attitude. For example, raising the nose to the correct climb attitude, and as the airspeed decreases, either overcontrol and further increase the pitch attitude or allow the nose to lower. As control pressures change with airspeed changes, cross-check must be increased and pressures readjusted.

4. Failure to trim off pressures. Unless the airplane is trimmed, there is difficulty in determining whether control pressure changes are induced by aerodynamic changes or by the pilot's own movements.

5. Failure to learn and use proper power settings.

6. Failure to cross-check both airspeed and vertical speed before making pitch or power adjustments.

7. Improper pitch and power coordination on slow-speed level offs due to slow cross-check of airspeed and altimeter indications.

8. Failure to cross-check the VSI against the other pitch control instruments, resulting in chasing the vertical speed.

9. Failure to note the rate of climb or descent to determine the lead for level offs, resulting in overshooting or undershooting the desired altitude.

10. Ballooning (allowing the nose to pitch up) on level offs from descents, resulting from failure to maintain descending attitude with forward-elevator pressure as power is increased to the level flight cruise setting.

11. Failure to recognize the approaching straight-and-level flight indications as level off is completed. Maintain an accelerated cross-check until positively established in straight-and-level flight.

Turns

Standard Rate Turns

A standard rate turn is one in which the pilot will do a complete 360° circle in 2 minutes or 3 degrees per second. A standard rate turn, although always 3 degrees per second, requires higher angles of bank as airspeed increases. To enter a standard rate level turn, apply coordinated aileron and rudder pressures in the desired direction of turn. Pilots commonly roll into turns at a much too rapid rate. During initial training in turns, base control pressures on the rate of cross-check and interpretation. Maneuvering an airplane faster than the capability to keep up with the changes in instrument indications only creates the need to make corrections.

A rule of thumb to determine the approximate angle of bank required for a standard rate turn is to use 15 percent of the true airspeed. A simple way to determine this amount is to divide the airspeed by 10 and add one-half the result. For example, at 100 knots, approximately 15° of bank is required (100 ÷ 10 = 10 + 5 = 15); at 120 knots, approximately 18° of bank is needed for a standard rate turn.

On the roll-in, use the attitude indicator to establish the approximate angle of bank, and then check the turn coordinator's miniature aircraft for a standard rate turn indication or the aircraft's turn-and-bank indicator. Maintain the bank for this rate of turn, using the turn coordinator's miniature aircraft as the primary bank reference and the attitude indicator as the supporting bank instrument. *[Figure 7-33]* Note the exact angle of bank shown on the banking scale of the attitude indicator when the turn coordinator indicates a standard rate turn.

During the roll-in, check the altimeter, VSI, and attitude indicator for the necessary pitch adjustments as the vertical lift component decreases with an increase in bank. If constant airspeed is to be maintained, the ASI becomes primary for power, and the throttle must be adjusted as drag increases. As the bank is established, trim off the pressures applied during pitch and power changes.

To recover to straight-and-level flight, apply coordinated aileron and rudder pressures opposite to the direction of the turn. Strive for the same rate of roll-out used to roll into the turn; fewer problems are encountered in estimating the lead necessary for roll-out on exact headings, especially on partial panel maneuvers. Upon initiation of the turn recovery, the attitude indicator becomes the primary bank instrument. When the airplane is approximately level, the heading indicator is the primary bank instrument as in straight-and-level flight. Pitch, power, and trim adjustments are made as changes in vertical lift component and airspeed occur. The ball should be checked throughout the turn, especially if control pressures are held rather than trimmed off.

Some airplanes are very stable during turns, requiring only slight trim adjustments that permit hands-off flight while the airplane remains in the established attitude. Other airplanes require constant, rapid cross-check and control during turns to correct overbanking tendencies. Due to the interrelationship of pitch, bank, and airspeed deviations during turns, cross-check must be fast in order to prevent an accumulation of errors.

Turns to Predetermined Headings

As long as an airplane is in a coordinated bank, it continues to turn. Thus, the roll-out to a desired heading must be started before the heading is reached. The amount of lead varies with the relationship between the rate of turn, angle of bank, and rate of recovery. For small heading changes, use a bank angle that does not exceed the number of degrees to be turned. Lead the desired heading by one-half the number of degrees of bank used. For example, if a 10° bank is used during a change in heading, start the roll-out 5 degrees before reaching the desired heading. For larger changes in heading, the amount

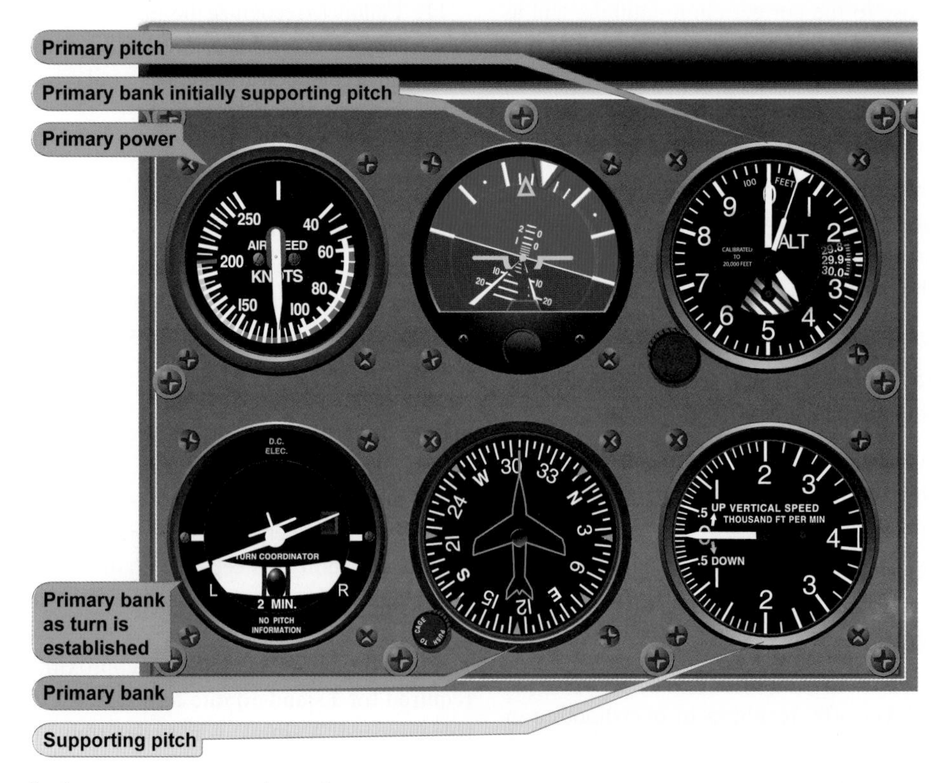

Figure 7-33. *Standard rate turn, constant airspeed.*

of lead varies since the angle of bank for a standard rate turn varies with the true airspeed.

Practice with a lead of one-half the angle of bank until the precise lead a given technique requires is determined. If rates of roll-in and roll-out are consistent, the precise amount of lead suitable to a particular roll-out technique can be determined.

Timed Turns

A timed turn is a turn in which the clock and the turn coordinator are used to change heading by a specific number of degrees in a given time. For example, in a standard rate turn (3 degrees per second), an airplane turns 45° in 15 seconds; in a half standard rate turn, the airplane turns 45° in 30 seconds.

Prior to performing timed turns, the turn coordinator should be calibrated to determine the accuracy of its indications. *[Figure 7-34]* Establish a standard rate turn as indicated by the turn coordinator, and as the sweep-second hand of the clock passes a cardinal point (12, 3, 6, 9), check the heading on the heading indicator. While holding the indicated rate of turn constant, note the indicated heading changes at 10 second intervals. If the airplane turns more than or less than 30° in that interval, a respectively larger or smaller deflection of the miniature aircraft of the turn coordinator is necessary to produce a standard rate turn. After calibrating the turn coordinator during turns in each direction, note the corrected deflections, if any, and apply them during all timed turns.

The same cross-check and control technique is used in making a timed turn that is used to execute turns to predetermined headings, except the clock is substituted for the heading indicator. The miniature aircraft of the turn coordinator is primary for bank control, the altimeter is primary for pitch control, and the ASI is primary for power control. Start the roll-in when the clock's second hand passes a cardinal point, hold the turn at the calibrated standard rate indication (or half-standard rate for small heading changes), and begin the roll-out when the computed number of seconds has elapsed. If the rates of roll-in and roll-out are the same, the time taken during entry and recovery does not need to be considered in the time computation.

Practice timed turns with a full instrument panel and check the heading indicator for the accuracy of turns. If the turns are executed without the gyro heading indicator, use the magnetic compass at the completion of the turn to check turn accuracy, taking compass deviation errors into consideration.

Compass Turns

In most small airplanes, the magnetic compass is the only direction-indicating instrument independent of other airplane instruments and power sources. Because of its operating characteristics, called compass errors, pilots are prone to use it only as a reference for setting the heading indicator, but knowledge of magnetic compass characteristics permits full use of the instrument to turn the airplane to correct and maintain headings.

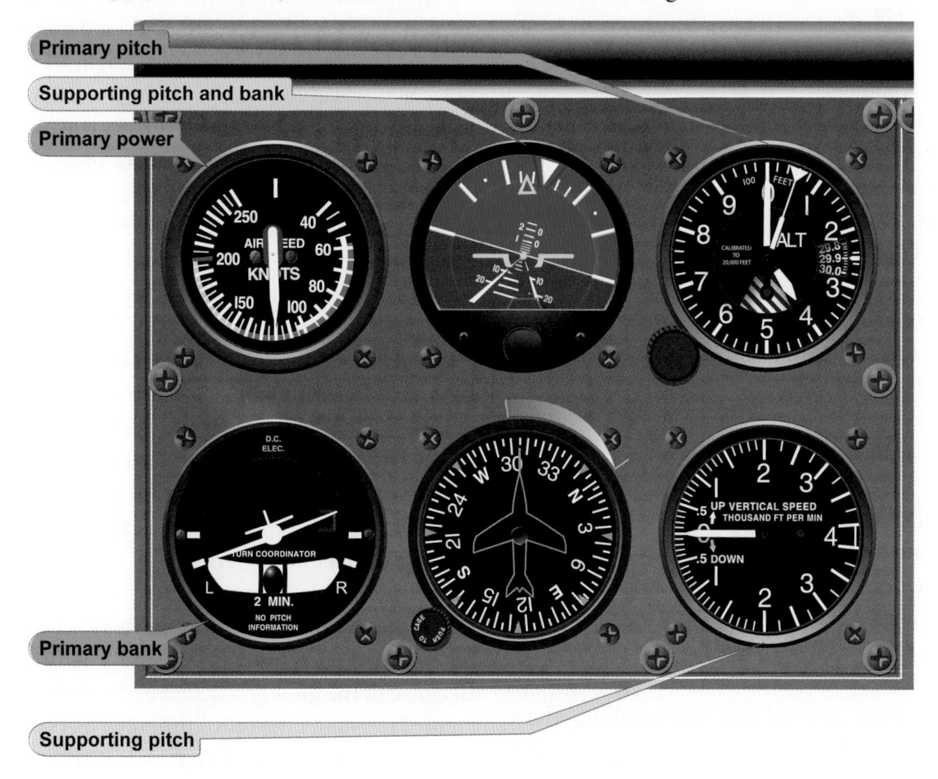

Figure 7-34. *Turn coordinator calibration.*

Remember the following points when making turns to magnetic compass headings or when using the magnetic compass as a reference for setting the heading indicator:

1. If on a north heading and a turn is started to the east or west, the compass indication lags or indicates a turn in the opposite direction.
2. If on a south heading and a turn is started toward the east or west, the compass indication precedes the turn, indicating a greater amount of turn than is actually occurring.
3. When on an east or west heading, the compass indicates correctly when starting a turn in either direction.
4. If on an east or west heading, acceleration results in a north turn indication; deceleration results in a south turn indication.
5. When maintaining a north or south heading, no error results from diving, climbing, or changing airspeed.

With an angle of bank between 15° and 18°, the amount of lead or lag to be used when turning to northerly or southerly headings varies with, and is approximately equal to, the latitude of the locality over which the turn is being made. When turning to a heading of north, the lead for roll-out must include the number of degrees of change of latitude, plus the lead normally used in recovery from turns. During a turn to a south heading, maintain the turn until the compass passes south the number of degrees of latitude, minus normal roll-out lead. *[Figure 7-35]*

For example, when turning from an easterly direction to north, where the latitude is 30°, start the roll-out when the compass reads 37° (30° plus one-half the 15° angle of bank, or whatever amount is appropriate for the rate of roll-out). When turning from an easterly direction to south, start the roll-out when the magnetic compass reads 203° (180° plus 30° minus one-half the angle of bank). When making similar turns from a westerly direction, the appropriate points at which to begin the roll-out would be 323° for a turn to north and 157° for a turn to south.

When turning to a heading of east or west from a northerly direction, start the roll-out approximately 10° to 12° before the east or west indication is reached. When turning to an east or west heading from a southerly direction, start the rollout approximately 5 degrees before the east or west indication is reached. When turning to other headings, the lead or lag must be interpolated.

Abrupt changes in attitude or airspeed and the resulting erratic movements of the compass card make accurate interpretations of the instrument very difficult. Proficiency in compass turns depends on knowledge of compass characteristics, smooth control technique, and accurate bank-and-pitch control.

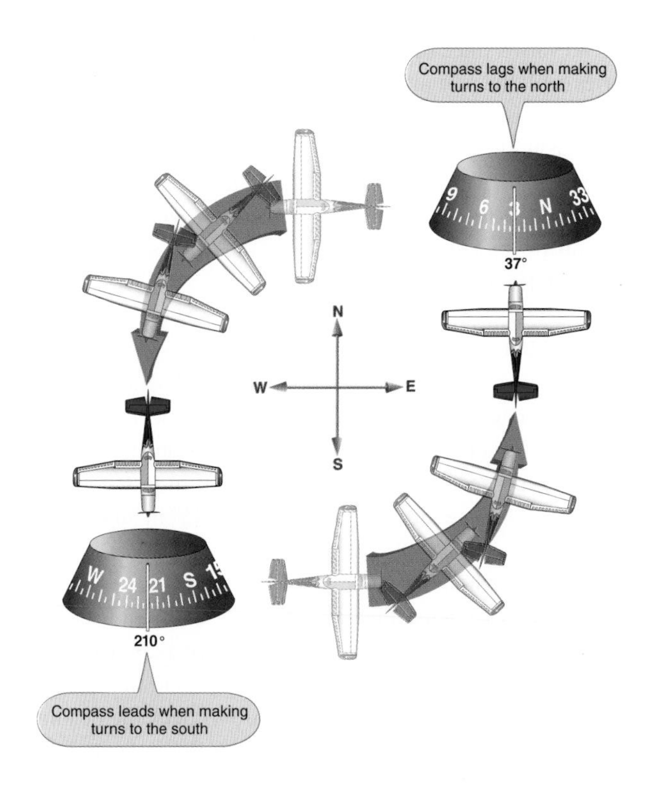

Figure 7-35. *North and south turn error.*

Steep Turns

For purposes of instrument flight training in conventional airplanes, any turn greater than a standard rate is considered steep. *[Figure 7-36]* The exact angle of bank at which a normal turn becomes steep is unimportant. What is important is learning to control the airplane with bank attitudes in excess of those normally used on instruments. Practicing steep turns will not only increase proficiency in the basic instrument flying skills, but also enable smooth, quick, and confident reactions to unexpected abnormal flight attitudes under instrument flight conditions.

Pronounced changes occur in the effects of aerodynamic forces on aircraft control at progressively greater bank attitudes. Skill in cross-check, interpretation, and control is increasingly necessary in proportion to the amount of these changes, though the techniques for entering, maintaining, and recovering from the turn are the same in principle for steep turns as for shallower turns.

Enter a steep turn in the same way as a shallower turn, but prepare to cross-check rapidly as the turn steepens. Because of the greatly reduced vertical lift component, pitch control is usually the most difficult aspect of this maneuver. Unless immediately noted and corrected with a pitch

Figure 7-36. *Steep left turn.*

increase, the loss of vertical lift results in rapid movement of the altimeter, vertical speed, and airspeed needles. The faster the rate of bank change, the more suddenly the lift changes occur. If a cross-check is fast enough to note the immediate need for pitch changes, smooth, steady back-elevator pressure will maintain constant altitude. However, overbanking to excessively steep angles without adjusting pitch as the bank changes occur requires increasingly stronger elevator pressure. The loss of vertical lift and increase in wing loading finally reach a point at which further application of back-elevator pressure tightens the turn without raising the nose.

How does a pilot recognize overbanking and low pitch attitude? What should a pilot do to correct them? If a rapid downward movement of the altimeter needle or vertical speed needle, together with an increase in airspeed, is observed despite application of back elevator pressure, the airplane is in a diving spiral. *[Figure 7-37]* Immediately shallow the bank with smooth and coordinated aileron and rudder pressures, hold or slightly relax elevator pressure, and increase the cross-check of the attitude indicator, altimeter, and VSI. Reduce power if the airspeed increase is rapid. When the vertical speed trends upward, the altimeter needle moves slower as the vertical lift increases. When the elevator is effective in raising the nose, hold the bank attitude shown on the attitude indicator and adjust elevator control pressures smoothly for the nose-high attitude appropriate to the bank maintained. If pitch control is consistently late on entries to steep turns, rollout immediately to straight-and-level flight and analyze possible errors. Practice shallower turns initially and learn the attitude changes and control responses required, then increase the banks as a quicker and more accurate cross-check and control techniques are developed.

The power necessary to maintain constant airspeed increases as the bank and drag increase. With practice, the power

Figure 7-37. *Diving spiral.*

settings appropriate to specific bank attitudes are learned, and adjustments can be made without undue attention to airspeed and power instruments. During training in steep turns, as in any other maneuver, attend to the most important tasks first. Keep the pitch attitude relatively constant, and more time can be devoted to cross-check and instrument interpretation.

During recovery from steep turns to straight-and-level flight, elevator and power control must be coordinated with bank control in proportion to the changes in aerodynamic forces. Back elevator pressures must be released and power decreased. The common errors associated with steep turns are the same as those discussed later in this section. Remember, errors are more exaggerated, more difficult to correct, and more difficult to analyze unless rates of entry and recovery are consistent with the level of proficiency in the three basic instrument flying skills.

Climbing and Descending Turns

To execute climbing and descending turns, combine the technique used in straight climbs and descents with the various turn techniques. The aerodynamic factors affecting lift and power control must be considered in determining power settings, and the rate of cross-check and interpretation must be increased to enable control of bank as well as pitch changes.

Change of Airspeed During Turns

Changing airspeed during turns is an effective maneuver for increasing proficiency in all three basic instrument skills. Since the maneuver involves simultaneous changes in all components of control, proper execution requires rapid cross-check and interpretation as well as smooth control. Proficiency in the maneuver also contributes to confidence in the instruments during attitude and power changes involved in more complex maneuvers. Pitch and power control techniques are the same as those used during changes in airspeed in straight-and-level flight.

The angle of bank necessary for a given rate of turn is proportional to the true airspeed. Since the turns are executed at a standard rate, the angle of bank must be varied in direct proportion to the airspeed change in order to maintain a constant rate of turn. During a reduction of airspeed, decrease the angle of bank and increase the pitch attitude to maintain altitude and a standard rate turn.

The altimeter and turn coordinator indications should remain constant throughout the turn. The altimeter is primary for pitch control and the miniature aircraft of the turn coordinator is primary for bank control. The manifold pressure gauge (or tachometer) is primary for power control while the airspeed is changing. As the airspeed approaches the new indication, the ASI becomes primary for power control.

Two methods of changing airspeed in turns may be used. In the first method, airspeed is changed after the turn is established. *[Figure 7-38]* In the second method, the airspeed change is initiated simultaneously with the turn entry. The first method is easier, but regardless of the method used, the rate of cross-check must be increased as power is reduced. As the airplane decelerates, check the altimeter and VSI for necessary pitch changes and the bank instruments for required bank changes.

Figure 7-38. *Change of airspeed during turn.*

If the miniature aircraft of the turn coordinator indicates a deviation from the desired deflection, adjust the bank. Adjust pitch attitude to maintain altitude. When approaching the desired airspeed, pitch attitude becomes primary for power control and the manifold pressure gauge (or tachometer) is adjusted to maintain the desired airspeed. Trim is important throughout the maneuver to relieve control pressures.

Until control technique is very smooth, frequent cross-check of the attitude indicator is essential to prevent overcontrolling and to provide approximate bank angles appropriate to the changing airspeeds.

Common Errors in Turns

Pitch

Pitch errors result from the following faults:

1. Preoccupation with bank control during turn entry and recovery. If 5 seconds are required to roll into a turn, check the pitch instruments as bank pressures are initiated. If bank control pressure and rate of bank change are consistent, a sense of the time required for an attitude change is developed. During the interval, check pitch, power, and trim—as well as bank—controlling the total attitude instead of one factor at a time.
2. Failure to understand or remember the need for changing the pitch attitude as the vertical lift component changes, resulting in consistent loss of altitude during entries.
3. Changing the pitch attitude before it is necessary. This fault is very likely if a cross-check is slow and rate of entry too rapid. The error occurs during the turn entry due to a mechanical and premature application of back-elevator control pressure.
4. Overcontrolling the pitch changes. This fault commonly occurs with the previous error.
5. Failure to properly adjust the pitch attitude as the vertical lift component increases during the roll-out, resulting in consistent gain in altitude on recovery to headings.
6. Failure to trim during turn entry and following turn recovery (if turn is prolonged).
7. Failure to maintain straight-and-level cross-check after roll-out. This error commonly follows a perfectly executed turn.
8. Erratic rates of bank change on entry and recovery, resulting from failure to cross-check the pitch instruments with a consistent technique appropriate to the changes in lift.

Bank

Bank and heading errors result from the following faults:

1. Overcontrolling, resulting in overbanking upon turn entry, overshooting and undershooting headings, as well as aggravated pitch, airspeed, and trim errors.
2. Fixation on a single bank instrument. On a 90° change of heading, for example, leave the heading indicator out of the cross-check for approximately 20 seconds after establishing a standard rate turn, since at 3° per second the turn will not approach the lead point until that time has elapsed. Make the cross-check selective, checking only what needs to be checked at the appropriate time.
3. Failure to check for precession of the horizon bar following recovery from a turn. If the heading indicator shows a change in heading when the attitude indicator shows level flight, the airplane is turning. If the ball is centered, the attitude gyro has precessed; if the ball is not centered, the airplane may be in a slipping or skidding turn. Center the ball with rudder pressure, check the attitude indicator and heading indicator, stop the heading change if it continues, and retrim.
4. Failure to use the proper degree of bank for the amount of heading change desired. Rolling into a 20° bank for a heading change of 10° will normally overshoot the heading. Use the bank attitude appropriate to the amount of heading change desired.
5. Failure to remember the heading to which the aircraft is being turned. This fault is likely when rushing the maneuver.
6. Turning in the wrong direction, due to misreading or misinterpreting the heading indicator, or to confusion regarding the location of points on the compass. Turn in the shortest direction to reach a given heading, unless there is a specific reason to turn the long way around. Study the compass rose and visualize at least the positions of the eight major points around the azimuth. A number of methods can be used to make quick computations for heading changes. For example, to turn from a heading of 305° to a heading of 110°, would a pilot turn right or left for the shortest way around? Subtracting 200 from 305 and adding 20, gives 125° as the reciprocal of 305°; therefore, execute the turn to the right. Likewise, to figure the reciprocal of a heading less than 180°, add 200 and subtract 20. Computations are done more quickly using multiples of 100s and 10s than by adding or subtracting 180° from the actual heading; therefore, the method suggested above may save time and confusion.

7. Failure to check the ball of the turn coordinator when interpreting the instrument for bank information. If the roll rate is reduced to zero, the miniature aircraft of the turn coordinator indicates only direction and rate of turn. Unless the ball is centered, do not assume the turn is resulting from a banked attitude.

Power

Power and airspeed errors result from the following faults:

1. Failure to cross-check the ASI as pitch changes are made.
2. Erratic use of power control. This may be due to improper throttle friction control, inaccurate throttle settings, chasing the airspeed readings, abrupt or overcontrolled pitch-and-bank changes, or failure to recheck the airspeed to note the effect of a power adjustment.
3. Poor coordination of throttle control with pitch-and-bank changes associated with slow cross-check or failure to understand the aerodynamic factors related to turns.

Trim

Trim errors result from the following faults:

1. Failure to recognize the need for a trim change due to slow cross-check and interpretation. For example, a turn entry at a rate too rapid for a cross-check leads to confusion in cross-check and interpretation with resulting tension on the controls.
2. Failure to understand the relationship between trim and attitude/power changes.
3. Chasing the vertical speed needle. Overcontrolling leads to tension and prevents sensing the pressures to be trimmed off.
4. Failure to trim following power changes.

Errors During Compass Turns

In addition to the faults discussed above, the following errors connected with compass turns should be noted:

1. Faulty understanding or computation of lead and lag.
2. Fixation on the compass during the roll-out. Until the airplane is in straight-and-level unaccelerated flight, it is unnecessary to read the indicated heading. Accordingly, after the roll-out, cross-check for straight-and-level flight before checking the accuracy of the turn.

Approach to Stall

Practicing approach to stall recoveries in various airplane configurations should build confidence in a pilot's ability to control the airplane in unexpected situations. Approach to stall should be practiced from straight flight and from shallow banks. The objective is to practice recognition and recovery from the approach to a stall.

Prior to stall recovery practice, select a safe altitude above the terrain, an area free of conflicting air traffic, appropriate weather, and the availability of radar traffic advisory service.

Approaches to stalls are accomplished in the following configurations:

1. Takeoff configuration—should begin from level flight near liftoff speed. Power should be applied while simultaneously increasing the angle of attack to induce an indication of a stall.
2. Clean configuration—should begin from a reduced airspeed, such as pattern airspeed, in level flight. Power should be applied while simultaneously increasing the angle of attack to induce an indication of a stall.
3. Approach or landing configuration—should be initiated at the appropriate approach or landing airspeed. The angle of attack should be smoothly increased to induce an indication of a stall.

Recoveries should be prompt in response to a stall warning device or an aerodynamic indication by smoothly reducing the angle of attack and applying maximum power or as recommended by the POH/AFM. The recovery should be completed without an excessive loss of altitude and on a predetermined heading, altitude, and airspeed.

Unusual Attitudes and Recoveries

An unusual attitude is an airplane attitude not normally required for instrument flight. Unusual attitudes may result from a number of conditions, such as turbulence, disorientation, instrument failure, confusion, preoccupation with flight deck duties, carelessness in cross-checking, errors in instrument interpretation, or lack of proficiency in aircraft control. Since unusual attitudes are not intentional maneuvers during instrument flight, except in training, they are often unexpected, and the reaction of an inexperienced or inadequately trained pilot to an unexpected abnormal flight attitude is usually instinctive rather than intelligent and deliberate. This individual reacts with abrupt muscular effort, which is purposeless and even hazardous in turbulent conditions, at excessive speeds, or at low altitudes. However, with practice, the techniques for rapid and safe recovery from unusual attitudes can be mastered.

When an unusual attitude is noted during the cross-check, the immediate problem is not how the airplane got there, but what it is doing and how to get it back to straight-and-level flight as quickly as possible.

Recognizing Unusual Attitudes

As a general rule, any time an instrument rate of movement or indication other than those associated with the basic instrument flight maneuvers is noted, assume an unusual attitude and increase the speed of cross-check to confirm the attitude, instrument error, or instrument malfunction.

Nose-high attitudes are shown by the rate and direction of movement of the altimeter needle, vertical speed needle, and airspeed needle, as well as the immediately recognizable indication of the attitude indicator (except in extreme attitudes). *[Figure 7-39]* Nose-low attitudes are shown by the same instruments, but in the opposite direction. *[Figure 7-40]*

Recovery from Unusual Attitudes

In moderate unusual attitudes, the pilot can normally reorient by establishing a level flight indication on the attitude indicator. However, the pilot should not depend on this instrument if the attitude indicator is the spillable type, because its upset limits may have been exceeded or it may have become inoperative due to mechanical malfunction. If it is the nonspillable-type instrument and is operating properly, errors up to 5 degrees of pitch-and-bank may result and its indications are very difficult to interpret in extreme attitudes. As soon as the unusual attitude is detected, the recommended recovery procedures stated in the POH/AFM should be initiated. If there are no recommended procedures stated in the POH/AFM, the recovery should be initiated by reference to the ASI, altimeter, VSI, and turn coordinator.

Nose-High Attitudes

If the airspeed is decreasing, or below the desired airspeed, increase power (as necessary in proportion to the observed deceleration), apply forward elevator pressure to lower the nose and prevent a stall, and correct the bank by applying coordinated aileron and rudder pressure to level the miniature aircraft and center the ball of the turn coordinator. The corrective control applications are made almost simultaneously, but in the sequence given above. A level pitch attitude is indicated by the reversal and stabilization

Figure 7-39. *Unusual attitude—nose-high.*

Figure 7-40. *Unusual attitude—nose-low.*

of the ASI and altimeter needles. Straight coordinated flight is indicated by the level miniature aircraft and centered ball of the turn coordinator.

Nose-Low Attitudes

If the airspeed is increasing, or is above the desired airspeed, reduce power to prevent excessive airspeed and loss of altitude. Correct the bank attitude with coordinated aileron and rudder pressure to straight flight by referring to the turn coordinator. Raise the nose to level flight attitude by applying smooth back elevator pressure. All components of control should be changed simultaneously for a smooth, proficient recovery. However, during initial training a positive, confident recovery should be made by the numbers, in the sequence given above. A very important point to remember is that the instinctive reaction to a nose-down attitude is to pull back on the elevator control.

After initial control has been applied, continue with a fast cross-check for possible overcontrolling, since the necessary initial control pressures may be large. As the rate of movement of altimeter and ASI needles decreases, the attitude is approaching level flight. When the needles stop and reverse direction, the aircraft is passing through level flight. As the indications of the ASI, altimeter, and turn coordinator stabilize, incorporate the attitude indicator into the cross-check.

The attitude indicator and turn coordinator should be checked to determine bank attitude and then corrective aileron and rudder pressures should be applied. The ball should be centered. If it is not, skidding and slipping sensations can easily aggravate disorientation and retard recovery. If entering the unusual attitude from an assigned altitude (either by an instructor or by air traffic control (ATC) if operating under instrument flight rules (IFR)), return to the original altitude after stabilizing in straight-and-level flight.

Common Errors in Unusual Attitudes

Common errors associated with unusual attitudes include the following faults:

1. Failure to keep the airplane properly trimmed. A flight deck interruption when holding pressures can easily lead to inadvertent entry into unusual attitudes.
2. Disorganized flight deck. Hunting for charts, logs, computers, etc., can seriously distract attention from the instruments.
3. Slow cross-check and fixations. The impulse is to stop and stare when noting an instrument discrepancy unless a pilot has trained enough to develop the skill required for immediate recognition.

4. Attempting to recover by sensory sensations other than sight. The discussion of disorientation in Chapter 3, Human Factors, indicates the importance of trusting the instruments.
5. Failure to practice basic instrument skills. All of the errors noted in connection with basic instrument skills are aggravated during unusual attitude recoveries until the elementary skills have been mastered.

Instrument Takeoff

Competency in instrument takeoffs will provide the proficiency and confidence necessary for use of flight instruments during departures under conditions of low visibility, rain, low ceilings, or disorientation at night. A sudden rapid transition from "visual" to "instrument" flight can result in serious disorientation and control problems.

Instrument takeoff techniques vary with different types of airplanes, but the method described below is applicable whether the airplane is single- or multiengine; tricycle gear or conventional gear.

Align the airplane with the centerline of the runway with the nosewheel or tailwheel straight. Lock the tailwheel, if so equipped, and hold the brakes firmly to avoid creeping while preparing for takeoff. Set the heading indicator with the nose index on the 5 degree mark nearest the published runway heading to allow instant detection of slight changes in heading during the takeoff. Make certain that the instrument is uncaged (if it has a caging feature) by rotating the knob after uncaging and checking for constant heading indication. If using an electric heading indicator with a rotatable needle, rotate the needle so that it points to the nose position, under the top index. Advance the throttle to an rpm that will provide partial rudder control. Release the brakes, advancing the power smoothly to takeoff setting.

During the takeoff roll, hold the heading constant on the heading indicator by using the rudder. In multiengine, propeller-driven airplanes, also use differential throttle to maintain direction. The use of brakes should be avoided, except as a last resort, as it usually results in overcontrolling and extending the takeoff roll. Once the brakes are released, any deviation in heading must be corrected instantly.

As the airplane accelerates, cross-check both heading indicator and ASI rapidly. The attitude indicator may precess to a slight nose-up attitude. As flying speed is approached (approximately 15–25 knots below takeoff speed), smoothly apply elevator control for the desired takeoff attitude on the attitude indicator. This is approximately a two bar width climb indication for most small airplanes.

Continue with a rapid cross-check of heading indicator and attitude indicator as the airplane leaves the ground. Do not pull it off; let it fly off while holding the selected attitude constant. Maintain pitch-and-bank control by referencing the attitude indicator, and make coordinated corrections in heading when indicated on the heading indicator. Cross-check the altimeter and VSI for a positive rate of climb (steady clockwise rotation of the altimeter needle, and the VSI showing a stable rate of climb appropriate to the airplane).

When the altimeter shows a safe altitude (approximately 100 feet), raise the landing gear and flaps, maintaining attitude by referencing the attitude indicator. Because of control pressure changes during gear and flap operation, overcontrolling is likely unless the pilot notes pitch indications accurately and quickly. Trim off control pressures necessary to hold the stable climb attitude. Check the altimeter, VSI, and airspeed for a smooth acceleration to the predetermined climb speed (altimeter and airspeed increasing, vertical speed stable). At climb speed, reduce power to climb setting (unless full power is recommended for climb by the POH/AFM and trim).

Throughout the instrument takeoff, cross-check and interpretation must be rapid and control positive and smooth. During liftoff, gear and flap retraction, power reduction, and the changing control reactions demand rapid cross-check, adjustment of control pressures, and accurate trim changes.

Common Errors in Instrument Takeoffs

Common errors during the instrument takeoff include the following:

1. Failure to perform an adequate flight deck check before the takeoff. Pilots have attempted instrument takeoffs with inoperative airspeed indicators (pitot tube obstructed), gyros caged, controls locked, and numerous other oversights due to haste or carelessness.
2. Improper alignment on the runway. This may result from improper brake application, allowing the airplane to creep after alignment or from alignment with the nosewheel or tailwheel cocked. In any case, the result is a built-in directional control problem as the takeoff starts.
3. Improper application of power. Abrupt application of power complicates directional control. Add power with a smooth, uninterrupted motion.
4. Improper use of brakes. Incorrect seat or rudder pedal adjustment, with feet in an uncomfortable position, frequently cause inadvertent application of brakes and excessive heading changes.

5. Overcontrolling rudder pedals. This fault may be caused by late recognition of heading changes, tension on the controls, misinterpretation of the heading indicator (and correcting in the wrong direction), failure to appreciate changing effectiveness of rudder control as the aircraft accelerates, and other factors. If heading changes are observed and corrected instantly with small movement of the rudder pedals, swerving tendencies can be reduced.
6. Failure to maintain attitude after becoming airborne. If the pilot reacts to seat-of-the-pants sensations when the airplane lifts off, pitch control is guesswork. The pilot may either allow excessive pitch or apply excessive forward elevator pressure, depending on the reaction to trim changes.
7. Inadequate cross-check. Fixations are likely during trim changes, attitude changes, gear and flap retractions, and power changes. Once an instrument or a control input is applied, continue the cross-check and note the effect during the next cross-check sequence.
8. Inadequate interpretation of instruments. Failure to understand instrument indications immediately indicates that further study of the maneuver is necessary.

Basic Instrument Flight Patterns

Flight patterns are basic maneuvers, flown by sole reference to the instruments rather than outside visual clues, for the purpose of practicing basic attitude flying. The patterns simulate maneuvers encountered on instrument flights, such as holding patterns, procedure turns, and approaches. After attaining a reasonable degree of proficiency in basic maneuvers, apply these skills to the various combinations of individual maneuvers. The following practice flight patterns are directly applicable to operational instrument flying.

Racetrack Pattern

1. Time 3 minutes straight-and-level flight from A to B. *[Figure 7-41]* During this interval, reduce airspeed to the holding speed appropriate for the aircraft.
2. Start a 180° standard rate turn to the right at B. Roll-out at C on the reciprocal of the heading originally used at A.
3. Time a 1 minute straight-and-level flight from C to D.
4. Start a 180° standard rate turn to the right at D, rolling-out on the original heading.
5. Fly 1 minute on the original heading, adjusting the outbound leg so that the inbound segment is 1 minute.

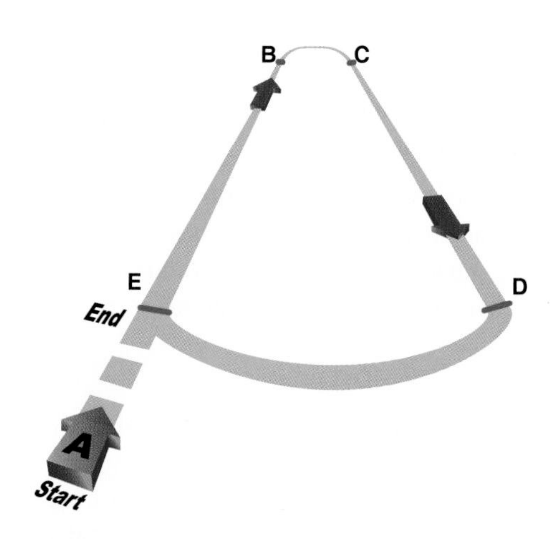

Figure 7-41. *Racetrack pattern (entire pattern in level flight).*

NOTE*:* This pattern is an exercise combining use of the clock with basic maneuvers.

Procedure Turn

A procedure turn is a maneuver that facilitates:

- A reversal in flight direction.
- A descent from an initial approach fix or assigned altitude to a permissible altitude (usually the procedure turn altitude).
- An interception of the inbound course at a sufficient distance allowing the aircraft to become aligned with the final approach.

Procedure turn types include the 45° turn, the 80/260 turn, and the teardrop turn. All of these turns are normally conducted no more than 10 nautical miles (NM) from the primary airport. The procedure turn altitude generally provides a minimum of 1,000' obstacle clearance in the procedure turn area (not necessarily within the 10 NM arc around the primary airport). Turns may have to be increased or decreased but should not exceed 30° of a bank angle.

Standard 45° Procedure Turn

1. Start timing at point A (usually identified on approach procedures by a fix). For example, fly outbound on a heading of 360° for a given time (2 minutes, in this example). *[Figure 7-42]*
2. After flying outbound for 2 minutes (point B), turn left 45° to a heading of 315° using a standard rate turn. After roll-out and stabilizing, fly this new heading of 315° for 40 seconds and the aircraft will be at the approximate position of C.

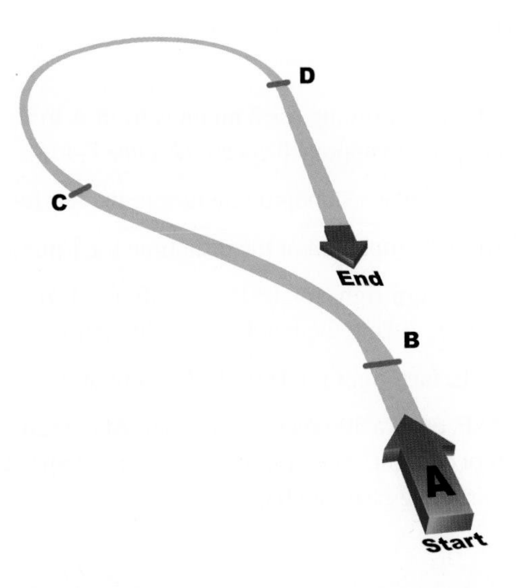

Figure 7-42. *Standard procedure turn (entire pattern in level flight).*

3. At point C, turn 225° right (using a standard rate turn) which will provide a heading of 180°. The timing is such that in a no wind environment, the pilot will be aligned with the final approach course of 180° at D. Wind conditions, however must be considered during the execution of the procedure turn. Compensating for wind may result in changes to outbound time, procedure turn heading and/or time and minor changes in the inbound turn.

80/260 Procedure Turn

1. Start timing at point A (usually identified on approach procedures by a fix). For example, fly outbound on a heading of 360° for 2 minutes. *[Figure 7-43]*

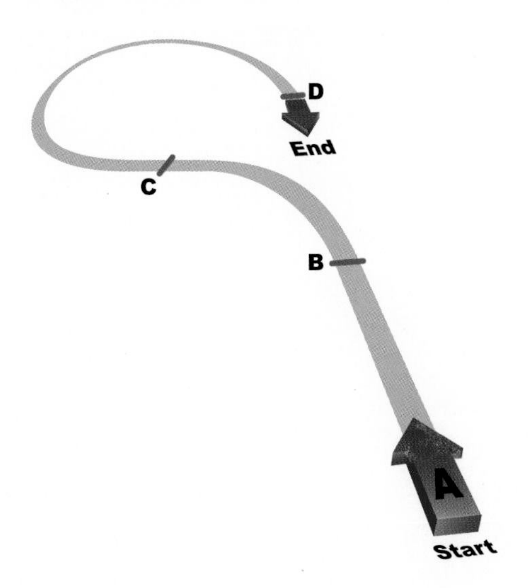

Figure 7-43. *80/260 procedure turn (entire pattern in level flight).*

2. At B, enter a left standard rate turn of 80° to a heading of 280°.
3. At the completion of the 80° turn to 280° (Point C), immediately turn right 260°, rolling-out on a heading of 180° (Point D) and also the reciprocal of the entry heading.

Teardrop Patterns

There are three typical teardrop procedure turns. A 30°, 20°, and a 10° teardrop pattern. The below steps indicate actions for all three starting on a heading of 360°. *[Figure 7-44]*

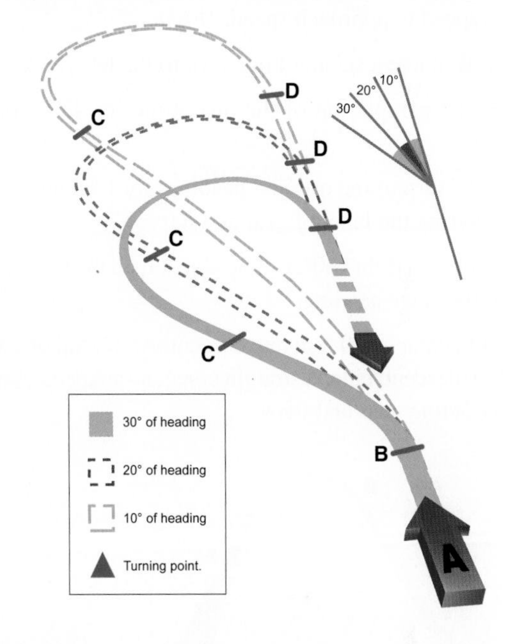

Figure 7-44. *Teardrop pattern (entire pattern in level flight).*

1. At point B (after stabilizing on the outbound course) turn left:
   - 30° to a heading of 330° and time for 1 minute
   - 20° to a heading of 340° and time for 2 minutes
   - 10° to a heading of 350° and time for 3 minutes
2. After the appropriate time above (Point C), make a standard rate turn to the right for:
   - 30° teardrop—210° to the final course heading of 180° (Point D)
   - 20° teardrop—200° to the final course heading of 180° (Point D)
   - 10° teardrop—190° to the final course heading of 180° (Point D)

By using the different teardrop patterns, a pilot is afforded the ability to manage time more efficiently. For instance, a 10° pattern for 3 minutes provides about three times the distance (and time) than a 30° pattern. Pattern selection should be based upon an individual assessment of the procedure turn requirements to include wind, complexity, the individual preparedness, etc.

Circling Approach Patterns

Pattern I

1. At A, start timing for 2 minutes from A to B; reduce airspeed to approach speed. *[Figure 7-45]*
2. At B, make a standard rate turn to the left for 45°.
3. At the completion of the turn, time for 45 seconds to C.
4. At C, turn to the original heading; fly 1 minute to D, lowering the landing gear and flaps.
5. At D, turn right 180°, rolling-out at E on the reciprocal of the entry heading.
6. At E, enter a 500 fpm rate descent. At the end of a 500 foot descent, enter a straight constant-airspeed climb, retracting gear and flaps.

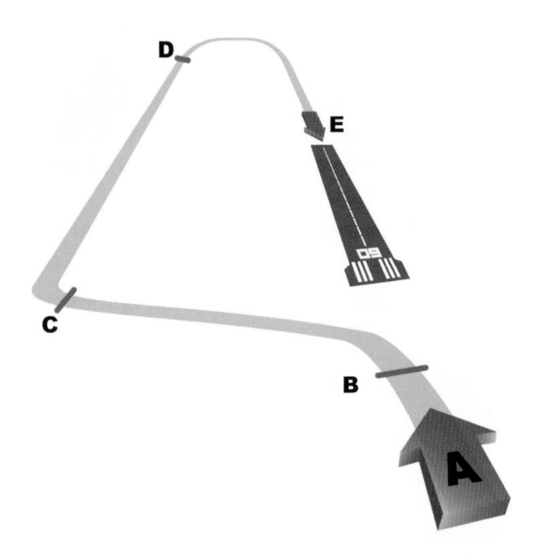

Figure 7-45. *Circling approach pattern I (imaginary runway).*

Pattern II

Steps:

1. At A, start timing for 2 minutes from A to B; reduce airspeed to approach speed. *[Figure 7-46]*
2. At B, make a standard rate turn to the left for 45°.
3. At the completion of the turn, time for 1 minute to C.
4. At C, turn right for 180° to D; fly for 1-1/2 minutes to E, lowering the landing gear and flaps.
5. At E, turn right for 180°, rolling-out at F.
6. At F, enter a 500 fpm rate descent. At the end of a 500 foot descent, enter a straight constant-airspeed climb, retracting gear and flaps.

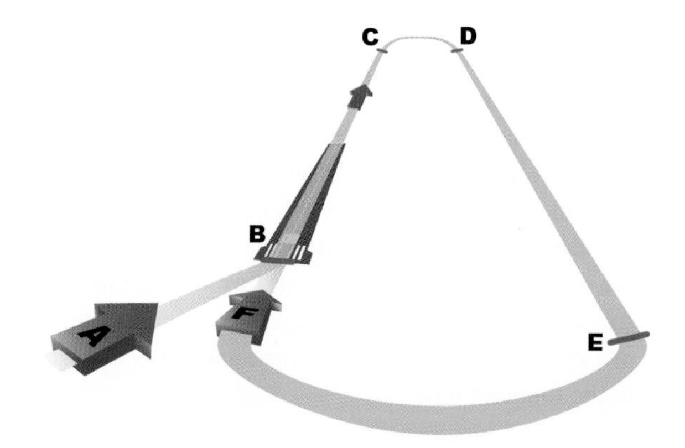

Figure 7-46. *Circling approach pattern II (imaginary runway).*

Chapter 7, Section II

Airplane Basic Flight Maneuvers

Using an Electronic Flight Display

Introduction

The previous chapters have laid the foundation for instrument flying. The pilot's ability to use and interpret the information displayed and apply corrective action is required to maneuver the aircraft and maintain safe flight. A pilot must recognize that each aircraft make and model flown may require a different technique. Aircraft weight, speed, and configuration changes require the pilot to vary his or her technique in order to perform successful attitude instrument flying. A pilot must become familiar with all sections of the Pilot's Operating Handbook/Airplane Flight Manual (POH/AFM) prior to performing any flight maneuver.

Chapter 7, Section II describes basic attitude instrument flight maneuvers and explains how to perform each one by interpreting the indications presented on the electronic flight display (EFD). In addition to normal flight maneuvers, "partial panel" flight is addressed. With the exception of the instrument takeoff, all flight maneuvers can be performed on "partial panel" with the Attitude Heading Reference System (AHRS) unit simulated or rendered inoperative.

Straight-and-Level Flight

Pitch Control

The pitch attitude of an airplane is the angle between the longitudinal axis of the airplane and the actual horizon. In level flight, the pitch attitude varies with airspeed and load. For training purposes, the latter factor can normally be disregarded in small airplanes. At a constant airspeed, there is only one specific pitch attitude for level flight. At slow cruise speeds, the level flight attitude is nose-high with indications as in *Figure 7-47*; at fast cruise speeds, the level flight attitude is nose-low. *[Figure 7-48] Figure 7-49* shows the indications for the attitude at normal cruise speeds.

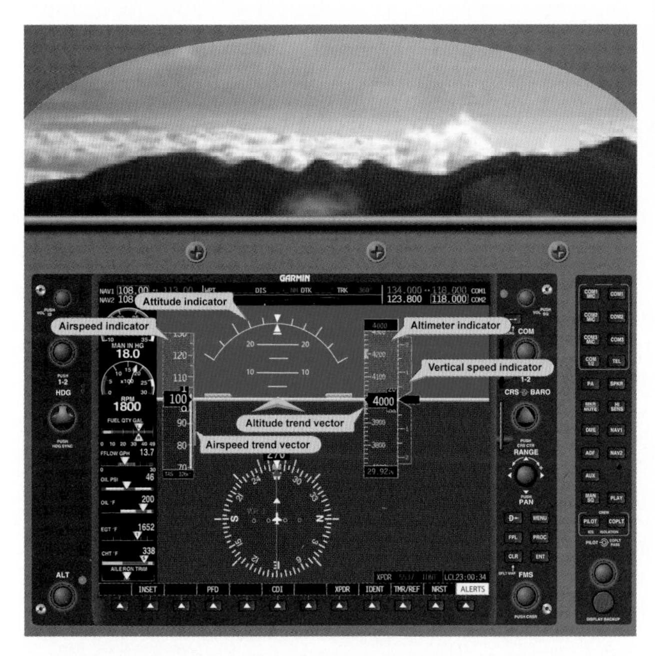

Figure 7-47. *Pitch attitude and airspeed in level flight, slow cruise speed.*

Figure 7-49. *Various pitch attitudes (right), aircraft shown in level flight.*

The instruments that directly or indirectly indicate pitch on the primary flight display (PFD) are the attitude indicator, altimeter, vertical speed indicator (VSI), airspeed indicator (ASI), and both airspeed and altitude trend indicators.

Attitude Indicator

The attitude indicator gives the pilot a direct indication of the pitch attitude. The increased size of the attitude display on the EFD system greatly increases situational awareness for the pilot. Most attitude indicators span the entire width of the PFD screen.

Figure 7-48. *Pitch attitude decreasing and airspeed increasing—indicates need to increase pitch.*

The aircraft pitch attitude is controlled by changing the deflection of the elevator. As the pilot pulls back on the control yoke causing the elevator to rise, the yellow chevron begins to show a displacement up from the artificial horizon line. This is caused by the AHRS unit sensing the changing angle between the longitudinal plane of the earth and the longitudinal axis of the aircraft.

The attitude indicator displayed on the PFD screen is a representation of outside visual cues. Rather than rely on the natural horizon visible during visual flight rules (VFR) flight, the pilot must rely on the artificial horizon of the PFD screen.

During normal cruise airspeed, the point of the yellow chevron (aircraft symbol) is positioned on the artificial horizon. Unlike conventional attitude indicators, the EFD attitude indicator does not allow for manipulating the position of the chevron in relationship to the artificial horizon. The position is fixed and therefore always display the pitch angle as calculated by the AHRS unit.

The attitude indicator only shows pitch attitude and does not indicate altitude. A pilot should not attempt to maintain level flight using the attitude indicator alone. It is important for the pilot to understand how small displacements both up and down can affect the altitude of the aircraft. To achieve this, the pilot should practice increasing the pitch attitude incrementally to become familiar with how each degree of pitch changes the altitude. *[Figures 7-50* and *7-51]* In both cases, the aircraft will slow and gain altitude.

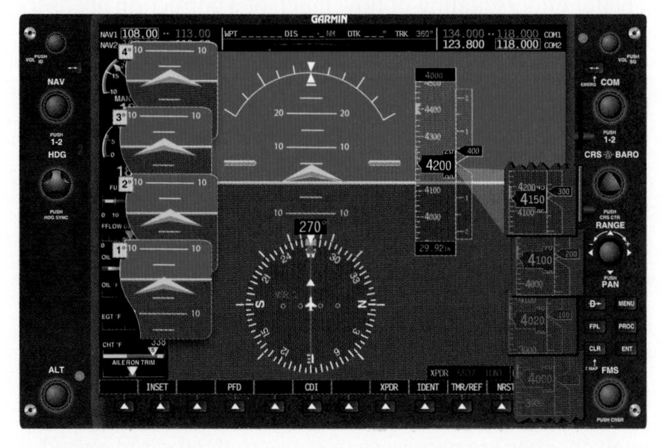

Figure 7-50. *Pitch indications for various attitudes (1° through 5°).*

The full height of the chevron is approximately 5 degrees and provides an accurate reference for pitch adjustment. It is imperative that the pilot make the desired changes to pitch by referencing the attitude indicator and then trimming off any excess control pressures. Relieving these pressures allow for a more stabilized flight and reduces pilot work load. Once the aircraft is trimmed for level flight, the pilot must smoothly and precisely manipulate the elevator control forces in order to change the pitch attitude.

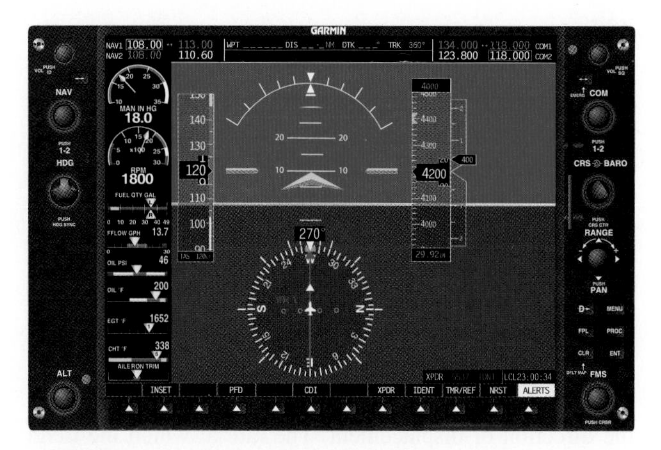

Figure 7-51. *Pitch illustrated at 10°.*

To master the ability to smoothly control the elevator, a pilot must develop a very light touch on the control yoke. The thumb and two fingers are normally sufficient to move the control yoke. The pilot should avoid griping the yoke with a full fist. When a pilot grips the yoke with a full fist, there is a tendency to apply excess pressures, thus changing the aircraft attitude.

Practice making smooth, small pitch changes both up and down until precise corrections can be made. With practice, a pilot is able to make pitch changes in 1 degree increments, smoothly controlling the attitude of the aircraft.

The last step in mastering elevator control is trim. Trimming the aircraft to relieve any control pressures is essential for smooth attitude instrument flight. To accomplish this, momentarily release the control yoke. Note which way the aircraft pitch attitude wants to move. Grasp the control yoke again and then reapply the pressure to return the attitude to the previous position. Apply trim in the direction of the control pressure. Small applications of trim make large changes in the pitch attitude. Be patient and make multiple changes to trim, if necessary.

Once the aircraft is in trim, relax on the control yoke as much as practicable. When pressure is held on the yoke, unconscious pressures are applied to the elevator and ailerons, which displaces the aircraft from its desired flightpath. If the aircraft is in trim, in calm, non-turbulent air, a pilot should be able to release the control yoke and maintain level flight for extended periods of time. This is one of the hardest skills to learn prior to successfully flying in instrument meteorological conditions (IMC).

Altimeter

At constant power, any deviation from level flight (except in turbulent air) must be the result of a pitch change. If the power is constant, the altimeter gives an indirect indication of the pitch attitude in level flight. Since the altitude should remain constant when the airplane is in level flight, any deviation from the desired altitude signals the need for a pitch change. For example, if the aircraft is gaining altitude, the nose must be lowered.

In the PFD, as the pitch starts to change, the altitude trend indicator on the altitude tape begins to show a change in the direction of displacement. The rate at which the trend indicator grows and the altimeter numbers change aids the pilot in determining how much of a pitch change is necessary to stop the trend.

As a pilot becomes familiar with a specific aircraft's instruments, he or she learns to correlate pitch changes, altimeter tapes, and altitude trend indicators. By adding the altitude tape display and the altitude trend indicator into the scan along with the attitude indicator, a pilot starts to develop the instrument cross-check.

Partial Panel Flight

One important skill to practice is partial panel flight by referencing the altimeter as the primary pitch indicator. Practice controlling the pitch by referencing the altitude tape and trend indicator alone without the use of the attitude indicator. Pilots need to learn to make corrections to altitude deviations by referencing the rate of change of the altitude tape and trend indicator. When operating in IMC and in a partial panel configuration, the pilot should avoid abrupt changes to the control yoke. Reacting abruptly to altitude changes can lead to large pitch changes and thus a larger divergence from the initial altitude.

When a pilot is controlling pitch by the altitude tape and altitude trend indicators alone, it is possible to overcontrol the aircraft by making a larger than necessary pitch correction. Overcontrolling causes the pilot to move from a nose-high attitude to a nose-low attitude and vice versa. Small changes to pitch are required to insure prompt corrective actions are taken to return the aircraft to its original altitude with less confusion.

When an altitude deviation occurs, two actions need to be accomplished. First, make a smooth control input to stop the needle movement. Once the altitude tape has stopped moving, make a change to the pitch attitude to start back to the entry altitude.

During instrument flight with limited instrumentation, it is imperative that only small and precise control inputs are made. Once a needle movement is indicated denoting a deviation in altitude, the pilot needs to make small control inputs to stop the deviation. Rapid control movements only compound the deviation by causing an oscillation effect. This type of oscillation can quickly cause the pilot to become disoriented and begin to fixate on the altitude. Fixation on the altimeter can lead to a loss of directional control as well as airspeed control.

As a general rule of thumb, for altitude deviations less than 100 feet, utilize a pitch change of 1 degree, which equates to ⅕ of the thickness of the chevron. Small incremental pitch changes allow the performance to be evaluated and eliminate overcontrolling of the aircraft.

Instrumentation needs to be utilized collectively, but failures will occur that leave the pilot with only limited instrumentation. That is why partial panel flying training is important. If the pilot understands how to utilize each instrument independently, no significant change is encountered in carrying out the flight when other instruments fail.

VSI Tape

The VSI tape provides for an indirect indication of pitch attitude and gives the pilot a more immediate indication of a pending altitude deviation. In addition to trend information, the vertical speed also gives a rate indication. By using the VSI tape in conjunction with the altitude trend tape, a pilot has a better understanding of how much of a correction needs to be made. With practice, the pilot will learn the performance of a particular aircraft and know how much pitch change is required in order to correct for a specific rate indication.

Unlike older analog VSIs, new glass panel displays have instantaneous VSIs. Older units had a lag designed into the system that was utilized to indicate rate information. The new glass panel displays utilize a digital air data computer that does not indicate a lag. Altitude changes are shown immediately and can be corrected for quickly.

The VSI tape should be used to assist in determining what pitch changes are necessary to return to the desired altitude. A good rule of thumb is to use a vertical speed rate of change that is double the altitude deviation. However, at no time should the rate of change be more than the optimum rate of climb or descent for the specific aircraft being flown. For example, if the altitude is off by 200 feet from the desired altitude, then a 400 feet per minute (fpm) rate of change would be sufficient to get the aircraft back to the original altitude. If the altitude has changed by 700 feet, then doubling that would necessitate a 1,400 fpm change. Most aircraft are not capable of that, so restrict changes to no more than optimum climb and descent. An optimum rate of change would vary between 500 and 1,000 fpm.

One error the instrument pilot encounters is overcontrolling. Overcontrolling occurs when a deviation of more than 200 fpm is indicated over the optimum rate of change. For example, an altitude deviation of 200 feet is indicated on the altimeter, a vertical speed rate of 400 feet should be indicated on the gauge. If the vertical speed rate showed 600 fpm (200 more than optimum), the pilot would be overcontrolling the aircraft.

When returning to altitude, the primary pitch instrument is the VSI tape. If any deviation from the desired vertical speed is indicated, make the appropriate pitch change using the attitude indicator.

As the aircraft approaches the target altitude, the vertical speed rate can be slowed in order to capture the altitude in a more stabilized fashion. Normally within 10 percent of the rate of climb or descent from the target altitude, begin to slow the vertical speed rate in order to level off at the target altitude. This allows the pilot to level at the desired altitude without rapid control inputs or experiencing discomfort due to G-load.

Airspeed Indicator (ASI)

The ASI presents an indirect indication of the pitch attitude. At a constant power setting and pitch attitude, airspeed remains constant. As the pitch attitude lowers, airspeed increases, and the nose should be raised.

As the pitch attitude is increased, the nose of the aircraft raises, which results in an increase in the angle of attack as well as an increase in induced drag. The increased drag begins to slow the momentum of the aircraft, which is indicated on the ASI. The airspeed trend indicator shows a trend as to where the airspeed will be in 6 seconds. Conversely, if the nose of the aircraft should begin to fall, the angle of attack, as well as induced drag, decreases.

There is a lag associated with the ASI when using it as a pitch instrument. It is not a lag associated with the construction of the ASI, but a lag associated with momentum change. Depending on the rate of momentum change, the ASI may not indicate a pitch change in a timely fashion. If the ASI is being used as the sole reference for pitch change, it may not allow for a prompt correction. However, if smooth pitch changes are executed, modern glass panel displays are capable of indicating 1 knot changes in airspeed and also capable of projecting airspeed trends.

When flying by reference to flight instruments alone, it is imperative that all of the flight instruments be cross-checked for pitch control. By cross-checking all pitch related instruments, the pilot can better visualize the aircraft attitude at all times.

As previously stated, the primary instrument for pitch is the instrument that gives the pilot the most pertinent information for a specific parameter. When in level flight and maintaining a constant altitude, what instrument shows a direct indication of altitude? The only instrument that is capable of showing altitude is the altimeter. The other instruments are supporting instruments that are capable of showing a trend away from altitude, but do not directly indicate an altitude.

The supporting instruments forewarn of an impending altitude deviation. With an efficient cross-check, a proficient pilot is better able to maintain altitude.

Bank Control

This discussion assumes the aircraft is being flown in coordinated flight, which means the longitudinal axis of the aircraft is aligned with the relative wind. On the PFD, the attitude indicator shows if the wings are level. The turn rate indicator, slip/skid indicator, and the heading indicator also indicate whether or not the aircraft is maintaining a straight (zero bank) flightpath.

Attitude Indicator

The attitude indicator is the only instrument on the PFD that has the capability of displaying the precise bank angle of the aircraft. This is made possible by the display of the roll scale depicted as part of the attitude indicator.

Figure 7-52 identifies the components that make up the attitude indicator display. Note that the top of the display is blue, representing sky, the bottom is brown, depicting dirt, and the white line separating them is the horizon. The lines parallel to the horizon line are the pitch scale, which is marked in 5 degree increments and labeled every 10 degrees. The pitch scale always remains parallel to the horizon.

The curved line in the blue area is the roll scale. The triangle on the top of the scale is the zero index. The hash marks on the scale represent the degree of bank. *[Figure 7-53]* The roll scale always remains in the same position relative to the horizon line.

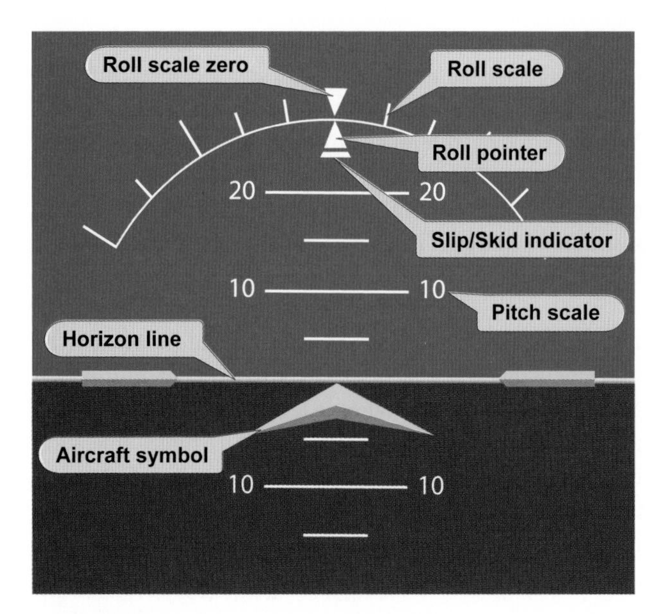

Figure 7-52. *Attitude indicator.*

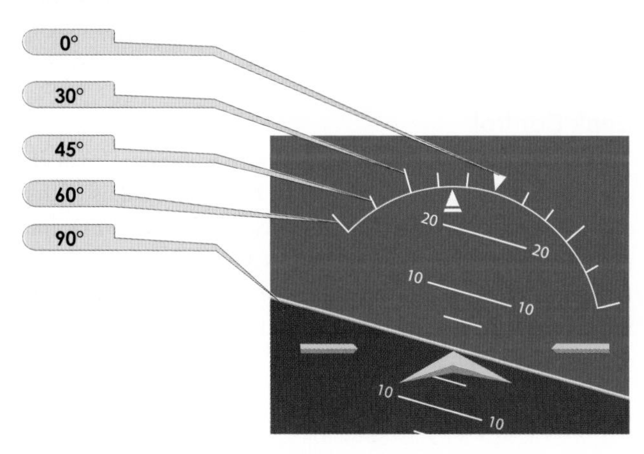

Figure 7-53. *Attitude indicator showing a 15° left bank.*

The roll pointer indicates the direction and degree of bank. *[Figure 7-53]* The roll pointer is aligned with the aircraft symbol. The roll pointer indicates the angle of the lateral axis of the aircraft compared to the natural horizon. The slip/skid indicator will show if the longitudinal axis of the aircraft is aligned with the relative wind, which is coordinated flight. With the roll index and the slip/skid indicator aligned, any deflection, either right or left of the roll index causes the aircraft to turn in that direction. With the small graduations on the roll scale, it is easy to determine the bank angle within approximately 1 degree. In coordinated flight, if the roll index is aligned with the roll pointer, the aircraft is achieving straight flight.

An advantage of EFDs is the elimination of the precession error. Precession error in analog gauges is caused by forces being applied to a spinning gyro. With the new solid state instruments, precession error has been eliminated.

Since the attitude indicator is capable of showing precise pitch and bank angles, the only time that the attitude indicator is a primary instrument is when attempting to fly at a specific bank angle or pitch angle. Other times, the attitude instrument can be thought of as a control instrument.

Horizontal Situation Indicator (HSI)

The horizontal situation indicator (HSI) is a rotating 360° compass card that indicates magnetic heading. The HSI is the only instrument that is capable of showing exact headings. The magnetic compass can be used as a backup instrument in case of an HSI failure; however, due to erratic, unstable movements, it is more likely to be used a supporting instrument.

In order for the pilot to achieve the desired rate of change, it is important for him or her to understand the relationship between the rate at which the HSI changes heading displays and the amount of bank angle required to meet that rate of change. A very small rate of heading change means the bank angle is small, and it takes more time to deviate from the desired straight flightpath. A larger rate of heading change means a greater bank angle happens at a faster rate.

Heading Indicator

The heading indicator is the large black box with a white number that indicates the magnetic heading of the aircraft. *[Figure 7-54]* The aircraft heading is displayed to the nearest degree. When this number begins to change, the pilot should be aware that straight flight is no longer being achieved.

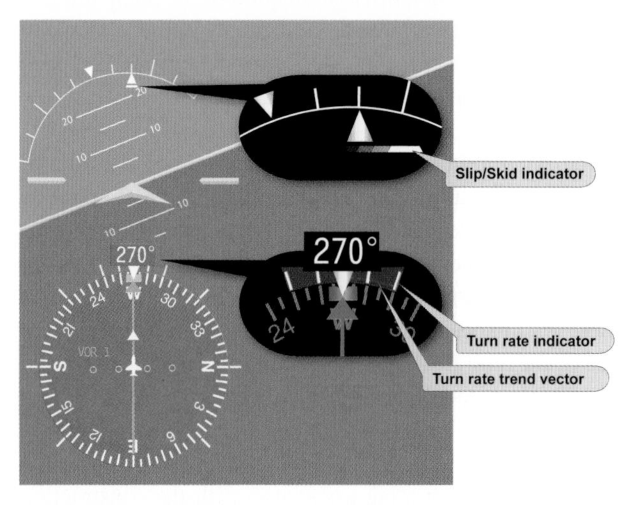

Figure 7-54. *Slip/skid and turn rate indicator.*

Turn Rate Indicator

The turn rate indicator gives an indirect indication of bank. It is a magenta trend indicator capable of displaying half-standard as well as standard rate turns to both the left and right. *[Figure 7-54]* The turn indicator is capable of indicating turns up to 4 degrees per second by extending the magenta line outward from the standard rate mark. If the rate of turn has exceeded 4 degrees per second, the magenta line can not precisely indicate where the heading will be in the next 6 seconds; the magenta line freezes and an arrowhead will be displayed. This alerts the pilot to the fact that the normal range of operation has been exceeded.

Slip/Skid Indicator

The slip/skid indicator is the small portion of the lower segmented triangle displayed on the attitude indicator. This instrument depicts whether the aircraft's longitudinal axis is aligned with the relative wind. *[Figure 7-54]*

The pilot must always remember to cross-check the roll index to the roll pointer when attempting to maintain straight flight. Any time the heading remains constant and the roll pointer and the roll index are not aligned, the aircraft is in uncoordinated flight. To make a correction, the pilot should apply rudder pressure to bring the aircraft back to coordinated flight.

Power Control

Power produces thrust which, with the appropriate angle of attack of the wing, overcomes the forces of gravity, drag, and inertia to determine airplane performance.

Power control must be related to its effect on altitude and airspeed, since any change in power setting results in a change in the airspeed or the altitude of the airplane. At any given airspeed, the power setting determines whether the airplane is in level flight, in a climb, or in a descent. If the power is increased in straight-and-level flight and the airspeed held constant, the airplane climbs; if power is decreased while the airspeed is held constant, the airplane descends. On the other hand, if altitude is held constant, the power applied determines the airspeed.

The relationship between altitude and airspeed determines the need for a change in pitch or power. If the airspeed is off the desired value, always check the altimeter before deciding that a power change is necessary. Think of altitude and airspeed as interchangeable; altitude can be traded for airspeed by lowering the nose, or convert airspeed to altitude by raising the nose. If altitude is higher than desired and airspeed is low, or vice versa, a change in pitch alone may return the airplane to the desired altitude and airspeed. *[Figure 7-55]* If both airspeed and altitude are high or if both are low, then a change in both pitch and power is necessary in order to return to the desired airspeed and altitude. *[Figure 7-56]*

For changes in airspeed in straight-and-level flight, pitch, bank, and power must be coordinated in order to maintain constant altitude and heading. When power is changed to vary airspeed in straight-and-level flight, a single-engine, propeller-driven airplane tends to change attitude around all axes of movement. Therefore, to maintain constant altitude and heading, apply various control pressures in proportion to the change in power. When power is added to increase airspeed, the pitch instruments indicate a climb unless forward-elevator control pressure is applied as the airspeed changes. With an increase in power, the airplane tends to yaw and roll to the left unless counteracting aileron and rudder pressures are applied. Keeping ahead of these changes requires increasing cross-check speed, which varies with the type of airplane and its torque characteristics, the extent of power and speed change involved.

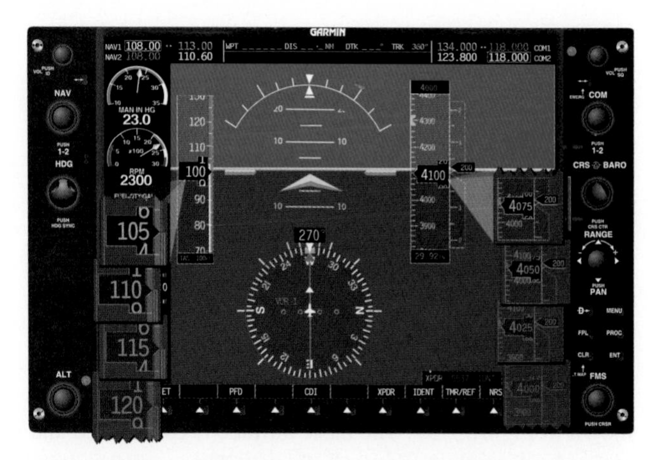

Figure 7-55. *An aircraft decreasing in airspeed while gaining altitude. In this case, the pilot has decreased pitch.*

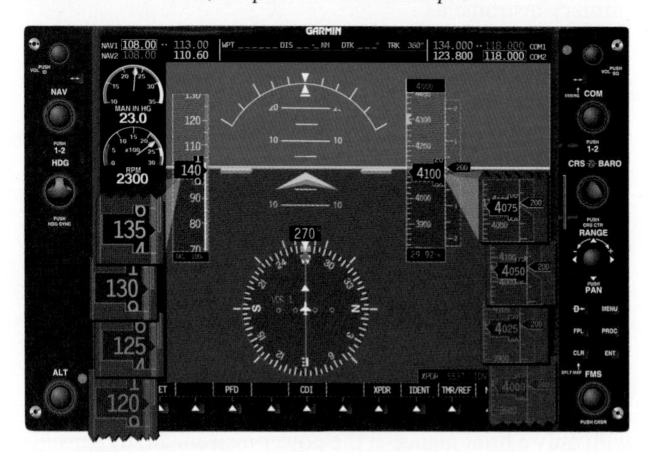

Figure 7-56. *Figure shows both an increase in speed and altitude where pitch adjustment alone is insufficient. In this situation, a reduction of power is also necessary.*

Power Settings

Power control and airspeed changes are much easier when approximate power settings necessary to maintain various airspeeds in straight-and-level flight are known in advance. However, to change airspeed by any appreciable amount, the common procedure is to underpower or overpower on initial power changes to accelerate the rate of airspeed change. (For small speed changes, or in airplanes that decelerate or accelerate rapidly, overpowering or underpowering is not necessary.)

Consider the example of an airplane that requires 23 inches of mercury ("Hg) to maintain a normal cruising airspeed of 120 knots, and 18 "Hg to maintain an airspeed of 100 knots. The reduction in airspeed from 120 knots to 100 knots while maintaining straight-and-level flight is discussed below and illustrated in *Figures 7-57, 7-58,* and *7-59.*

Instrument indications, prior to the power reduction, are shown in *Figure 7-57*. The basic attitude is established and maintained on the attitude indicator. The specific pitch, bank, and power control requirements are detected on these primary instruments:

Altimeter—Primary Pitch
Heading Indicator—Primary Bank
Airspeed Indicator—Primary Power

Supporting pitch and bank instruments are shown in *Figure 7-57*. Note that the supporting power instrument is the manifold pressure gauge (or tachometer if the propeller is fixed pitch). However, when a smooth power reduction to approximately 15 "Hg (underpower) is made, the manifold pressure gauge becomes the primary power instrument. *[Figure 7-58]* With practice, power setting can be changed with only a brief glance at the power instrument, by sensing the movement of the throttle, the change in sound, and the changes in the feel of control pressures.

As the thrust decreases, increase the speed of the cross-check and be ready to apply left rudder, back-elevator, and aileron control pressure the instant the pitch and bank instruments show a deviation from altitude and heading. As proficiency is obtained, a pilot will learn to cross-check, interpret, and control the changes with no deviation of heading and altitude. Assuming smooth air and ideal control technique, as airspeed decreases, a proportionate increase in airplane pitch attitude is required to maintain altitude. Similarly, effective torque control means counteracting yaw with rudder pressure.

As the power is reduced, the altimeter is primary for pitch, the heading indicator is primary for bank, and the manifold pressure gauge is momentarily primary for power (at 15 "Hg in *Figure 7-58*). Control pressures should be trimmed off as the airplane decelerates. As the airspeed approaches the desired airspeed of 100 knots, the manifold pressure is adjusted to approximately 18 "Hg and becomes the supporting power instrument. The ASI again becomes primary for power. *[Figure 7-59]*

Airspeed Changes in Straight-and-Level Flight

Practice of airspeed changes in straight-and-level flight provides an excellent means of developing increased proficiency in all three basic instrument skills and brings out some common errors to be expected during training in straight-and-level flight. Having learned to control the airplane in a clean configuration (minimum drag conditions), increase proficiency in cross-check and control by practicing speed changes while extending or retracting the flaps and landing gear. While practicing, be sure to comply with the

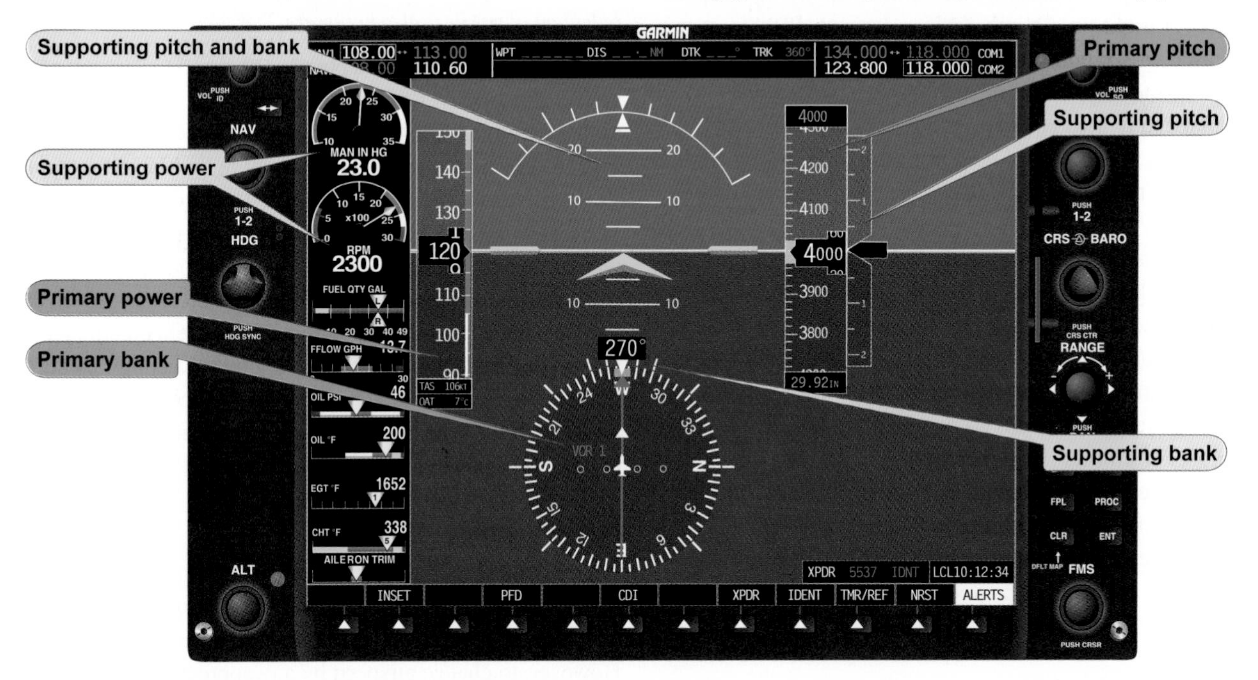

Figure 7-57. *Straight-and-level flight (normal cruising speed).*

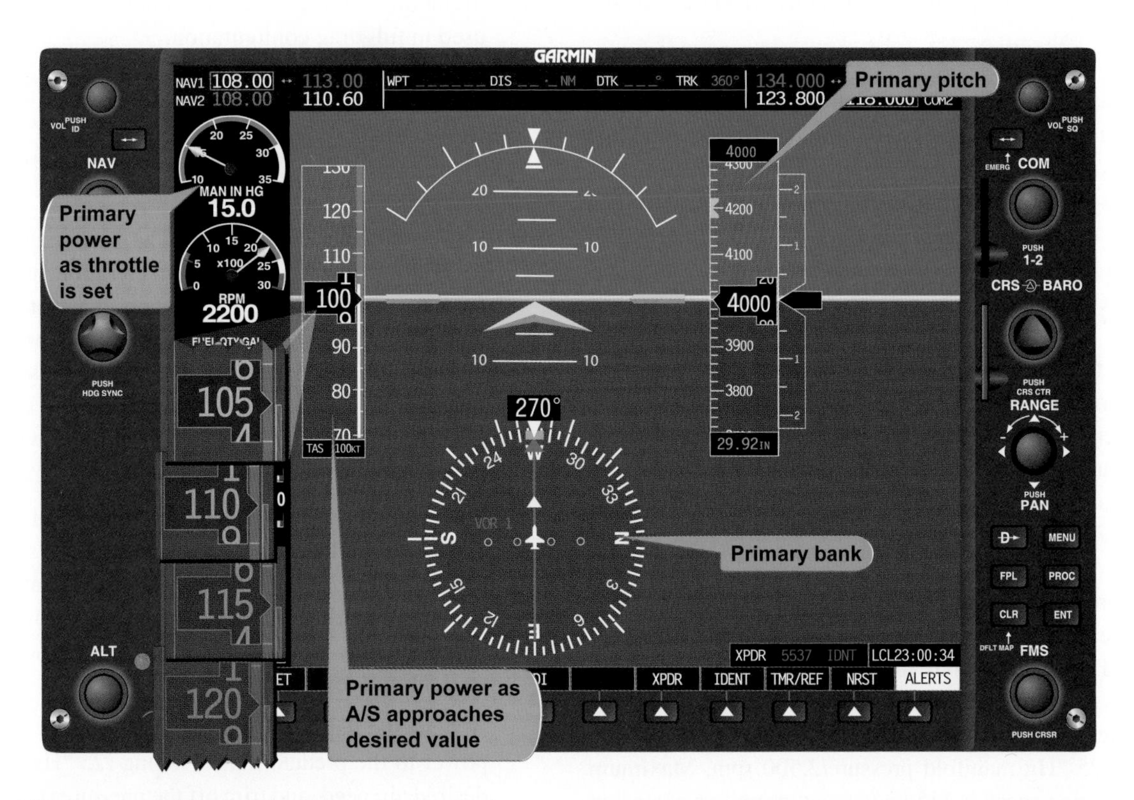

Figure 7-58. *Straight-and-level flight (airspeed decreasing).*

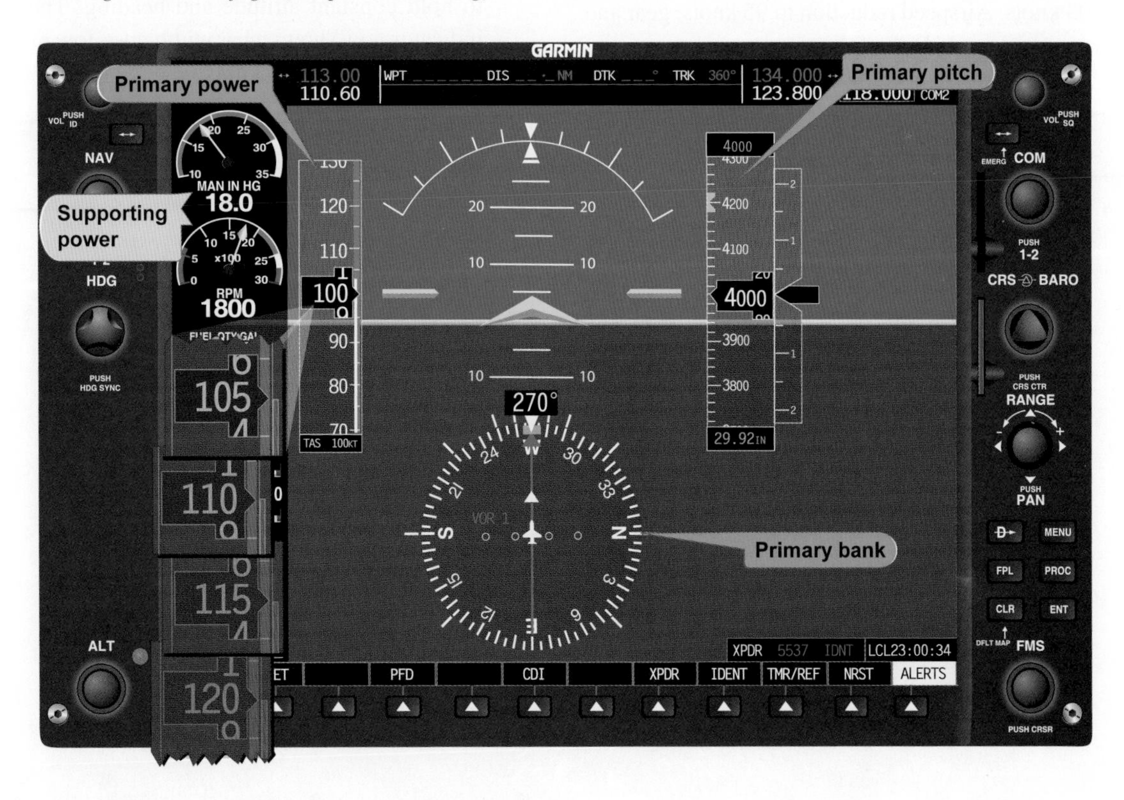

Figure 7-59. *Straight-and-level flight (reduced airspeed stabilized).*

airspeed limitations specified in the POH/AFM for gear and flap operation.

Sudden and exaggerated attitude changes may be necessary in order to maintain straight-and-level flight as the landing gear is extended and the flaps are lowered in some airplanes. The nose tends to pitch down with gear extension, and when flaps are lowered, lift increases momentarily (at partial flap settings) followed by a marked increase in drag as the flaps near maximum extension.

Control technique varies according to the lift and drag characteristics of each airplane. Accordingly, knowledge of the power settings and trim changes associated with different combinations of airspeed, gear, and flap configurations reduces instrument cross-check and interpretation problems. *[Figure 7-60]*

For example, assume that in straight-and-level flight instruments indicate 120 knots with power at 23 "Hg manifold pressure/2,300 revolutions per minute (rpm), gear and flaps up. After reduction in airspeed, with gear and flaps fully extended, straight-and-level flight at the same altitude requires 25 "Hg manifold pressure/2,500 rpm. Maximum gear extension speed is 115 knots; maximum flap extension speed is 105 knots. Airspeed reduction to 95 knots, gear and flaps down, can be made in the following manner:

1. Maintain rpm at 2,500, since a high power setting is used in full drag configuration.
2. Reduce manifold pressure to 10 "Hg. As the airspeed decreases, increase cross-check speed.
3. Make trim adjustments for an increased angle of attack and decrease in torque.
4. Lower the gear at 115 knots. The nose may tend to pitch down and the rate of deceleration increases. Increase pitch attitude to maintain constant altitude and trim off some of the back-elevator pressures. If full flaps are lowered at 105 knots, cross-check, interpretation, and control must be very rapid. A simpler technique is to stabilize attitude with gear down before lowering the flaps.
5. Since 18 "Hg manifold pressure holds level flight at 100 knots with the gear down, increase power smoothly to that setting as the ASI shows approximately 105 knots, and retrim. The attitude indicator now shows approximately two-and-a-half bar width nose-high in straight-and-level flight.
6. Actuate the flap control and simultaneously increase power to the predetermined setting (25 "Hg) for the desired airspeed, and trim off the pressures necessary to hold constant altitude and heading. The attitude indicator now shows a bar width nose-low in straight-and-level flight at 95 knots.

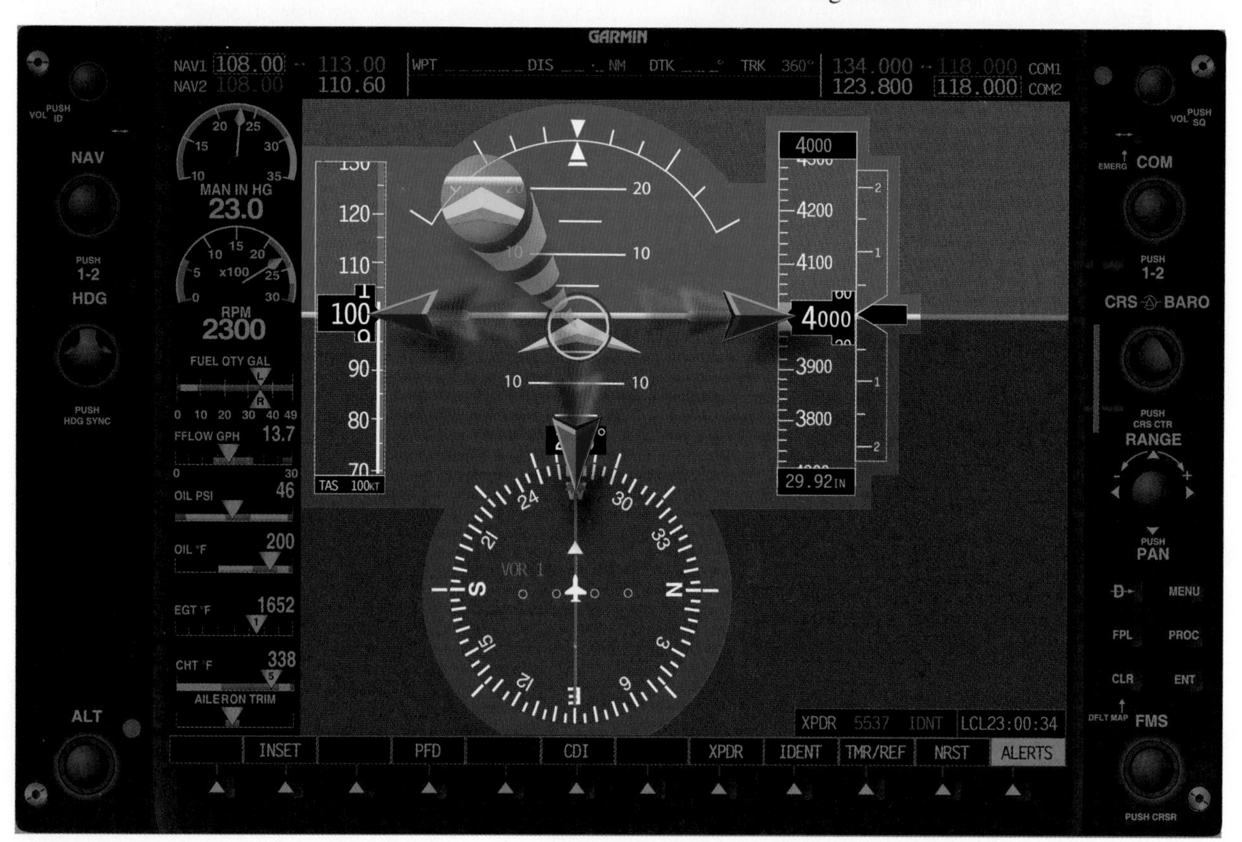

Figure 7-60. *Cross-check supporting instruments.*

Trim Technique

Trim control is one of the most important flight habits to cultivate. Trimming refers to relieving any control pressures that need to be applied by the pilot to the control surfaces to maintain a desired flight attitude. The desired result is for the pilot to be able to take his or her hands off the control surfaces and have the aircraft remain in the current attitude. Once the aircraft is trimmed for hands-off flight, the pilot is able to devote more time to monitoring the flight instruments and other aircraft systems.

In order to trim the aircraft, apply pressure to the control surface that needs trimming and roll the trim wheel in the direction pressure is being held. Relax the pressure that is being applied to the control surface and monitor the primary instrument for that attitude. If the desired performance is achieved, fly hands off. If additional trimming is required, redo the trimming steps.

An aircraft is trimmed for a specific airspeed, not pitch attitude or altitude. Any time an aircraft changes airspeed, there is a need to re-trim. For example, an aircraft is flying at 100 knots straight-and-level. An increase of 50 rpm causes the airspeed to increase. As the airspeed increases, additional lift is generated and the aircraft climbs. Once the additional thrust has stabilized at some higher altitude, the airspeed will again stabilize at 100 knots.

This demonstrates how trim is associated with airspeed and not altitude. If the initial altitude is to be maintained, forward pressure would need to be applied to the control wheel while the trim wheel needs to be rolled forward to eliminate any control pressures. Rolling forward on the trim wheel is equal to increasing for a trimmed airspeed. Any time the airspeed is changed, re-trimming is required. Trimming can be accomplished during any transitional period; however, prior to final trimming, the airspeed must be held constant. If the airspeed is allowed to change, the trim is not adjusted properly and the altitude varies until the airspeed for which the aircraft is trimmed is achieved.

Common Errors in Straight-and-Level Flight

Pitch

Pitch errors usually result from the following errors:

1. Improper adjustment of the yellow chevron (aircraft symbol) on the attitude indicator.

   Corrective Action: Once the aircraft has leveled off and the airspeed has stabilized, make small corrections to the pitch attitude to achieve the desired performance. Cross-check the supporting instruments for validation.

2. Insufficient cross-check and interpretation of pitch instruments. *[Figure 7-61]*

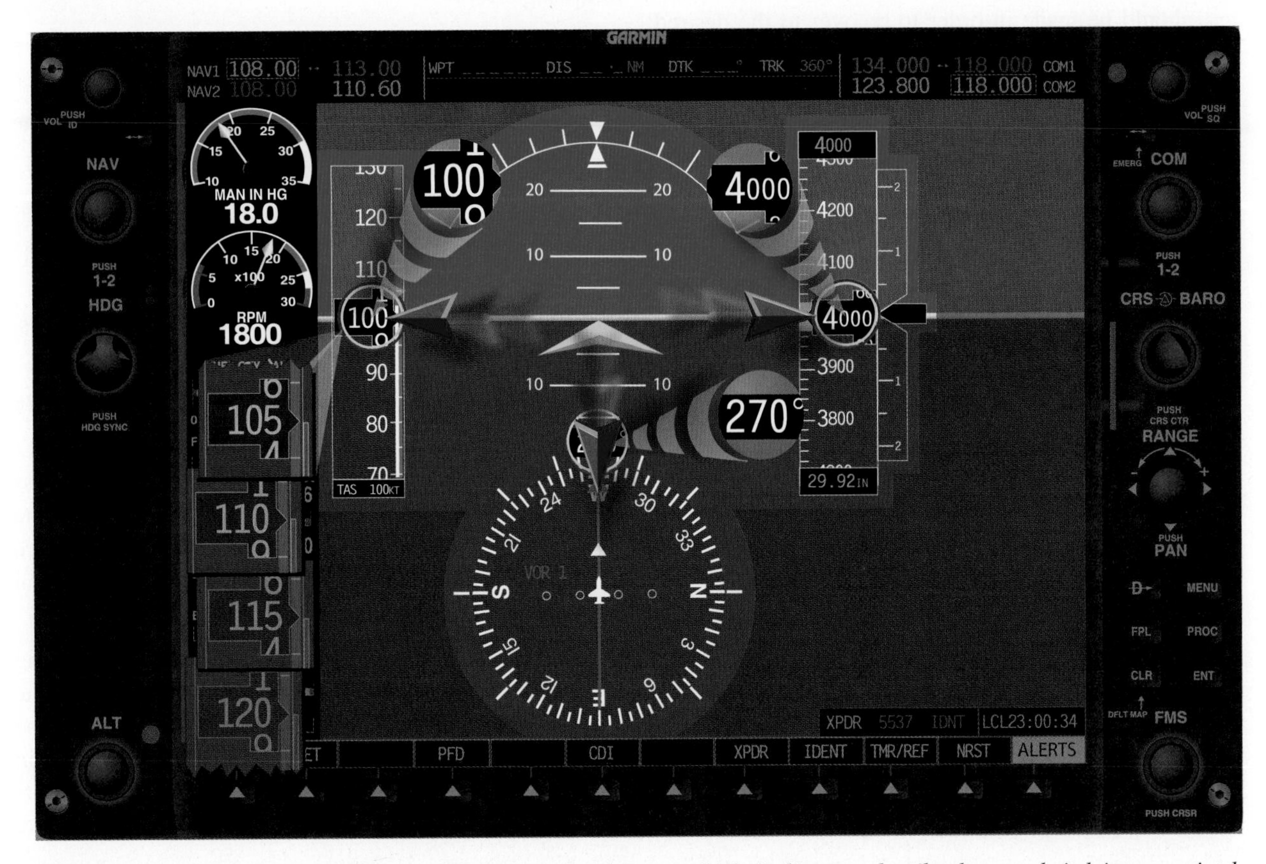

Figure 7-61. *Insufficient cross-check. The problem is power and not nose-high. In this case, the pilot decreased pitch inappropriately.*

Example: The airspeed indication is low. The pilot, believing a nose-high pitch attitude exists, applies forward pressure without noting that a low power setting is the cause of the airspeed discrepancy.

Corrective Action: Increase the rate of cross-check of all the supporting flight instruments. Airspeed and altitude should be stabilized before making a control input.

3. Acceptance of deviations.

   Example: A pilot has an altitude range of ±100 feet according to the practical test standards for straight-and level-flight. When the pilot notices that the altitude has deviated by 60 feet, no correction is made because the altitude is holding steady and is within the standards.

   Corrective Action: The pilot should cross-check the instruments and, when a deviation is noted, prompt corrective actions should be taken in order to bring the aircraft back to the desired altitude. Deviations from altitude should be expected but not accepted.

4. Overcontrolling—excessive pitch changes.

   Example: A pilot notices a deviation in altitude. In an attempt to quickly return to altitude, the pilot makes a large pitch change. The large pitch change destabilizes the attitude and compounds the error.

   Corrective Action: Small, smooth corrections should be made in order to recover to the desired altitude (0.5° to 2° depending on the severity of the deviation). Instrument flying is comprised of small corrections to maintain the aircraft attitude. When flying in IMC, a pilot should avoid making large attitude changes in order to avoid loss of aircraft control and spatial disorientation.

5. Failure to maintain pitch corrections.

   Pitch changes need to be made promptly and held for validation. Many times pilots make corrections and allow the pitch attitude to change due to not trimming the aircraft. It is imperative that any time a pitch change is made; the trim is readjusted in order to eliminate any control pressures that are being held. A rapid cross-check aids in avoiding any deviations from the desired pitch attitude.

   Example: A pilot notices a deviation in altitude. A change in the pitch attitude is accomplished but no adjustment to the trim is made. Distractions cause the pilot to slow the cross-check and an inadvertent reduction in the pressure to the control column commences. The pitch attitude then changes, thus complicating recovery to the desired altitude.

   Corrective Action: The pilot should initiate a pitch change and then immediately trim the aircraft to relieve any control pressures. A rapid cross-check should be established in order to validate the desired performance is being achieved.

6. Fixation during cross-check.

   Devoting an unequal amount of time to one instrument either for interpretation or assigning too much importance to an instrument. Equal amounts of time should be spent during the cross-check to avoid an unnoticed deviation in one of the aircraft attitudes.

   Example: A pilot makes a correction to the pitch attitude and then devotes all of the attention to the altimeter to determine if the pitch correction is valid. During this time, no attention is paid to the heading indicator, which shows a turn to the left. *[Figure 7-62]*

   Corrective Action: The pilot should monitor all instrumentation during the cross-check. Do not fixate on one instrument waiting for validation. Continue to scan all instruments to avoid allowing the aircraft to begin a deviation in another attitude.

Heading

Heading errors usually result from but are not limited to the following errors:

1. Failure to cross-check the heading indicator, especially during changes in power or pitch attitude.
2. Misinterpretation of changes in heading, with resulting corrections in the wrong direction.
3. Failure to note and remember a preselected heading.
4. Failure to observe the rate of heading change and its relation to bank attitude.
5. Overcontrolling in response to heading changes, especially during changes in power settings.
6. Anticipating heading changes with premature application of rudder pressure.
7. Failure to correct small heading deviations. Unless zero error in heading is the goal, a pilot will tolerate larger and larger deviations. Correction of a 1 degree error takes far less time and concentration than correction of a 20° error.
8. Correcting with improper bank attitude. If correcting a 10° heading error with a 20° bank correction, the aircraft will roll past the desired heading before the bank is established, requiring another correction in the opposite direction. Do not multiply existing errors with errors in corrective technique.

Figure 7-62. *The pilot has fixated on pitch and altitude, leaving bank indications unattended. Note the trend line to the left.*

9. Failure to note the cause of a previous heading error and thus repeating the same error. For example, the airplane is out of trim with a left wing low tendency. Repeated corrections for a slight left turn are made, yet trim is ignored.

Power

Power errors usually result from but are not limited to the following errors:

1. Failure to become familiar with the aircraft's specific power settings and pitch attitudes.
2. Abrupt use of throttle.
3. Failure to lead the airspeed when making power changes, climbs, or descents.

   Example: When leveling off from a descent, increase the power in order to avoid the airspeed from bleeding off due to the decrease in momentum of the aircraft. If the pilot waits to bring in the power until after the aircraft is established in the level pitch attitude, the aircraft will have already decreased below the speed desired, which will require additional adjustment in the power setting.
4. Fixation on airspeed tape or manifold pressure indications during airspeed changes, resulting in erratic control of airspeed, power, as well as pitch and bank attitudes.

Trim

Trim errors usually result from the following faults:

1. Improper adjustment of seat or rudder pedals for comfortable position of legs and feet. Tension in the ankles makes it difficult to relax rudder pressures.
2. Confusion about the operation of trim devices, which differ among various airplane types. Some trim wheels are aligned appropriately with the airplane's axes; others are not. Some rotate in a direction contrary to expectations.
3. Failure to understand the principles of trim and that the aircraft is being trimmed for airspeed, not a pitch attitude.
4. Faulty sequence in trim techniques. Trim should be utilized to relieve control pressures, not to change pitch attitudes. The proper trim technique has the pilot holding the control wheel first and then trimming to relieve any control pressures. Continuous trim changes are required as the power setting is changed. Utilize the trim continuously, but in small amounts.

Straight Climbs and Descents

Each aircraft has a specific pitch attitude and airspeed that corresponds to the most efficient climb rate for a specified weight. The POH/AFM contains the speeds that produce the desired climb. These numbers are based on maximum gross weight. Pilots must be familiar with how the speeds vary with weight so they can compensate during flight.

Entry

Constant Airspeed Climb From Cruise Airspeed

To enter a constant airspeed climb from cruise airspeed, slowly and smoothly apply aft elevator pressure in order to raise the yellow chevron (aircraft symbol) until the tip points to the desired degree of pitch. *[Figure 7-63]* Hold the aft control pressure and smoothly increase the power to the climb power setting. This increase in power may be initiated either prior to initiating the pitch change or after having established the desired pitch setting. Consult the POH/AFM for specific climb power settings if anything other than a full power climb is desired. Pitch attitudes vary depending on the type of aircraft being flown. As airspeed decreases, control forces need to be increased in order to compensate for the additional elevator deflection required to maintain attitude. Utilize trim to eliminate any control pressures. By effectively using trim, the pilot is better able to maintain the desired pitch without constant attention. The pilot is thus able to devote more time to maintaining an effective scan of all instrumentation.

The VSI should be utilized to monitor the performance of the aircraft. With a smooth pitch transition, the VSI tape should begin to show an immediate trend upward and stabilize on a

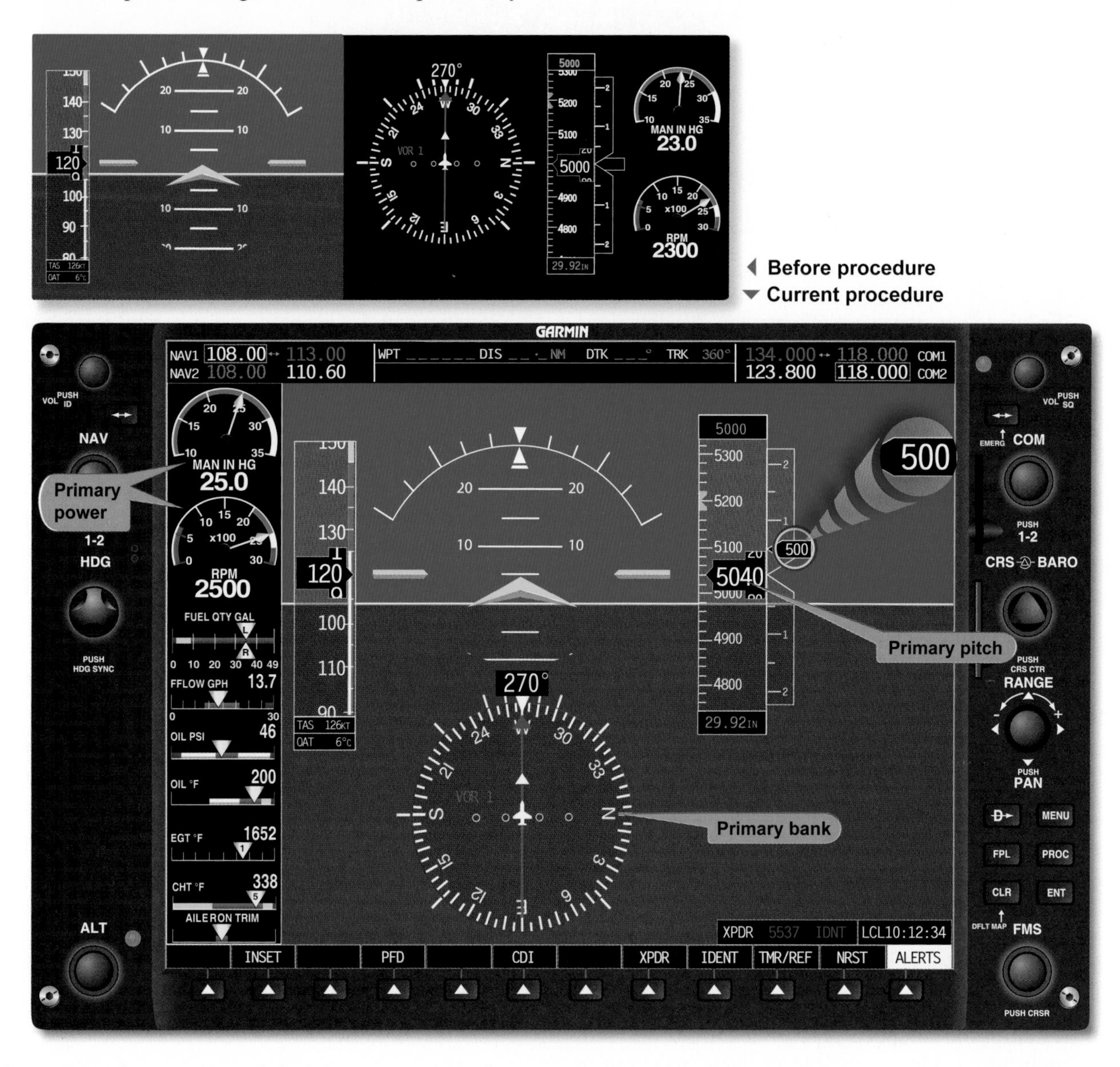

Figure 7-63. *Constant airspeed climb from cruise airspeed.*

rate of climb equivalent to the pitch and power setting being utilized. Depending on current weight and atmospheric conditions, this rate will be different. This requires the pilot to be knowledgeable of how weight and atmospheric conditions affect aircraft performance.

Once the aircraft is stabilized at a constant airspeed and pitch attitude, the primary flight instrument for pitch will be the ASI and the primary bank instrument will be the heading indicator. The primary power instrument will be the tachometer or the manifold pressure gauge depending on the aircraft type. If the pitch attitude is correct, the airspeed should slowly decrease to the desired speed. If there is any variation in airspeed, make small pitch changes until the aircraft is stabilized at the desired speed. Any change in airspeed requires a trim adjustment.

Constant Airspeed Climb from Established Airspeed

In order to enter a constant airspeed climb, first complete the airspeed reduction from cruise airspeed to climb airspeed. Maintain straight-and-level flight as the airspeed is reduced. The entry to the climb is similar to the entry from cruise airspeed with the exception that the power must be increased when the pitch attitude is raised. *[Figure 7-64]* Power added after the pitch change shows a decrease in airspeed due to the increased drag encountered. Power added prior to a pitch change causes the airspeed to increase due to the excess thrust.

Constant Rate Climbs

Constant rate climbs are very similar to the constant airspeed climbs in the way the entry is made. As power is added, smoothly apply elevator pressure to raise the yellow chevron

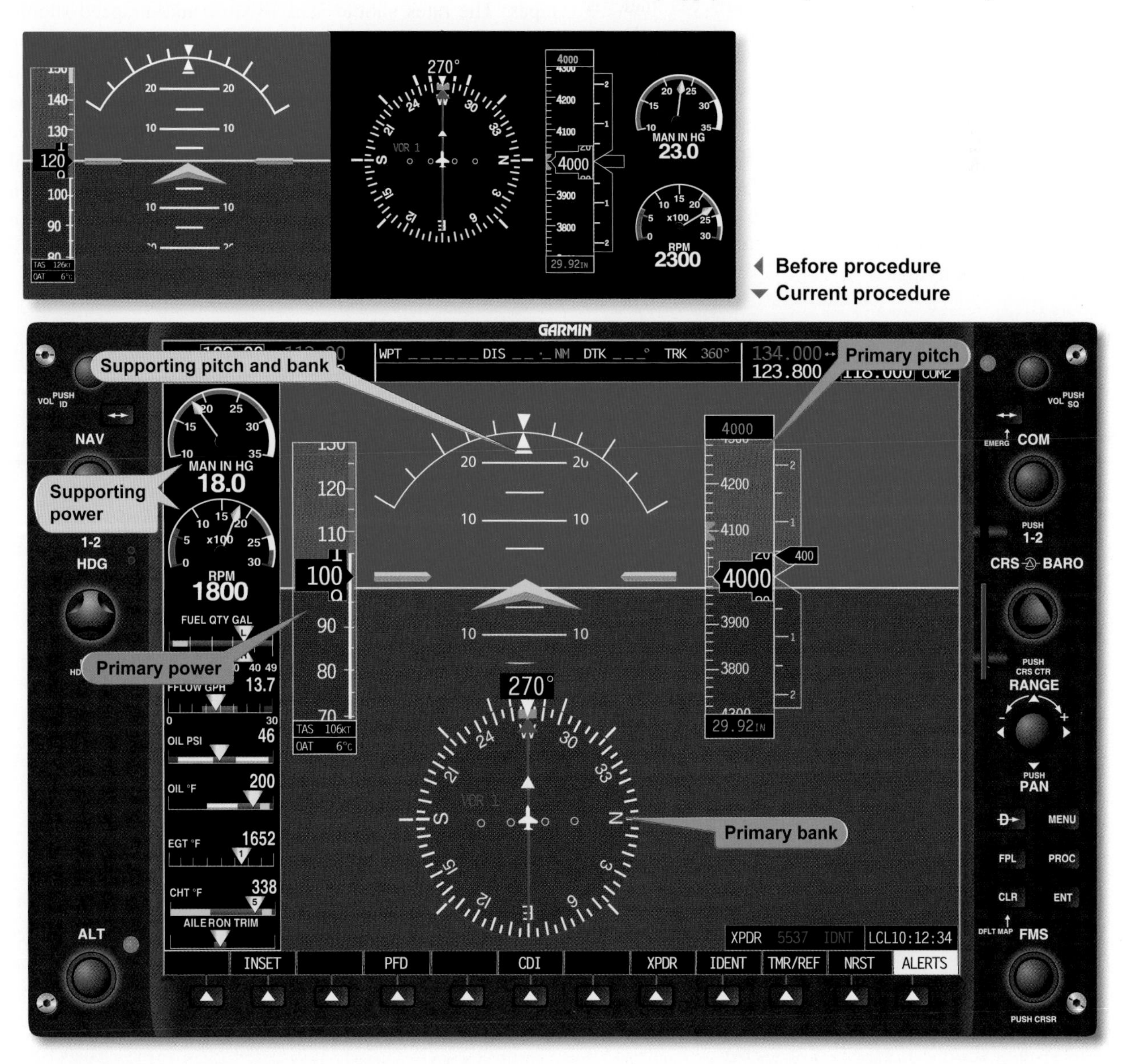

Figure 7-64. *Constant airspeed climb from established airspeed.*

to the desired pitch attitude that equates to the desired vertical speed rate. The primary instrument for pitch during the initial portion of the maneuver is the ASI until the vertical speed rate stabilizes and then the VSI tape becomes primary. The ASI then becomes the primary instrument for power. If any deviation from the desired vertical speed is noted, small pitch changes will be required in order to achieve the desired vertical speed. *[Figure 7-65]*

When making changes to compensate for deviations in performance, pitch, and power, pilot inputs need to be coordinated to maintain a stable flight attitude. For instance, if the vertical speed is lower than desired but the airspeed is correct, an increase in pitch momentarily increases the vertical speed. However, the increased drag quickly starts to degrade the airspeed if no increase in power is made. A change to any one variable mandates a coordinated change in the other.

Conversely, if the airspeed is low and the pitch is high, a reduction in the pitch attitude alone may solve the problem. Lower the nose of the aircraft very slightly to see if a power reduction is necessary. Being familiar with the pitch and power settings for the aircraft aids in achieving precise attitude instrument flying.

Leveling Off

Leveling off from a climb requires a reduction in the pitch prior to reaching the desired altitude. If no change in pitch is made until reaching the desired altitude, the momentum of the aircraft causes the aircraft to continue past the desired altitude throughout the transition to a level pitch attitude. The amount of lead to be applied depends on the vertical speed rate. A higher vertical speed requires a larger lead for level off. A good rule of thumb to utilize is to lead the level off by 10 percent of the vertical speed rate (1,000 fpm ÷ 10 = 100 feet lead).

To level off at the desired altitude, refer to the attitude display and apply smooth forward elevator pressure toward the desired level pitch attitude while monitoring the VSI and altimeter tapes. The rates should start to slow and airspeed should begin to increase. Maintain the climb power setting until the airspeed approaches the desired cruise airspeed. Continue to monitor the altimeter to maintain the desired altitude as the airspeed increases. Prior to reaching the cruise airspeed, the power must be reduced to avoid overshooting the desired speed. The amount of lead time that is required depends on the speed at which the aircraft accelerates. Utilization of the airspeed trend indicator can assist by showing how quickly the aircraft will arrive at the desired speed.

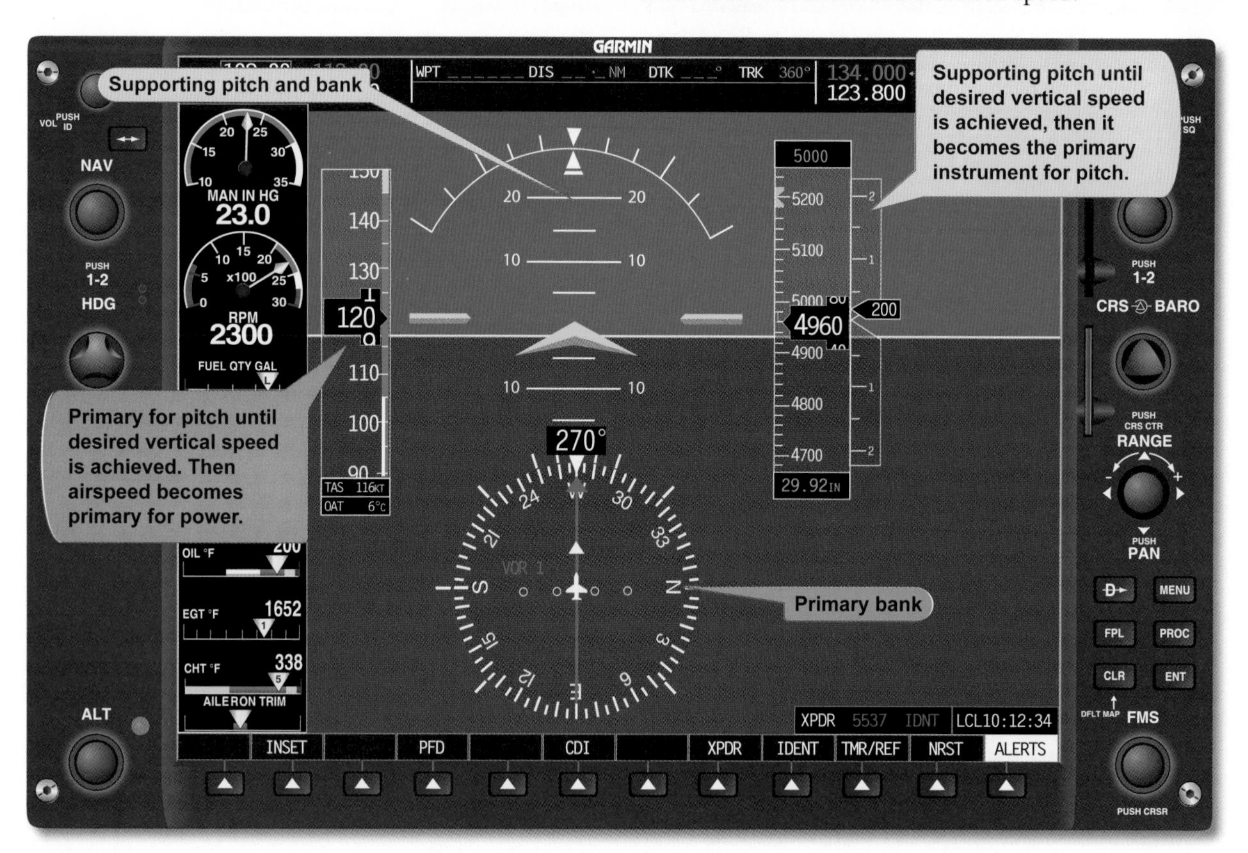

Figure 7-65. *Constant rate climbs.*

To level off at climbing airspeed, lower the nose to the appropriate pitch attitude for level flight with a simultaneous reduction in power to a setting that maintains the desired speed. With a coordinated reduction in pitch and power, there should be no change in the airspeed.

Descents

Descending flight can be accomplished at various airspeeds and pitch attitudes by reducing power, lowering the nose to a pitch attitude lower than the level flight attitude, or adding drag. Once any of these changes have been made, the airspeed eventually stabilizes During this transitional phase, the only instrument that displays an accurate indication of pitch is the attitude indicator. Without the use of the attitude indicator (such as in partial panel flight), the ASI tape, the VSI tape, and the altimeter tape shows changing values until the aircraft stabilizes at a constant airspeed and constant rate of descent. The altimeter tape continues to show a descent. Hold pitch constant and allow the aircraft to stabilize. During any change in attitude or airspeed, continuous application of trim is required to eliminate any control pressures that need to be applied to the control yoke. An increase in the scan rate during the transition is important since changes are being made to the aircraft flightpath and speed. *[Figure 7-66]*

Entry

Descents can be accomplished with a constant rate, constant airspeed, or a combination. The following method can accomplish any of these with or without an attitude indicator. Reduce the power to allow the aircraft to decelerate to the desired airspeed while maintaining straight-and-level flight. As the aircraft approaches the desired airspeed, reduce the

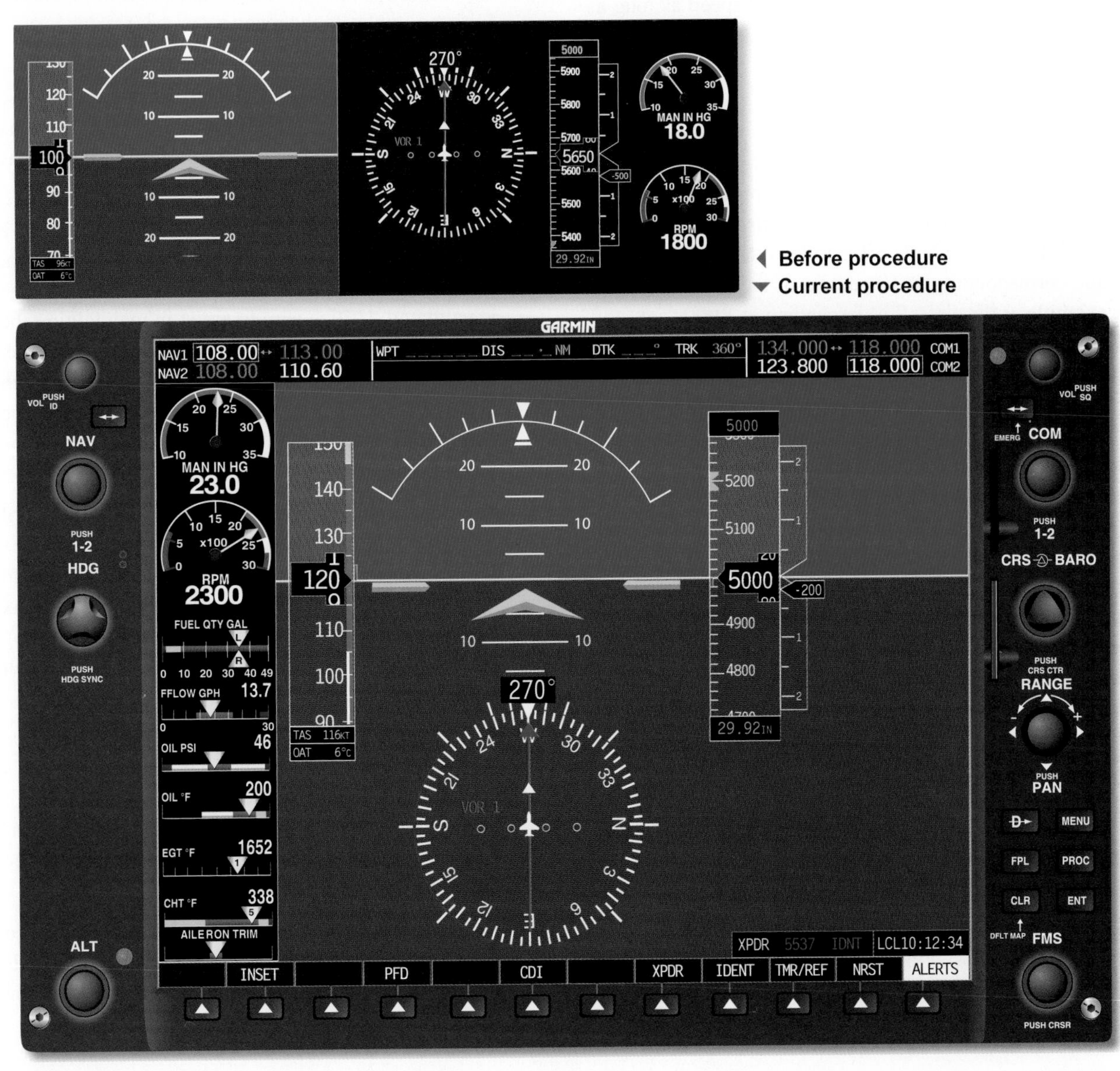

Figure 7-66. *The top image illustrates a reduction of power and descending at 500 fpm to an altitude of 5,000 feet. The bottom image illustrates an increase in power and the initiation of leveling off.*

power to a predetermined value. The airspeed continues to decrease below the desired airspeed unless a simultaneous reduction in pitch is performed. The primary instrument for pitch is the ASI tape. If any deviation from the desired speed is noted, make small pitch corrections by referencing the attitude indicator and validate the changes made with the airspeed tape. Utilize the airspeed trend indicator to judge if the airspeed is increasing and at what rate. Remember to trim off any control pressures.

The entry procedure for a constant rate descent is the same except the primary instrument for pitch is the VSI tape. The primary instrument for power is the ASI. When performing a constant rate descent while maintaining a specific airspeed, coordinated use of pitch and power is required. Any change in pitch directly affects the airspeed. Conversely, any change in airspeed has a direct impact on vertical speed as long as the pitch is being held constant.

Leveling Off

When leveling off from a descent with the intention of returning to cruise airspeed, first start by increasing the power to cruise prior to increasing the pitch back toward the level flight attitude. A technique used to determine how soon to start the level off is to lead the level off by an altitude corresponding to 10 percent of the rate of descent. For example, if the aircraft is descending at 1,000 fpm, start the level off 100 feet above the level off altitude. If the pitch attitude change is started late, there is a tendency to overshoot the desired altitude unless the pitch change is made with a rapid movement. Avoid making any rapid changes that could lead to control issues or spatial disorientation. Once in level pitch attitude, allow the aircraft to accelerate to the desired speed. Monitor the performance on the airspeed and altitude tapes. Make adjustments to the power in order to correct any deviations in the airspeed. Verify that the aircraft is maintaining level flight by cross-checking the altimeter tape. If deviations are noticed, make an appropriate smooth pitch change in order to arrive back at desired altitude. Any change in pitch requires a smooth coordinated change to the power setting. Monitor the airspeed in order to maintain the desired cruise airspeed.

To level off at a constant airspeed, the pilot must again determine when to start to increase the pitch attitude toward the level attitude. If pitch is the only item that is changing, airspeed varies due to the increase in drag as the aircraft's pitch increases. A smooth coordinated increase in power needs to be made to a predetermined value in order to maintain speed. Trim the aircraft to relieve any control pressure that may have to be applied.

Common Errors in Straight Climbs and Descents

Climbing and descending errors usually result from but are not limited to the following errors:

1. Overcontrolling pitch on beginning the climb. Aircraft familiarization is the key to achieving precise attitude instrument flying. Until the pilot becomes familiar with the pitch attitudes associated with specific airspeeds, the pilot must make corrections to the initial pitch settings. Changes do not produce instantaneous and stabilized results; patience must be maintained while the new speeds and vertical speed rates stabilize. Avoid the temptations to make a change and then rush into making another change until the first one is validated. Small changes produce more expeditious results and allow for a more stabilized flightpath. Large changes to pitch and power are more difficult to control and can further complicate the recovery process.
2. Failure to increase the rate of instrument cross-check. Any time a pitch or power change is made, an increase in the rate a pilot cross-checks the instrument is required. A slow cross-check can lead to deviations in other flight attitudes.
3. Failure to maintain new pitch attitudes. Once a pitch change is made to correct for a deviation, that pitch attitude must be maintained until the change is validated. Utilize trim to assist in maintaining the new pitch attitude. If the pitch is allowed to change, it is impossible to validate whether the initial pitch change was sufficient to correct the deviation. The continuous changing of the pitch attitude delays the recovery process.
4. Failure to utilize effective trim techniques. If control pressures have to be held by the pilot, validation of the initial correction is impossible if the pitch is allowed to vary. Pilots have the tendency to either apply or relax additional control pressures when manually holding pitch attitudes. Trim allows the pilot to fly without holding pressure on the control yoke.
5. Failure to learn and utilize proper power settings. Any time a pilot is not familiar with an aircraft's specific pitch and power settings, or does not utilize them, a change in flightpaths takes longer. Learn pitch and power settings in order to expedite changing the flightpath.
6. Failure to cross-check both airspeed and vertical speed prior to making adjustments to pitch and or power. It is possible that a change in one may correct a deviation in the other.

7. Uncoordinated use of pitch and power during level offs. During level offs, both pitch and power settings need to be made in unison in order to achieve the desired results. If pitch is increased before adding power, additional drag is generated thereby reducing airspeed below the desired value.
8. Failure to utilize supporting pitch instruments leads to chasing the VSI. Always utilize the attitude indicator as the control instrument on which to change the pitch.
9. Failure to determine a proper lead time for level off from a climb or descent. Waiting too long can lead to overshooting the altitude.
10. Ballooning—Failure to maintain forward control pressure during level off as power is increased. Additional lift is generated causing the nose of the aircraft to pitch up.

Turns

Standard Rate Turns

The previous sections have addressed flying straight-and-level as well as climbs and descents. However, attitude instrument flying is not accomplished solely by flying in a straight line. At some point, the aircraft needs to be turned to maneuver along victor airways, global positioning system (GPS) courses, and instrument approaches. The key to instrument flying is smooth, controlled changes to pitch and bank. Instrument flying should be a slow but deliberate process that takes the pilot from departure airport to destination airport without any radical flight maneuvers.

A turn to specific heading should be made at standard rate. Standard rate is defined as a turning rate of 3 degrees per second, which yields a complete 360° turn in 2 minutes. A turning rate of 3 degrees per second allows for a timely heading change, as well as allowing the pilot sufficient time to cross-check the flight instruments and avoid drastic changes to the aerodynamic forces being exerted on the aircraft. At no time should the aircraft be maneuvered faster than the pilot is comfortable cross-checking the flight instruments. Most autopilots are programmed to turn at standard rate.

Establishing a Standard Rate Turn

In order to initiate a standard rate turn, approximate the bank angle and then establish that bank angle on the attitude indicator. A rule of thumb to determine the approximate angle of bank is to use 15 percent of the true airspeed. A simple way to determine this amount is to divide the airspeed by 10 and add one-half the result. For example, at 100 knots, approximately 15° of bank is required (100/10 = 10 + 5 = 15); at 120 knots, approximately 18° of bank is needed for a standard-rate turn. Cross-check the turn rate indicator, located on the HSI, to determine if that bank angle is sufficient to deliver a standard rate turn. Slight modifications may need to be made to the bank angle in order to achieve the desired performance. The primary bank instrument in this case is the turn rate indicator since the goal is to achieve a standard rate turn. The turn rate indicator is the only instrument that can specifically indicate a standard rate turn. The attitude indicator is used only to establish a bank angle (control instrument) but can be utilized as a supporting instrument by cross-checking the bank angle to determine if the bank is greater or less than what was calculated.

As the aircraft rolls into the bank, the vertical component of lift begins to decrease. *[Figure 7-67]* As this happens, additional lift must be generated to maintain level flight. Apply aft control pressure on the yoke sufficient to stop any altitude loss trend. With the increase in lift that needs to be generated, additional induced drag is also generated. This additional drag causes the aircraft to start to decelerate. To counteract this, apply additional thrust by adding power to the power lever. Once altitude and airspeed is being maintained, utilize the trim wheel to eliminate any control forces that need to be held on the control column.

When rolling out from a standard rate turn, the pilot needs to utilize coordinated aileron and rudder and roll-out to a wings level attitude utilizing smooth control inputs. The roll-out rate should be the same as the roll-in rate in order to estimate the lead necessary to arrive at the desired heading without over- or undershooting.

During the transition from the turn back to straight flight, the attitude indicator becomes the primary instrument for bank. Once the wings are level, the heading indicator becomes the primary instrument for bank. As bank decreases, the vertical component increases if the pitch attitude is not decreased sufficiently to maintain level flight. An aggressive cross-check keeps the altimeter stationary if forward control pressure is applied to the control column. As the bank angle is decreased, the pitch attitude should be decreased accordingly in order to arrive at the level pitch attitude when the aircraft reaches zero bank. Remember to utilize the trim wheel to eliminate any excess control forces that would otherwise need to be held.

Common Errors

1. One common error associated with standard rate turns is due to pilot inability to hold the appropriate bank angle that equates to a standard rate. The primary bank instrument during the turn is the turn rate indicator; however, the bank angle varies slightly. With an

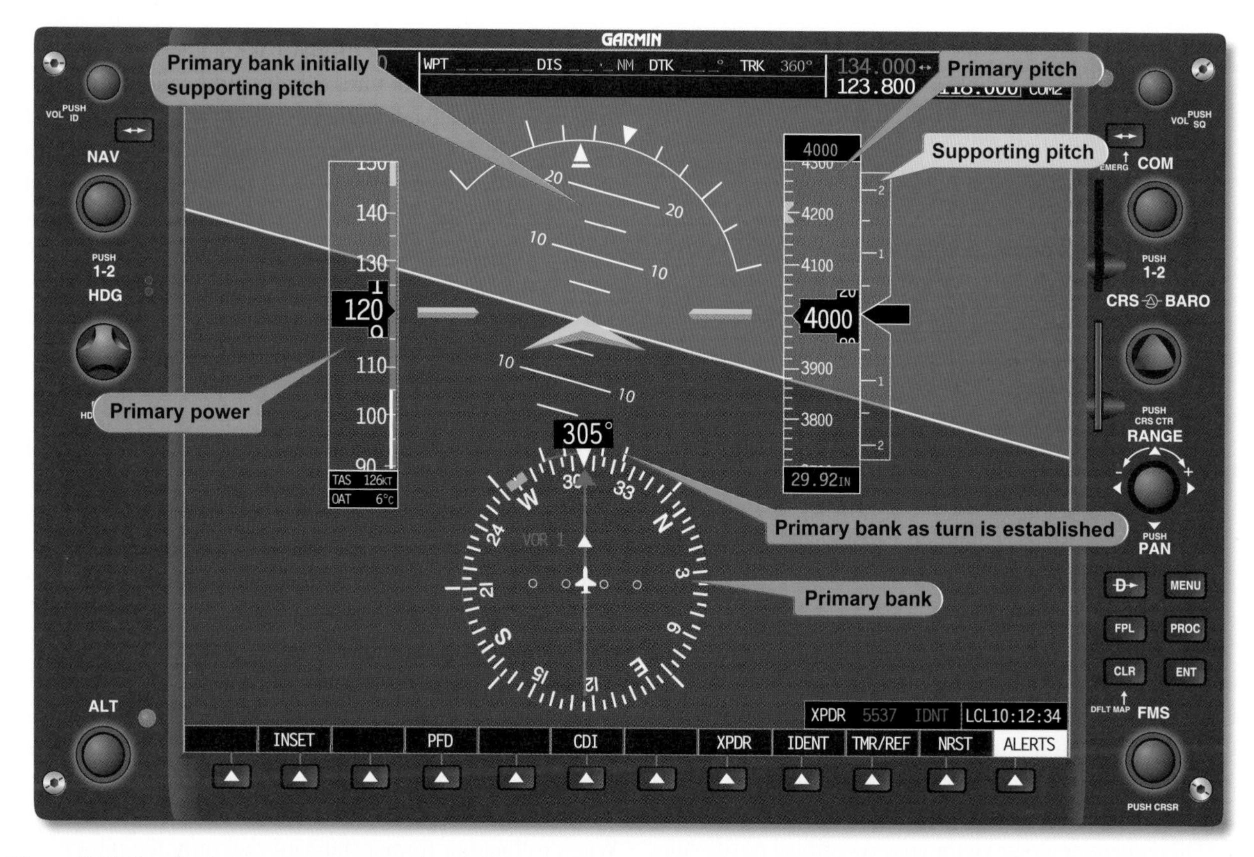

Figure 7-67. *Standard rate turn—constant airspeed.*

aggressive cross-check, a pilot should be able to minimize errors arising from over- or underbanking.

2. Another error normally encountered during standard rate turns is inefficient or lack of adequate cross-checking. Pilots need to establish an aggressive cross-check in order to detect and eliminate all deviations from altitude, airspeed, and bank angle during a maneuver.
3. Fixation is a major error associated with attitude instrument flying in general. Pilots training for their instrument rating tend to focus on what they perceive to be the most important task at hand and abandon their cross-check by applying all of their attention to the turn rate indicator. A modified radial scan works well to provide the pilot with adequate scanning of all instrumentation during the maneuver.

Turns to Predetermined Headings

Turning the aircraft is one of the most basic maneuvers that a pilot learns during initial flight training. Learning to control the aircraft, maintaining coordination, and smoothly rolling out on a desired heading are all keys to proficient attitude instrument flying.

EFDs allow the pilot to better utilize all instrumentation during all phases of attitude instrument flying by consolidating all traditional instrumentation onto the PFD. The increased size of the attitude indicator, which stretches the entire width of the PFD, allows the pilot to maintain better pitch control while the introduction of the turn rate indicator positioned directly on the compass rose aids the pilot in determining when to begin a roll-out for the desired heading.

When determining what bank angle to utilize when making a heading change, a general rule states that for a small heading change, do not use a bank angle that is greater than the total number of degrees of change needed. For instance, if a heading change of 20° is needed, a bank angle of not more than 20° is required. Another rule of thumb that better defines the bank angle is half the total number of degrees of heading change required, but never greater than standard rate. The exact bank angle that equates to a standard rate turn varies due to true airspeed.

With this in mind and the angle of bank calculated, the next step is determining when to start the roll-out process. For example:

An aircraft begins a turn from a heading of 030° to a heading of 120°. With the given airspeed, a standard rate turn has yielded a 15° bank. The pilot wants to begin a smooth coordinated roll-out to the desired heading when the heading indicator displays approximately 112°. The necessary calculations are:

15° bank (standard rate) ÷ 2 = 7.5°
120° – 7.5° = 112.5°

By utilizing this technique, the pilot is better able to judge if any modifications need to be made to the amount of lead once the amount of over- or undershooting is established.

Timed Turns

Timed turns to headings are performed in the same fashion with an EFD as with an analog equipped aircraft. The instrumentation used to perform this maneuver is the turn rate indicator as well as the clock. The purpose of this maneuver is to allow the pilot to gain proficiency in scanning as well as to further develop the pilot's ability to control the aircraft without standard instrumentation.

Timed turns become essential when controlling the aircraft with a loss of the heading indicator. This may become necessary due to a loss of the AHRS unit or the magnetometer. In any case, the magnetic compass is still available for navigation. The reason for timed turns instead of magnetic compass turns is the simplicity of the maneuver. Magnetic compass turns require the pilot to take into account various errors associated with the compass; timed turns do not.

Prior to initiating a turn, determine if the standard rate indication on the turn rate indicator actually delivers a 3 degrees per second turn. To accomplish this, a calibration must be made. Establish a turn in either direction at the indicated standard rate. Start the digital timer as the compass rolls past a cardinal heading. Stop the timer once the compass card rolls through another cardinal heading. Roll wings level and compute the rate of turn. If the turn rate indicator is calibrated and indicating correctly, 90° of heading change should take 30 seconds. If the time taken to change heading by 90° is more or less than 30 seconds, then a deflection above or below the standard rate line needs to be made to compensate for the difference. Once the calibration has been completed in one direction, proceed to the opposite direction. When both directions have been calibrated, apply the calibrated calculations to all timed turns.

In order to accomplish a timed turn, the amount of heading change needs to be established. For a change in heading from 120° to a heading of 360°, the pilot calculates the difference and divides that number by 3. In this case, 120° divided by 3° per second equals 40 seconds. This means that it would take 40 seconds for an aircraft to change heading 120° if that aircraft were held in a perfect standard rate turn. Timing for the maneuver should start as the aircraft begins rolling into the standard rate turn. Monitor all flight instruments during this maneuver. The primary pitch instrument is the altimeter. The primary power instrument is the ASI and the primary bank instrument is the turn rate indicator.

Once the calculated time expires, start a smooth coordinated roll-out. As long as the pilot utilizes the same rate of roll-in as roll-out, the time it takes for both will not need to be included in the calculations. With practice, the pilot should level the wings on the desired heading. If any deviation has occurred, make small corrections to establish the correct heading.

Compass Turns

The magnetic compass is the only instrument that requires no other source of power for operation. In the event of an AHRS or magnetometer failure, the magnetic compass is the instrument the pilot uses to determine aircraft heading. For a more detailed explanation on the use of the magnetic compass, see page 7-21.

Steep Turns

For the purpose of instrument flight training, a steep turn is defined as any turn in excess of standard rate. A standard rate turn is defined as 3 degrees per second. The bank angle that equates to a turn rate of 3 degrees per second varies according to airspeed. As airspeed increases, the bank angle must be increased. The exact bank angle that equates to a standard rate turn is unimportant. Normal standard rate turn bank angles range from 10° to 20°. The goal of training in steep turn maneuvers is pilot proficiency in controlling the aircraft with excessive bank angles.

Training in excessive bank angles challenges the pilot in honing cross-checking skills and improves altitude control throughout a wider range of flight attitudes. Although the current instrument flight check practical test standards (PTS) do not call for a demonstration of steep turns on the certification check flight, this does not eliminate the need for the instrument pilot-in-training to demonstrate proficiency to an instructor.

Training in steep turns teaches the pilot to recognize and to adapt to rapidly changing aerodynamic forces that necessitate an increase in the rate of cross-checking all flight instruments. The procedures for entering, maintaining, and exiting a steep turn are the same as for shallower turns. Proficiency in instrument cross-check and interpretation is increased due to the higher aerodynamic forces and increased speed at which the forces are changing.

Performing the Maneuver

To enter a steep turn to the left, roll into a coordinated 45° bank turn to the left. An advantage that glass panel displays have over analog instrumentation is a 45° bank indication on the roll scale. This additional index on the roll scale allows the pilot to precisely roll into the desired bank angle instead of having to approximate it as is necessary with analog instrumentation. *[Figure 7-68]*

Figure 7-68. *Steep left turn.*

As soon as the bank angle increases from level flight, the vertical component of lift begins to decrease. If the vertical component of lift is allowed to continue to decrease, a pronounced loss of altitude is indicated on the altimeter along with the VSI tape, as well as the altitude trend indicator. Additionally, the airspeed begins to increase due to the lowered pitch attitude. It is very important to have a comprehensive scan developed prior to training in steep turns. Utilization of all of the trend indicators, as well the VSI, altimeter, and ASI, is essential in learning to fly steep turns by reference to instruments alone.

In order to avoid a loss of altitude, the pilot begins to slowly increase back pressure on the control yoke in order to increase the pitch attitude. The pitch change required is usually no more than 3 degrees to 5 degrees, depending on the type of aircraft. As the pilot increases back pressure, the angle of attack increases, thus increasing the vertical component of lift. When a deviation in altitude is indicated, proper control force corrections need to be made. During initial training of steep turns, pilots have a tendency to overbank. Over banking is when the bank angle exceeds 50°. As the outboard wing begins to travel faster through the air, it begins to generate a greater and greater differential in lift compared to the inboard wing. As the bank angle continues to progress more and more steeply past 45°, the two components of lift (vertical and horizontal) become inversely proportionate.

Once the angle has exceeded 45°, the horizontal component of lift is now the greater force. If altitude should continue to decrease and the pilot only applies back yoke pressure, the aircraft's turn radius begins to tighten due to the increased horizontal force. If aft control pressure continues to increase, there comes a point where the loss of the vertical component of lift and aerodynamic wing loading prohibits the nose of the aircraft from being raised. Any increase in pitch only tightens the turning radius.

The key to successfully performing a steep turn by reference to instruments alone is the thorough understanding of the aerodynamics involved, as well as a quick and reliable cross-check. The pilot should utilize the trim to avoid holding control forces for any period of time. With time and practice, a flight instructor can demonstrate how to successfully fly steep turns with and without the use of trim. Once the aircraft is trimmed for the maneuver, accomplishing the maneuver is virtually a hands-off effort. This allows additional time for cross-checking and interpreting the instruments.

It is imperative when correcting for a deviation in altitude, that the pilot modify the bank angle ±5° in order to vary the vertical component of lift, not just adjust back pressure. These two actions should be accomplished simultaneously.

During the recovery from steep turns to straight-and-level flight, aft control forces must be varied with the power control to arrive back at entry altitude, heading and airspeed.

Steps:

1. Perform clearing turns.
2. Roll left into a 45° bank turn and immediately begin to increase the pitch attitude by approximately 3° to 5°.
3. As the bank rolls past 30°, increase power to maintain the entry airspeed.
4. Apply trim to eliminate any aft control wheel forces.
5. Begin rolling out of the steep turn approximately 20° prior to the desired heading.
6. Apply forward control pressure and place the pitch attitude in the level cruise pitch attitude.
7. Reduce power to the entry power setting to maintain the desired airspeed.
8. Re-trim the aircraft as soon as practical or continue into a right hand steep turn and continue from step 3.

9. Once the maneuver is complete, establish cruise flight and accomplish all appropriate checklist items.

Unusual Attitude Recovery Protection

Unusual attitudes are some of the most hazardous situations for a pilot to be in. Without proper recovery training on instrument interpretation and aircraft control, a pilot can quickly aggravate an abnormal flight attitude into a potentially fatal accident.

Analog gauges require the pilot to scan between instruments to deduce the aircraft attitude. Individually, these gauges lack the necessary information needed for a successful recovery.

EFDs have additional features to aid in recognition and recovery from unusual flight attitudes. The PFD displays all the flight instruments on one screen. Each instrument is superimposed over a full-screen representation of the attitude indicator. With this configuration, the pilot no longer needs to transition from one instrument to another.

The new unusual attitude recovery protection allows the pilot to be able to quickly determine the aircraft's attitude and make a safe, proper, and prompt recovery. Situational awareness is increased by the introduction of the large full-width artificial horizon depicted on the PFD. This now allows for the attitude indicator to be in view during all portions of the scan.

One problem with analog gauges is that the attitude indicator displays a complete blue or brown segment when the pitch attitude is increased toward 90° nose-up or nose-down.

With the EFDs, the attitude indicator is designed to retain a portion of both sky and land representation at all times. This improvement allows the pilot to always know the quickest way to return to the horizon. Situational awareness is greatly increased.

NOTE: The horizon line starts moving downward at approximately 47° pitch up. From this point on, the brown segment remains visible to show the pilot the quickest way to return to the level pitch attitude. *[Figure 7-69]*

NOTE: The horizon line starts moving upward at approximately 27° pitch down. From this point on, the blue segment remains visible to show the pilot the quickest way to return to the level pitch attitude. *[Figure 7-70]*

It is imperative to understand that the white line on the attitude indicator is the horizon line. The break between the blue and brown symbols is only a reference and should not be thought of as the artificial horizon.

Another important advancement is the development of the unusual attitude recovery protection that is built into the PFD software and made capable by the AHRS. In the case of a nose-high unusual attitude, the unusual attitude recovery protection displays red chevrons that point back to the horizon line. These

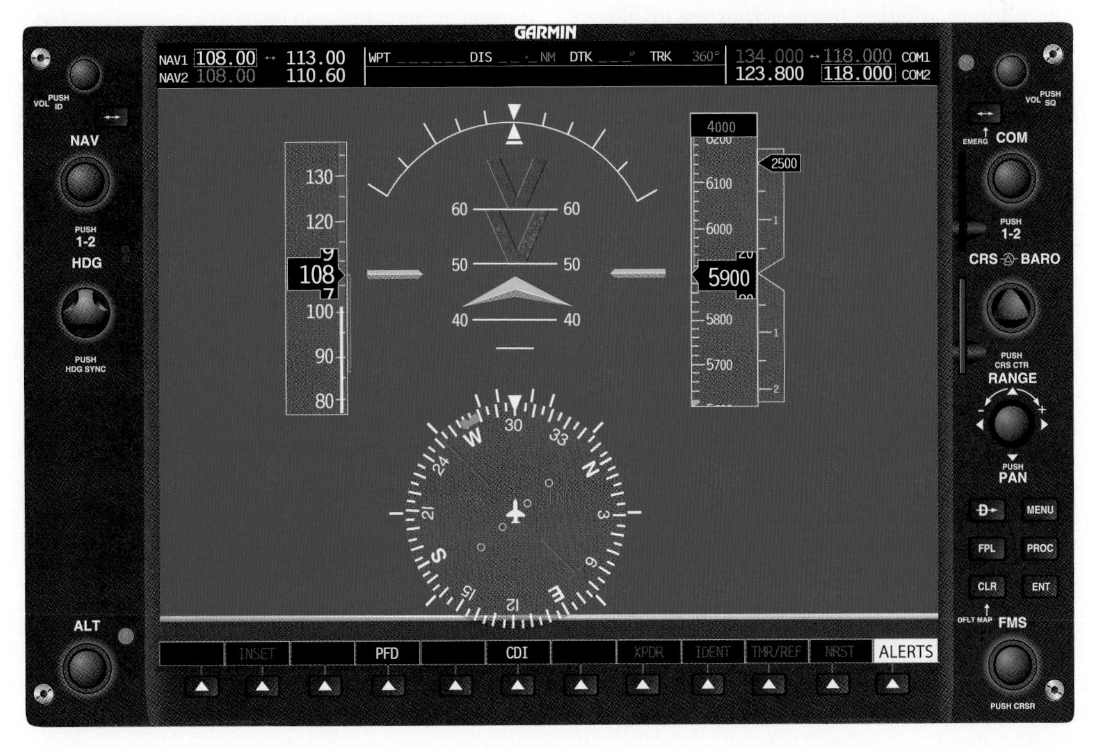

Figure 7-69. *Unusual attitude recovery protection. Note the brown horizon line is visible at the bottom.*

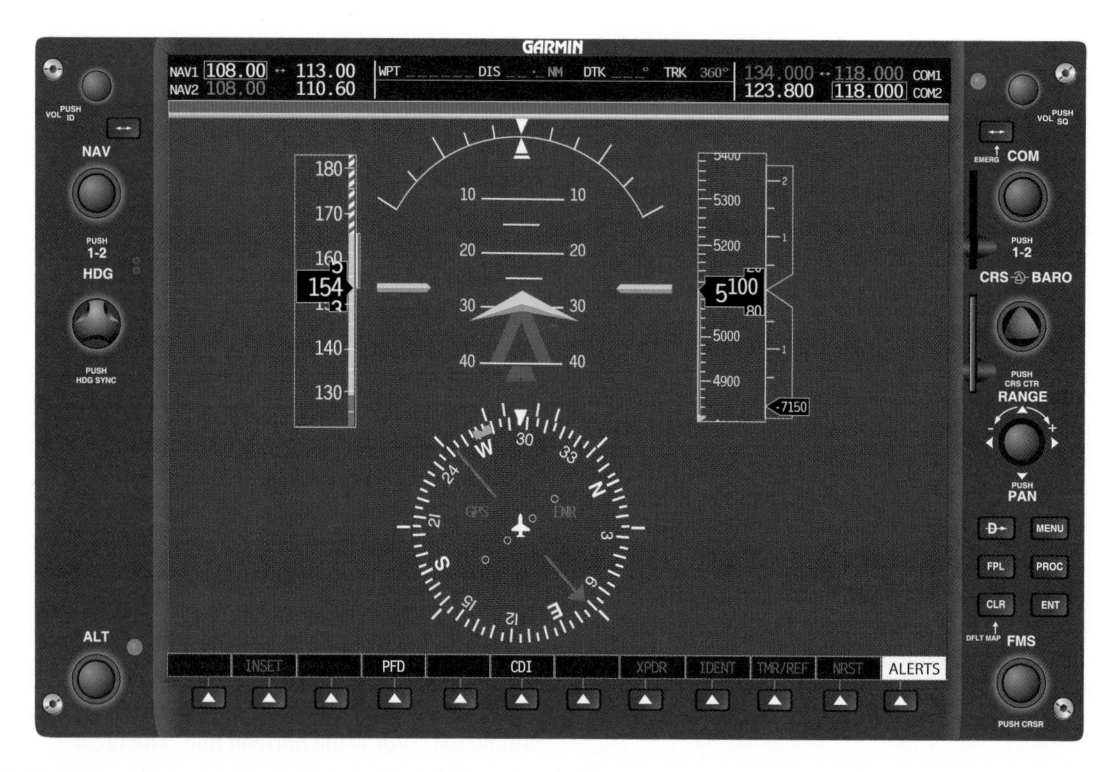

Figure 7-70. *Horizon line starts moving upward at 27°. Note that the blue sky remains visible at 17° nose-down.*

chevrons are positioned at 50° up on the attitude indicator. The chevrons appear when the aircraft approaches a nose-high attitude of 30°. The software automatically declutters the PFD leaving only airspeed, heading, attitude, altimeter, VSI tape, and the trend vectors. The decluttered information reappears when the pitch attitude falls below 25°.

For nose-low unusual attitudes, the chevrons are displayed when the pitch exceeds 15° nose-down. If the pitch continues to decrease, the unusual attitude recovery protection de-clutters the screen at 20° nose-down. The decluttered information reappears when the pitch increases above 15°.

Additionally, there are bank limits that trigger the unusual attitude protection. If the aircraft's bank increases beyond 60°, a continuation of the roll index occurs to indicate the shortest direction to roll the wings back to level. At 65°, the PFD de-clutters. All information reappears when the bank decreases below 60°.

In *Figure 7-71*, the aircraft has rolled past 60°. Observe the white line that continues from the end of the bank index. This line appears to indicate the shortest distance back to wings level.

When experiencing a failure of the AHRS unit, all unusual attitude protection is lost. The failure of the AHRS results in the loss of all heading and attitude indications on the PFD. In addition, all modes of the autopilot, except for roll and altitude hold, are lost.

The following picture series represents how important this technology is in increasing situational awareness, and how critical it is in improving safety.

Figure 7-72 shows the unusual attitude protection with valid AHRS and air data computer (ADC) inputs. The bright red chevrons pointing down to the horizon indicate a nose-high unusual attitude that can be easily recognized and corrected.

NOTE: The red chevrons point back to the level pitch attitude. The trend indicators show where the airspeed and altitude will be in 6 seconds. The trend indicator on the heading indicator shows which direction the aircraft is turning. The slip/skid indicator clearly shows if the aircraft is coordinated. This information helps the pilot determine which type of unusual attitude the aircraft has taken.

Now look at *Figure 7-73*. The display shows the same airspeed as the picture above; however, the AHRS unit has failed. The altimeter and the VSI tape are the only clear indications that the aircraft is in a nose-high attitude. The one key instrument that is no longer present is the slip/skid indicator. There is not a standby turn coordinator installed in the aircraft for the pilot to reference.

The magnetic compass indicates a heading is being maintained; however, it is not as useful as a turn coordinator or slip/skid indicator.

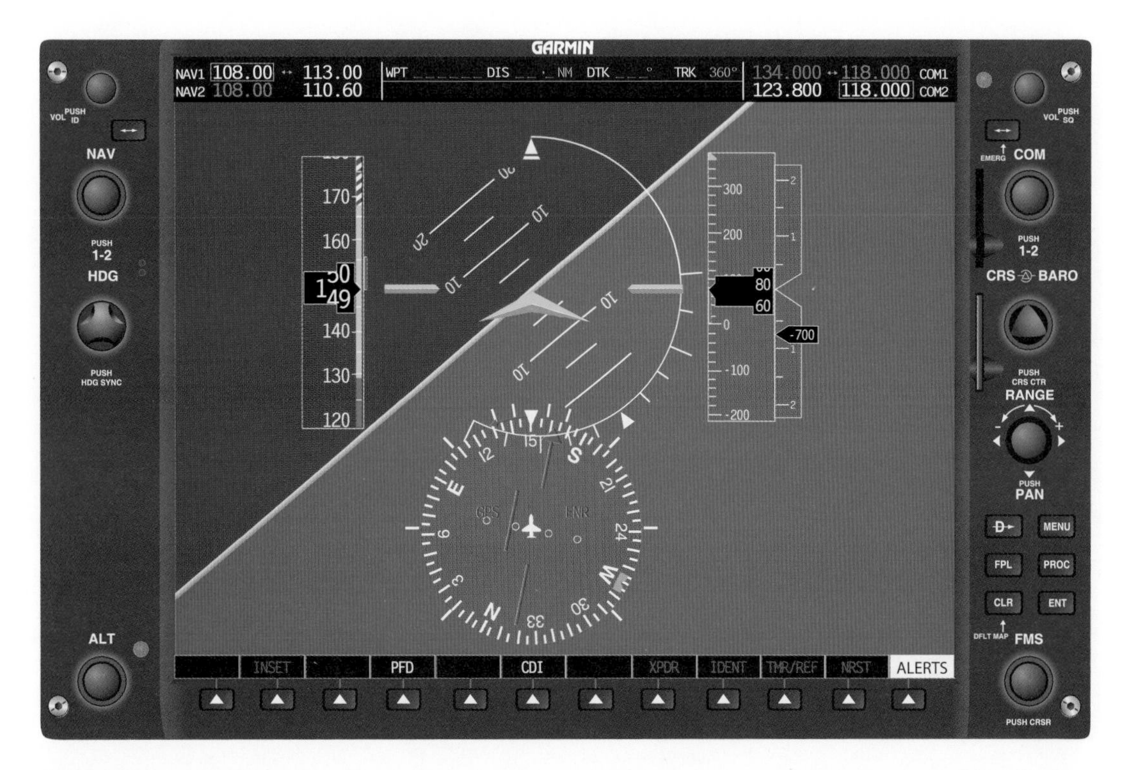

Figure 7-71. *Aircraft rolled past 60°.*

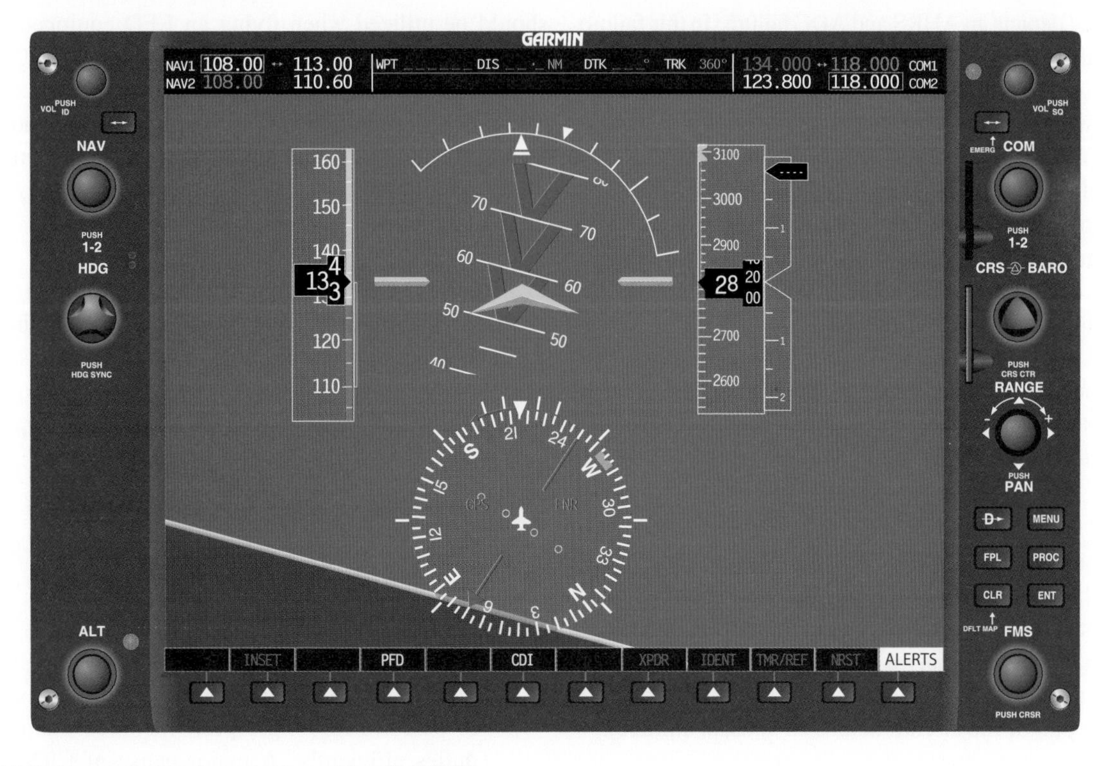

Figure 7-72. *Unusual attitude protection with valid AHRS.*

Figure 7-73. *AHRS unit failed.*

Figure 7-74 depicts an AHRS and ADC failure. In this failure scenario, there are no indications of the aircraft's attitude. The manufacturer recommends turning on the autopilot, which is simply a wing leveler.

With a failure of the primary instrumentation on the PFD, the only references available are the standby instruments. The standby instrumentation consists of an analog ASI, attitude indicator, altimeter, and magnetic compass. There is no standby turn coordinator installed.

In extreme nose-high or nose-low pitch attitudes, as well as high bank angles, the analog attitude indicator has the potential to tumble, rendering it unusable.

Autopilot Usage

The autopilot is equipped with inputs from a turn coordinator installed behind the MFD screen. This turn coordinator is installed solely for the use of the autopilot to facilitate the roll mode, which is simply a wing leveler. This protection is always available, barring a failure of the turn coordinator (to aid the pilot if the aircraft attains an unusual attitude).

NOTE: The pilot is not able to gain access to the turn coordinator. This instrument is installed behind the MFD panel. *[Figure 7-75]*

Most EFD equipped aircraft are coming from the factory with autopilots installed. However, the purchaser of the aircraft can specify if an autopilot is to be installed. Extreme caution should be utilized when flying an EFD equipped aircraft without an autopilot in IMC with an AHRS and ADC failure.

The autopilot should be utilized to reduce workload, which affords the pilot more time to monitor the flight. Utilization of the autopilot also decreases the chances of entry into an unusual attitude.

Flying an EFD-equipped aircraft without the use of an autopilot has been shown to increase workload and decrease situational awareness for pilots first learning to flying the new system.

Common Errors Leading to Unusual Attitudes

The following errors have the potential to disrupt a pilot's situational awareness and lead to unusual attitudes.

1. Improper trimming techniques. A failure to keep the aircraft trimmed for level flight at all times can turn a momentary distraction into an emergency situation if the pilot stops cross-checking.
2. Poor crew resource management (CRM) skills. Failure to perform all single-pilot resource management duties efficiently. A major cause of CRM-related accidents comes from the failure of the pilot to maintain an organized flight deck. Items that are being utilized for the flight portion should be neatly arranged for easy access. A disorganized flight deck can lead to a distraction that causes the pilot to cease cross-checking the instruments long enough to enter an unusual attitude.

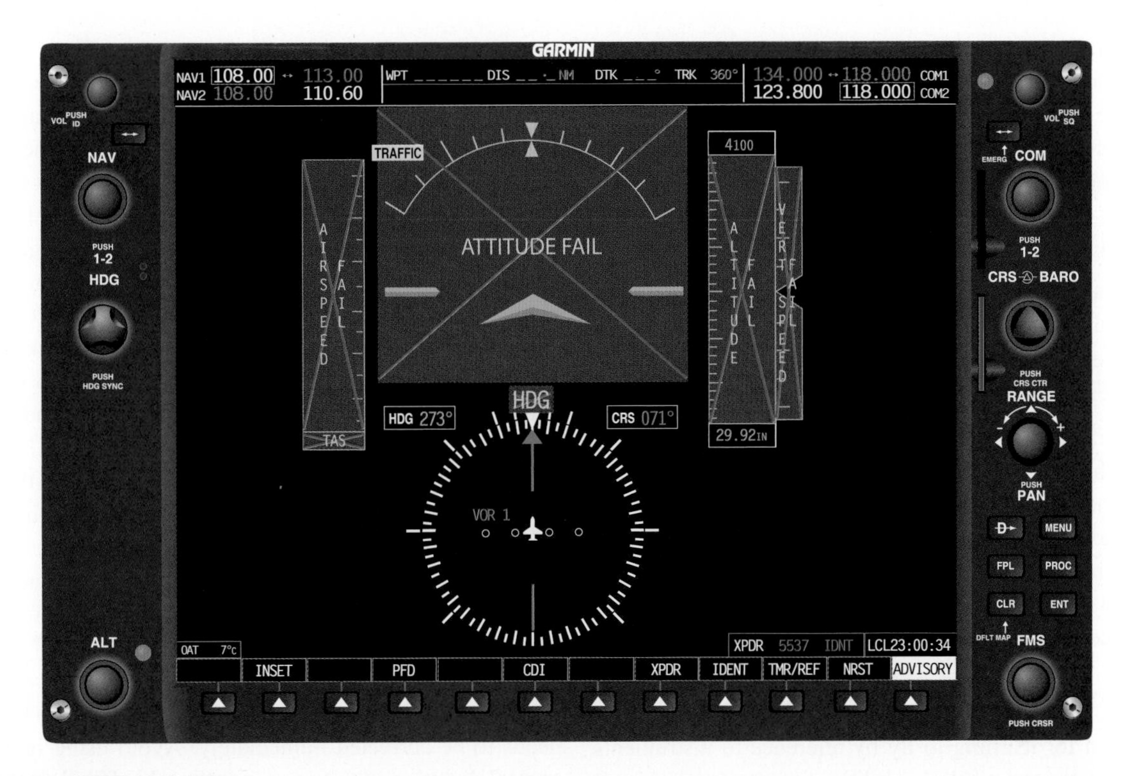

Figure 7-74. *AHRS ADC failure.*

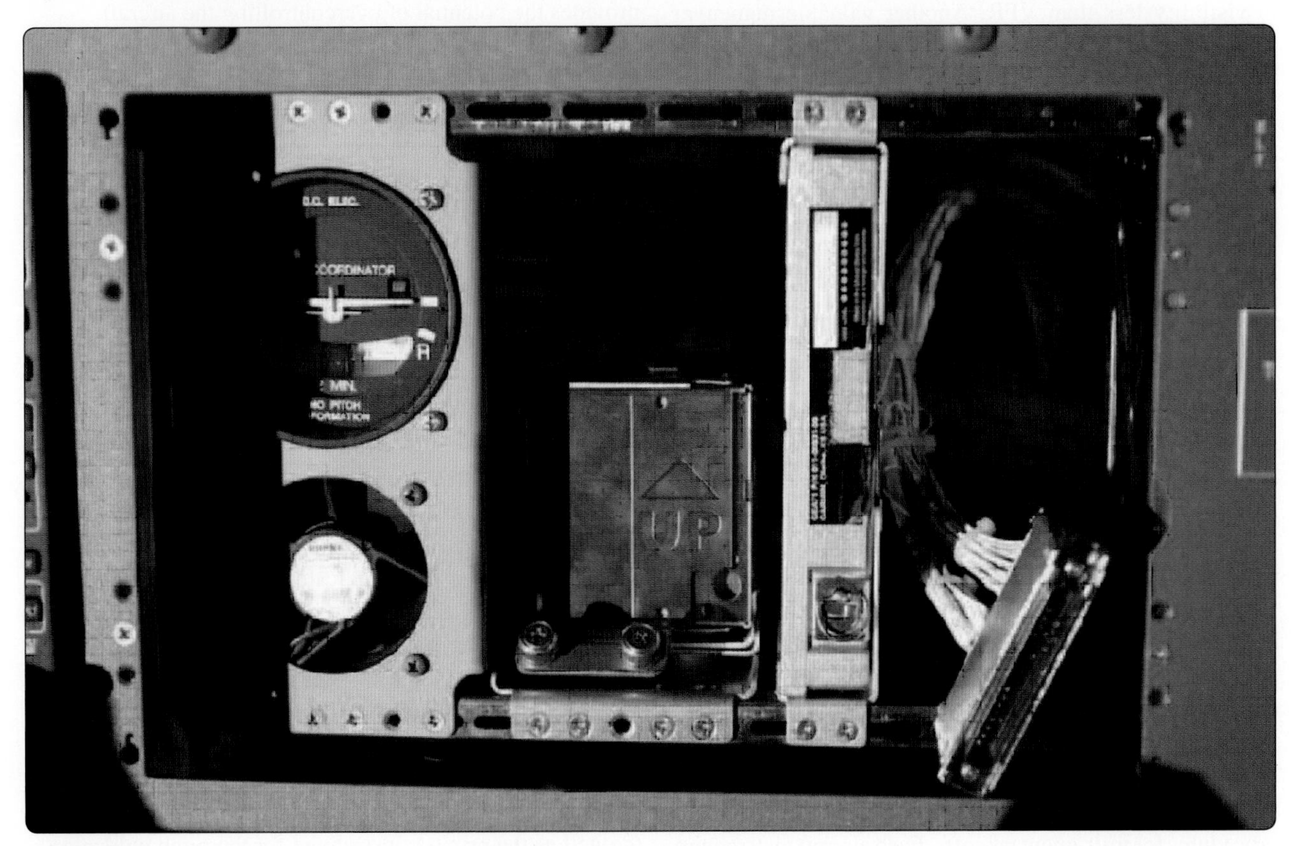

Figure 7-75. *This autopilot requires roll information from a turn coordinator.*

3. Fixation is displayed when a pilot focuses far too much attention on one instrument because he or she perceives something is wrong or a deviation is occurring. It is important for the instrument pilot to remember that a cross-check of several instruments for corroboration is more valuable than checking a single instrument.
4. Attempting to recover by sensory sensations other than sight. Recovery by instinct almost always leads to erroneous corrections due to the illusions that are prevalent during instrument flight.
5. Failure to practice basic attitude instrument flying. When a pilot does not fly instrument approach procedures or even basic attitude instrument flying maneuvers for long periods of time, skill levels diminish. Pilots should avoid flying in IMC if they are not proficient. They should seek a qualified instructor to receive additional instruction prior to entry into IMC.

Instrument Takeoff

The reason for learning to fly by reference to instruments alone is to expand a pilot's abilities to operate an aircraft in visibility less than VFR. Another valuable maneuver to learn is the instrument takeoff. This maneuver requires the pilot to maneuver the aircraft during the takeoff roll by reference to flight instruments alone with no outside visual reference. With practice, this maneuver becomes as routine as a standard rate turn.

The reason behind practicing instrument takeoffs is to reduce the disorientation that can occur during the transitional phase of quickly moving the eyes from the outside references inside to the flight instruments.

One EFD system currently offers what is trademarked as synthetic vision. Synthetic vision is a three-dimensional computer-generated representation of the terrain that lies ahead of the aircraft. The display shows runways as well as a depiction of the terrain features based on a GPS terrain database. Similar to a video game, the display generates a runway the pilot can maneuver down in order to maintain directional control. As long as the pilot tracks down the computer-generated runway, the aircraft remains aligned with the actual runway.

Not all EFD systems have such an advanced visioning system. With all other systems, the pilot needs to revert to the standard procedures for instrument takeoffs. Each aircraft may require a modification to the maneuver; therefore, always obtain training on any new equipment to be used.

In order to accomplish an instrument takeoff, the aircraft needs to be maneuvered on the centerline of the runway facing the direction of departure with the nose or tail wheel straight. Assistance from the instructor may be necessary if the pilot has been taxiing while wearing a view limiting device. Lock the tail wheel, if so equipped, and hold the brakes firmly to prevent the aircraft from creeping. Cross-check the heading indicator on the PFD with the magnetic compass and adjust for any deviations noted on the compass card. Set the heading to the nearest 5 degree mark closest to the runway heading. This allows the pilot to quickly detect any deviations from the desired heading and allows prompt corrective actions during the takeoff roll. Using the omnibearing select (OBS) mode on the GPS, rotate the OBS selector until the needle points to the runway heading. This adds additional situational awareness during the takeoff roll. Smoothly apply power to generate sufficient rudder authority for directional control. Release the brakes and continue to advance the power to the takeoff setting.

As soon as the brakes are released, any deviation in heading needs to be corrected immediately. Avoid using brakes to control direction as this increases the takeoff roll, as well as provides the potential of overcontrolling the aircraft.

Continuously cross-check the ASI and the heading indicator as the aircraft accelerates. As the aircraft approaches 15-25 knots below the rotation speed, smoothly apply aft elevator pressure to increase the pitch attitude to the desired takeoff attitude (approximately 7° for most small airplanes). With the pitch attitude held constant, continue to cross-check the flight instruments and allow the aircraft to fly off of the runway. Do not pull the aircraft off of the runway. Pulling the aircraft off of the runway imposes left turning tendencies due to P-Factor, which will yaw the aircraft to the left and destabilize the takeoff.

Maintain the desired pitch and bank attitudes by referencing the attitude indicator and cross-check the VSI tape for an indication of a positive rate of climb. Take note of the magenta 6-second altimeter trend indicator. The trend should show positive. Barring turbulence, all trend indications should be stabilized. The airspeed trend indicator should not be visible at this point if the airspeed is being held constant. An activation of the airspeed trend indicator shows that the pitch attitude is not being held at the desired value and, therefore, the airspeed is changing. The desired performance is to be climbing at a constant airspeed and vertical speed rate. Use the ASI as the primary instrument for the pitch indication.

Once the aircraft has reached a safe altitude (approximately 100 feet for insufficient runway available for landing should an engine failure occur) retract the landing gear and flaps while referencing the ASI and attitude indicator to maintain the desired pitch. As the configuration is changed, an increase in aft control pressure is needed in order to maintain the desired pitch attitude. Smoothly increase the aft control pressure to compensate for the change in configuration. Anticipate the changes and increase the rate of cross-check. The airspeed tape and altitude tape increases while the VSI tape is held constant. Allow the aircraft to accelerate to the desired climb speed. Once the desired climb speed is reached, reduce the power to the climb power setting as printed in the POH/AFM. Trim the aircraft to eliminate any control pressures.

Common Errors in Instrument Takeoffs

Common errors associated with the instrument takeoff include, but are not limited to, the following:

1. Failure to perform an adequate flight deck check before the takeoff. Pilots have attempted instrument takeoff with inoperative airspeed indicators (pitot tube obstructed), controls locked, and numerous other oversights due to haste or carelessness. It is imperative to cross-check the ASI as soon as possible. No airspeed is indicated until 20 knots of true airspeed is generated in some systems.
2. Improper alignment on the runway. This may result from improper brake applications, allowing the airplane to creep after alignment, or from alignment with the nosewheel or tailwheel cocked. In any case, the result is a built-in directional control problem as the takeoff starts.
3. Improper application of power. Abrupt applications of power complicate directional control. Power should be applied in a smooth and continuous manner to arrive at the takeoff power setting within approximately 3 seconds.
4. Improper use of brakes. Incorrect seat or rudder pedal adjustment, with feet in an uncomfortable position, frequently causes inadvertent application of brakes and excessive heading changes.
5. Overcontrolling rudder pedals. This fault may be caused by late recognition of heading changes, tension on the controls, misinterpretation of the heading indicator (and correcting in the wrong direction), failure to appreciate changing effectiveness of rudder control as the aircraft accelerates, and other factors. If heading changes are observed and corrected instantly with small movement of the rudder pedals, swerving tendencies can be reduced.
6. Failure to maintain attitude after becoming airborne. If the pilot reacts to seat-of-the-pants sensations when the airplane lifts off, pitch control is guesswork. The pilot may either allow excessive pitch or apply excessive forward-elevator pressure, depending on the reaction to trim changes.
7. Inadequate cross-check. Fixations are likely during the trim changes, attitude changes, gear and flap retractions, and power changes. Once an instrument or a control input is applied, continue the cross-check and note the effect control during the next cross-check sequence.
8. Inadequate interpretation of instruments. Failure to understand instrument indications immediately indicates that further study of the maneuver is necessary.

Basic Instrument Flight Patterns

After attaining a reasonable degree of proficiency in basic maneuvers, apply these skills to the various combinations of individual maneuvers. The practice flight patterns, beginning on page 7-30, are directly applicable to operational instrument flying.

Chapter 8

Helicopter Attitude Instrument Flying

Introduction

Attitude instrument flying in helicopters is essentially visual flying with the flight instruments substituted for the various reference points on the helicopter and the natural horizon. Control changes, required to produce a given attitude by reference to instruments, are identical to those used in helicopter visual flight rules (VFR) flight, and pilot thought processes are the same. Basic instrument training is intended to be a building block toward attaining an instrument rating.

Flight Instruments

When flying a helicopter with reference to the flight instruments, proper instrument interpretation is the basis for aircraft control. Skill, in part, depends on understanding how a particular instrument or system functions, including its indications and limitations (see Chapter 5, Flight Instruments). With this knowledge, a pilot can quickly interpret an instrument indication and translate that information into a control response.

Instrument Flight

To achieve smooth, positive control of the helicopter during instrument flight, three fundamental skills must be developed. They are instrument cross-check, instrument interpretation, and aircraft control.

Instrument Cross-Check

Cross-checking, sometimes referred to as scanning, is the continuous and logical observation of instruments for attitude and performance information. In attitude instrument flying, an attitude is maintained by reference to the instruments, which produces the desired result in performance. Due to human error, instrument error, and helicopter performance differences in various atmospheric and loading conditions, it is difficult to establish an attitude and have performance remain constant for a long period of time. These variables make it necessary to constantly check the instruments and make appropriate changes in the helicopter's attitude. The actual technique may vary depending on what instruments are installed and where they are installed, as well as pilot experience and proficiency level. This discussion concentrates on the six basic flight instruments. *[Figure 8-1]*

At first, there may be a tendency to cross-check rapidly, looking directly at the instruments without knowing exactly what information is needed. However, with familiarity and practice, the instrument cross-check reveals definite trends during specific flight conditions. These trends help a pilot control the helicopter as it makes a transition from one flight condition to another.

When full concentration is applied to a single instrument, a problem called fixation is encountered. This results from a natural human inclination to observe a specific instrument carefully and accurately, often to the exclusion of other instruments. Fixation on a single instrument usually results in poor control. For example, while performing a turn, there is a tendency to watch only the turn-and-slip indicator instead of including other instruments in the cross-check. This fixation on the turn-and-slip indicator often leads to a loss of altitude through poor pitch-and-bank control. Look at each

Figure 8-1. *A radial scan pattern of the flight instruments enables the helicopter pilot to fully comprehend the condition and direction of the helicopter.*

instrument only long enough to understand the information it presents, and then proceed to the next one. Similarly, too much emphasis can be placed on a single instrument, instead of relying on a combination of instruments necessary for helicopter performance information. This differs from fixation in that other instruments are included in a cross-check, but too much attention is placed on one particular instrument.

During performance of a maneuver, there is sometimes a failure to anticipate significant instrument indications following attitude changes. For example, during level off from a climb or descent, a pilot may concentrate on pitch control, while forgetting about heading or roll information. This error, called omission, results in erratic control of heading and bank.

In spite of these common errors, most pilots can adapt well to flight by instrument reference after instruction and practice. Many find that they can control the helicopter more easily and precisely by instruments.

Instrument Interpretation

The flight instruments together give a picture of what is happening. No one instrument is more important than the next; however, during certain maneuvers or conditions, those instruments that provide the most pertinent and useful information are termed primary instruments. Those which back up and supplement the primary instruments are termed supporting instruments. For example, since the attitude indicator is the only instrument that provides instant and direct aircraft attitude information, it should be considered primary during any change in pitch or bank attitude. After the new attitude is established, other instruments become primary, and the attitude indicator usually becomes the supporting instrument.

Aircraft Control

Controlling a helicopter is the result of accurately interpreting the flight instruments and translating these readings into correct control responses. Aircraft control involves adjustment to pitch, bank, power, and trim in order to achieve a desired flight path.

Pitch attitude control is controlling the movement of the helicopter about its lateral axis. After interpreting the helicopter's pitch attitude by reference to the pitch instruments (attitude indicator, altimeter, airspeed indicator, and vertical speed indicator (VSI)), cyclic control adjustments are made to affect the desired pitch attitude. In this chapter, the pitch attitudes depicted are approximate and vary with different helicopters.

Bank attitude control is controlling the angle made by the lateral tilt of the rotor and the natural horizon or the movement of the helicopter about its longitudinal axis. After interpreting the helicopter's bank instruments (attitude indicator, heading indicator, and turn indicator), cyclic control adjustments are made to attain the desired bank attitude.

Power control is the application of collective pitch with corresponding throttle control, where applicable. In straight-and-level flight, changes of collective pitch are made to correct for altitude deviation if the error is more than 100 feet or the airspeed is off by more than 10 knots. If the error is less than that amount, a pilot should use a slight cyclic climb or descent.

In order to fly a helicopter by reference to the instruments, it is important to know the approximate power settings required for a particular helicopter in various load configurations and flight conditions.

Trim, in helicopters, refers to the use of the cyclic centering button, if the helicopter is so equipped, to relieve all possible cyclic pressures. Trim also refers to the use of pedal adjustment to center the ball of the turn indicator. Pedal trim is required during all power changes.

The proper adjustment of collective pitch and cyclic friction helps a pilot relax during instrument flight. Friction should be adjusted to minimize overcontrolling and to prevent creeping, but not applied to such a degree that control movement is limited. In addition, many helicopters equipped for instrument flight contain stability augmentation systems or an autopilot to help relieve pilot workload.

Straight-and-Level Flight

Straight-and-level unaccelerated flight consists of maintaining the desired altitude, heading, airspeed, and pedal trim.

Pitch Control

The pitch attitude of a helicopter is the angular relation of its longitudinal axis to the natural horizon. If available, the attitude indicator is used to establish the desired pitch attitude. In level flight, pitch attitude varies with airspeed and center of gravity (CG). At a constant altitude and a stabilized airspeed, the pitch attitude is approximately level. *[Figure 8-2]*

Attitude Indicator

The attitude indicator gives a direct indication of the pitch attitude of the helicopter. In visual flight, attain the desired pitch attitude by using the cyclic to raise and lower the nose

Figure 8-2. *The flight instruments for pitch control are the airspeed indicator, attitude indicator, altimeter, and vertical speed indicator.*

of the helicopter in relation to the natural horizon. During instrument flight, follow exactly the same procedure in raising or lowering the miniature aircraft in relation to the horizon bar.

There is some delay between control application and resultant instrument change. This is the normal control lag in the helicopter and should not be confused with instrument lag. The attitude indicator may show small misrepresentations of pitch attitude during maneuvers involving acceleration, deceleration, or turns. This precession error can be detected quickly by cross-checking the other pitch instruments.

If the miniature aircraft is properly adjusted on the ground, it may not require readjustment in flight. If the miniature aircraft is not on the horizon bar after level off at normal cruising airspeed, adjust it as necessary while maintaining level flight with the other pitch instruments. Once the miniature aircraft has been adjusted in level flight at normal cruising airspeed, leave it unchanged so it gives an accurate picture of pitch attitude at all times.

When making initial pitch attitude corrections to maintain altitude, the changes of attitude should be small and smoothly applied. The initial movement of the horizon bar should not exceed one bar width high or low. *[Figure 8-3]* If a further adjustment is required, an additional correction of one-half bar normally corrects any deviation from the desired altitude. This one-and-one-half bar correction is normally the maximum pitch attitude correction from level flight attitude.

After making the correction, cross-check the other pitch instruments to determine whether the pitch attitude change is sufficient. If additional correction is needed to return to altitude, or if the airspeed varies more than 10 knots from that desired, adjust the power.

Figure 8-3. *The initial pitch correction at normal cruise is one bar width or less.*

Altimeter

The altimeter gives an indirect indication of the pitch attitude of the helicopter in straight-and-level flight. Since the altitude should remain constant in level flight, deviation from the desired altitude indicates a need for a change in pitch attitude and power as necessary. When losing altitude, raise the pitch attitude and adjust power as necessary. When gaining altitude, lower the pitch attitude and adjust power as necessary. Indications for power changes are explained in the next paragraph.

The rate at which the altimeter moves helps to determine pitch attitude. A very slow movement of the altimeter indicates a small deviation from the desired pitch attitude, while a fast movement of the altimeter indicates a large deviation from the desired pitch attitude. Make any corrective action promptly with small control changes. Also, remember that movement of the altimeter should always be corrected by two distinct changes. The first is a change of attitude to stop the altimeter movement; the second is a change of attitude to return smoothly to the desired altitude. If altitude and airspeed are more than 100 feet and 10 knots low, respectively, apply power in addition to an increase of pitch attitude. If the altitude and airspeed are high by more than 100 feet and 10 knots, reduce power and lower the pitch attitude.

There is a small lag in the movement of the altimeter; however, for all practical purposes, consider that the altimeter gives an immediate indication of a change or a need for change in pitch attitude. Since the altimeter provides the most pertinent information regarding pitch in level flight, it is considered primary for pitch.

Vertical Speed Indicator (VSI)

The VSI gives an indirect indication of the pitch attitude of the helicopter and should be used in conjunction with the other pitch instruments to attain a high degree of accuracy and precision. The instrument indicates zero when in level flight. Any movement of the needle from the zero position shows a need for an immediate change in pitch attitude to return it to zero. Always use the VSI in conjunction with the altimeter in level flight. If a movement of the VSI is detected, immediately use the proper corrective measures to return it to zero. If the correction is made promptly, there is usually little or no change in altitude. If the needle of the VSI does not indicate zero, the altimeter indicates a gain or loss of altitude.

The initial movement of the vertical speed needle is instantaneous and indicates the trend of the vertical movement of the helicopter. A period of time is necessary for the VSI to reach its maximum point of deflection after a correction has been made. This time element is commonly referred to as instrument lag. The lag is directly proportional to the speed and magnitude of the pitch change. When employing smooth control techniques and small adjustments in pitch attitude are made, lag is minimized, and the VSI is easy to interpret.

Overcontrolling can be minimized by first neutralizing the controls and allowing the pitch attitude to stabilize, then readjusting the pitch attitude by noting the indications of the other pitch instruments.

Occasionally, the VSI may be slightly out of calibration. This could result in the instrument indicating a slight climb or descent even when the helicopter is in level flight. If the instrument cannot be calibrated properly, this error must be taken into consideration when using the VSI for pitch control. For example, if a descent of 100 feet per minute (fpm) is the vertical speed indication when the helicopter is in level flight, use that indication as level flight. Any deviation from that reading would indicate a change in attitude.

Airspeed Indicator

The airspeed indicator gives an indirect indication of helicopter pitch attitude. With a given power setting and pitch attitude, the airspeed remains constant. If the airspeed increases, the nose is too low and should be raised. If the airspeed decreases, the nose is too high and should be lowered. A rapid change in airspeed indicates a large change in pitch attitude, and a slow change in airspeed indicates a small change in pitch attitude. There is very little lag in the indications of the airspeed indicator. If, while making attitude changes, there is some lag between control application and change of airspeed, it is most likely due to cyclic control lag. Generally, a departure from the desired airspeed, due to an inadvertent pitch attitude change, also results in a change in altitude. For example, an increase in airspeed due to a low pitch attitude results in a decrease in altitude. A correction in the pitch attitude regains both airspeed and altitude.

Bank Control

The bank attitude of a helicopter is the angular relation of its lateral axis to the natural horizon. To maintain a straight course in visual flight, keep the lateral axis of the helicopter level with the natural horizon. Assuming the helicopter is in coordinated flight, any deviation from a laterally level attitude produces a turn. *[Figure 8-4]*

Attitude Indicator

The attitude indicator gives a direct indication of the bank attitude of the helicopter. For instrument flight, the miniature aircraft and the horizon bar of the attitude indicator are substituted for the actual helicopter and the natural horizon. Any change in bank attitude of the helicopter is indicated instantly by the miniature aircraft. For proper interpretation of this instrument, imagine being in the miniature aircraft. If the helicopter is properly trimmed and the rotor tilts, a turn begins. The turn can be stopped by leveling the miniature aircraft with the horizon bar. The ball in the turn-and-slip indicator should always be kept centered through proper pedal trim.

The angle of bank is indicated by the pointer on the banking scale at the top of the instrument. Small bank angles, which may not be seen by observing the miniature aircraft, can easily be determined by referring to the banking scale pointer.

Figure 8-4. *The flight instruments used for bank control are the attitude, heading, and turn indicators.*

Pitch-and-bank attitudes can be determined simultaneously on the attitude indicator. Even though the miniature aircraft is not level with the horizon bar, pitch attitude can be established by observing the relative position of the miniature aircraft and the horizon bar. *[Figure 8-5]*

The attitude indicator may show small misrepresentations of bank attitude during maneuvers that involve turns. This precession error can be detected immediately by closely cross-checking the other bank instruments during these maneuvers. Precession is normally noticed when rolling out of a turn. If, upon completion of a turn, the miniature aircraft is level and the helicopter is still turning, make a small change of bank attitude to center the turn needle and stop the movement of the heading indicator.

Heading Indicator

In coordinated flight, the heading indicator gives an indirect indication of a helicopter's bank attitude. When a helicopter is banked, it turns. When the lateral axis of a helicopter is level, it flies straight. Therefore, in coordinated flight when the heading indicator shows a constant heading, the helicopter is level laterally. A deviation from the desired heading indicates a bank in the direction the helicopter is turning. A small angle of bank is indicated by a slow change of heading; a large angle of bank is indicated by a rapid change of heading. If a turn is noticed, apply opposite cyclic until the heading indicator

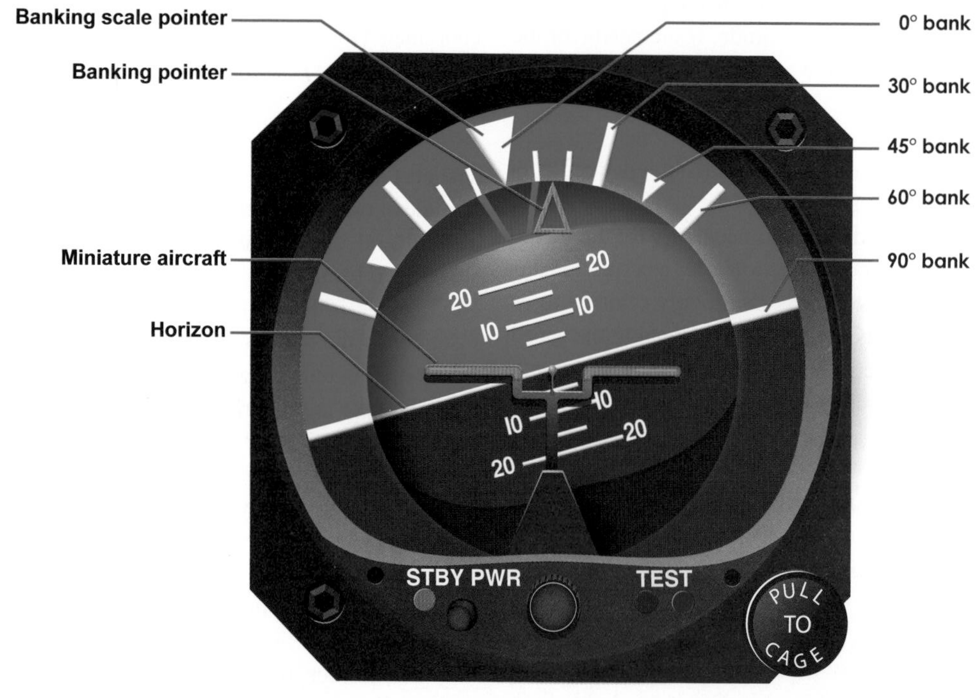

Figure 8-5. *The banking scale at the top of the attitude indicator indicates varying degrees of bank. In this example, the helicopter is banked approximately 15° to the right.*

indicates the desired heading, simultaneously ensuring the ball is centered. When making the correction to the desired heading, do not use a bank angle greater than that required to achieve a standard rate turn. In addition, if the number of degrees of change is small, limit the bank angle to the number of degrees to be turned. Bank angles greater than these require more skill and precision in attaining the desired results. During straight-and-level flight, the heading indicator is the primary reference for bank control.

Turn Indicator

During coordinated flight, the needle of the turn-and-slip indicator gives an indirect indication of the bank attitude of the helicopter. When the needle is displaced from the vertical position, the helicopter is turning in the direction of the displacement. Thus, if the needle is displaced to the left, the helicopter is turning left. Bringing the needle back to the vertical position with the cyclic produces straight flight. A close observation of the needle is necessary to accurately interpret small deviations from the desired position.

Cross-check the ball of the turn-and-slip indicator to determine if the helicopter is in coordinated flight. *[Figure 8-6]* If the rotor is laterally level and pedal pressure properly compensates for torque, the ball remains in the center. To center the ball, level the helicopter laterally by reference to the other bank instruments, then center the ball with pedal trim. Torque correction pressures vary as power changes are made. Always check the ball after such changes.

Common Errors During Straight-and-Level Flight

1. Failure to maintain altitude
2. Failure to maintain heading
3. Overcontrolling pitch and bank during corrections
4. Failure to maintain proper pedal trim
5. Failure to cross-check all available instruments

Power Control During Straight-and-Level Flight

Establishing specific power settings is accomplished through collective pitch adjustments and throttle control, where necessary. For reciprocating-powered helicopters, power indication is observed on the manifold pressure gauge. For turbine-powered helicopters, power is observed on the torque gauge. (Although most instrument flight rules (IFR)-certified helicopters are turbine powered, depictions within this chapter use a reciprocating-powered helicopter as this is where training is most likely conducted.)

At any given airspeed, a specific power setting determines whether the helicopter is in level flight, in a climb, or in a descent. For example, cruising airspeed maintained with cruising power results in level flight. If a pilot increases the power setting and holds the airspeed constant, the helicopter climbs. Conversely, if the pilot decreases power and holds the airspeed constant, the helicopter descends.

Figure 8-6. *Coordinated flight is indicated by centering of the ball.*

If the altitude is held constant, power determines the airspeed. For example, at a constant altitude, cruising power results in cruising airspeed. Any deviation from the cruising power setting results in a change of airspeed. When power is added to increase airspeed, the nose of the helicopter pitches up and yaws to the right in a helicopter with a counterclockwise main rotor blade rotation. *[Figure 8-7]* When power is reduced to decrease airspeed, the nose pitches down and yaws to the left. *[Figure 8-8]* The yawing effect is most pronounced in single-rotor helicopters and is absent in helicopters with counter-rotating rotors. To counteract the yawing tendency of the helicopter, apply pedal trim during power changes.

To maintain a constant altitude and airspeed in level flight, coordinate pitch attitude and power control. The relationship between altitude and airspeed determines the need for a change in power and/or pitch attitude. If the altitude is constant and the airspeed is high or low, change the power to obtain the desired airspeed. During the change in power, make an accurate interpretation of the altimeter, then counteract any deviation from the desired altitude by an appropriate change of pitch attitude. If the altitude is low and the airspeed is high, or vice versa, a change in pitch attitude alone may return the helicopter to the proper altitude and airspeed. If both airspeed and altitude are low, or if both are high, changes in both power and pitch attitude are necessary.

To make power control easy when changing airspeed, it is necessary to know the approximate power settings for the various airspeeds at which the helicopter is flown. When the airspeed is to be changed by any appreciable amount, adjust the power so that it is over or under that setting necessary to maintain the new airspeed. As the power approaches the desired setting, include the manifold pressure in the cross-check to determine when the proper adjustment has been accomplished. As the airspeed is changing, adjust the pitch attitude to maintain a constant altitude. A constant heading should be maintained throughout the change. As the desired airspeed is approached, adjust power to the new cruising power setting and further adjust pitch attitude to maintain altitude. The instrument indications for straight-and-level flight at normal cruise and during the transition from normal cruise to slow cruise are illustrated in *Figures 8-9* and *8-10*. After the airspeed stabilizes at slow cruise, the attitude indicator shows an approximate level pitch attitude.

The altimeter is the primary pitch instrument during level flight, whether flying at a constant airspeed or during a change in airspeed. Altitude should not change during airspeed transitions, and the heading indicator remains the primary bank instrument. Whenever the airspeed is changed by an appreciable amount, the manifold pressure gauge is momentarily the primary instrument for power control. When the airspeed approaches the desired reading, the airspeed indicator again becomes the primary instrument for power control.

Figure 8-7. *Flight instrument indications in straight-and-level flight with power increasing.*

Figure 8-8. *Flight instrument indications in straight-and-level flight with power decreasing.*

Figure 8-9. *Flight instrument indications in straight-and-level flight at normal cruise speed.*

Figure 8-10. *Flight instrument indications in straight-and-level flight with airspeed decreasing.*

To produce straight-and-level flight, the cross-check of the pitch-and-bank instruments should be combined with the power control instruments. With a constant power setting, a normal cross-check should be satisfactory. When changing power, the speed of the cross-check must be increased to cover the pitch and bank instruments adequately. This is necessary to counteract any deviations immediately.

Common Errors During Airspeed Changes

1. Improper use of power
2. Overcontrolling pitch attitude
3. Failure to maintain heading
4. Failure to maintain altitude
5. Improper pedal trim

Straight Climbs (Constant Airspeed and Constant Rate)

For any power setting and load condition, there is only one airspeed that gives the most efficient rate of climb. To determine this, consult the climb data for the type of helicopter being flown. The technique varies according to the airspeed on entry and whether a constant airspeed or constant rate climb is made.

Entry

To enter a constant airspeed climb from cruise airspeed when the climb speed is lower than cruise speed, simultaneously increase power to the climb power setting and adjust pitch attitude to the approximate climb attitude. A helicopter may or may not have an exact "climb attitude." To slow down to climb (versus cruise) airspeed, the nose must be raised. Depending on power and horizontal stabilizer configuration and effectiveness, the nose may be level during an established climb or slightly nose high. Many helicopters are very capable of climbing and never raising the nose. A short deceleration period may be necessary to slow to a more efficient climb airspeed, but the attitude indicator is often level after the climb is stabilized. The increase in power causes the helicopter to start climbing and only very slight back cyclic pressure is needed to complete the change from level to climb attitude. The attitude indicator should be used to accomplish the pitch change. If the transition from level flight to a climb is smooth, the VSI shows an immediate upward trend and then stops at a rate appropriate to the stabilized airspeed and attitude. Primary and supporting instruments for climb entry are illustrated in *Figure 8-11*.

When the helicopter stabilizes at a constant airspeed and attitude, the airspeed indicator becomes primary for pitch. The manifold pressure continues to be primary for power and should be monitored closely to determine if the proper climb power setting is being maintained. Primary and supporting instruments for a stabilized constant airspeed climb are shown in *Figure 8-12*.

Figure 8-11. *Flight instrument indications during climb entry for a constant-airspeed climb.*

Figure 8-12. *Flight instrument indications in a stabilized constant-airspeed climb.*

The technique and procedures for entering a constant rate climb are very similar to those previously described for a constant airspeed climb. For training purposes, a constant rate climb is entered from climb airspeed. Use the rate appropriate for the particular helicopter being flown. Normally, in helicopters with low climb rates, 500 fpm is appropriate. In helicopters capable of high climb rates, use a rate of 1,000 fpm.

To enter a constant rate climb, increase power to the approximate setting for the desired rate. As power is applied, the airspeed indicator is primary for pitch until the vertical speed approaches the desired rate. At this time, the VSI becomes primary for pitch. Change pitch attitude by reference to the attitude indicator to maintain the desired vertical speed. When the VSI becomes primary for pitch, the airspeed indicator becomes primary for power. *[Figure 8-13]* Adjust power to maintain desired airspeed. Pitch attitude and power corrections should be closely coordinated. To illustrate this, if the vertical speed is correct but the airspeed is low, add power. As power is increased, it may be necessary to lower the pitch attitude slightly to avoid increasing the vertical rate. Adjust the pitch attitude smoothly to avoid overcontrolling. Small power corrections are usually sufficient to bring the airspeed back to the desired indication.

Level Off

The level off from a constant airspeed climb must be started before reaching the desired altitude. Although the amount of lead varies with the type of helicopter being flown and pilot technique, the most important factor is vertical speed. As a rule of thumb, use 10 percent of the vertical velocity as the lead point. For example, if the rate of climb is 500 fpm, initiate the level off approximately 50 feet before the desired altitude. When the proper lead altitude is reached, the altimeter becomes primary for pitch. Adjust the pitch attitude to the level flight attitude for that airspeed. Cross-check the altimeter and VSI to determine when level flight has been attained at the desired altitude. If cruise airspeed is higher than climb airspeed, leave the power at the climb power setting until the airspeed approaches cruise airspeed, and then reduce it to the cruise power setting. The level off from a constant rate climb is accomplished in the same manner as the level off from a constant airspeed climb.

Straight Descents (Constant Airspeed and Constant Rate)

A descent may be performed at any normal airspeed the helicopter can attain, but the airspeed must be determined prior to entry. The technique is determined by the type of descent, a constant airspeed, or a constant rate.

Entry

If airspeed is higher than descending airspeed, and a constant airspeed descent is desired, reduce power to a descent power setting and maintain a constant altitude using cyclic pitch control. This slows the helicopter. As the helicopter

Figure 8-13. *Flight instrument indications in a stabilized constant-rate climb.*

approaches the descending airspeed, the airspeed indicator becomes primary for pitch and the manifold pressure is primary for power. Holding the airspeed constant causes the helicopter to descend. For a constant rate descent, reduce the power to the approximate setting for the desired rate. If the descent is started at the descending airspeed, the airspeed indicator is primary for pitch until the VSI approaches the desired rate. At this time, the VSI becomes primary for pitch, and the airspeed indicator becomes primary for power. Coordinate power and pitch attitude control as previously described on page 8-10 for constant rate climbs.

Level Off

The level off from a constant airspeed descent may be made at descending airspeed or at cruise airspeed, if this is higher than descending airspeed. As in a climb level off, the amount of lead depends on the rate of descent and control technique. For a level off at descending airspeed, the lead should be approximately 10 percent of the vertical speed. At the lead altitude, simultaneously increase power to the setting necessary to maintain descending airspeed in level flight. At this point, the altimeter becomes primary for pitch, and the airspeed indicator becomes primary for power.

To level off at an airspeed higher than descending airspeed, increase the power approximately 100 to 150 feet prior to reaching the desired altitude. The power setting should be that which is necessary to maintain the desired airspeed in level flight. Hold the vertical speed constant until approximately 50 feet above the desired altitude. At this point, the altimeter becomes primary for pitch and the airspeed indicator becomes primary for power. The level off from a constant rate descent should be accomplished in the same manner as the level off from a constant airspeed descent.

Common Errors During Straight Climbs and Descents

1. Failure to maintain heading
2. Improper use of power
3. Poor control of pitch attitude
4. Failure to maintain proper pedal trim
5. Failure to level off on desired altitude

Turns

Turns made by reference to the flight instruments should be made at a precise rate. Turns described in this chapter are those not exceeding a standard rate of 3° per second as indicated on the turn-and-slip indicator. True airspeed determines the angle of bank necessary to maintain a standard rate turn. A rule of thumb to determine the approximate angle of bank required for a standard rate turn is to use 15 percent of the airspeed. A simple way to determine this amount is to divide the airspeed by 10 and add one-half the result. For example, at 60 knots approximately 9° of bank is required ($60 \div 10 = 6$, $6 + 3 = 9$); at 80 knots approximately 12° of bank is needed for a standard rate turn.

To enter a turn, apply lateral cyclic in the direction of the desired turn. The entry should be accomplished smoothly, using the attitude indicator to establish the approximate bank angle. When the turn indicator indicates a standard rate turn, it becomes primary for bank. The attitude indicator now becomes a supporting instrument. During level turns, the altimeter is primary for pitch, and the airspeed indicator is primary for power. Primary and supporting instruments for a stabilized standard rate turn are illustrated in *Figure 8-14*. If an increase in power is required to maintain airspeed, slight forward cyclic pressure may be required since the helicopter tends to pitch up as collective pitch is increased. Apply pedal trim, as required, to keep the ball centered.

To recover to straight-and-level flight, apply cyclic in the direction opposite the turn. The rate of roll-out should be the same as the rate used when rolling into the turn. As the turn recovery is initiated, the attitude indicator becomes primary for bank. When the helicopter is approximately level, the heading indicator becomes primary for bank as in straight-and-level flight. Cross-check the airspeed indicator and ball closely to maintain the desired airspeed and pedal trim.

Turn to a Predetermined Heading

A helicopter turns as long as its lateral axis is tilted; therefore, the recovery must start before the desired heading is reached. The amount of lead varies with the rate of turn and piloting technique.

As a guide, when making a 3° per second rate of turn, use a lead of one-half the bank angle. For example, if using a 12° bank angle, use half of that, or 6°, as the lead point prior to the desired heading. Use this lead until the exact amount required by a particular technique can be determined. The bank angle should never exceed the number of degrees to be turned. As in any standard rate turn, the rate of recovery should be the same as the rate of entry. During turns to predetermined headings, cross-check the primary and supporting pitch, bank, and power instruments closely.

Timed Turns

A timed turn is a turn in which the clock and turn-and-slip indicator are used to change heading a definite number of degrees in a given time. For example, using a standard rate turn, a helicopter turns 45° in 15 seconds. Using a half-standard rate turn, the helicopter turns 45° in 30 seconds. Timed turns can be used if the heading indicator becomes inoperative.

Figure 8-14. *Flight instrument indications in a standard-rate turn to the left.*

Prior to performing timed turns, the turn coordinator should be calibrated to determine the accuracy of its indications. To do this, establish a standard rate turn by referring to the turn-and-slip indicator. Then, as the sweep second hand of the clock passes a cardinal point (12, 3, 6, or 9), check the heading on the heading indicator. While holding the indicated rate of turn constant, note the heading changes at 10-second intervals. If the helicopter turns more or less than 30° in that interval, a smaller or larger deflection of the needle is necessary to produce a standard rate turn. After the turn-and-slip indicator has been calibrated during turns in each direction, note the corrected deflections, if any, and apply them during all timed turns.

Use the same cross-check and control technique in making timed turns that is used to make turns to a predetermined heading, but substitute the clock for the heading indicator. The needle of the turn-and-slip indicator is primary for bank control, the altimeter is primary for pitch control, and the airspeed indicator is primary for power control. Begin the roll-in when the clock's second hand passes a cardinal point; hold the turn at the calibrated standard rate indication or half-standard rate for small changes in heading; then begin the roll-out when the computed number of seconds has elapsed. If the roll-in and roll-out rates are the same, the time taken during entry and recovery need not be considered in the time computation.

If practicing timed turns with a full instrument panel, check the heading indicator for the accuracy of the turns. If executing turns without the heading indicator, use the magnetic compass at the completion of the turn to check turn accuracy, taking compass deviation errors into consideration.

Change of Airspeed in Turns

Changing airspeed in turns is an effective maneuver for increasing proficiency in all three basic instrument skills. Since the maneuver involves simultaneous changes in all components of control, proper execution requires a rapid cross-check and interpretation, as well as smooth control. Proficiency in the maneuver also contributes to confidence in the instruments during attitude and power changes involved in more complex maneuvers.

Pitch and power control techniques are the same as those used during airspeed changes in straight-and-level flight. As discussed previously, the angle of bank necessary for a given rate of turn is proportional to the true airspeed. Since the turns are executed at standard rate, the angle of bank must be varied in direct proportion to the airspeed change in order to maintain a constant rate of turn. During a reduction of airspeed, decrease the angle of bank and increase the pitch attitude to maintain altitude and a standard rate turn.

Altimeter and turn indicator readings should remain constant throughout the turn. The altimeter is primary for pitch control, and the turn needle is primary for bank control. Manifold pressure is primary for power control while the airspeed is changing. As the airspeed approaches the new indication, the airspeed indicator becomes primary for power control.

Two methods of changing airspeed in turns may be used. In the first method, airspeed is changed after the turn is established. In the second method, the airspeed change is initiated simultaneously with the turn entry. The first method is easier, but regardless of the method used, the rate of cross-check must be increased as power is reduced. As the helicopter decelerates, check the altimeter and VSI for needed pitch changes and the bank instruments for needed bank changes. If the needle of the turn-and-slip indicator shows a deviation from the desired deflection, change the bank. Adjust pitch attitude to maintain altitude. When the airspeed approaches that desired, the airspeed indicator becomes primary for power control. Adjust the power to maintain the desired airspeed. Use pedal trim to ensure the maneuver is coordinated.

Until control technique is very smooth, frequently cross-check the attitude indicator to keep from overcontrolling and to provide approximate bank angles appropriate for the changing airspeeds.

Compass Turns

The use of gyroscopic heading indicators makes heading control very easy. However, if the heading indicator fails or the helicopter is not equipped with one, use the magnetic compass for heading reference. When making compass-only turns, a pilot needs to adjust for the lead or lag created by acceleration and deceleration errors so that the helicopter rolls out on the desired heading. When turning to a heading of north, the lead for the roll-out must include the number of degrees of latitude plus the lead normally used in recovery from turns. During a turn to a south heading, maintain the turn until the compass passes south the number of degrees of latitude, minus the normal roll-out lead. For example, when turning from an easterly direction to north, where the latitude is 30°, start the roll-out when the compass reads 37° (30° plus one-half the 15° angle of bank or whatever amount is appropriate for the rate of roll-out). When turning from an easterly direction to south, start the roll-out when the magnetic compass reads 203° (180° plus 30° minus one-half the angle of bank). When making similar turns from a westerly direction, the appropriate points at which to begin the roll-out would be 323° for a turn to north and 157° for a turn to south.

30° Bank Turn

A turn using 30° of bank is seldom necessary or advisable in instrument meteorological conditions (IMC) and is considered an unusual attitude in a helicopter. However, it is an excellent maneuver to practice to increase the ability to react quickly and smoothly to rapid changes of attitude. Even though the entry and recovery techniques are the same as for any other turn, it is more difficult to control pitch because of the decrease in vertical lift as the bank increases. Also, because of the decrease in vertical lift, there is a tendency to lose altitude and/or airspeed. Therefore, to maintain a constant altitude and airspeed, additional power is required. Do not initiate a correction, however, until the instruments indicate the need for one. During the maneuver, note the need for a correction on the altimeter and VSI, check the attitude indicator, and then make the necessary adjustments. After making a change, check the altimeter and VSI again to determine whether or not the correction was adequate.

Climbing and Descending Turns

For climbing and descending turns, the techniques described previously for straight climbs, descents, and standard rate turns are combined. For practice, simultaneously turn and start the climb or descent. The primary and supporting instruments for a stabilized constant airspeed left climbing turn are illustrated in *Figure 8-15*. The level off from a climbing or descending turn is the same as the level off from a straight climb or descent. To return to straight-and-level flight, stop the turn and then level off, or level off and then stop the turn, or simultaneously level off and stop the turn. During climbing and descending turns, keep the ball of the turn indicator centered with pedal trim.

Common Errors During Turns

1. Failure to maintain desired turn rate
2. Failure to maintain altitude in level turns
3. Failure to maintain desired airspeed
4. Variation in the rate of entry and recovery
5. Failure to use proper lead in turns to a heading
6. Failure to properly compute time during timed turns
7. Failure to use proper leads and lags during the compass turns
8. Improper use of power
9. Failure to use proper pedal trim

Figure 8-15. *Flight instrument indications for a stabilized left climbing turn at a constant airspeed.*

Unusual Attitudes

Any maneuver not required for normal helicopter instrument flight is an unusual attitude and may be caused by any one or combination of factors, such as turbulence, disorientation, instrument failure, confusion, preoccupation with flight deck duties, carelessness in cross-checking, errors in instrument interpretation, or lack of proficiency in aircraft control. Due to the instability characteristics of the helicopter, unusual attitudes can be extremely critical. As soon as an unusual attitude is detected, make a recovery to straight-and-level flight as soon as possible with a minimum loss of altitude.

To recover from an unusual attitude, a pilot should correct bank-and-pitch attitude and adjust power as necessary. All components are changed almost simultaneously, with little lead of one over the other. A pilot must be able to perform this task with and without the attitude indicator. If the helicopter is in a climbing or descending turn, adjust bank, pitch, and power. The bank attitude should be corrected by referring to the turn-and-slip indicator and attitude indicator. Pitch attitude should be corrected by reference to the altimeter, airspeed indicator, VSI, and attitude indicator. Adjust power by referring to the airspeed indicator and manifold pressure.

Since the displacement of the controls used in recovery from unusual attitudes may be greater than those used for normal flight, make careful adjustments as straight-and-level flight is approached. Cross-check the other instruments closely to avoid overcontrolling.

Common Errors During Unusual Attitude Recoveries

1. Failure to make proper pitch correction
2. Failure to make proper bank correction
3. Failure to make proper power correction
4. Overcontrolling pitch and/or bank attitude
5. Overcontrolling power
6. Excessive loss of altitude

Emergencies

Emergencies during instrument flight are handled similarly to those occurring during VFR flight. A thorough knowledge of the helicopter and its systems, as well as good aeronautical knowledge and judgment, is the best preparation for emergency situations. Safe operations begin with preflight planning and a thorough preflight inspection. Plan a route of flight to include adequate landing sites in the event of an emergency landing. Make sure all resources, such as maps, publications, flashlights, and fire extinguishers, are readily available for use in an emergency.

During any emergency, first fly the aircraft. This means ensure the helicopter is under control, and determine emergency landing sites. Then perform the emergency checklist memory items, followed by items written in the rotorcraft flight manual (RFM). When all these items are under control, notify air traffic control (ATC). Declare any emergency on the last assigned ATC frequency. If one was not issued, transmit on the emergency frequency 121.5. Set the transponder to the emergency squawk code 7700. This code triggers an alarm or special indicator in radar facilities.

When experiencing most in-flight emergencies, such as low fuel or complete electrical failure, land as soon as possible. In the event of an electrical fire, turn off all nonessential equipment and land immediately. Some essential electrical instruments, such as the attitude indicator, may be required for a safe landing. A navigation radio failure may not require an immediate landing if the flight can continue safely. In this case, land as soon as practical. ATC may be able to provide vectors to a safe landing area. For specific details on what to do during an emergency, refer to the RFM for the helicopter.

Autorotations

Both straight-ahead and turning autorotations should be practiced by reference to instruments. This training ensures prompt corrective action to maintain positive aircraft control in the event of an engine failure.

To enter autorotation, reduce collective pitch smoothly to maintain a safe rotor RPM and apply pedal trim to keep the ball of the turn-and-slip indicator centered. The pitch attitude of the helicopter should be approximately level as shown by the attitude indicator. The airspeed indicator is the primary pitch instrument and should be adjusted to the recommended autorotation speed. The heading indicator is primary for bank in a straight-ahead autorotation. In a turning autorotation, a standard rate turn should be maintained by reference to the needle of the turn-and-slip indicator.

Common Errors During Autorotations

1. Uncoordinated entry due to improper pedal trim
2. Poor airspeed control due to improper pitch attitude
3. Poor heading control in straight-ahead autorotations
4. Failure to maintain proper rotor RPM
5. Failure to maintain a standard rate turn during turning autorotations

Servo Failure

Most helicopters certified for single-pilot IFR flight are required to have autopilots, which greatly reduces pilot workload. If an autopilot servo fails, however, resume manual control of the helicopter. The amount of workload increase depends on which servo fails. If a cyclic servo fails, a pilot may want to land immediately because the workload increases tremendously. If an antitorque or collective servo fails, continuing to the next suitable landing site might be possible.

Instrument Takeoff

The procedures and techniques described here should be modified as necessary to conform to those set forth in the operating instructions for the particular helicopter being flown. During training, instrument takeoffs should not be attempted except when receiving instruction from an appropriately certificated, proficient flight instructor pilot.

Adjust the miniature aircraft in the attitude indicator, as appropriate, for the aircraft being flown. After the helicopter is aligned with the runway or takeoff pad, to prevent forward movement of a helicopter equipped with a wheel-type landing gear, set the parking brakes or apply the toe brakes. If the parking brake is used, it must be unlocked after the takeoff has been completed. Apply sufficient friction to the collective pitch control to minimize overcontrolling and to prevent creeping. Excessive friction should be avoided since it limits collective pitch movement.

After checking all instruments for proper indications, start the takeoff by applying collective pitch and a predetermined power setting. Add power smoothly and steadily to gain airspeed and altitude simultaneously and to prevent settling to the ground. As power is applied and the helicopter becomes airborne, use the antitorque pedals initially to maintain the desired heading. At the same time, apply forward cyclic to begin accelerating to climbing airspeed. During the initial acceleration, the pitch attitude of the helicopter, as read on the attitude indicator, should be one- to two-bar widths low. The primary and supporting instruments after becoming airborne are illustrated in *Figure 8-16*. As the airspeed increases to the appropriate climb airspeed, adjust pitch gradually to climb attitude. As climb airspeed is reached, reduce power to the climb power setting and transition to a fully coordinated straight climb.

During the initial climb out, minor heading corrections should be made with pedals only until sufficient airspeed is attained to transition to fully coordinated flight. Throughout the instrument takeoff, instrument cross-check and interpretations must be rapid and accurate and aircraft control positive and smooth.

Figure 8-16. *Flight instrument indications during an instrument takeoff.*

Common Errors During Instrument Takeoffs

1. Failure to maintain heading
2. Overcontrolling pedals
3. Failure to use required power
4. Failure to adjust pitch attitude as climbing airspeed is reached

Changing Technology

Advances in technology have brought about changes in the instrumentation found in all types of aircraft, including helicopters. Electronic displays commonly referred to as "glass cockpits" are becoming more common. Primary flight displays (PFDs) and multi-function displays (MFDs) are changing not only what information is available to a pilot but also how that information is displayed.

Illustrations of technological advancements in instrumentation are described as follows. In *Figure 8-17,* a typical PFD depicts an aircraft flying straight-and-level at 3,000 feet and 100 knots. *Figure 8-18* illustrates a nose-low pitch attitude in a right turn. MFDs can be configured to provide navigation information, such as the moving map in *Figure 8-19* or information pertaining to aircraft systems as in *Figure 8-20.*

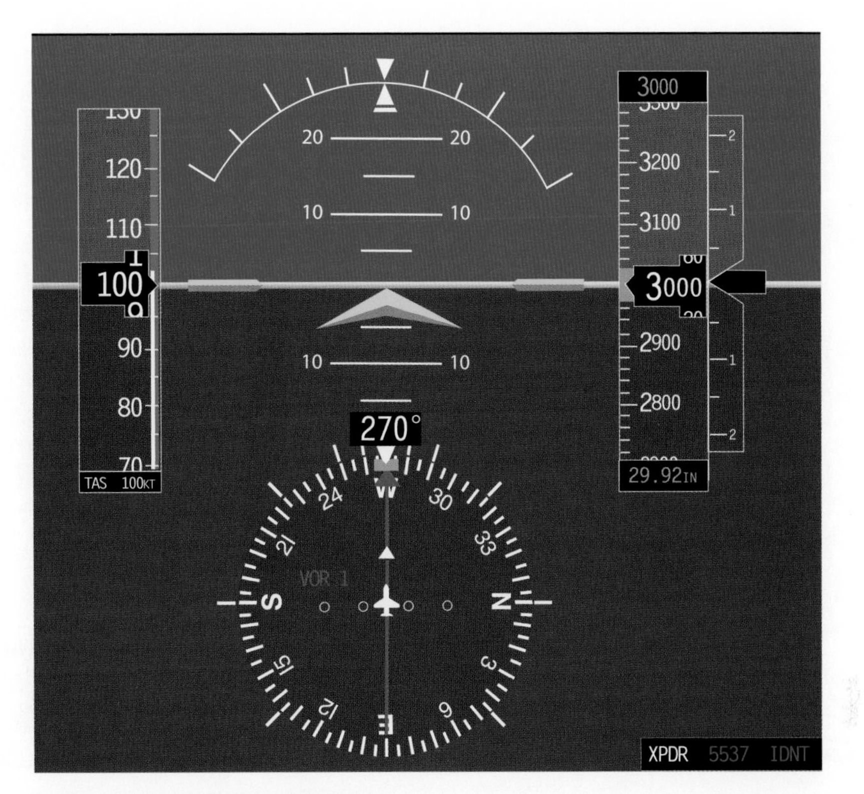

Figure 8-17. *PFD indications during straight-and-level flight.*

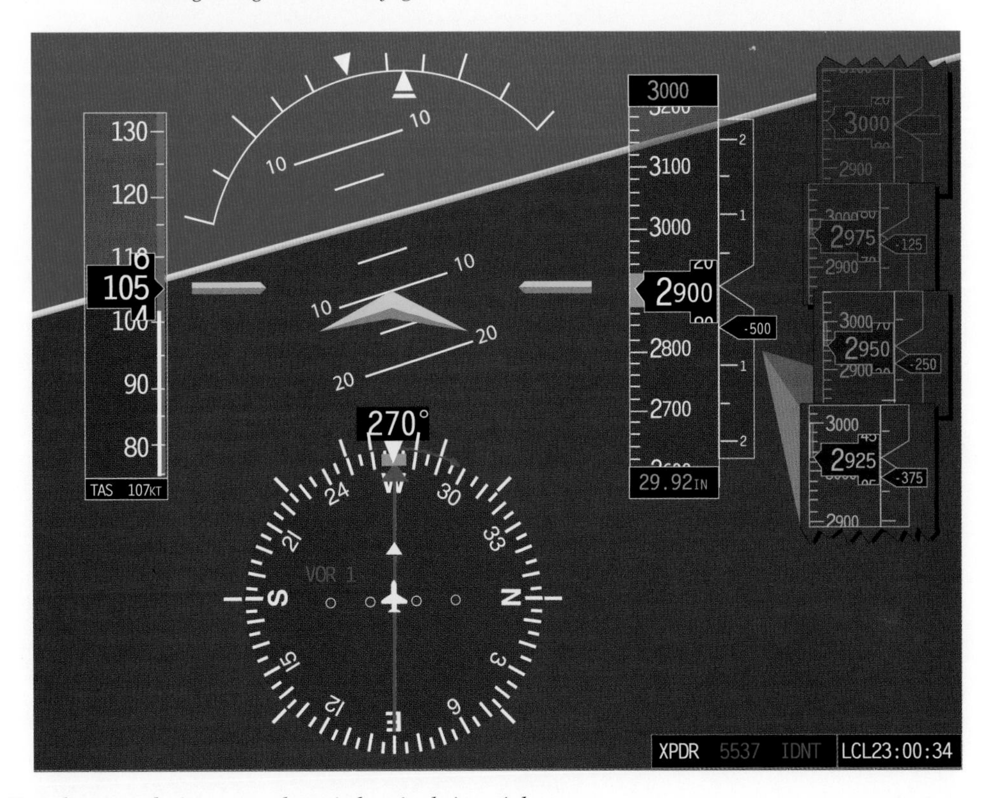

Figure 8-18. *PFD indications during a nose-low pitch attitude in a right turn.*

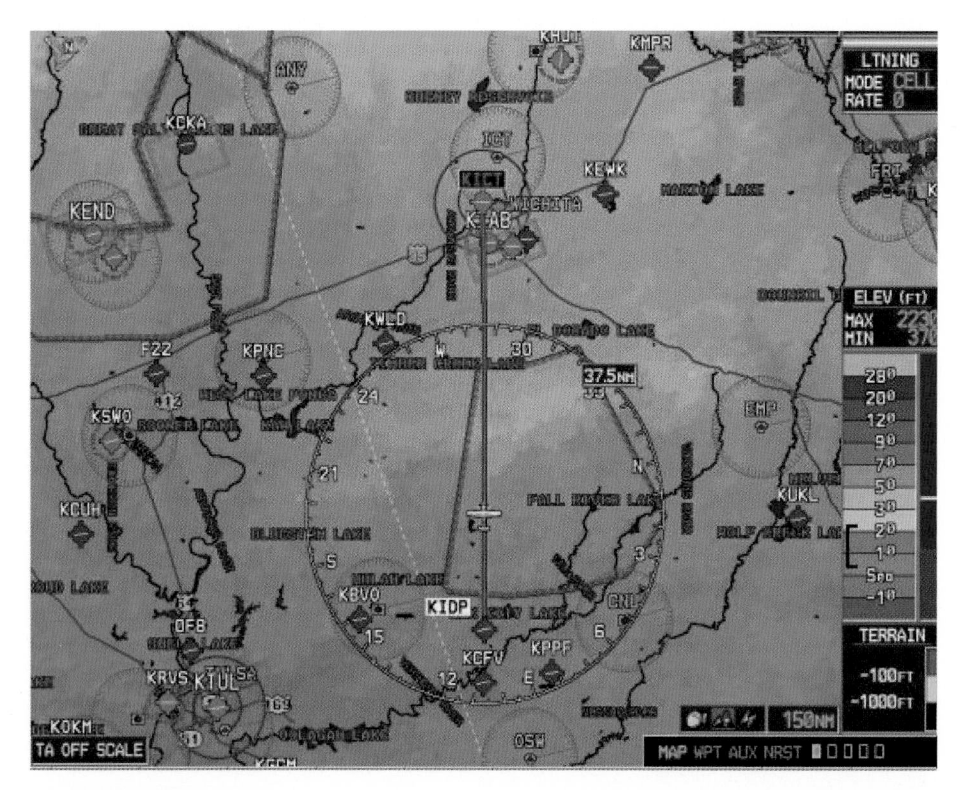

Figure 8-19. *MFD display of a moving map.*

Figure 8-20. *MFD display of aircraft systems.*

Chapter 9

Navigation Systems

Introduction

This chapter provides the basic radio principles applicable to navigation equipment, as well as an operational knowledge of how to use these systems in instrument flight. This information provides the framework for all instrument procedures, including standard instrument departure procedures (SIDS), departure procedures (DPs), holding patterns, and approaches, because each of these maneuvers consists mainly of accurate attitude instrument flying and accurate tracking using navigation systems.

Basic Radio Principles

A radio wave is an electromagnetic (EM) wave with frequency characteristics that make it useful. The wave travels long distances through space (in or out of the atmosphere) without losing too much strength. An antenna is used to convert electric current into a radio wave so it can travel through space to the receiving antenna, which converts it back into an electric current for use by a receiver.

How Radio Waves Propagate

All matter has a varying degree of conductivity or resistance to radio waves. The Earth itself acts as the greatest resistor to radio waves. Radiated energy that travels near the ground induces a voltage in the ground that subtracts energy from the wave, decreasing the strength of the wave as the distance from the antenna becomes greater. Trees, buildings, and mineral deposits affect the strength to varying degrees. Radiated energy in the upper atmosphere is likewise affected as the energy of radiation is absorbed by molecules of air, water, and dust. The characteristics of radio wave propagation vary according to the signal frequency and the design, use, and limitations of the equipment.

Ground Wave

A ground wave travels across the surface of the Earth. You can best imagine a ground wave's path as being in a tunnel or alley bounded by the surface of the Earth and by the ionosphere, which keeps the ground wave from going out into space. Generally, the lower the frequency, the farther the signal travels.

Ground waves are usable for navigation purposes because they travel reliably and predictably along the same route day after day and are not influenced by too many outside factors. The ground wave frequency range is generally from the lowest frequencies in the radio range (perhaps as low as 100 Hz) up to approximately 1,000 kHz (1 MHz). Although there is a ground wave component to frequencies above this, up to 30 MHz, the ground wave at these higher frequencies loses strength over very short distances.

Sky Wave

The sky wave, at frequencies of 1 to 30 MHz, is good for long distances because these frequencies are refracted or "bent" by the ionosphere, causing the signal to be sent back to Earth from high in the sky and received great distances away. *[Figure 9-1]* Used by high frequency (HF) radios in aircraft, messages can be sent across oceans using only 50 to 100 watts of power. Frequencies that produce a sky wave are not used for navigation because the pathway of the signal from transmitter to receiver is highly variable. The wave is "bounced" off of the ionosphere, which is always changing due to the varying amount of the sun's radiation reaching it (night/day and seasonal variations, sunspot activity, etc.). The sky wave is, therefore, unreliable for navigation purposes.

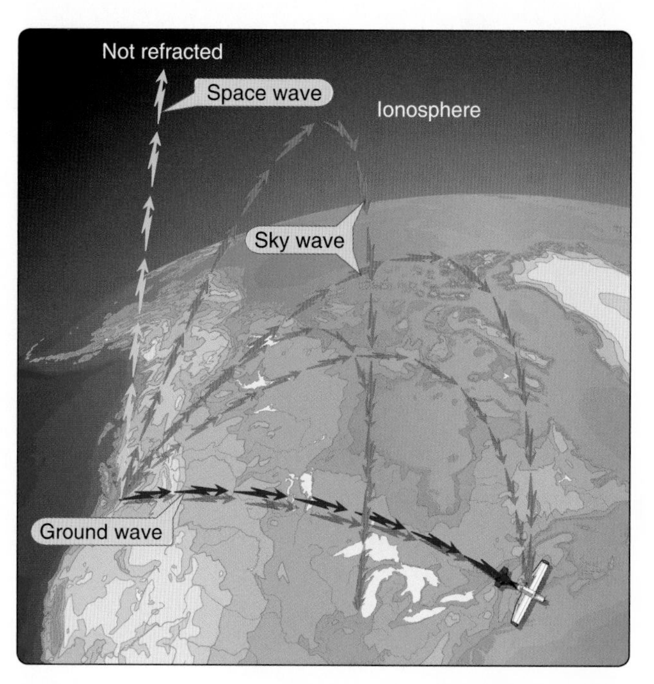

Figure 9-1. *Ground, space, and sky wave propogation.*

For aeronautical communication purposes, the sky wave (HF) is about 80 to 90 percent reliable. HF is being gradually replaced by more reliable satellite communication.

Space Wave

When able to pass through the ionosphere, radio waves of 15 MHz and above (all the way up to many GHz), are considered space waves. Most navigation systems operate with signals propagating as space waves. Frequencies above 100 MHz have nearly no ground or sky wave components. They are space waves, but (except for global positioning system (GPS)) the navigation signal is used before it reaches the ionosphere so the effect of the ionosphere, which can cause some propagation errors, is minimal. GPS errors caused by passage through the ionosphere are significant and are corrected for by the GPS receiver system.

Space waves have another characteristic of concern to users. Space waves reflect off hard objects and may be blocked if the object is between the transmitter and the receiver. Site and terrain error, as well as propeller/rotor modulation error in very high omnidirectional range (VOR) systems, is caused by this bounce. Instrument landing system (ILS) course distortion is also the result of this phenomenon, which led to the need for establishment of ILS critical areas.

Generally, space waves are "line of sight" receivable, but those of lower frequencies "bend" somewhat over the horizon. The VOR signal at 108 to 118 MHz is a lower frequency than distance measuring equipment (DME) at 962 to 1213 MHz. Therefore, when an aircraft is flown "over the horizon" from a VOR/DME station, the DME is normally the first to stop functioning.

Disturbances to Radio Wave Reception

Static distorts the radio wave and interferes with normal reception of communications and navigation signals. Low-frequency airborne equipment, such as automatic direction finder (ADF) and LORAN (LOng RAnge Navigation) ,are particularly subject to static disturbance. Using very high frequency (VHF) and ultra-high frequency (UHF) frequencies avoids many of the discharge noise effects. Static noise heard on navigation or communication radio frequencies may be a warning of interference with navigation instrument displays. Some of the problems caused by precipitation static (P-static) are:

- Complete loss of VHF communications.
- Erroneous magnetic compass readings.
- Aircraft flying with one wing low while using the autopilot.
- High-pitched squeal on audio.
- Motorboat sound on audio.
- Loss of all avionics.
- Inoperative very-low frequency (VLF) navigation system.
- Erratic instrument readouts.
- Weak transmissions and poor radio reception.
- St. Elmo's Fire.

Traditional Navigation Systems

Nondirectional Radio Beacon (NDB)

The nondirectional radio beacon (NDB) is a ground-based radio transmitter that transmits radio energy in all directions. The ADF, when used with an NDB, determines the bearing from the aircraft to the transmitting station. The indicator may be mounted in a separate instrument in the aircraft panel. *[Figure 9-2]* The ADF needle points to the NDB ground station to determine the relative bearing (RB) to the transmitting station. It is the number of degrees measured clockwise between the aircraft's heading and the direction from which the bearing is taken. The aircraft's magnetic heading (MH) is the direction the aircraft is pointed with respect to magnetic north. The magnetic bearing (MB) is the direction to or from a radio transmitting station measured relative to magnetic north.

Figure 9-2. *ADF indicator instrument and receiver.*

NDB Components

The ground equipment, the NDB, transmits in the frequency range of 190 to 535 kHz. Most ADFs also tune the AM broadcast band frequencies above the NDB band (550 to 1650 kHz). However, these frequencies are not approved for navigation because stations do not continuously identify themselves, and they are much more susceptible to sky wave propagation especially from dusk to dawn. NDB stations are capable of voice transmission and are often used for transmitting the Automated Weather Observing System (AWOS). The aircraft must be in operational range of the NDB. Coverage depends on the strength of the transmitting station. Before relying on ADF indications, identify the station by listening to the Morse code identifier. NDB stations are usually two letters or an alpha-numeric combination.

ADF Components

The airborne equipment includes two antennas: a receiver and the indicator instrument. The "sense" antenna (non-directional) receives signals with nearly equal efficiency from all directions. The "loop" antenna receives signals better from two directions (bidirectional). When the loop and sense antenna inputs are processed together in the ADF radio, the result is the ability to receive a radio signal well in all directions but one, thus resolving all directional ambiguity. The indicator instrument can be one of four kinds: fixed-card ADF, rotatable compass-card ADF, or radio magnetic indicator (RMI) with either one needle or dual needle. Fixed-card ADF (also known as the relative bearing indicator (RBI)) always indicates zero at the top of the instrument, with the needle indicating the RB to the station. *Figure 9-3* indicates an RB of 135°; if the MH is 045°, the MB to the station is 180°. (MH + RB = MB to the station.)

Figure 9-3. *Relative bearing (RB) on a fixed-card indicator. Note that the card always indicates 360° or north. In this case, the RB to the station is 135° to the right. If the aircraft were on a magnetic heading of 360°, then the magnetic bearing (MB) would also be 135°.*

The movable-card ADF allows the pilot to rotate the aircraft's present heading to the top of the instrument so that the head of the needle indicates MB to the station and the tail indicates MB from the station. *Figure 9-4* indicates a heading of 045°, MB to the station of 180°, and MB from the station of 360°.

The RMI differs from the movable-card ADF in that it automatically rotates the azimuth card (remotely controlled by a gyrocompass) to represent aircraft heading. The RMI has two needles, which can be used to indicate navigation information from either the ADF or the VOR receiver. When a needle is being driven by the ADF, the head of the needle indicates the MB TO the station tuned on the ADF receiver. The tail of the needle is the bearing FROM the station. When a needle of the RMI is driven by a VOR receiver, the needle indicates where the aircraft is radially with respect to the VOR station. The needle points the bearing TO the station as read on the azimuth card. The tail of the needle points to the radial of the VOR the aircraft is currently on or crossing. *Figure 9-5* indicates a heading of 360°, the MB to the station is 005°, and the MB from the station is 185°.

Figure 9-4. *Relative bearing (RB) on a movable-card indicator. By placing the aircraft's magnetic heading (MH) of 045° under the top index, the RB of 135° to the right is also the magnetic bearing (no wind conditions), which takes you to the transmitting station.*

Figure 9-5. *Radio magnetic indicator (RMI). Because the aircraft's magnetic heading (MH) is automatically changed, the relative bearing (RB), in this case 095°, indicates the magnetic bearing (095°) to the station (no wind conditions) and the MH that takes you there.*

Function of ADF

The ADF can be used to plot your position, track inbound and outbound, and intercept a bearing. These procedures are used to execute holding patterns and nonprecision instrument approaches.

Orientation

The ADF needle points TO the station, regardless of aircraft heading or position. The RB indicated is thus the angular relationship between the aircraft heading and the station, measured clockwise from the nose of the aircraft. Think of the nose/tail and left/right needle indications, visualizing the ADF dial in terms of the longitudinal axis of the aircraft. When the needle points to 0°, the nose of the aircraft points directly to the station; with the pointer on 210°, the station is 30° to the left of the tail; with the pointer on 090°, the station is off the right wingtip. The RB alone does not indicate aircraft position. The RB must be related to aircraft heading in order to determine direction to or from the station.

Station Passage

When you are near the station, slight deviations from the desired track result in large deflections of the needle. Therefore, it is important to establish the correct drift correction angle as soon as possible. Make small heading corrections (not over 5°) as soon as the needle shows a deviation from course, until it begins to rotate steadily toward a wingtip position or shows erratic left/right oscillations. You are abeam a station when the needle points 90° off your track. Hold your last corrected heading constant and time station passage when the needle shows either wingtip position or settles at or near the 180° position. The time interval from the first indications of station proximity to positive station passage varies with altitude—a few seconds at low levels to 3 minutes at high altitude.

Homing

The ADF may be used to "home" in on a station. Homing is flying the aircraft on any heading required to keep the needle pointing directly to the 0° RB position. To home in on a station, tune the station, identify the Morse code signal, and then turn the aircraft to bring the ADF azimuth needle to the 0° RB position. Turns should be made using the heading indicator. When the turn is complete, check the ADF needle and make small corrections as necessary.

Figure 9-6 illustrates homing starting from an initial MH of 050° and an RB of 310°, indicating a 50° left turn is needed to produce an RB of zero. Turn left, rolling out at 50° minus 50° equals 360°. Small heading corrections are then made to zero the ADF needle.

If there is no wind, the aircraft homes to the station on a direct track over the ground. With a crosswind, the aircraft follows a circuitous path to the station on the downwind side of the direct track to the station.

Tracking

Tracking uses a heading that maintains the desired track to or from the station regardless of crosswind conditions. Interpretation of the heading indicator and needle is done to maintain a constant MB to or from the station.

To track inbound, turn to the heading that produces a zero RB. Maintain this heading until off-course drift is indicated by displacement of the needle, which occurs if there is a crosswind (needle moving left = wind from the left; needle moving right = wind from the right). A rapid rate of bearing change with a constant heading indicates either a strong crosswind or close proximity to the station or both. When there is a definite (2° to 5°) change in needle reading, turn in the direction of needle deflection to intercept the initial MB. The angle of interception must be greater than the number of degrees of drift, otherwise the aircraft slowly drifts due to the wind pushing the aircraft. If repeated often enough, the track to the station appears circular and the distance greatly increased as compared to a straight track. The intercept angle depends on the rate of drift, the aircraft speed, and station proximity. Initially, it is standard to double the RB when turning toward your course.

For example, if your heading equals your course and the needle points 10° left, turn 20° left, twice the initial RB. *[Figure 9-7]* This is your intercept angle to capture the RB. Hold this heading until the needle is deflected 20° in the opposite direction. That is, the deflection of the needle equals the interception angle (in this case 20°). The track has been intercepted, and the aircraft remains on track as long as the RB remains the same number of degrees as the wind correction angle (WCA), the angle between the desired track and the heading of the aircraft necessary to keep the aircraft tracking over the desired track. Lead the interception to avoid overshooting the track. Turn 10° toward the inbound course. You are now inbound with a 10° left correction angle.

NOTE: In *Figure 9-7*, for the aircraft closest to the station, the WCA is 10° left and the RB is 10° right. If those values do not change, the aircraft tracks directly to the station. If you observe off-course deflection in the original direction, turn again to the original interception heading. When the desired course has been re-intercepted, turn 5° toward the inbound course, proceeding inbound with a 15° drift correction. If the initial 10° drift correction is excessive, as shown by needle deflection away from the wind, turn to parallel the desired course and let the wind drift you back on course. When the needle is again zeroed, turn into the wind with a reduced drift correction angle.

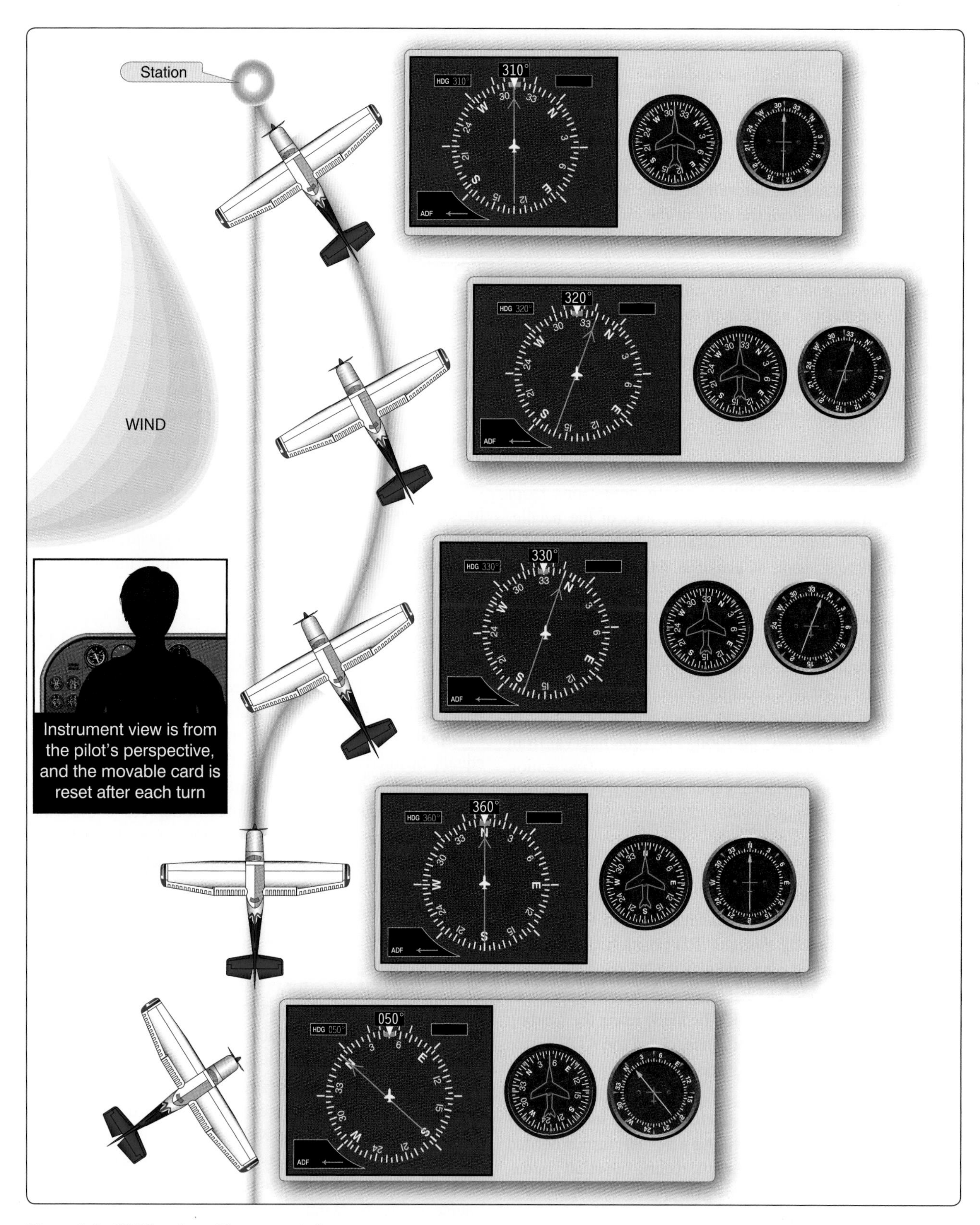

Figure 9-6. *ADF homing with a crosswind.*

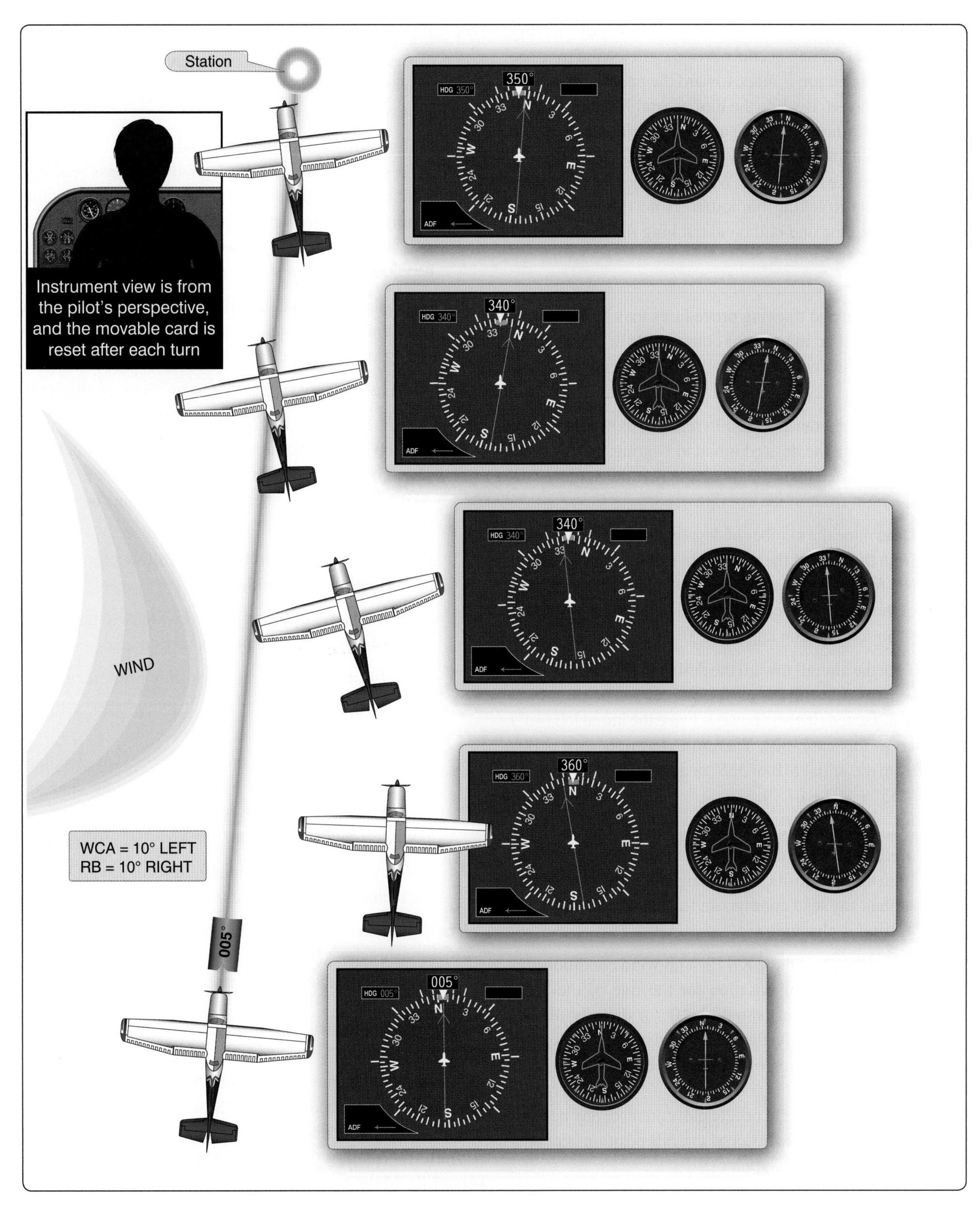

Figure 9-7. *ADF tracking inbound.*

To track outbound, the same principles apply: needle moving left = wind from the left, needle moving right = wind from the right. Wind correction is made toward the needle deflection. The only exception is while the turn to establish the WCA is being made, the direction of the azimuth needle deflections is reversed. When tracking inbound, needle deflection decreases while turning to establish the WCA, and needle deflection increases when tracking outbound. Note the example of course interception and outbound tracking in *Figure 9-8.*

Intercepting Bearings

ADF orientation and tracking procedures may be applied to intercept a specified inbound or outbound MB. To intercept an inbound bearing of 355°, the following steps may be used. *[Figure 9-9]*

1. Determine your position in relation to the station by paralleling the desired inbound bearing. In this case, turn to a heading of 355°. Note that the station is to the right front of the aircraft.
2. Determine the number of degrees of needle deflection from the nose of the aircraft. In this case, the needle's RB from the aircraft's nose is 40° to the right. A rule of thumb for interception is to double this RB amount as an interception angle (80°).
3. Turn the aircraft toward the desired MB the number of degrees determined for the interception angle, which as indicated (in two above) is twice the initial RB (40°) or, in this case, 80°. Therefore, the right turn is 80° from the initial MB of 355° or a turn to 075° magnetic (355° + 80° + 075°).
4. Maintain this interception heading of 075° until the needle is deflected the same number of degrees "left" from the zero position as the angle of interception 080° (minus any lead appropriate for the rate at which the bearing is changing).
5. Turn left 80° and the RB (in a no wind condition and with proper compensation for the rate of the ADF needle movement) should be 0° or directly off the nose. Additionally, the MB should be 355° indicating proper interception of the desired course.

NOTE: The rate of an ADF needle movement, or any bearing pointer for that matter, is faster as aircraft position becomes closer to the station or waypoint (WP).

Interception of an outbound MB can be accomplished by the same procedures as for the inbound intercept, except that it is necessary to substitute the 180° position for the zero position on the needle.

Operational Errors of ADF

Some of the common pilot-induced errors associated with ADF navigation are listed below to help you avoid making the same mistakes. The errors are:

1. Failure to keep the heading indicator set so that it agrees with the corrected magnetic compass reading. Initiating an ADF approach without verifying that the heading indicator agrees with the corrected compass indicator reading may cause the pilot to believe that he is on course but still impact the terrain (CFIT).
2. Improper tuning and station identification. Many pilots have made the mistake of homing or tracking to the wrong station.
3. Positively identifying any malfunctions of the RMI slaving system or ignoring the warning flag.
4. Dependence on homing rather than proper tracking. This commonly results from sole reliance on the ADF indications rather than correlating them with heading indications.
5. Poor orientation due to failure to follow proper steps in orientation and tracking.
6. Careless interception angles, very likely to happen if you rush the initial orientation procedure.
7. Overshooting and undershooting predetermined MBs, often due to forgetting the course interception angles used.
8. Failure to maintain selected headings. Any heading change is accompanied by an ADF needle change. The instruments must be read in combination before any interpretation is made.
9. Failure to understand the limitations of the ADF and the factors that affect its use.
10. Overcontrolling track corrections close to the station (chasing the ADF needle) due to failure to understand or recognize station approach.

Very High Frequency Omnidirectional Range (VOR)

VOR is the primary navigational aid (NAVAID) used by civil aviation in the National Airspace System (NAS). The VOR ground station is oriented to magnetic north and transmits azimuth information to the aircraft, providing 360 courses TO or FROM the VOR station. When DME is installed with the VOR, it is referred to as a VOR/DME and provides both azimuth and distance information. When military tactical air navigation (TACAN) equipment is installed with the VOR, it is known as a VORTAC and provides both azimuth and distance information.

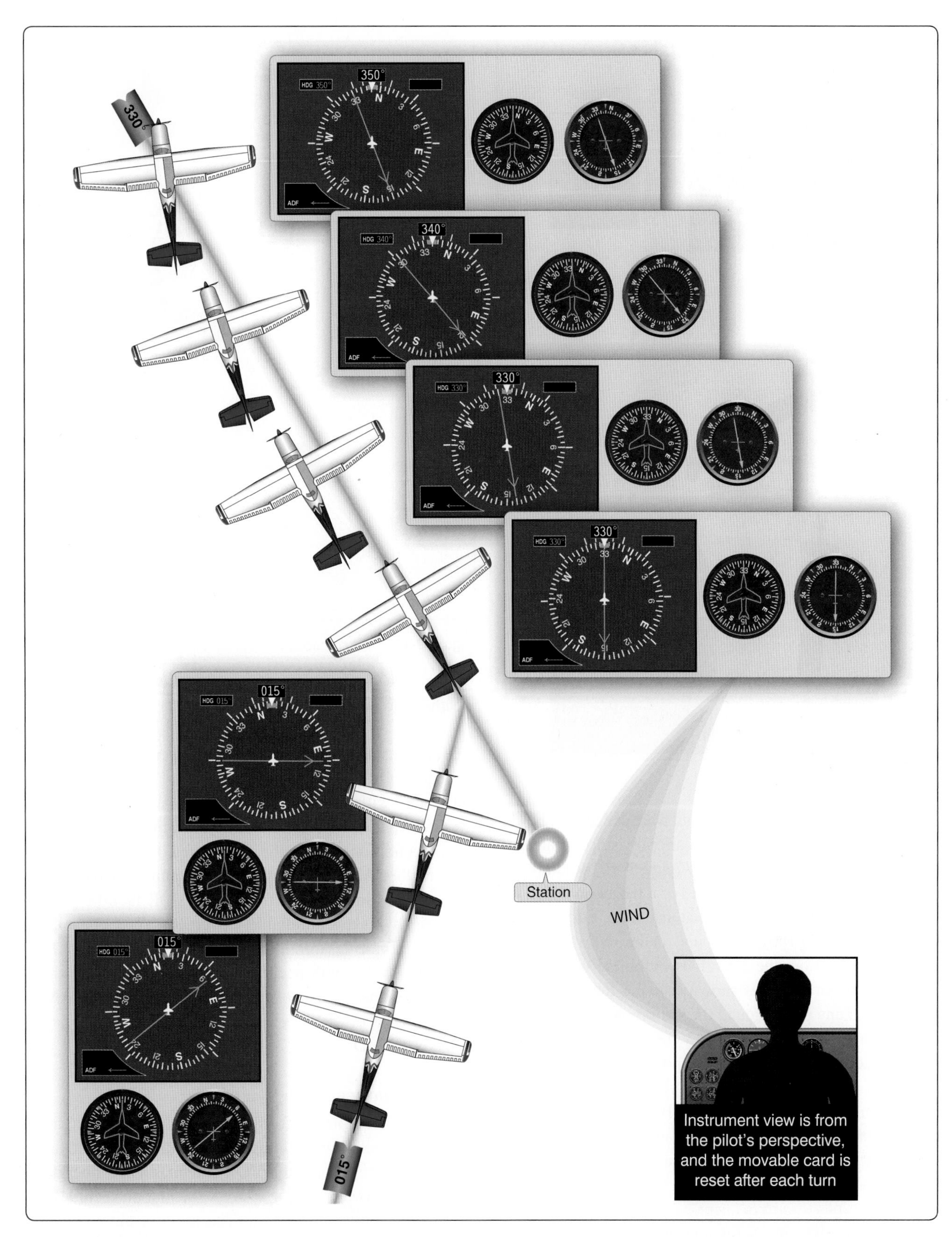

Figure 9-8. *ADF interception and tracking outbound.*

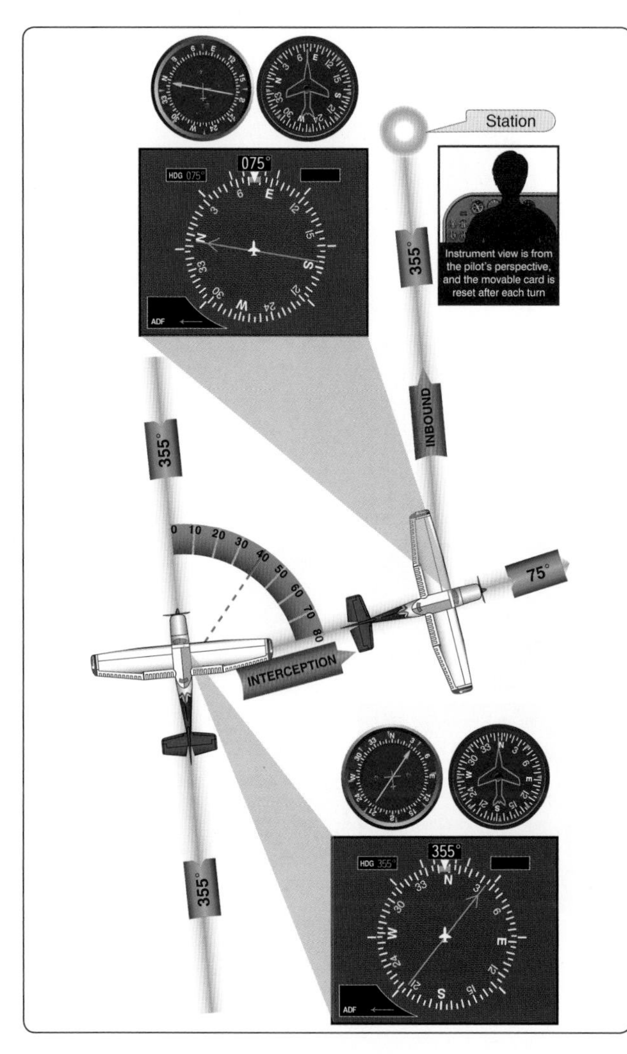

Figure 9-9. *Interception of bearing.*

The courses oriented FROM the station are called radials. The VOR information received by an aircraft is not influenced by aircraft attitude or heading. *[Figure 9-10]* Radials can be envisioned to be like the spokes of a wheel on which the aircraft is on one specific radial at any time. For example, aircraft A (heading 180°) is inbound on the 360° radial; after crossing the station, the aircraft is outbound on the 180° radial at A1. Aircraft B is shown crossing the 225° radial. Similarly, at any point around the station, an aircraft can be located somewhere on a specific VOR radial. Additionally, a VOR needle on an RMI always points to the course that takes you to the VOR station where conversely the ADF needle points to the station as a RB from the aircraft. In the example above, the ADF needle at position A would be pointed straight ahead, at A1 to the aircraft's 180° position (tail) and at B to the aircraft's right.

The VOR receiver measures and presents information to indicate bearing TO or FROM the station. In addition to the navigation signals transmitted by the VOR, a Morse code signal is transmitted concurrently to identify the facility, as

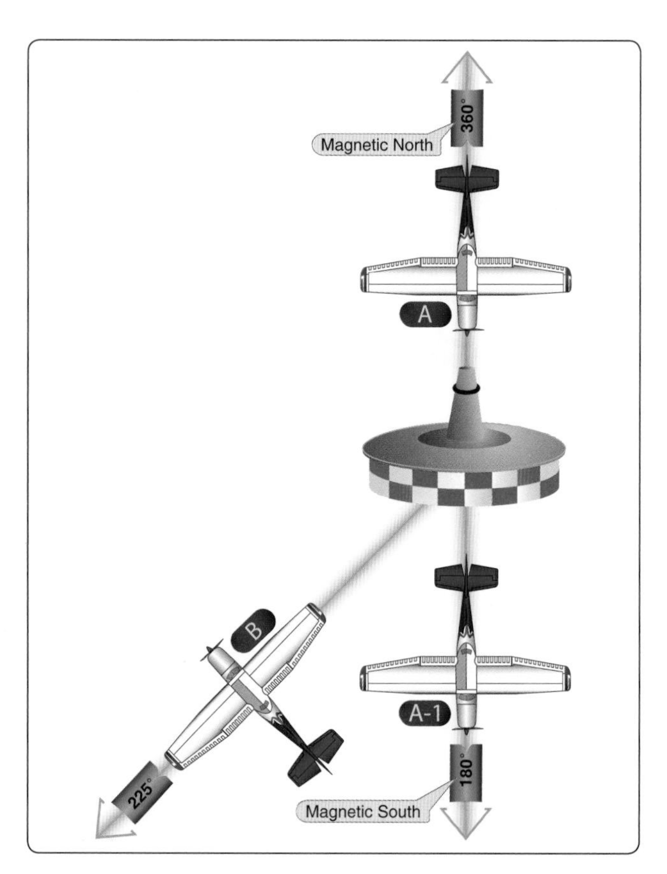

Figure 9-10. *VOR radials.*

well as voice transmissions for communication and relay of weather and other information.

VORs are classified according to their operational uses. The standard VOR facility has a power output of approximately 200 watts, with a maximum usable range depending upon the aircraft altitude, class of facility, location of the facility, terrain conditions within the usable area of the facility, and other factors. Above and beyond certain altitude and distance limits, signal interference from other VOR facilities and a weak signal make it unreliable. Coverage is typically at least 40 miles at normal minimum instrument flight rules (IFR) altitudes. VORs with accuracy problems in parts of their service volume are listed in Notices to Airmen (NOTAMs) and in the Airport/Facility Directory (A/FD) under the name of the NAVAID.

VOR Components

The ground equipment consists of a VOR ground station, which is a small, low building topped with a flat white disc, upon which are located the VOR antennas and a fiberglass cone-shaped tower. *[Figure 9-11]* The station includes an automatic monitoring system. The monitor automatically turns off defective equipment and turns on the standby transmitter. Generally, the accuracy of the signal from the ground station is within 1°.

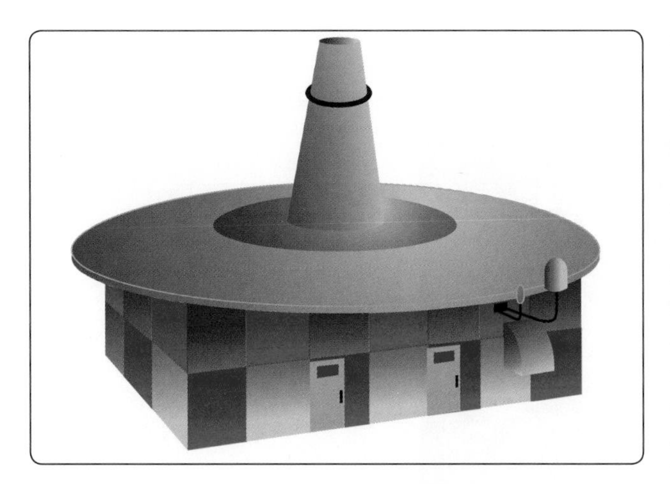

Figure 9-11. *VOR transmitter (ground station).*

VOR facilities are aurally identified by Morse code, or voice, or both. The VOR can be used for ground-to-air communication without interference with the navigation signal. VOR facilities operate within the 108.0 to 117.95 MHz frequency band and assignment between 108.0 and 112.0 MHz is in even-tenth increments to preclude any conflict with ILS localizer frequency assignment, which uses the odd tenths in this range.

The airborne equipment includes an antenna, a receiver, and the indicator instrument. The receiver has a frequency knob to select any of the frequencies between 108.0 to 117.95 MHz. The ON/OFF/volume control turns on the navigation receiver and controls the audio volume. The volume has no effect on the operation of the receiver. You should listen to the station identifier before relying on the instrument for navigation.

VOR indicator instruments have at least the essential components shown in the instrument illustrated in *Figure 9-12.*

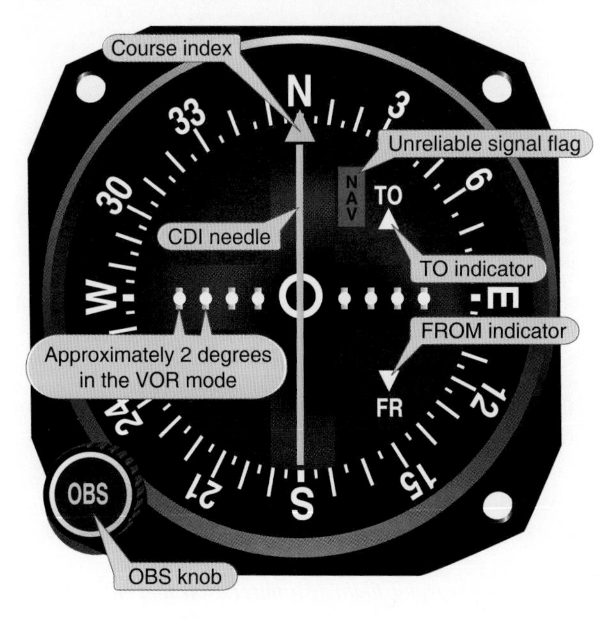

Figure 9-12. *The VOR indicator instrument.*

Omnibearing Selector (OBS)

The desired course is selected by turning the omnibearing selector (OBS) knob until the course is aligned with the course index mark or displayed in the course window.

Course Deviation Indicator (CDI)

The course deviation indicator (CDI) is composed of an instrument face and a needle hinged to move laterally across the instrument face. The needle centers when the aircraft is on the selected radial or its reciprocal. Full needle deflection from the center position to either side of the dial indicates the aircraft is 12° or more off course, assuming normal needle sensitivity. The outer edge of the center circle is 2° off course; with each dot representing an additional 2°.

TO/FROM Indicator

The TO/FROM indicator shows whether the selected course, if intercepted and flown, takes the aircraft TO or FROM the station. It does not indicate whether the aircraft is heading to or from the station.

Flags or Other Signal Strength Indicators

The device that indicates a usable or an unreliable signal may be an "OFF" flag. It retracts from view when signal strength is sufficient for reliable instrument indications. Alternately, insufficient signal strength may be indicated by a blank or OFF in the TO/FROM window.

The indicator instrument may also be a horizontal situation indicator (HSI), which combines the heading indicator and CDI. *[Figure 9-13]* The combination of navigation information from VOR/Localizer (LOC) with aircraft heading information provides a visual picture of the aircraft's location and direction. This decreases pilot workload especially with tasks such as course intercepts, flying a back-course approach, or holding pattern entry. (See Chapter 5, Flight Instruments, for operational characteristics.) *[Figure 9-14]*

Function of VOR

Orientation

The VOR does not account for the aircraft heading. It only relays the aircraft direction from the station and has the same indications regardless of which way the nose is pointing. Tune the VOR receiver to the appropriate frequency of the selected VOR ground station, turn up the audio volume, and identify the station's signal audibly. Then, rotate the OBS to center the CDI needle and read the course under or over the index.

In *Figure 9-12*, 360° TO is the course indicated, while in *Figure 9-15*, 180° TO is the course. The latter indicates that the aircraft (which may be heading in any direction) is, at this moment, located at any point on the 360° radial (line from the station) except directly over the station or very

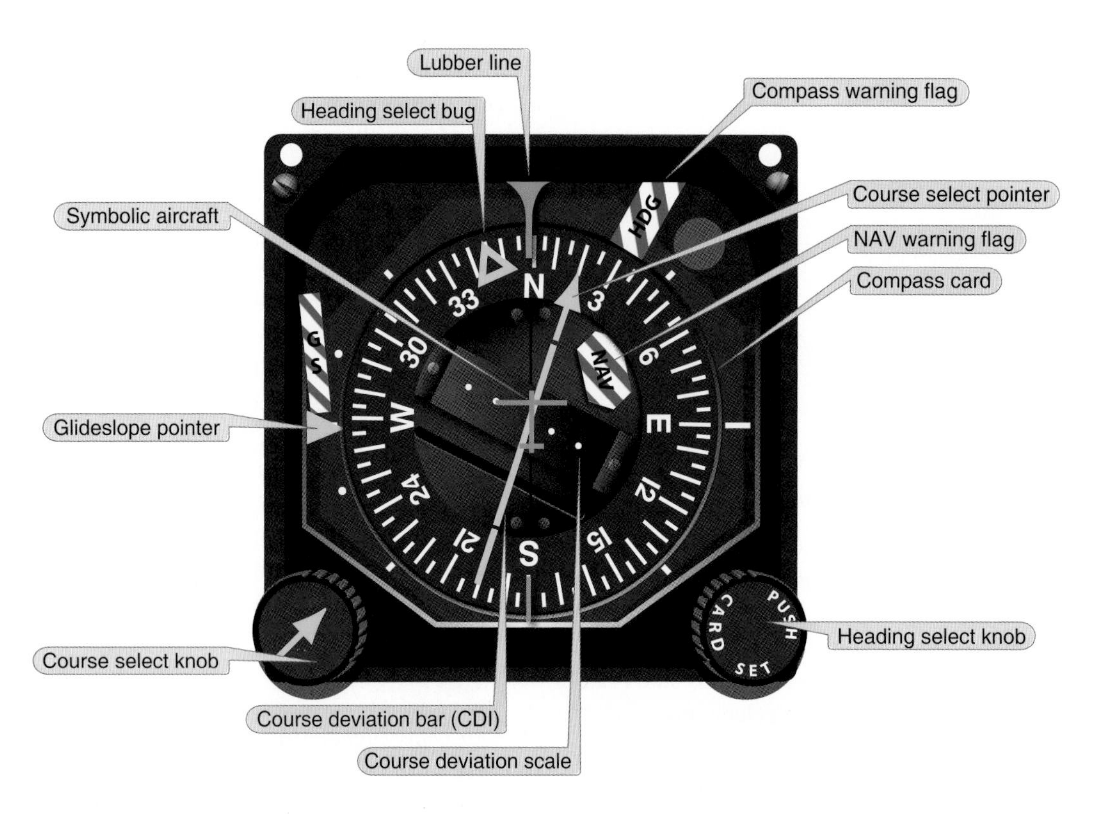

Figure 9-13. *A typical horizontal situation indicator (HSI).*

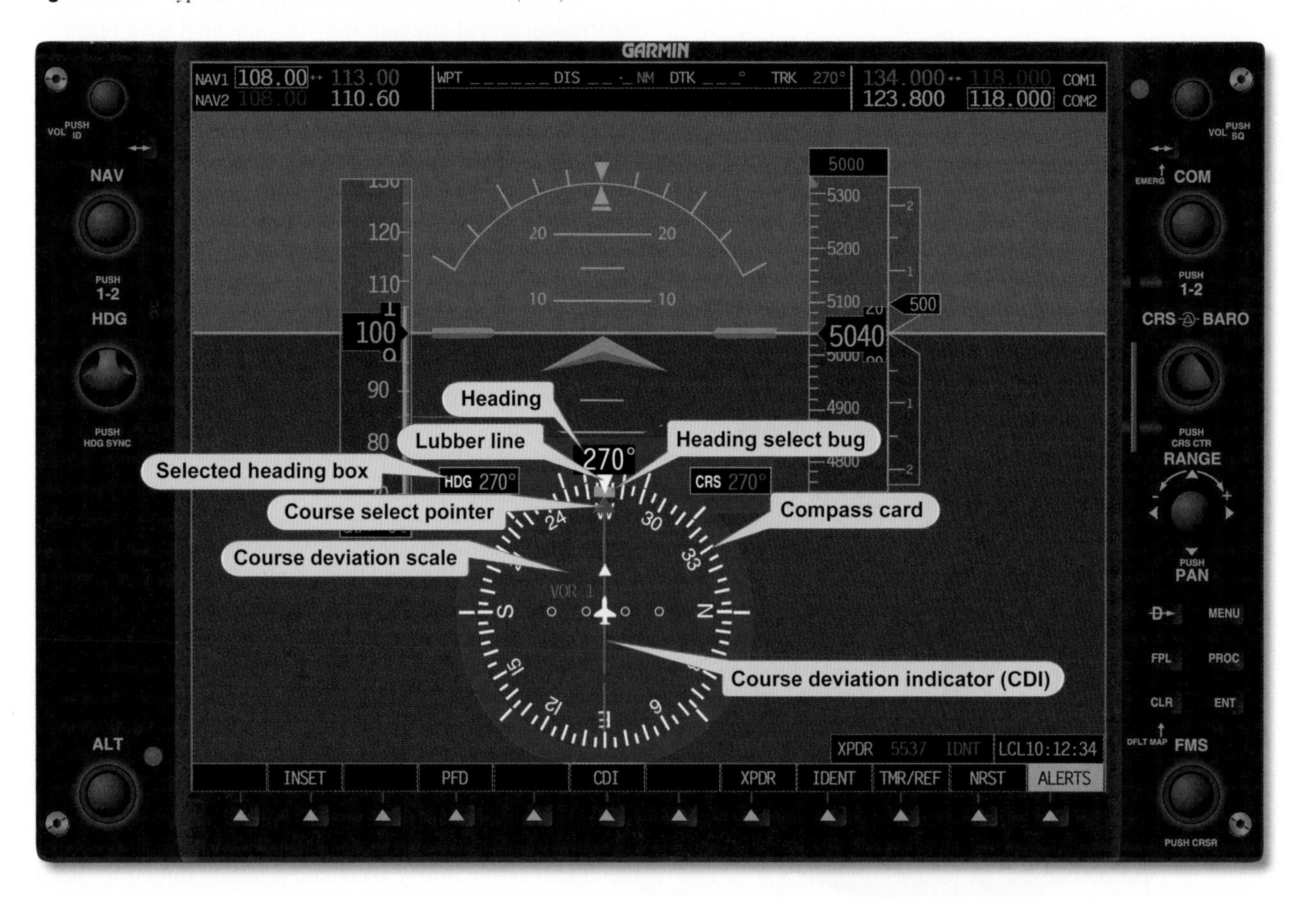

Figure 9-14. *An HSI display as seen on the pilot's primary flight display (PFD) on an electronic flight instrument. Note that only attributes related to the HSI are labeled.*

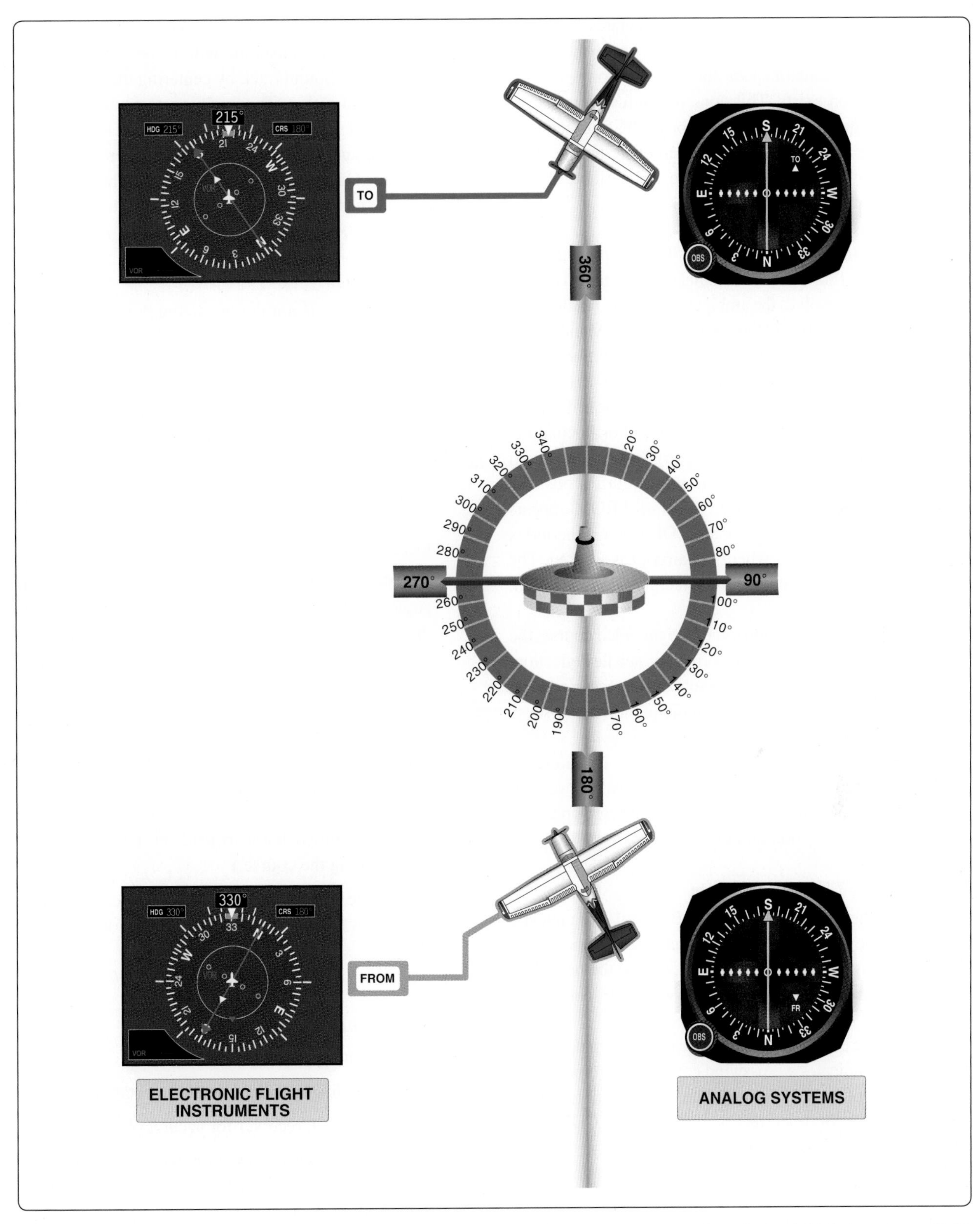

Figure 9-15. *CDI interpretation. The CDI as typically found on analog systems (right) and as found on electronic flight instruments (left).*

close to it, as in *Figure 9-15*. The CDI deviates from side to side as the aircraft passes over or nearly over the station because of the volume of space above the station where the zone of confusion exists. This zone of confusion is caused by lack of adequate signal directly above the station due to the radiation pattern of the station's antenna, and because the resultant of the opposing reference and variable signals is small and constantly changing.

The CDI in *Figure 9-15* indicates 180°, meaning that the aircraft is on the 180° or the 360° radial of the station. The TO/FROM indicator resolves the ambiguity. If the TO indicator is showing, then it is 180° TO the station. The FROM indication indicates the radial of the station the aircraft is presently on. Movement of the CDI from center, if it occurs at a relatively constant rate, indicates the aircraft is moving or drifting off the 180°/360° line. If the movement is rapid or fluctuating, this is an indication of impending station passage (the aircraft is near the station). To determine the aircraft's position relative to the station, rotate the OBS until FROM appears in the window, and then center the CDI needle. The index indicates the VOR radial where the aircraft is located. The inbound (to the station) course is the reciprocal of the radial.

If the VOR is set to the reciprocal of the intended course, the CDI reflects reverse sensing. To correct for needle deflection, turn away from the needle. To avoid this reverse sensing situation, set the VOR to agree with the intended course.

A single NAVAID allows a pilot to determine the aircraft's position relative to a radial. Indications from a second NAVAID are needed in order to narrow the aircraft's position down to an exact location on this radial.

Tracking TO and FROM the Station

To track to the station, rotate the OBS until TO appears, then center the CDI. Fly the course indicated by the index. If the CDI moves off center to the left, follow the needle by correcting course to the left, beginning with a 20° correction.

When flying the course indicated on the index, a left deflection of the needle indicates a crosswind component from the left. If the amount of correction brings the needle back to center, decrease the left course correction by half. If the CDI moves left or right now, it should do so much more slowly, and smaller heading corrections can be made for the next iteration.

Keeping the CDI centered takes the aircraft to the station. To track to the station, the OBS value at the index is not changed. To home to the station, the CDI needle is periodically centered, and the new course under the index is used for the aircraft heading. Homing follows a circuitous route to the station, just as with ADF homing.

To track FROM the station on a VOR radial, you should first orient the aircraft's location with respect to the station and the desired outbound track by centering the CDI needle with a FROM indication. The track is intercepted by either flying over the station or establishing an intercept heading. The magnetic course of the desired radial is entered under the index using the OBS and the intercept heading held until the CDI centers. Then the procedure for tracking to the station is used to fly outbound on the specified radial.

Course Interception

If the desired course is not the one being flown, first orient the aircraft's position with respect to the VOR station and the course to be flown, and then establish an intercept heading. The following steps may be used to intercept a predetermined course, either inbound or outbound. Steps 1–3 may be omitted when turning directly to intercept the course without initially turning to parallel the desired course.

1. Determine the difference between the radial to be intercepted and the radial on which the aircraft is located (205° – 160° = 045°).
2. Double the difference to determine the interception angle, which will not be less than 20° nor greater than 90° (45° × 2 = 090°). 205° + 090° = 295° for the intercept).
3. Rotate the OBS to the desired radial or inbound course.
4. Turn to the interception heading.
5. Hold this heading constant until the CDI center, which indicates the aircraft is on course. (With practice in judging the varying rates of closure with the course centerline, pilots learn to lead the turn to prevent overshooting the course.)
6. Turn to the MH corresponding to the selected course, and follow tracking procedures inbound or outbound.

Course interception is illustrated in *Figure 9-16*.

VOR Operational Errors

Typical pilot-induced errors include:

1. Careless tuning and identification of station.
2. Failure to check receiver for accuracy/sensitivity.
3. Turning in the wrong direction during an orientation. This error is common until visualizing position rather than heading.
4. Failure to check the ambiguity (TO/FROM) indicator, particularly during course reversals, resulting in reverse sensing and corrections in the wrong direction.

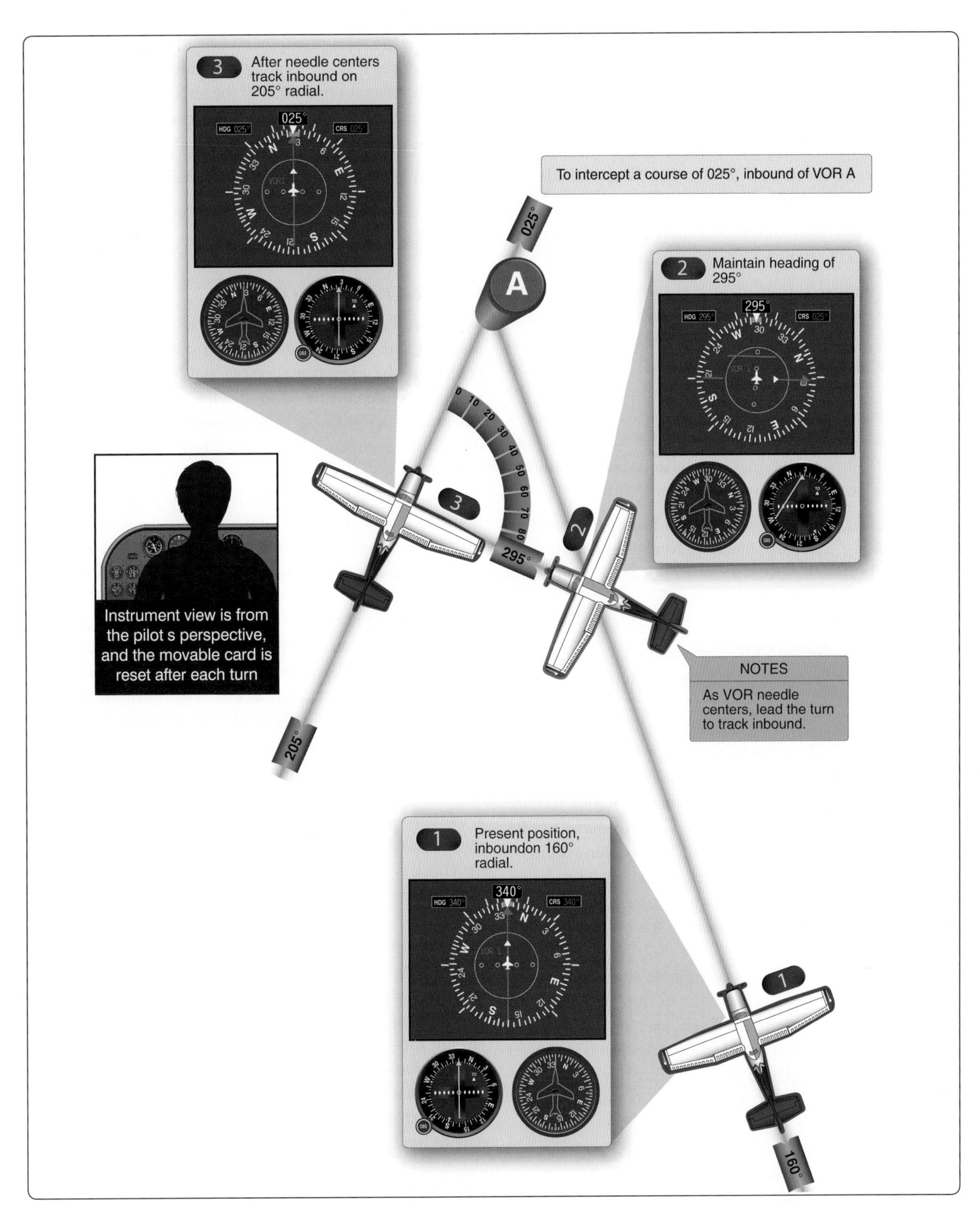

Figure 9-16. *Course interception (VOR).*

5. Failure to parallel the desired radial on a track interception problem. Without this step, orientation to the desired radial can be confusing. Since pilots think in terms of left and right of course, aligning the aircraft position to the radial/course is essential.
6. Overshooting and undershooting radials on interception problems.
7. Overcontrolling corrections during tracking, especially close to the station.
8. Misinterpretation of station passage. On VOR receivers not equipped with an ON/OFF flag, a voice transmission on the combined communication and navigation radio (NAV/COM) in use for VOR may cause the same TO/FROM fluctuations on the ambiguity meter as shown during station passage. Read the whole receiver—TO/FROM, CDI, and OBS—before you make a decision. Do not utilize a VOR reading observed while transmitting.
9. Chasing the CDI, resulting in homing instead of tracking. Careless heading control and failure to bracket wind corrections make this error common.

VOR Accuracy

The effectiveness of the VOR depends upon proper use and adjustment of both ground and airborne equipment.

The accuracy of course alignment of the VOR is generally plus or minus 1°. On some VORs, minor course roughness may be observed, evidenced by course needle or brief flag alarm. At a few stations, usually in mountainous terrain, the pilot may occasionally observe a brief course needle oscillation similar to the indication of "approaching station." Pilots flying over unfamiliar routes are cautioned to be on the alert for these vagaries, and in particular, to use the TO/FROM indicator to determine positive station passage.

Certain propeller revolutions per minute (rpm) settings or helicopter rotor speeds can cause the VOR CDI to fluctuate as much as plus or minus 6°. Slight changes to the RPM setting normally smooths out this roughness. Pilots are urged to check for this modulation phenomenon prior to reporting a VOR station or aircraft equipment for unsatisfactory operation.

VOR Receiver Accuracy Check

VOR system course sensitivity may be checked by noting the number of degrees of change as the OBS is rotated to move the CDI from center to the last dot on either side. The course selected should not exceed 10° or 12° either side. In addition, Title 14 of the Code of Federal Regulations (14 CFR) part 91 provides for certain VOR equipment accuracy checks, and an appropriate endorsement, within 30 days prior to flight under IFR. To comply with this requirement and to ensure satisfactory operation of the airborne system, use the following means for checking VOR receiver accuracy:

1. VOR test facility (VOT) or a radiated test signal from an appropriately rated radio repair station.
2. Certified checkpoints on the airport surface.
3. Certified airborne checkpoints.

VOR Test Facility (VOT)

The Federal Aviation Administration (FAA) VOT transmits a test signal that provides users a convenient means to determine the operational status and accuracy of a VOR receiver while on the ground where a VOT is located. Locations of VOTs are published in the A/FD. Two means of identification are used: one is a series of dots and the other is a continuous tone. Information concerning an individual test signal can be obtained from the local flight service station (FSS.) The airborne use of VOT is permitted; however, its use is strictly limited to those areas/altitudes specifically authorized in the A/FD or appropriate supplement.

To use the VOT service, tune in the VOT frequency 108.0 MHz on the VOR receiver. With the CDI centered, the OBS should read 0° with the TO/FROM indication showing FROM or the OBS should read 180° with the TO/FROM indication showing TO. Should the VOR receiver operate an RMI, it would indicate 180° on any OBS setting.

A radiated VOT from an appropriately rated radio repair station serves the same purpose as an FAA VOT signal, and the check is made in much the same manner as a VOT with some differences.

The frequency normally approved by the Federal Communications Commission (FCC) is 108.0 MHz; however, repair stations are not permitted to radiate the VOR test signal continuously. The owner or operator of the aircraft must make arrangements with the repair station to have the test signal transmitted. A representative of the repair station must make an entry into the aircraft logbook or other permanent record certifying to the radial accuracy and the date of transmission.

Certified Checkpoints

Airborne and ground checkpoints consist of certified radials that should be received at specific points on the airport surface or over specific landmarks while airborne in the immediate vicinity of the airport. Locations of these checkpoints are published in the A/FD.

Should an error in excess of ±4° be indicated through use of a ground check, or ±6° using the airborne check, IFR flight shall not be attempted without first correcting the source of the error. No correction other than the correction card figures supplied by the manufacturer should be applied in making these VOR receiver checks.

If a dual system VOR (units independent of each other except for the antenna) is installed in the aircraft, one system may be checked against the other. Turn both systems to the same VOR ground facility and note the indicated bearing to that station. The maximum permissible variation between the two indicated bearings is 4°.

Distance Measuring Equipment (DME)

When used in conjunction with the VOR system, DME makes it possible for pilots to determine an accurate geographic position of the aircraft, including the bearing and distance TO or FROM the station. The aircraft DME transmits interrogating radio frequency (RF) pulses, which are received by the DME antenna at the ground facility. The signal triggers ground receiver equipment to respond to the interrogating aircraft. The airborne DME equipment measures the elapsed time between the interrogation signal sent by the aircraft and reception of the reply pulses from the ground station. This time measurement is converted into distance in nautical miles (NM) from the station.

Some DME receivers provide a groundspeed in knots by monitoring the rate of change of the aircraft's position relative to the ground station. Groundspeed values are accurate only when tracking directly to or from the station.

DME Components

VOR/DME, VORTAC, ILS/DME, and LOC/DME navigation facilities established by the FAA provide course and distance information from collocated components under a frequency pairing plan. DME operates on frequencies in the UHF spectrum between 962 MHz and 1213 MHz. Aircraft receiving equipment that provides for automatic DME selection assures reception of azimuth and distance information from a common source when designated VOR/DME, VORTAC, ILS/DME, and LOC/DME are selected. Some aircraft have separate VOR and DME receivers, each of which must be tuned to the appropriate navigation facility. The airborne equipment includes an antenna and a receiver.

The pilot-controllable features of the DME receiver include:

Channel (Frequency) Selector
Many DMEs are channeled by an associated VHF radio, or there may be a selector switch so a pilot can select which VHF radio is channeling the DME. For a DME with its own frequency selector, use the frequency of the associated VOR/DME or VORTAC station.

ON/OFF/Volume Switch
The DME identifier is heard as a Morse code identifier with a tone somewhat higher than that of the associated VOR or LOC. It is heard once for every three or four times the VOR or LOC identifier is heard. If only one identifier is heard about every 30 seconds, the DME is functional, but the associated VOR or LOC is not.

Mode Switch
The mode switch selects between distance (DIST) or distance in NMs, groundspeed, and time to station. There may also be one or more HOLD functions that permit the DME to stay channeled to the station that was selected before the switch was placed in the hold position. This is useful when you make an ILS approach at a facility that has no collocated DME, but there is a VOR/DME nearby.

Altitude
Some DMEs correct for slant-range error.

Function of DME

A DME is used for determining the distance from a ground DME transmitter. Compared to other VHF/UHF NAVAIDs, a DME is very accurate. The distance information can be used to determine the aircraft position or flying a track that is a constant distance from the station. This is referred to as a DME arc.

DME Arc

There are many instrument approach procedures (IAPs) that incorporate DME arcs. The procedures and techniques given here for intercepting and maintaining such arcs are applicable to any facility that provides DME information. Such a facility may or may not be collocated with the facility that provides final approach guidance.

As an example of flying a DME arc, refer to *Figure 9-17* and follow these steps:

1. Track inbound on the OKT 325° radial, frequently checking the DME mileage readout.
2. A 0.5 NM lead is satisfactory for groundspeeds of 150 knots or less; start the turn to the arc at 10.5 miles. At higher groundspeeds, use a proportionately greater lead.
3. Continue the turn for approximately 90°. The roll-out heading is 055° in a no wind condition.
4. During the last part of the intercepting turn, monitor the DME closely. If the arc is being overshot (more than 1.0 NM), continue through the originally planned roll-out heading. If the arc is being undershot, roll-out of the turn early.

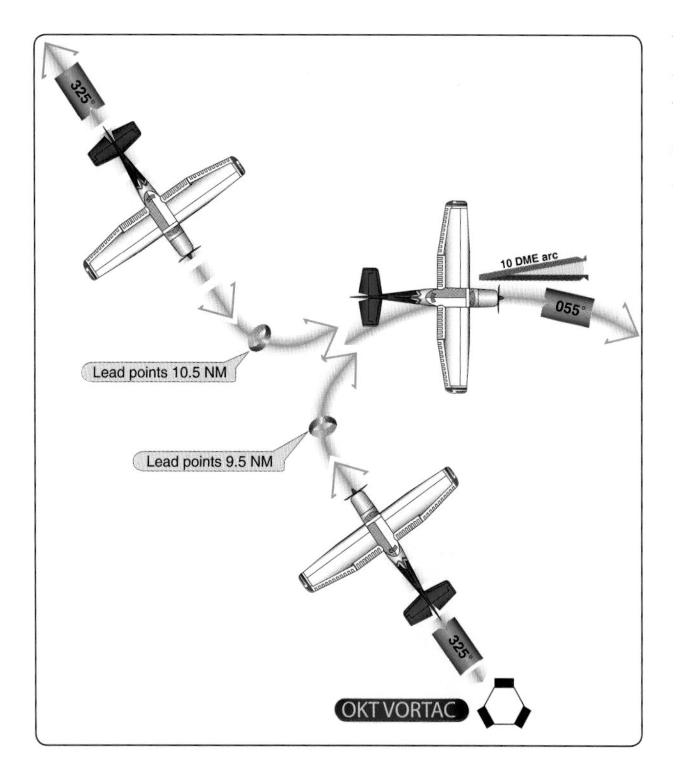

Figure 9-17. *DME arc interception.*

The procedure for intercepting the 10 DME when outbound is basically the same, the lead point being 10 NM minus 0.5 NM or 9.5 NM.

When flying a DME arc with wind, it is important to keep a continuous mental picture of the aircraft's position relative to the facility. Since the wind-drift correction angle is constantly changing throughout the arc, wind orientation is important. In some cases, wind can be used in returning to the desired track. High airspeeds require more pilot attention because of the higher rate of deviation and correction.

Maintaining the arc is simplified by keeping slightly inside the curve; thus, the arc is turning toward the aircraft and interception may be accomplished by holding a straight course. When outside the curve, the arc is "turning away" and a greater correction is required.

To fly the arc using the VOR CDI, center the CDI needle upon completion of the 90° turn to intercept the arc. The aircraft's heading is found very near the left or right side (270° or 90° reference points) of the instrument. The readings at that side location on the instrument give primary heading information while on the arc. Adjust the aircraft heading to compensate for wind and to correct for distance to maintain the correct arc distance. Recenter the CDI and note the new primary heading indicated whenever the CDI gets 2°– 4° from center.

With an RMI, in a no wind condition, pilots should theoretically be able to fly an exact circle around the facility by maintaining an RB of 90° or 270°. In actual practice, a series of short legs are flown. To maintain the arc in *Figure 9-18*, proceed as follows:

1. With the RMI bearing pointer on the wingtip reference (90° or 270° position) and the aircraft at the desired DME range, maintain a constant heading and allow the bearing pointer to move 5°– 10° behind the wingtip. This causes the range to increase slightly.
2. Turn toward the facility to place the bearing pointer 5°– 10° ahead of the wingtip reference, and then maintain heading until the bearing pointer is again behind the wingtip. Continue this procedure to maintain the approximate arc.
3. If a crosswind causes the aircraft to drift away from the facility, turn the aircraft until the bearing pointer is ahead of the wingtip reference. If a crosswind causes the aircraft to drift toward the facility, turn until the bearing is behind the wingtip.
4. As a guide in making range corrections, change the RB 10°– 20° for each half-mile deviation from the desired arc. For example, in no-wind conditions, if the aircraft is ½ to 1 mile outside the arc and the bearing pointer is on the wingtip reference, turn the aircraft 20° toward the facility to return to the arc.

Without an RMI, orientation is more difficult since there is no direct azimuth reference. However, the procedure can be flown using the OBS and CDI for azimuth information and the DME for arc distance.

Intercepting Lead Radials

A lead radial is the radial at which the turn from the arc to the inbound course is started. When intercepting a radial from a DME arc, the lead varies with arc radius and groundspeed. For the average general aviation aircraft, flying arcs such as those depicted on most approach charts at speeds of 150 knots or less, the lead is under 5°. There is no difference between intercepting a radial from an arc and intercepting it from a straight course.

With an RMI, the rate of bearing movement should be monitored closely while flying the arc. Set the course of the radial to be intercepted as soon as possible and determine the approximate lead. Upon reaching this point, start the intercepting turn. Without an RMI, the technique for radial interception is the same except for azimuth information, which is available only from the OBS and CDI.

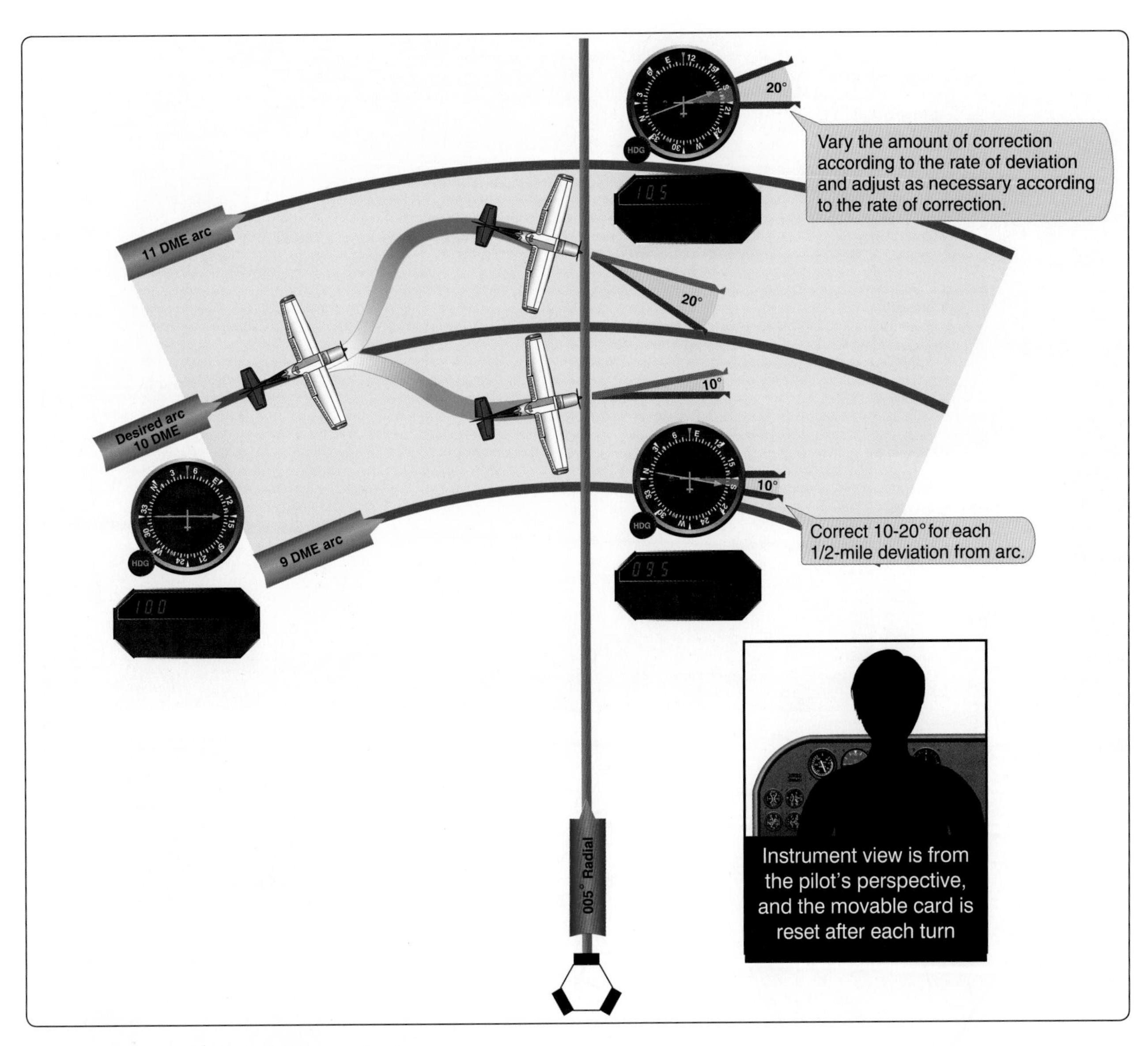

Figure 9-18. *Using DME and RMI to maintain an arc.*

The technique for intercepting a localizer from a DME arc is similar to intercepting a radial. At the depicted lead radial (LR 223 or LR 212 in *Figures 9-19, 9-20,* and *9-21*), a pilot having a single VOR/LOC receiver should set it to the localizer frequency. If the pilot has dual VOR/LOC receivers, one unit may be used to provide azimuth information and the other set to the localizer frequency. Since these lead radials provide 7° of lead, a half-standard rate turn should be used until the LOC needle starts to move toward center.

DME Errors

A DME/DME fix (a location based on two DME lines of position from two DME stations) provides a more accurate aircraft location than using a VOR and a DME fix.

DME signals are line-of-sight; the mileage readout is the straight line distance from the aircraft to the DME ground facility and is commonly referred to as slant range distance. Slant range refers to the distance from the aircraft's antenna to the ground station (A line at an angle to the ground transmitter. GPS systems provide distance as the horizontal measurement from the WP to the aircraft. Therefore, at 3,000 feet and 0.5 miles the DME (slant range) would read 0.6 NM while the GPS distance would show the actual horizontal distance of .5 DME. This error is smallest at low altitudes and/or at long ranges. It is greatest when the aircraft is closer to the facility, at which time the DME receiver displays altitude (in NM) above the facility. Slant range error is negligible if the aircraft is one mile or more from the ground facility for each 1,000 feet of altitude above the elevation of the facility.

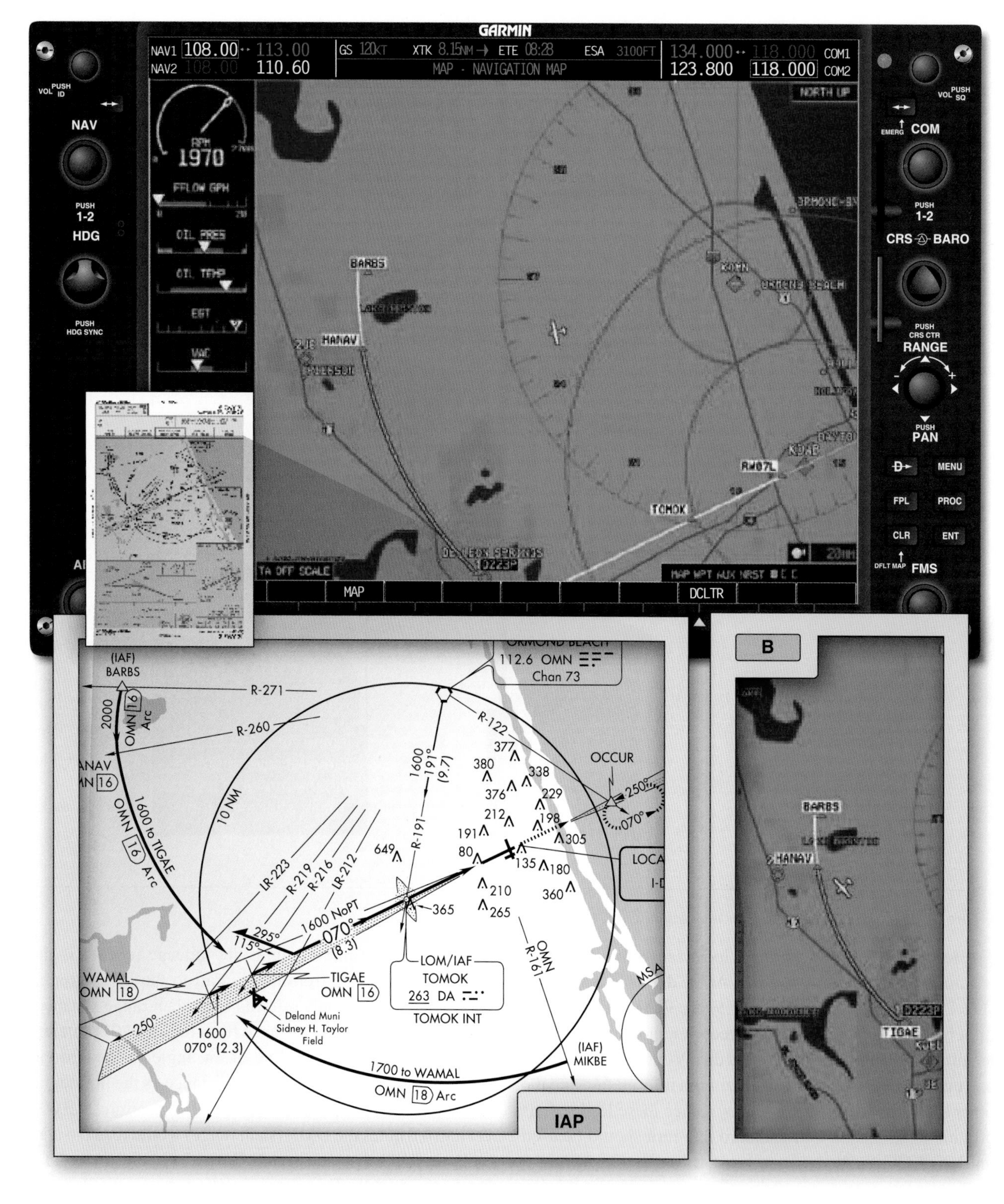

Figure 9-19. *An aircraft is displayed heading southwest to intercept the localizer approach, using the 16 NM DME arc off of ORM.*

Figure 9-20. *The same aircraft illustrated in Figure 9-19 shown on the ORM radial near TIGAE intersection turning inbound for the localizer.*

Figure 9-21. *Aircraft is illustrated inbound on the localizer course.*

Area Navigation (RNAV)

Area navigation (RNAV) equipment includes VOR/DME, LORAN, GPS, and inertial navigation systems (INS). RNAV equipment is capable of computing the aircraft position, actual track, groundspeed, and then presenting meaningful information to the pilot. This information may be in the form of distance, cross-track error, and time estimates relative to the selected track or WP. In addition, the RNAV equipment installations must be approved for use under IFR. The Pilot's Operating Handbook/Airplane Flight Manual (POH/AFM) should always be consulted to determine what equipment is installed, the operations that are approved, and the details of equipment use. Some aircraft may have equipment that allows input from more than one RNAV source, thereby providing a very accurate and reliable navigation source.

VOR/DME RNAV

VOR RNAV is based on information generated by the present VORTAC or VOR/DME system to create a WP using an airborne computer. As shown in *Figure 9-22*, the value of side A is the measured DME distance to the VOR/DME. Side B, the distance from the VOR/DME to the WP, and angle 1 (VOR radial or the bearing from the VORTAC to the WP) are values set in the flight deck control. The bearing from the VOR/DME to the aircraft, angle 2, is measured by the VOR receiver. The airborne computer continuously compares

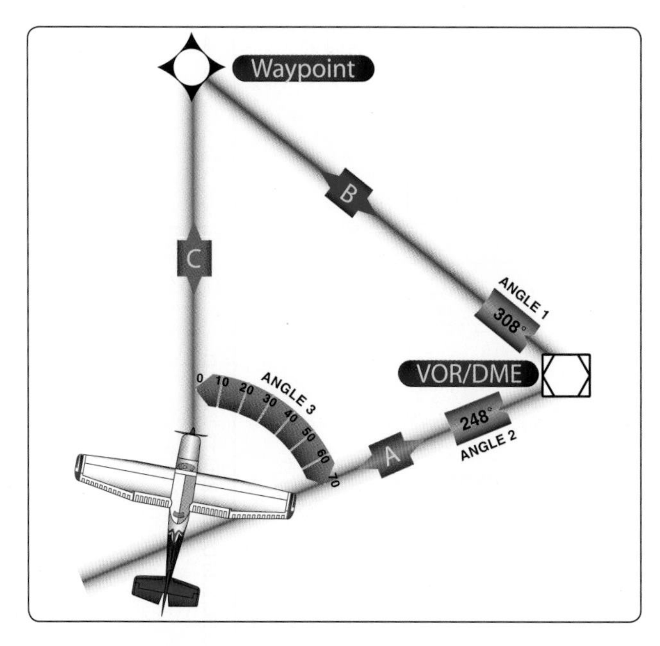

Figure 9-22. *RNAV computation.*

angles 1 and 2 and determines angle 3 and side C, which is the distance in NMs and magnetic course from the aircraft to the WP. This is presented as guidance information on the flight deck display.

VOR/DME RNAV Components

Although RNAV flight deck instrument displays vary among manufacturers, most are connected to the aircraft CDI with a switch or knob to select VOR or RNAV guidance. There is usually a light or indicator to inform the pilot whether VOR or RNAV is selected. *[Figure 9-23]* The display includes the WP, frequency, mode in use, WP radial and distance, DME distance, groundspeed, and time to station.

Figure 9-23. *Onboard RNAV receivers have changed significantly. Originally, RNAV receivers typically computed combined data from VOR, VORTAC, and/or DME. That is generally not the case now. Today, GPS such as the GNC 300 and the Bendix King KLS 88 LORAN receivers compute waypoints based upon embedded databases and aircraft positional information.*

Most VOR/DME RNAV systems have the following airborne controls:

1. OFF/ON/Volume control to select the frequency of the VOR/DME station to be used.
2. MODE select switch used to select VOR/DME mode, with:
   a. Angular course width deviation (standard VOR operation); or
   b. Linear cross-track deviation as standard (±5 NM full scale CDI).
3. RNAV mode, with direct to WP with linear cross-track deviation of ±5 NM.
4. RNAV/APPR (approach mode) with linear deviation of ±1.25 NM as full scale CDI deflection.
5. WP select control. Some units allow the storage of more than one WP; this control allows selection of any WP in storage.
6. Data input controls. These controls allow user input of WP number or ident, VOR or LOC frequency, WP radial and distance.

While DME groundspeed readout is accurate only when tracking directly to or from the station in VOR/DME mode, in RNAV mode the DME groundspeed readout is accurate on any track.

Function of VOR/DME RNAV

The advantages of the VOR/DME RNAV system stem from the ability of the airborne computer to locate a WP wherever it is convenient, as long as the aircraft is within reception range of both nearby VOR and DME facilities. A series of these WPs make up an RNAV route. In addition to the published routes, a random RNAV route may be flown under IFR if it is approved by air traffic control (ATC). RNAV DPs and standard terminal arrival routes (STARs) are contained in the DP and STAR booklets.

VOR/DME RNAV approach procedure charts are also available. Note in the VOR/DME RNAV chart excerpt shown in *Figure 9-24* that the WP identification boxes contain the following information: WP name, coordinates, frequency, identifier, radial distance (facility to WP), and reference facility elevation. The initial approach fix (IAF), final approach fix (FAF), and missed approach point (MAP) are labeled.

To fly a route or to execute an approach under IFR, the RNAV equipment installed in the aircraft must be approved for the appropriate IFR operations.

In vertical navigation (VNAV) mode, vertical guidance is provided, as well as horizontal guidance in some installations. A WP is selected at a point where the descent begins, and another WP is selected where the descent ends. The

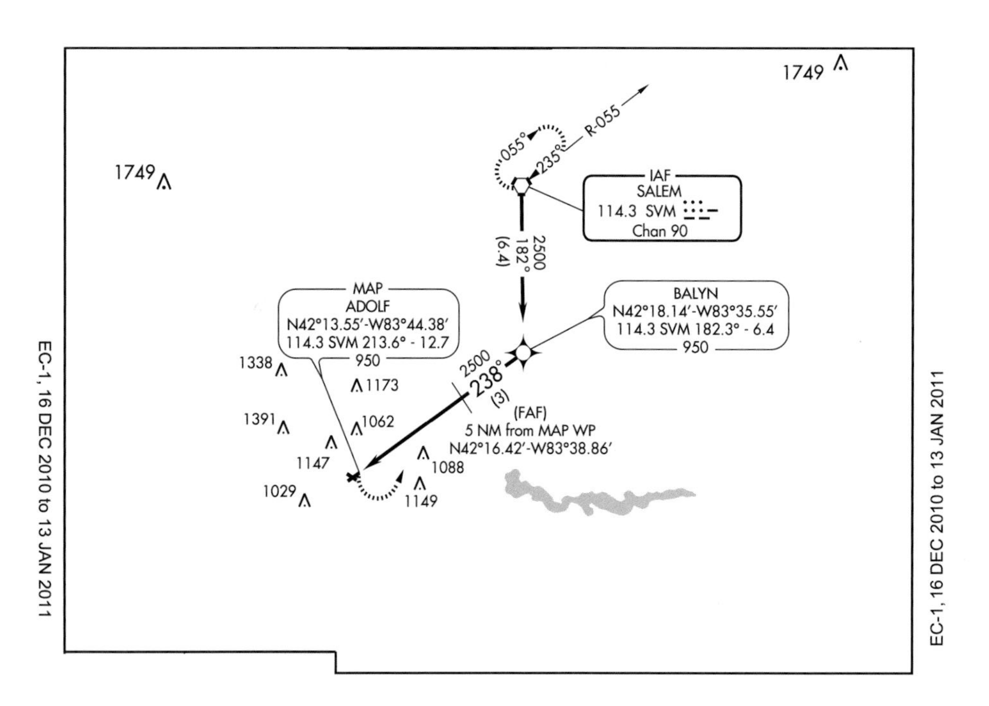

Figure 9-24. *VOR/DME RNAV RWY 25 approach (excerpt).*

RNAV equipment computes the rate of descent relative to the groundspeed; on some installations, it displays vertical guidance information on the GS indicator. When using this type of equipment during an instrument approach, the pilot must keep in mind that the vertical guidance information provided is not part of the nonprecision approach. Published nonprecision approach altitudes must be observed and complied with, unless otherwise directed by ATC.

To fly to a WP using RNAV, observe the following procedure *[Figure 9-25]*:

1. Select the VOR/DME frequency.
2. Select the RNAV mode.
3. Select the radial of the VOR that passes through the WP (225°).
4. Select the distance from the DME to the WP (12 NM).
5. Check and confirm all inputs, and center the CDI needle with the TO indicator showing.
6. Maneuver the aircraft to fly the indicated heading plus or minus wind correction to keep the CDI needle centered.
7. The CDI needle indicates distance off course of 1 NM per dot; the DME readout indicates distance in NM from the WP; the groundspeed reads closing speed (knots) to the WP; and the time to station (TTS) reads time to the WP.

VOR/DME RNAV Errors

The limitation of this system is the reception volume. Published approaches have been tested to ensure this is not a problem. Descents/approaches to airports distant from the VOR/DME facility may not be possible because, during the approach, the aircraft may descend below the reception altitude of the facility at that distance.

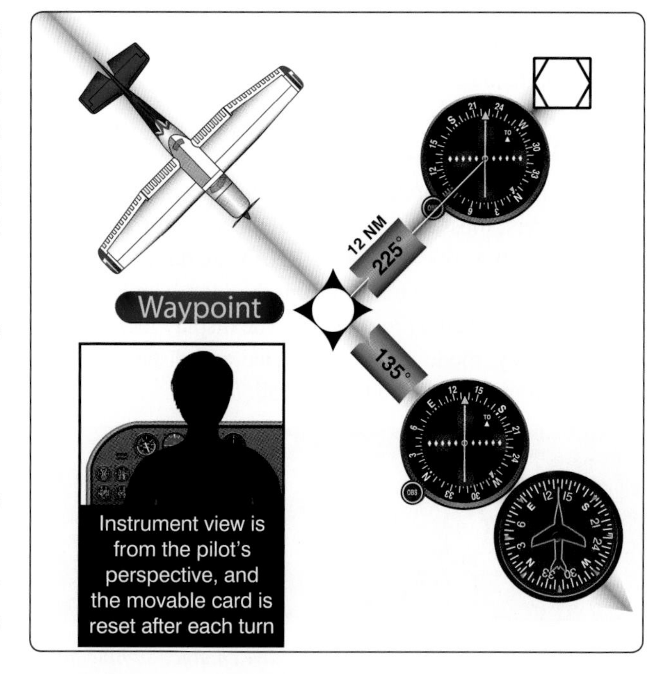

Figure 9-25. *Aircraft/DME/waypoint relationship.*

Advanced Technologies

Global Navigation Satellite System (GNSS)

The Global Navigation Satellite System (GNSS) is a constellation of satellites providing a high-frequency signal that contains time and distance that is picked up by a receiver thereby. *[Figure 9-26]* The receiver that picks up multiple signals from different satellites is able to triangulate its position from these satellites.

Figure 9-26. *A typical example (GNS 480) of a stand-alone GPS receiver and display.*

Three GNSSs exist today: the GPS, a United States system; the Russian GNSS (GLONASS); and Galileo, a European system.

1. GLONASS is a network of 24 satellites that can be picked up by any GLONASS receiver, allowing the user to pinpoint their position.
2. Galileo planned to be a network of 30 satellites that continuously transmit high-frequency radio signals containing time and distance data that can be picked up by a Galileo receiver with operational expectancy by 2013.
3. The GPS came on line in 1992 with 24 satellites and today utilizes 30 satellites.

Global Positioning System (GPS)

The GPS is a satellite-based radio navigation system that broadcasts a signal that is used by receivers to determine precise position anywhere in the world. The receiver tracks multiple satellites and determines a measurement that is then used to determine the user location. *[Figure 9-27]*

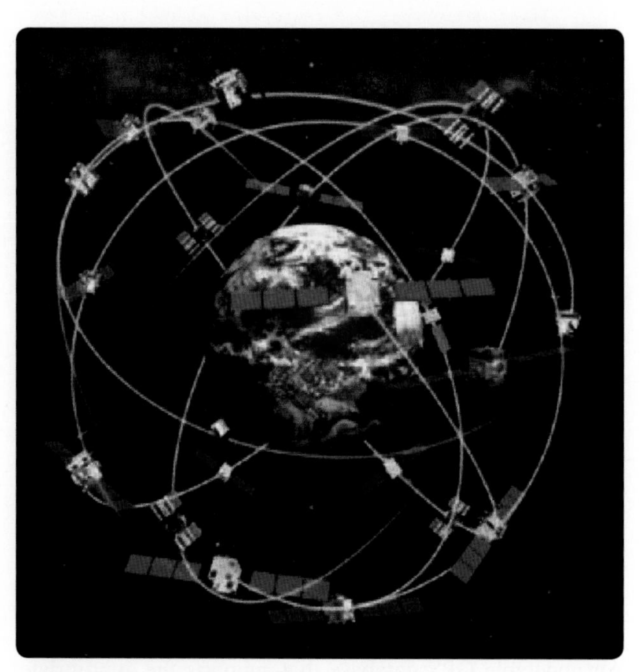

Figure 9-27. *Typical GPS satellite array.*

The Department of Defense (DOD) developed and deployed GPS as a space-based positioning, velocity, and time system. The DOD is responsible for operation of the GPS satellite constellation, and constantly monitors the satellites to ensure proper operation. The GPS system permits Earth-centered coordinates to be determined and provides aircraft position referenced to the DOD World Geodetic System of 1984 (WGS-84). Satellite navigation systems are unaffected by weather and provide global navigation coverage that fully meets the civil requirements for use as the primary means of navigation in oceanic airspace and certain remote areas. Properly certified GPS equipment may be used as a supplemental means of IFR navigation for domestic en route, terminal operations and certain IAPs. Navigational values, such as distance and bearing to a WP and groundspeed, are computed from the aircraft's current position (latitude and longitude) and the location of the next WP. Course guidance is provided as a linear deviation from the desired track of a Great Circle route between defined WPs.

GPS may not be approved for IFR use in other countries. Prior to its use, pilots should ensure that GPS is authorized by the appropriate countries.

GPS Components

GPS consists of three distinct functional elements: space, control, and user.

The space element consists of over 30 Navstar satellites. This group of satellites is called a constellation. The space element consists of 24 Navigation System using Timing and Ranging (NAVSTAR) satellites in 6 orbital planes. The satellites in each plane are spaced 60° apart for complete coverage and are located (nominally) at about 11,000 miles above the Earth. The planes are arranged so that there are always five satellites in view at any time on the Earth. Presently, there are at least 31 Block II/IIA/IIR and IIR-M satellites in orbit with the additional satellites representing replacement satellites (upgraded systems) and spares. Recently, the Air Force received funding for procurement of 31 Block IIF satellites. The GPS constellation broadcasts a pseudo-random code timing signal and data message that the aircraft equipment processes to obtain satellite position and status data. By knowing the precise location of each satellite and precisely matching timing with the atomic clocks on the satellites, the aircraft receiver/processor can accurately measure the time each signal takes to arrive at the receiver and, therefore, determine aircraft position.

The control element consists of a network of ground-based GPS monitoring and control stations that ensure the accuracy of satellite positions and their clocks. In its present form, it has five monitoring stations, three ground antennas, and a master control station.

The user element consists of antennas and receiver/processors on board the aircraft that provide positioning, velocity, and precise timing to the user. GPS equipment used while operating under IFR must meet the standards set forth in Technical Standard Order (TSO) C-129 (or equivalent); meet the airworthiness installation requirements; be "approved" for that type of IFR operation; and be operated in accordance with the applicable POH/AFM or flight manual supplement.

An updatable GPS database that supports the appropriate operations (e.g., en route, terminal, and instrument approaches) is required when operating under IFR. The aircraft GPS navigation database contains WPs from the geographic areas where GPS navigation has been approved for IFR operations. The pilot selects the desired WPs from the database and may add user-defined WPs for the flight.

Equipment approved in accordance with TSO C-115a, visual flight rules (VFR), and hand-held GPS systems do not meet the requirements of TSO C-129 and are not authorized for IFR navigation, instrument approaches, or as a principal instrument flight reference. During IFR operations, these units (TSO C-115a) may be considered only an aid to situational awareness.

Prior to GPS/WAAS IFR operation, the pilot must review appropriate NOTAMs and aeronautical information. This information is available on request from an flight service station (FSS). The FAA does provide NOTAMs to advise pilots of the status of the WAAS and level of service available.

Function of GPS

GPS operation is based on the concept of ranging and triangulation from a group of satellites in space that act as precise reference points. The receiver uses data from a minimum of four satellites above the mask angle (the lowest angle above the horizon at which it can use a satellite).

The aircraft GPS receiver measures distance from a satellite using the travel time of a radio signal. Each satellite transmits a specific code, called a course/acquisition (CA) code, which contains information about satellite position, the GPS system time, and the health and accuracy of the transmitted data. Knowing the speed at which the signal traveled (approximately 186,000 miles per second) and the exact broadcast time, the distance traveled by the signal can be computed from the arrival time. The distance derived from this method of computing distance is called a pseudo-range because it is not a direct measurement of distance, but a measurement based on time. In addition to knowing the distance to a satellite, a receiver needs to know the satellite's exact position in space, its ephemeris. Each satellite transmits information about its exact orbital location. The GPS receiver uses this information to establish the precise position of the satellite.

Using the calculated pseudo-range and position information supplied by the satellite, the GPS receiver/processor mathematically determines its position by triangulation from several satellites. The GPS receiver needs at least four satellites to yield a three-dimensional position (latitude, longitude, and altitude) and time solution. The GPS receiver computes navigational values (distance and bearing to a WP, groundspeed, etc.) by using the aircraft's known latitude/longitude and referencing these to a database built into the receiver.

The GPS receiver verifies the integrity (usability) of the signals received from the GPS constellation through receiver autonomous integrity monitoring (RAIM) to determine if a satellite is providing corrupted information. RAIM needs a minimum of five satellites in view or four satellites and a barometric altimeter baro-aiding to detect an integrity anomaly. For receivers capable of doing so, RAIM needs six satellites in view (or five satellites with baro-aiding) to isolate a corrupt satellite signal and remove it from the navigation solution.

Generally, there are two types of RAIM messages. One type indicates that there are not enough satellites available to provide RAIM and another type indicates that the RAIM has detected a potential error that exceeds the limit for the current phase of flight. Without RAIM capability, the pilot has no assurance of the accuracy of the GPS position.

Aircraft using GPS navigation equipment under IFR for domestic en route, terminal operations, and certain IAPs, must be equipped with an approved and operational alternate means of navigation appropriate to the flight. The avionics necessary to receive all of the ground-based facilities appropriate for the route to the destination airport and any required alternate airport must be installed and operational. Ground-based facilities necessary for these routes must also be operational. Active monitoring of alternative navigation equipment is not required if the GPS receiver uses RAIM for integrity monitoring. Active monitoring of an alternate means of navigation is required when the RAIM capability of the GPS equipment is lost. In situations where the loss of RAIM capability is predicted to occur, the flight must rely on other approved equipment, delay departure, or cancel the flight.

GPS Substitution

IFR En Route and Terminal Operations

GPS systems, certified for IFR en route and terminal operations, may be used as a substitute for ADF and DME receivers when conducting the following operations within the United States NAS.

1. Determining the aircraft position over a DME fix. This includes en route operations at and above 24,000 feet mean sea level (MSL) (FL 240) when using GPS for navigation.
2. Flying a DME arc.
3. Navigating TO/FROM an NDB/compass locator.
4. Determining the aircraft position over an NDB/compass locator.
5. Determining the aircraft position over a fix defined by an NDB/compass locator bearing crossing a VOR/LOC course.
6. Holding over an NDB/compass locator.

GPS Substitution for ADF or DME

Using GPS as a substitute for ADF or DME is subject to the following restrictions:

1. This equipment must be installed in accordance with appropriate airworthiness installation requirements and operated within the provisions of the applicable POH/AFM or supplement.
2. The required integrity for these operations must be provided by at least en route RAIM or equivalent.
3. WPs, fixes, intersections, and facility locations to be used for these operations must be retrieved from the GPS airborne database. The database must be current. If the required positions cannot be retrieved from the airborne database, the substitution of GPS for ADF and/or DME is not authorized
4. Procedures must be established for use when RAIM outages are predicted or occur. This may require the flight to rely on other approved equipment or require the aircraft to be equipped with operational NDB and/or DME receivers. Otherwise, the flight must be rerouted, delayed, canceled, or conducted under VFR.
5. The CDI must be set to terminal sensitivity (1 NM) when tracking GPS course guidance in the terminal area.
6. A non-GPS approach procedure must exist at the alternate airport when one is required. If the non-GPS approaches on which the pilot must rely require DME or ADF, the aircraft must be equipped with DME or ADF avionics as appropriate.
7. Charted requirements for ADF and/or DME can be met using the GPS system, except for use as the principal instrument approach navigation source.

NOTE: The following provides guidance that is not specific to any particular aircraft GPS system. For specific system guidance, refer to the POH/AFM, or supplement, or contact the system manufacturer.

To Determine Aircraft Position Over a DME Fix:

1. Verify aircraft GPS system integrity monitoring is functioning properly and indicates satisfactory integrity.
2. If the fix is identified by a five-letter name that is contained in the GPS airborne database, select either the named fix as the active GPS WP or the facility establishing the DME fix as the active GPS WP. When using a facility as the active WP, the only acceptable facility is the DME facility that is charted as the one used to establish the DME fix. If this facility is not in the airborne database, it is not authorized for use.
3. If the fix is identified by a five-letter name that is not contained in the GPS airborne database, or if the fix is not named, select the facility establishing the DME fix or another named DME fix as the active GPS WP.
4. When selecting the named fix as the active GPS WP, a pilot is over the fix when the GPS system indicates the active WP.
5. If selecting the DME providing facility as the active GPS WP, a pilot is over the fix when the GPS distance from the active WP equals the charted DME value, and the aircraft is established on the appropriate bearing or course.

To Fly a DME Arc:

1. Verify aircraft GPS system integrity monitoring is functioning properly and indicates satisfactory integrity.
2. Select from the airborne database the facility providing the DME arc as the active GPS WP. The only acceptable facility is the DME facility on which the arc is based. If this facility is not in your airborne database, you are not authorized to perform this operation.
3. Maintain position on the arc by reference to the GPS distance instead of a DME readout.

To Navigate TO or FROM an NDB/Compass Locator:

1. Verify aircraft GPS system integrity monitoring is functioning properly and indicates satisfactory integrity.
2. Select the NDB/compass locator facility from the airborne database as the active WP. If the chart depicts the compass locator collocated with a fix of the same name, use of that fix as the active WP in place of the compass locator facility is authorized.
3. Select and navigate on the appropriate course to or from the active WP.

To Determine Aircraft Position Over an NDB/Compass Locator:

1. Verify aircraft GPS system integrity monitoring is functioning properly and indicates satisfactory integrity.
2. Select the NDB/compass locator facility from the airborne database. When using an NDB/compass locator, the facility must be charted and be in the airborne database. If the facility is not in the airborne database, pilots are not authorized to use a facility WP for this operation.
3. A pilot is over the NDB/compass locator when the GPS system indicates arrival at the active WP.

To Determine Aircraft Position Over a Fix Made Up of an NDB/Compass Locator Bearing Crossing a VOR/LOC Course:

1. Verify aircraft GPS system integrity monitoring is functioning properly and indicates satisfactory integrity.
2. A fix made up by a crossing NDB/compass locator bearing is identified by a five-letter fix name. Pilots may select either the named fix or the NDB/compass locator facility providing the crossing bearing to establish the fix as the active GPS WP. When using an NDB/compass locator, that facility must be charted and be in the airborne database. If the facility is not in the airborne database, pilots are not authorized to use a facility WP for this operation.
3. When selecting the named fix as the active GPS WP, pilot is over the fix when the GPS system indicates the pilot is at the WP.
4. When selecting the NDB/compass locator facility as the active GPS WP, pilots are over the fix when the GPS bearing to the active WP is the same as the charted NDB/compass locator bearing for the fix flying the prescribed track from the non-GPS navigation source.

To Hold Over an NDB/Compass Locator:

1. Verify aircraft GPS system integrity monitoring is functioning properly and indicates satisfactory integrity.
2. Select the NDB/compass locator facility from the airborne database as the active WP. When using a facility as the active WP, the only acceptable facility is the NDB/compass locator facility which is charted. If this facility is not in the airborne database, its use is not authorized.
3. Select nonsequencing (e.g., "HOLD" or "OBS") mode and the appropriate course in accordance with the POH/AFM or supplement.
4. Hold using the GPS system in accordance with the POH/AFM or supplement.

IFR Flight Using GPS

Preflight preparations should ensure that the GPS is properly installed and certified with a current database for the type of operation. The GPS operation must be conducted in accordance with the FAA-approved POH/AFM or flight manual supplement. Flightcrew members must be thoroughly familiar with the particular GPS equipment installed in the aircraft, the receiver operation manual, and the POH/AFM or flight manual supplement. Unlike ILS and VOR, the basic operation, receiver presentation to the pilot and some capabilities of the equipment can vary greatly. Due to these differences, operation of different brands or even models of the same brand of GPS receiver under IFR should not be attempted without thorough study of the operation of that particular receiver and installation. Using the equipment in flight under VFR conditions prior to attempting IFR operation allows for further familiarization.

Required preflight preparations should include checking NOTAMs relating to the IFR flight when using GPS as a supplemental method of navigation. GPS satellite outages are issued as GPS NOTAMs both domestically and internationally. Pilots may obtain GPS RAIM availability information for an airport by specifically requesting GPS aeronautical information from an FSS during preflight briefings. GPS RAIM aeronautical information can be obtained for a 3-hour period: the estimated time of arrival (ETA), and 1 hour before to 1 hour after the ETA hour, or a 24-hour time frame for a specific airport. FAA briefers provide RAIM information for a period of 1 hour before to 1 hour after the ETA, unless a specific timeframe is requested by the pilot. If flying a published GPS departure, the pilot should also request a RAIM prediction for the departure airport. Some GPS receivers have the capability to predict RAIM availability. The pilot should also ensure that the required underlying ground-based navigation facilities and related aircraft equipment appropriate to the route of flight, terminal operations, instrument approaches for the destination, and alternate airports/heliports are operational for the ETA. If the required ground-based facilities and equipment are not available, the flight should be rerouted, rescheduled, canceled, or conducted under VFR.

Except for programming and retrieving information from the GPS receiver, planning the flight is accomplished in a similar manner to conventional NAVAIDs. Departure WP, DP, route, STAR, desired approach, IAF, and destination airport are entered into the GPS receiver according to the manufacturer's instructions. During preflight, additional information may be entered for functions such as ETA, fuel planning, winds aloft, etc.

When the GPS receiver is turned on, it begins an internal process of test and initialization. When the receiver is initialized, the user develops the route by selecting a WP or series of WPs, verifies the data, and selects the active flight plan. This procedure varies widely among receivers made by different manufacturers. GPS is a complex system, offering little standardization between receiver models. It is the pilot's responsibility to be familiar with the operation of the equipment in the aircraft.

The GPS receiver provides navigational values such as track, bearing, groundspeed, and distance. These are computed from the aircraft's present latitude and longitude to the location of the next WP. Course guidance is provided between WPs. The pilot has the advantage of knowing the aircraft's actual track over the ground. As long as track and bearing to the WP are matched up (by selecting the correct aircraft heading), the aircraft is going directly to the WP.

GPS Instrument Approaches

There is a mixture of GPS overlay approaches (approaches with "or GPS" in the title) and GPS stand-alone approaches in the United States.

NOTE: GPS instrument approach operations outside the United States must be authorized by the appropriate country authority.

While conducting these IAPs, ground-based NAVAIDs are not required to be operational and associated aircraft avionics need not be installed, operational, turned on, or monitored; however, monitoring backup navigation systems is always recommended when available.

Pilots should have a basic understanding of GPS approach procedures and practice GPS IAPs under visual meteorological conditions (VMC) until thoroughly proficient with all aspects of their equipment (receiver and installation) prior to attempting flight in instrument meteorological conditions (IMC). *[Figure 9-28]*

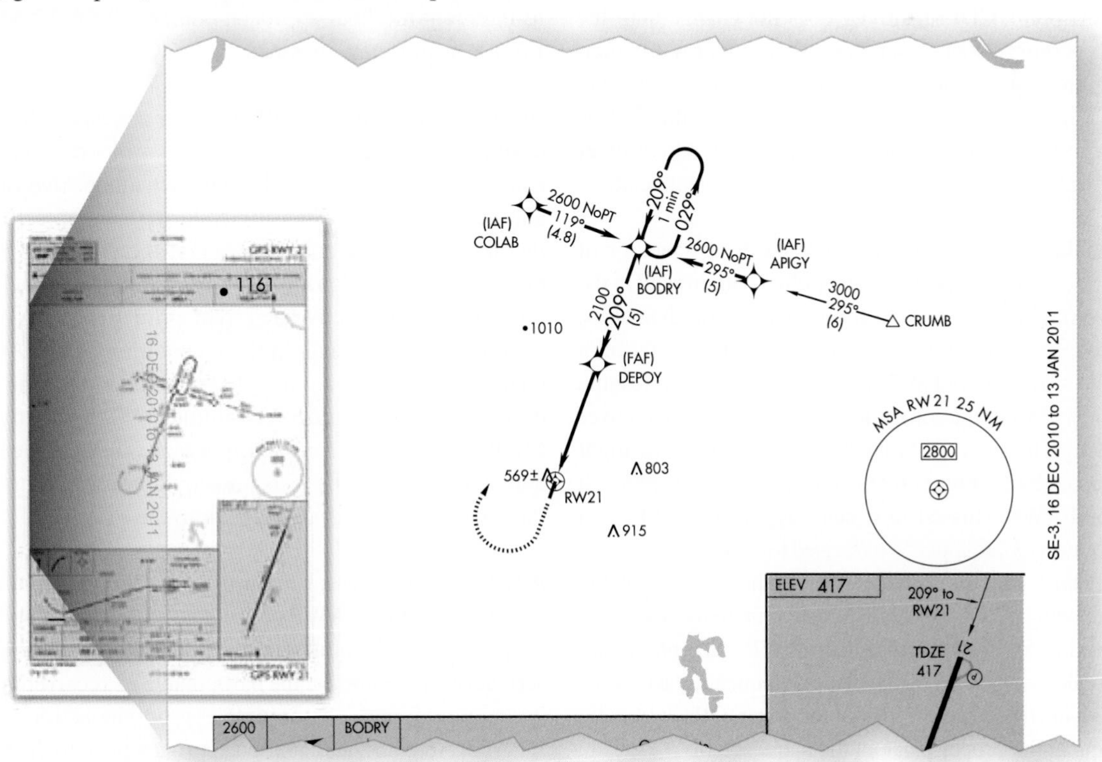

Figure 9-28. *A GPS stand-alone approach.*

All IAPs must be retrievable from the current GPS database supplied by the manufacturer or other FAA-approved source. Flying point to point on the approach does not assure compliance with the published approach procedure. The proper RAIM sensitivity is not available and the CDI sensitivity does not automatically change to 0.3 NM. Manually setting CDI sensitivity does not automatically change the RAIM sensitivity on some receivers. Some existing nonprecision approach procedures cannot be coded for use with GPS and are not available as overlays.

GPS approaches are requested and approved by ATC using the GPS title, such as "GPS RWY 24" or "RNAV RWY 35." Using the manufacturer's recommended procedures, the desired approach and the appropriate IAF are selected from the GPS receiver database. Pilots should fly the full approach from an initial approach waypoint (IAWP) or feeder fix unless specifically cleared otherwise. Randomly joining an approach at an intermediate fix does not ensure terrain clearance.

When an approach has been loaded in the flight plan, GPS receivers give an "arm" annunciation 30 NM straight line distance from the airport/heliport reference point. The approach mode should be "armed" when within 30 NM distance so the receiver changes from en route CDI (±5 NM) and RAIM (±2 NM) sensitivity to ±1 NM terminal sensitivity. Where the IAWP is within 30 NM, a CDI sensitivity change occurs once the approach mode is armed and the aircraft is within 30 NM. Where the IAWP is beyond the 30 NM point, CDI sensitivity does not change until the aircraft is within 30 NM even if the approach is armed earlier. Feeder route obstacle clearance is predicated on the receiver CDI and RAIM being in terminal CDI sensitivity within 30 NM of the airport/heliport reference point; therefore, the receiver should always be armed no later than the 30 NM annunciation.

Pilots should pay particular attention to the exact operation of their GPS receivers for performing holding patterns and in the case of overlay approaches, operations such as procedure turns. These procedures may require manual intervention by the pilot to stop the sequencing of WPs by the receiver and to resume automatic GPS navigation sequencing once the maneuver is complete. The same WP may appear in the route of flight more than once and consecutively (e.g., IAWP, final approach waypoint (FAWP), missed approach waypoint (MAWP) on a procedure turn). Care must be exercised to ensure the receiver is sequenced to the appropriate WP for the segment of the procedure being flown, especially if one or more fly-over WPs are skipped (e.g., FAWP rather than IAWP if the procedure turn is not flown). The pilot may need to sequence past one or more fly-overs of the same WP in order to start GPS automatic sequencing at the proper place in the sequence of WPs.

When receiving vectors to final, most receiver operating manuals suggest placing the receiver in the nonsequencing mode on the FAWP and manually setting the course. This provides an extended final approach course in cases where the aircraft is vectored onto the final approach course outside of any existing segment that is aligned with the runway. Assigned altitudes must be maintained until established on a published segment of the approach. Required altitudes at WPs outside the FAWP or step-down fixes must be considered. Calculating the distance to the FAWP may be required in order to descend at the proper location.

When within 2 NM of the FAWP with the approach mode armed, the approach mode switches to active, which results in RAIM and CDI sensitivity changing to the approach mode. Beginning 2 NM prior to the FAWP, the full scale CDI sensitivity changes smoothly from ±1 NM to ±0.3 NM at the FAWP. As sensitivity changes from ±1 NM to ±0.3 NM approaching the FAWP, and the CDI not centered, the corresponding increase in CDI displacement may give the impression the aircraft is moving further away from the intended course even though it is on an acceptable intercept heading. If digital track displacement information (cross-track error) is available in the approach mode, it may help the pilot remain position oriented in this situation. Being established on the final approach course prior to the beginning of the sensitivity change at 2 NM helps prevent problems in interpreting the CDI display during ramp-down. Requesting or accepting vectors, which causes the aircraft to intercept the final approach course within 2 NM of the FAWP, is not recommended.

Incorrect inputs into the GPS receiver are especially critical during approaches. In some cases, an incorrect entry can cause the receiver to leave the approach mode. Overriding an automatically selected sensitivity during an approach cancels the approach mode annunciation. If the approach mode is not armed by 2 NM prior to the FAWP, the approach mode does not become active at 2 NM prior to the FAWP and the equipment will flag. In these conditions, the RAIM and CDI sensitivity do not ramp down, and the pilot should not descend to minimum descent altitude (MDA) but fly to the MAWP and execute a missed approach. The approach active annunciator and/or the receiver should be checked to ensure the approach mode is active prior to the FAWP.

A GPS missed approach requires pilot action to sequence the receiver past the MAWP to the missed approach portion of the procedure. The pilot must be thoroughly familiar with the activation procedure for the particular GPS receiver installed in the aircraft and must initiate appropriate action after the MAWP. Activating the missed approach prior to the MAWP causes CDI sensitivity to change immediately to terminal (±1 NM) sensitivity, and the receiver continues to navigate to the MAWP. The receiver does not sequence past the MAWP. Turns should not begin prior to the MAWP. If the missed approach is not activated, the GPS receiver displays an extension of the inbound final approach course and the along track distance (ATD) increases from the MAWP until it is manually sequenced after crossing the MAWP.

Missed approach routings in which the first track is via a course rather than direct to the next WP require additional action by the pilot to set the course. Being familiar with all of the required inputs is especially critical during this phase of flight.

Departures and Instrument Departure Procedures (DPs)

The GPS receiver must be set to terminal (±1 NM) CDI sensitivity and the navigation routes contained in the database in order to fly published IFR charted departures and DPs. Terminal RAIM should be provided automatically by the receiver. (Terminal RAIM for departure may not be available unless the WPs are part of the active flight plan rather than proceeding direct to the first destination.) Certain segments of a DP may require some manual intervention by the pilot, especially when radar vectored to a course or required to intercept a specific course to a WP. The database may not contain all of the transitions or departures from all runways and some GPS receivers do not contain DPs in the database. It is necessary that helicopter procedures be flown at 70 knots or less since helicopter departure procedures and missed approaches use a 20:1 obstacle clearance surface (OCS), which is double the fixed-wing OCS. Turning areas are based on this speed also. Missed approach routings in which the first track is via a course rather than direct to the next WP require additional action by the pilot to set the course. Being familiar with all of the required inputs is especially critical during this phase of flight.

GPS Errors

Normally, with 30 satellites in operation, the GPS constellation is expected to be available continuously worldwide. Whenever there are fewer than 24 operational satellites, GPS navigational capability may not be available at certain geographic locations. Loss of signals may also occur in valleys surrounded by high terrain, and any time the aircraft's GPS antenna is "shadowed" by the aircraft's structure (e.g., when the aircraft is banked).

Certain receivers, transceivers, mobile radios, and portable receivers can cause signal interference. Some VHF transmissions may cause "harmonic interference." Pilots can isolate the interference by relocating nearby portable receivers, changing frequencies, or turning off suspected causes of the interference while monitoring the receiver's signal quality data page.

GPS position data can be affected by equipment characteristics and various geometric factors, which typically cause errors of less than 100 feet. Satellite atomic clock inaccuracies, receiver/processors, signals reflected from hard objects (multi-path), ionospheric and tropospheric delays, and satellite data transmission errors may cause small position errors or momentary loss of the GPS signal.

System Status

The status of GPS satellites is broadcast as part of the data message transmitted by the GPS satellites. GPS status information is also available by means of the United States Coast Guard navigation information service: (703) 313-5907 or on the internet at www.navcen.uscg.gov. Additionally, satellite status is available through the NOTAM system.

The GPS receiver verifies the integrity (usability) of the signals received from the GPS constellation through RAIM to determine if a satellite is providing corrupted information. At least one satellite, in addition to those required for navigation, must be in view for the receiver to perform the RAIM function; thus, RAIM needs a minimum of five satellites in view or four satellites and a barometric altimeter (baro-aiding) to detect an integrity anomaly. For receivers capable of doing so, RAIM needs six satellites in view (or five satellites with baro-aiding) to isolate the corrupt satellite signal and remove it from the navigation solution.

RAIM messages vary somewhat between receivers; however, there are two most commonly used types. One type indicates that there are not enough satellites available to provide RAIM integrity monitoring and another type indicates that the RAIM integrity monitor has detected a potential error that exceeds the limit for the current phase of flight. Without RAIM capability, the pilot has no assurance of the accuracy of the GPS position.

Selective Availability. Selective availability is a method by which the accuracy of GPS is intentionally degraded. This feature is designed to deny hostile use of precise GPS positioning data. Selective availability was discontinued on May 1, 2000, but many GPS receivers are designed to assume that selective availability is still active. New receivers may take advantage of the discontinuance of selective availability based on the performance values in ICAO Annex 10 and do not need to be designed to operate outside of that performance.

GPS Familiarization

Pilots should practice GPS approaches under VMC until thoroughly proficient with all aspects of their equipment (receiver and installation) prior to attempting flight by IFR in IMC. Some of the tasks which the pilot should practice are:

1. Utilizing the RAIM prediction function;
2. Inserting a DP into the flight plan, including setting terminal CDI sensitivity, if required, and the conditions under which terminal RAIM is available for departure (some receivers are not DP or STAR capable);
3. Programming the destination airport;
4. Programming and flying the overlay approaches (especially procedure turns and arcs);
5. Changing to another approach after selecting an approach;
6. Programming and flying "direct" missed approaches;
7. Programming and flying "routed" missed approaches;
8. Entering, flying, and exiting holding patterns, particularly on overlay approaches with a second WP in the holding pattern;
9. Programming and flying a "route" from a holding pattern;
10. Programming and flying an approach with radar vectors to the intermediate segment;
11. Indication of the actions required for RAIM failure both before and after the FAWP; and
12. Programming a radial and distance from a VOR (often used in departure instructions).

Differential Global Positioning Systems (DGPS)

Differential global positioning systems (DGPS) are designed to improve the accuracy of GNSS by measuring changes in variables to provide satellite positioning corrections.

Because multiple receivers receiving the same set of satellites produce similar errors, a reference receiver placed at a known location can compute its theoretical position accurately and can compare that value to the measurements provided by the navigation satellite signals. The difference in measurement between the two signals is an error that can be corrected by providing a reference signal correction.

As a result of this differential input accuracy of the satellite system can be increased to meters. The Wide Area Augmentation System (WAAS) and Local Area Augmentation System (LAAS) are examples of differential global positioning systems.

Wide Area Augmentation System (WAAS)

The WAAS is designed to improve the accuracy, integrity, and availability of GPS signals. WAAS allows GPS to be used as the aviation navigation system from takeoff through Category I precision approaches. ICAO has defined Standards for satellite-based augmentation systems (SBAS), and Japan and Europe are building similar systems that are planned to be interoperable with WAAS: EGNOS, the European Geostationary Navigation Overlay System, and MSAS, the Japanese Multifunctional Transport Satellite (MTSAT) Satellite-based Augmentation System. The result will be a worldwide seamless navigation capability similar to GPS but with greater accuracy, availability, and integrity.

Unlike traditional ground-based navigation aids, WAAS will cover a more extensive service area in which surveyed wide-area ground reference stations are linked to the WAAS network. Signals from the GPS satellites are monitored by these stations to determine satellite clock and ephemeris corrections. Each station in the network relays the data to a wide-area master station where the correction information is computed. A correction message is prepared and uplinked to a geostationary satellite (GEO) via a ground uplink and then broadcast on the same frequency as GPS to WAAS receivers within the broadcast coverage area. *[Figure 9-29]*

In addition to providing the correction signal, WAAS provides an additional measurement to the aircraft receiver, improving the availability of GPS by providing, in effect, an additional GPS satellite in view. The integrity of GPS is improved through real-time monitoring, and the accuracy is improved by providing differential corrections to reduce errors. *[Figure 9-30]* As a result, performance improvement is sufficient to enable approach procedures with GPS/WAAS glidepaths. At this time the FAA has completed installation of 25 wide area ground reference systems, two master stations, and four ground uplink stations.

General Requirements

WAAS avionics must be certified in accordance with TSO-C145A, Airborne Navigation Sensors Using the GPS Augmented by the WAAS; or TSO-146A for stand-alone systems. GPS/WAAS operation must be conducted in accordance with the FAA-approved aircraft flight manual (AFM) and flight manual supplements. Flight manual supplements must state the level of approach procedure that the receiver supports.

Instrument Approach Capabilities

WAAS receivers support all basic GPS approach functions and provide additional capabilities with the key benefit to generate an electronic glidepath, independent of ground equipment or barometric aiding. This eliminates several problems, such as cold temperature effects, incorrect altimeter setting, or lack of a local altimeter source, and allows approach procedures to be built without the cost

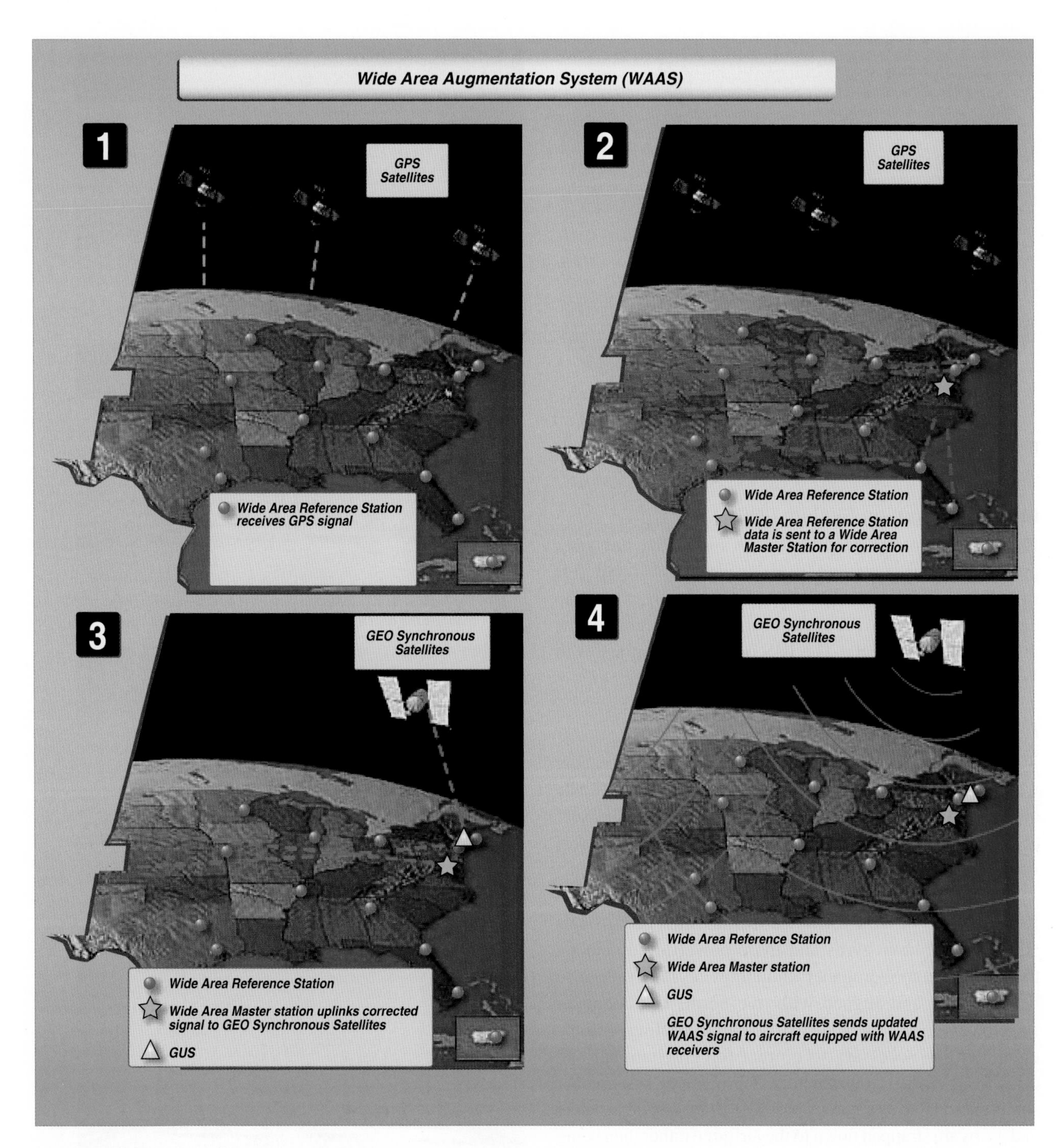

Figure 9-29. *WAAS satellite representation.*

of installing ground stations at each airport. A new class of approach procedures, which provide vertical guidance requirements for precision approaches, has been developed to support satellite navigation use for aviation applications. These new procedures called Approach with Vertical Guidance (APV) include approaches such as the LNAV/VNAV procedures presently being flown with barometric vertical navigation.

Local Area Augmentation System (LAAS)

LAAS is a ground-based augmentation system that uses a GPS-reference facility located on or in the vicinity of the airport being serviced. This facility has a reference receiver that measures GPS satellite pseudo-range and timing and retransmits the signal. Aircraft landing at LAAS-equipped airports are able to conduct approaches to Category I level and above for properly equipped aircraft. *[Figures 9-31* and *9-32]*

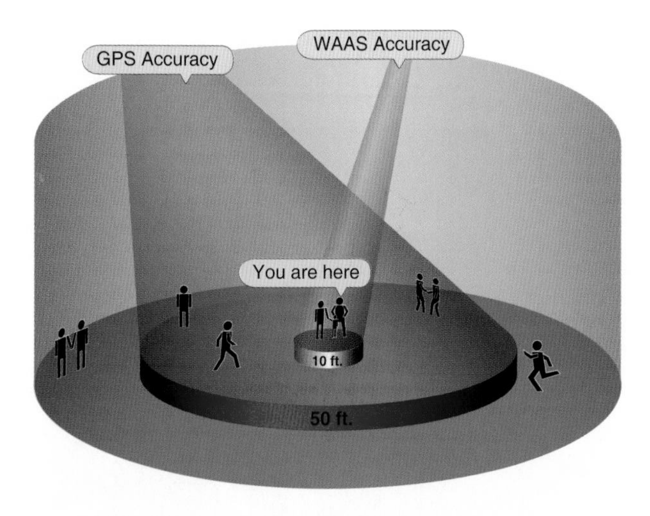

Figure 9-30. *WAAS provides performance enhancement for GPS approach procedures through real-time monitoring.*

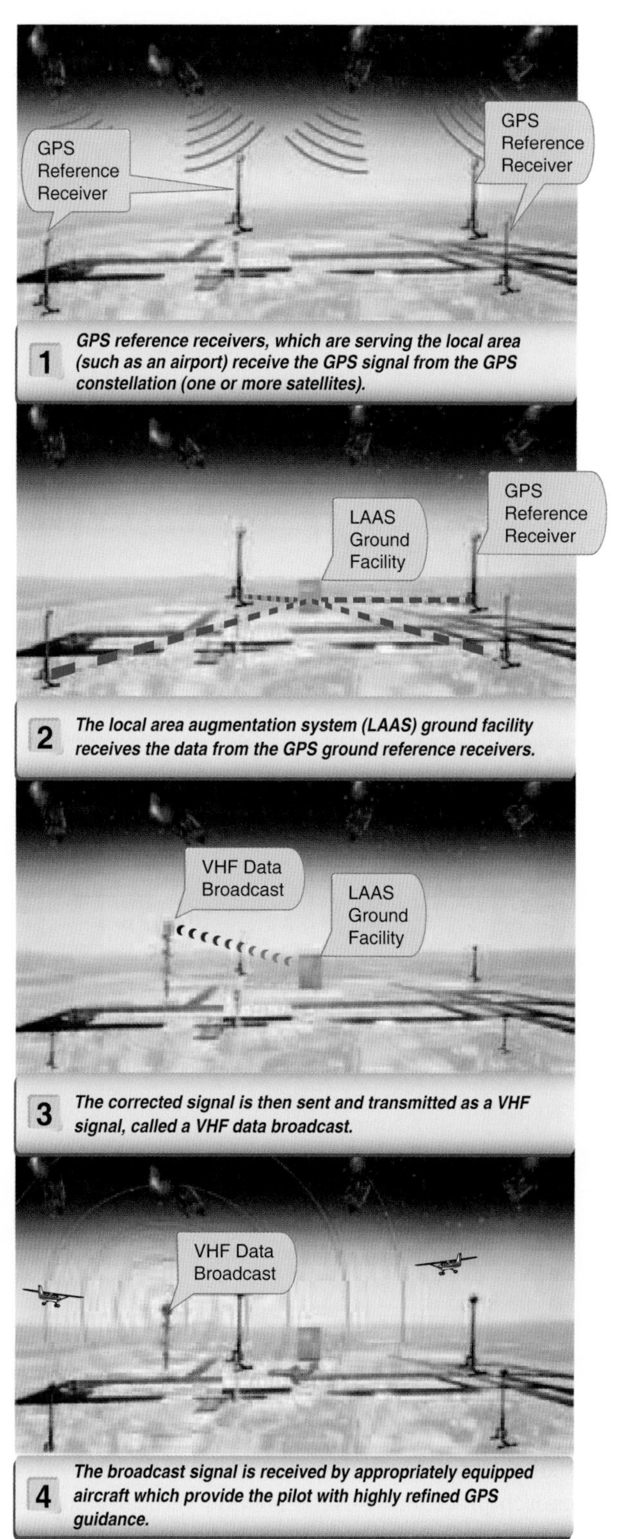

GPS Reference Receiver
GPS Reference Receiver
1 GPS reference receivers, which are serving the local area (such as an airport) receive the GPS signal from the GPS constellation (one or more satellites).
LAAS Ground Facility
GPS Reference Receiver
2 The local area augmentation system (LAAS) ground facility receives the data from the GPS ground reference receivers.
VHF Data Broadcast
LAAS Ground Facility
3 The corrected signal is then sent and transmitted as a VHF signal, called a VHF data broadcast.
VHF Data Broadcast
4 The broadcast signal is received by appropriately equipped aircraft which provide the pilot with highly refined GPS guidance.

Figure 9-31. *LAAS representation.*

Inertial Navigation System (INS)

Inertial Navigation System (INS) is a system that navigates precisely without any input from outside of the aircraft. It is fully self-contained. The INS is initialized by the pilot, who enters into the system the exact location of the aircraft on the ground before the flight. The INS is also programmed with WPs along the desired route of flight.

INS Components

INS is considered a stand-alone navigation system, especially when more than one independent unit is onboard. The airborne equipment consists of an accelerometer to measure acceleration—which, when integrated with time, gives velocity—and gyros to measure direction.

Later versions of the INS, called inertial reference systems (IRS), utilize laser gyros and more powerful computers; therefore, the accelerometer mountings no longer need to be kept level and aligned with true north. The computer system can handle the added workload of dealing with the computations necessary to correct for gravitational and directional errors. Consequently, these newer systems are sometimes called strap down systems, as the accelerometers and gyros are strapped down to the airframe rather than being mounted on a structure that stays fixed with respect to the horizon and true north.

INS Errors

The principal error associated with INS is degradation of position with time. INS computes position by starting with accurate position input which is changed continuously as accelerometers and gyros provide speed and direction inputs. Both accelerometers and gyros are subject to very small errors; as time passes, those errors probably accumulate.

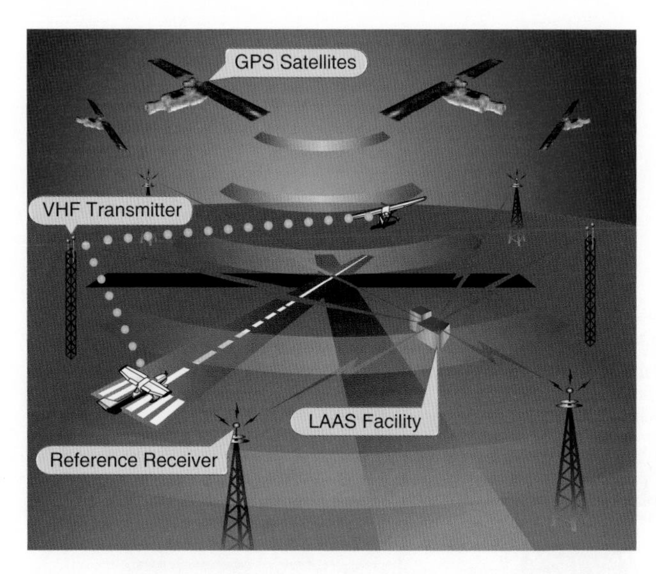

Figure 9-32. *The LAAS system working with GPS satellites, reference receivers and radio transmitters which are located on or in the vicinity of the airport.*

While the best INS/IRS display errors of 0.1 to 0.4 NM after flights across the North Atlantic of 4 to 6 hours, smaller and less expensive systems are being built that show errors of 1 to 2 NM per hour. This accuracy is more than sufficient for a navigation system that can be combined with and updated by GPS. The synergy of a navigation system consisting of an INS/IRS unit in combination with a GPS resolves the errors and weaknesses of both systems. GPS is accurate all the time it is working but may be subject to short and periodic outages. INS is made more accurate because it is continually updated and continues to function with good accuracy if the GPS has moments of lost signal.

Instrument Approach Systems

Most navigation systems approved for en route and terminal operations under IFR, such as VOR, NDB, and GPS, may also be approved to conduct IAPs. The most common systems in use in the United States are the ILS, simplified directional facility (SDF), localizer-type directional aid (LDA), and microwave landing system (MLS). These systems operate independently of other navigation systems. There are new systems being developed, such as WAAS and LAAS. Other systems have been developed for special use.

Instrument Landing Systems (ILS)

The ILS system provides both course and altitude guidance to a specific runway. The ILS system is used to execute a precision instrument approach procedure or precision approach. *[Figure 9-33]* The system consists of the following components:

1. A localizer providing horizontal (left/right) guidance along the extended centerline of the runway.
2. A glideslope (GS) providing vertical (up/down) guidance toward the runway touchdown point, usually at a 3° slope.
3. Marker beacons providing range information along the approach path.
4. Approach lights assisting in the transition from instrument to visual flight.

The following supplementary elements, though not specific components of the system, may be incorporated to increase safety and utility:

1. Compass locators providing transition from en route NAVAIDs to the ILS system and assisting in holding procedures, tracking the localizer course, identifying the marker beacon sites, and providing a FAF for ADF approaches.
2. DME collocated with the GS transmitter providing positive distance-to-touchdown information or DME associated with another nearby facility (VOR or stand-alone), if specified in the approach procedure.

ILS approaches are categorized into three different types of approaches based on the equipment at the airport and the experience level of the pilot. Category I approaches provide for approach height above touchdown of not less than 200 feet. Category II approaches provide for approach to a height above touchdown of not less than 100 feet. Category III approaches provide lower minimums for approaches without a decision height minimum. While pilots need only be instrument rated and the aircraft be equipped with the appropriate airborne equipment to execute Category I approaches, Category II and III approaches require special certification for the pilots, ground equipment, and airborne equipment.

ILS Components

Ground Components

The ILS uses a number of different ground facilities. These facilities may be used as a part of the ILS system, as well as part of another approach. For example, the compass locator may be used with NDB approaches.

Localizer

The localizer (LOC) ground antenna array is located on the extended centerline of the instrument runway of an airport, located at the departure end of the runway to prevent it from being a collision hazard. This unit radiates a field pattern, which develops a course down the centerline of the runway toward the middle markers (MMs) and outer markers (OMs) and a similar course along the runway centerline in the opposite direction. These are called the front and back

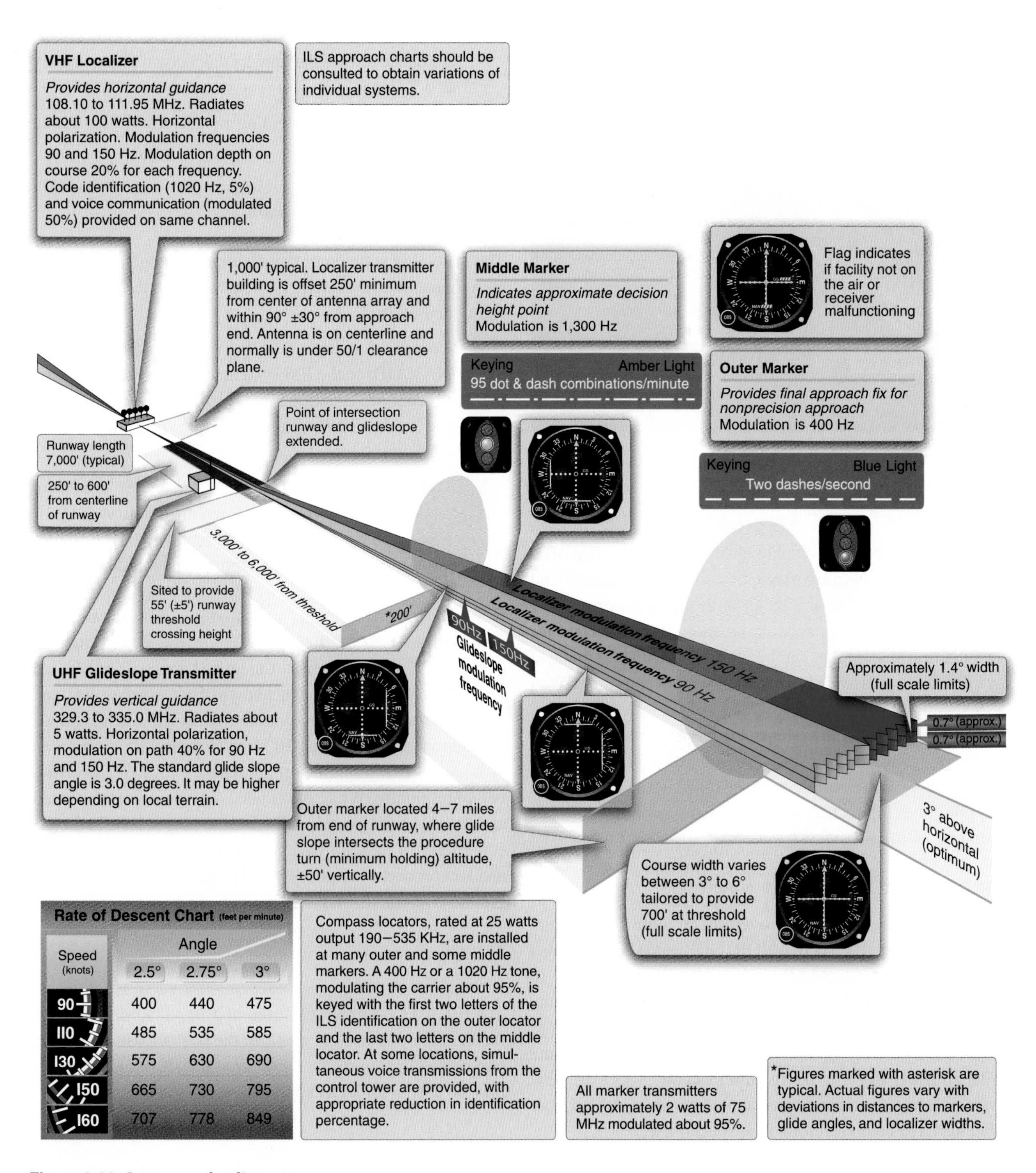
VHF Localizer
Provides horizontal guidance
108.10 to 111.95 MHz. Radiates about 100 watts. Horizontal polarization. Modulation frequencies 90 and 150 Hz. Modulation depth on course 20% for each frequency. Code identification (1020 Hz, 5%) and voice communication (modulated 50%) provided on same channel.
ILS approach charts should be consulted to obtain variations of individual systems.
1,000' typical. Localizer transmitter building is offset 250' minimum from center of antenna array and within 90° ±30° from approach end. Antenna is on centerline and normally is under 50/1 clearance plane.
Middle Marker
Indicates approximate decision height point
Modulation is 1,300 Hz
Flag indicates if facility not on the air or receiver malfunctioning
Keying
Amber Light
95 dot & dash combinations/minute
Outer Marker
Provides final approach fix for nonprecision approach
Modulation is 400 Hz
Keying
Blue Light
Two dashes/second
Point of intersection runway and glideslope extended.
Runway length 7,000' (typical)
250' to 600' from centerline of runway
3,000' to 6,000' from threshold
Sited to provide 55' (±5') runway threshold crossing height
*200'
90Hz
150Hz
Glideslope modulation frequency
Localizer modulation frequency 150 Hz
Localizer modulation frequency 90 Hz
Approximately 1.4° width (full scale limits)
0.7° (approx.)
0.7° (approx.)
UHF Glideslope Transmitter
Provides vertical guidance
329.3 to 335.0 MHz. Radiates about 5 watts. Horizontal polarization, modulation on path 40% for 90 Hz and 150 Hz. The standard glide slope angle is 3.0 degrees. It may be higher depending on local terrain.
Outer marker located 4–7 miles from end of runway, where glide slope intersects the procedure turn (minimum holding) altitude, ±50' vertically.
3° above horizontal (optimum)
Course width varies between 3° to 6° tailored to provide 700' at threshold (full scale limits)
Rate of Descent Chart (feet per minute)
Compass locators, rated at 25 watts output 190–535 KHz, are installed at many outer and some middle markers. A 400 Hz or a 1020 Hz tone, modulating the carrier about 95%, is keyed with the first two letters of the ILS identification on the outer locator and the last two letters on the middle locator. At some locations, simultaneous voice transmissions from the control tower are provided, with appropriate reduction in identification percentage.
All marker transmitters approximately 2 watts of 75 MHz modulated about 95%.
*Figures marked with asterisk are typical. Actual figures vary with deviations in distances to markers, glide angles, and localizer widths.

Speed (knots)	Angle 2.5°	2.75°	3°
90	400	440	475
110	485	535	585
130	575	630	690
150	665	730	795
160	707	778	849

Figure 9-33. *Instrument landing systems.*

courses, respectively. The localizer provides course guidance, transmitted at 108.1 to 111.95 MHz (odd tenths only), throughout the descent path to the runway threshold from a distance of 18 NM from the antenna to an altitude of 4,500 feet above the elevation of the antenna site. *[Figure 9-34]*

The localizer course width is defined as the angular displacement at any point along the course between a full "fly-left" (CDI needle fully deflected to the left) and a full "fly-right" indication (CDI needle fully deflected to the right). Each localizer facility is audibly identified by a three-letter designator transmitted at frequent regular intervals. The ILS identification is preceded by the letter "I" (two dots). For example, the ILS localizer at Springfield, Missouri, transmits the identifier ISGF. The localizer includes a voice feature on

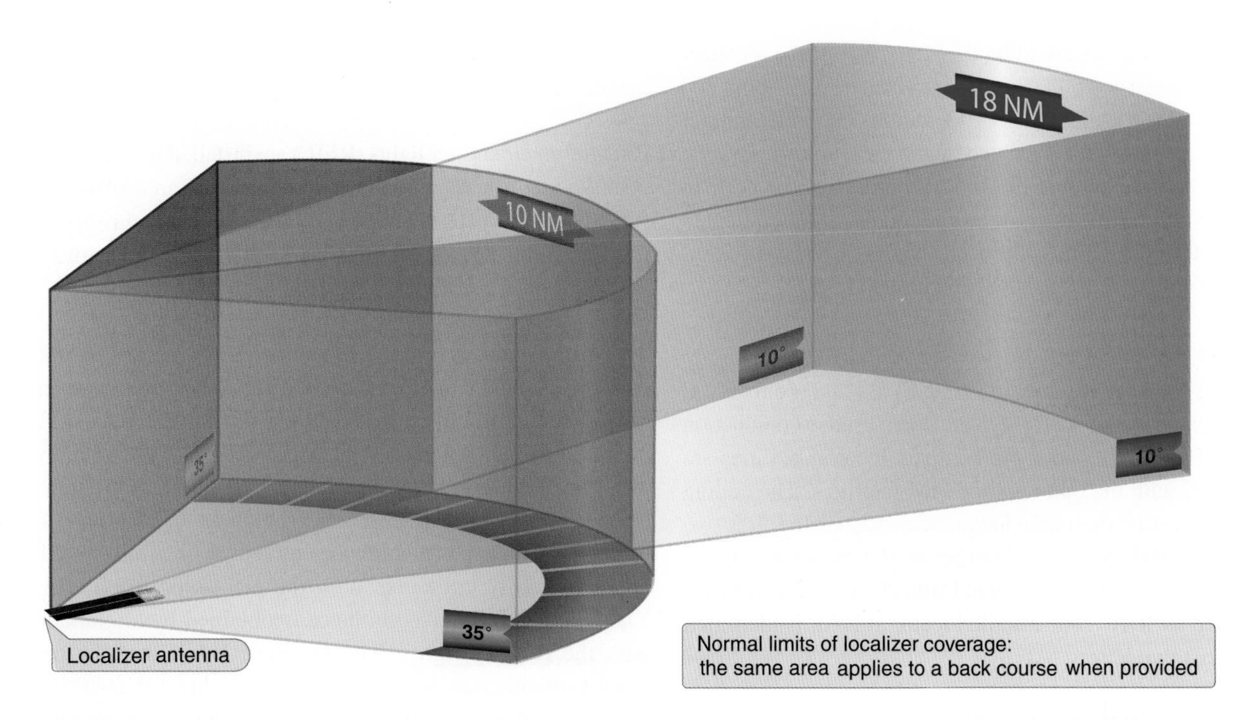

Figure 9-34. *Localizer coverage limits.*

its frequency for use by the associated ATC facility in issuing approach and landing instructions.

The localizer course is very narrow, normally 5°. This results in high needle sensitivity. With this course width, a full-scale deflection shows when the aircraft is 2.5° to either side of the centerline. This sensitivity permits accurate orientation to the landing runway. With no more than one-quarter scale deflection maintained, the aircraft will be aligned with the runway.

Glideslope (GS)

GS describes the systems that generate, receive, and indicate the ground facility radiation pattern. The glidepath is the straight, sloped line the aircraft should fly in its descent from where the GS intersects the altitude used for approaching the FAF to the runway touchdown zone. The GS equipment is housed in a building approximately 750 to 1,250 feet down the runway from the approach end of the runway and between 400 and 600 feet to one side of the centerline.

The course projected by the GS equipment is essentially the same as would be generated by a localizer operating on its side. The GS projection angle is normally adjusted to 2.5° to 3.5° above horizontal, so it intersects the MM at about 200 feet and the OM at about 1,400 feet above the runway elevation. At locations where standard minimum obstruction clearance cannot be obtained with the normal maximum GS angle, the GS equipment is displaced farther from the approach end of the runway if the length of the runway permits; or the GS angle may be increased up to 4°.

Unlike the localizer, the GS transmitter radiates signals only in the direction of the final approach on the front course. The system provides no vertical guidance for approaches on the back course. The glidepath is normally 1.4° thick. At 10 NM from the point of touchdown, this represents a vertical distance of approximately 1,500 feet, narrowing to a few feet at touchdown.

Marker Beacons

Two VHF marker beacons, outer and middle, are normally used in the ILS system. *[Figure 9-35]* A third beacon, the inner, is used where Category II operations are certified. A marker beacon may also be installed to indicate the FAF on the ILS back course.

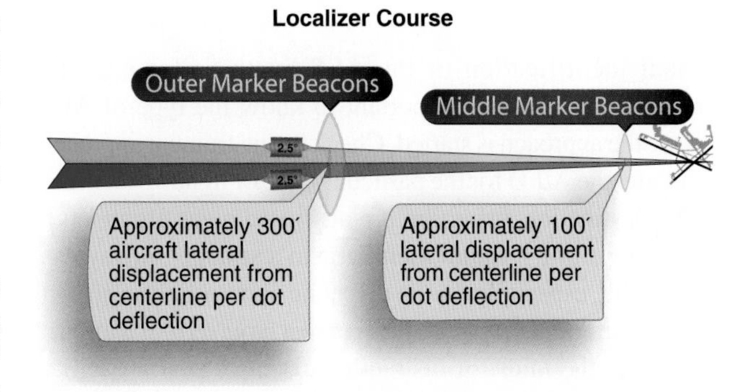

Figure 9-35. *Localizer receiver indications and aircraft displacement.*

The OM is located on the localizer front course 4–7 miles from the airport to indicate a position at which an aircraft, at the appropriate altitude on the localizer course, will intercept the glidepath. The MM is located approximately 3,500 feet from the landing threshold on the centerline of the localizer front course at a position where the GS centerline is about 200 feet above the touchdown zone elevation. The inner marker (IM), where installed, is located on the front course between the MM and the landing threshold. It indicates the point at which an aircraft is at the decision height on the glidepath during a Category II ILS approach. The back-course marker, where installed, indicates the back-course FAF.

Compass Locator

Compass locators are low-powered NDBs and are received and indicated by the ADF receiver. When used in conjunction with an ILS front course, the compass locator facilities are collocated with the outer and/or MM facilities. The coding identification of the outer locator consists of the first two letters of the three-letter identifier of the associated LOC. For example, the outer locator at Dallas/Love Field (DAL) is identified as "DA." The middle locator at DAL is identified by the last two letters "AL."

Approach Lighting Systems (ALS)

Normal approach and letdown on the ILS is divided into two distinct stages: the instrument approach stage using only radio guidance, and the visual stage, when visual contact with the ground runway environment is necessary for accuracy and safety. The most critical period of an instrument approach, particularly during low ceiling/visibility conditions, is the point at which the pilot must decide whether to land or execute a missed approach. As the runway threshold is approached, the visual glidepath separates into individual lights. At this point, the approach should be continued by reference to the runway touchdown zone markers. The approach lighting system (ALS) provides lights that will penetrate the atmosphere far enough from touchdown to give directional, distance, and glidepath information for safe visual transition.

Visual identification of the ALS by the pilot must be instantaneous, so it is important to know the type of ALS before the approach is started. Check the instrument approach chart and the A/FD for the particular type of lighting facilities at the destination airport before any instrument flight. With reduced visibility, rapid orientation to a strange runway can be difficult, especially during a circling approach to an airport with minimum lighting facilities or to a large terminal airport located in the midst of distracting city and ground facility lights. Some of the most common ALS systems are shown in *Figure 9-36*.

A high-intensity flasher system, often referred to as "the rabbit," is installed at many large airports. The flashers consist of a series of brilliant blue-white bursts of light flashing in sequence along the approach lights, giving the effect of a ball of light traveling towards the runway. Typically, "the rabbit" makes two trips toward the runway per second.

Runway end identifier lights (REIL) are installed for rapid and positive identification of the approach end of an instrument runway. The system consists of a pair of synchronized flashing lights placed laterally on each side of the runway threshold facing the approach area.

The visual approach slope indicator (VASI) gives visual descent guidance information during the approach to a runway. The standard VASI consists of light bars that project a visual glidepath, which provides safe obstruction clearance within the approach zone. The normal GS angle is 3°; however, the angle may be as high as 4.5° for proper obstacle clearance. On runways served by ILS, the VASI angle normally coincides with the electronic GS angle. Visual left/right course guidance is obtained by alignment with the runway lights. The standard VASI installation consists of either 2-, 3-, 4-, 6-, 12-, or 16-light units arranged in downwind and upwind light bars. Some airports serving long-bodied aircraft have three-bar VASIs that provide two visual glidepaths to the same runway. The first glidepath encountered is the same as provided by the standard VASI. The second glidepath is about 25 percent higher than the first and is designed for the use of pilots of long-bodied aircraft.

The basic principle of VASI is that of color differentiation between red and white. Each light projects a beam having a white segment in the upper part and a red segment in the lower part of the beam. From a position above the glidepath the pilot sees both bars as white. Lowering the aircraft with respect to the glidepath, the color of the upwind bars changes from white to pink to red. When on the proper glidepath, the landing aircraft will overshoot the downwind bars and undershoot the upwind bars. Thus the downwind (closer) bars are seen as white and the upwind bars as red. From a position below the glidepath, both light bars are seen as red. Moving up to the glidepath, the color of the downwind bars changes from red to pink to white. When below the glidepath, as indicated by a distinct all-red signal, a safe obstruction clearance might not exist. A standard two-bar VASI is illustrated in *Figure 9-37*.

ILS Airborne Components

Airborne equipment for the ILS system includes receivers for the localizer, GS, marker beacons, ADF, DME, and the respective indicator instruments.

The typical VOR receiver is also a localizer receiver with common tuning and indicating equipment. Some receivers have separate function selector switches, but most switch between VOR and LOC automatically by sensing if odd

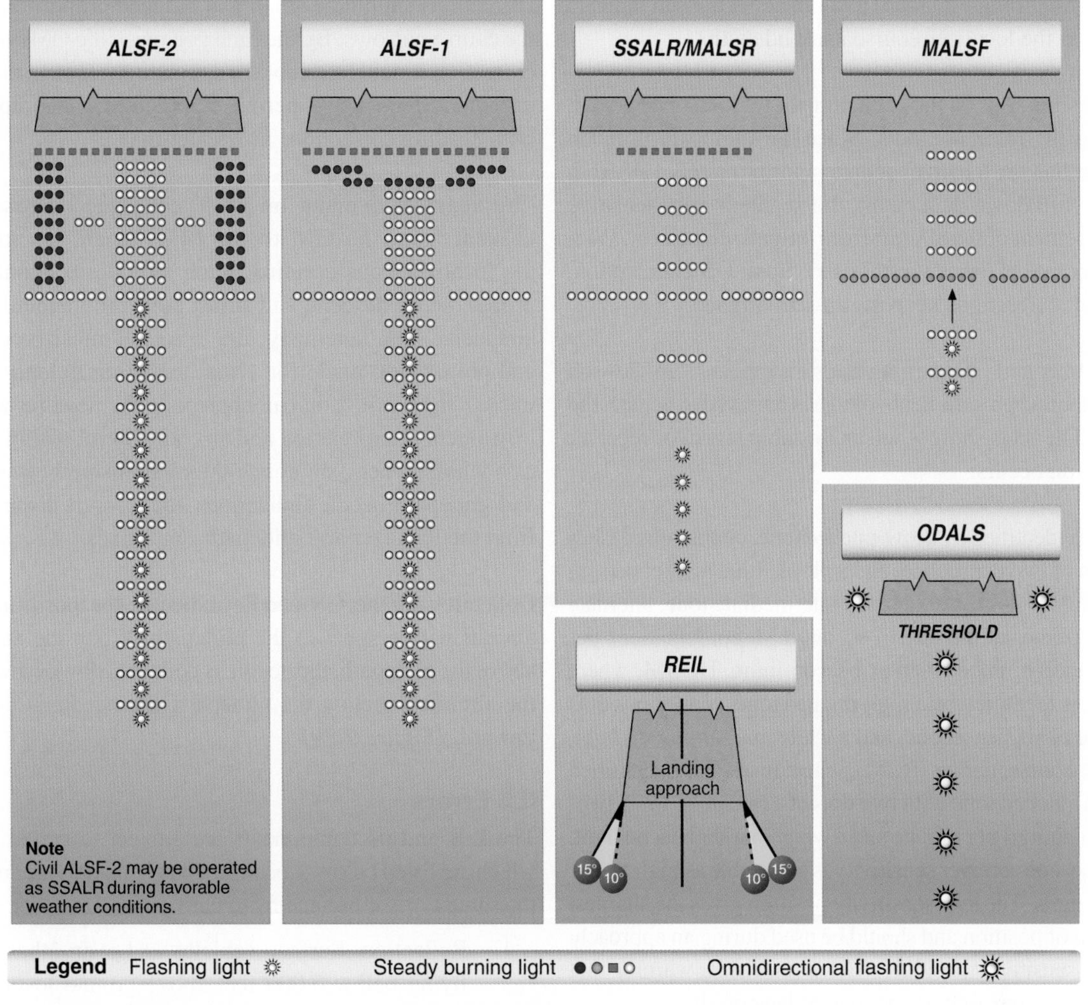

ALSF—Approach light system with sequenced flashing lights

SSALR—Simplified short approach light system with runway alignment indicator lights

MALSR—Medium intensity approach light system with runway alignment indicator lights

REIL—Runway end identification lights

MALSF—Medium intensity approach light system with sequenced flashing lights (and runway alignment)

ODALS—Omnidirectional approach light system

Figure 9-36. *Precision and nonprecision ALS configuration.*

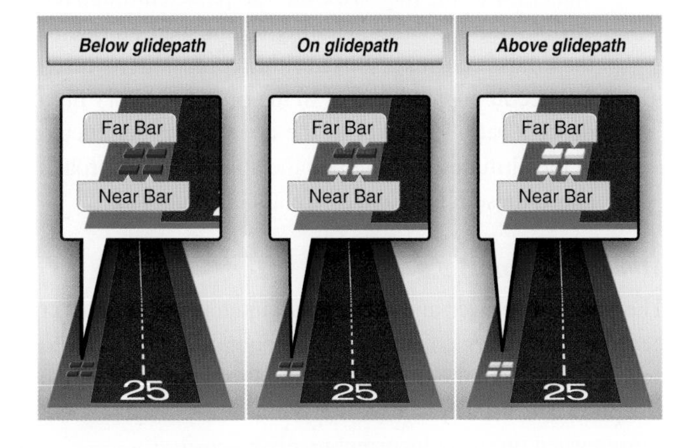

Figure 9-37. *Standard two-bar VASI.*

tenths between 108 and 111.95 MHz have been selected. Otherwise, tuning of VOR and localizer frequencies is accomplished with the same knobs and switches, and the CDI indicates "on course" as it does on a VOR radial.

Though some GS receivers are tuned separately, in a typical installation the GS is tuned automatically to the proper frequency when the localizer is tuned. Each of the 40 localizer channels in the 108.10 to 111.95 MHz band is paired with a corresponding GS frequency.

When the localizer indicator also includes a GS needle, the instrument is often called a cross-pointer indicator. The crossed horizontal (GS) and vertical (localizer) needles are free to move through standard five-dot deflections to indicate position on the localizer course and glidepath.

When the aircraft is on the glidepath, the needle is horizontal, overlying the reference dots. Since the glidepath is much narrower than the localizer course (approximately 1.4° from full up to full down deflection), the needle is very sensitive to displacement of the aircraft from on-path alignment. With the proper rate of descent established upon GS interception, very small corrections keep the aircraft aligned.

The localizer and GS warning flags disappear from view on the indicator when sufficient voltage is received to actuate the needles. The flags show when an unstable signal or receiver malfunction occurs.

The OM is identified by a low-pitched tone, continuous dashes at the rate of two per second, and a purple/blue marker beacon light. The MM is identified by an intermediate tone, alternate dots and dashes at the rate of 95 dot/dash combinations per minute, and an amber marker beacon light. The IM, where installed, is identified by a high-pitched tone, continuous dots at the rate of six per second, and a white marker beacon light. The back-course marker (BCM), where installed, is identified by a high-pitched tone with two dots at a rate of 72 to 75 two-dot combinations per minute and a white marker beacon light. Marker beacon receiver sensitivity is selectable as high or low on many units. The low-sensitivity position gives the sharpest indication of position and should be used during an approach. The high-sensitivity position provides an earlier warning that the aircraft is approaching the marker beacon site.

ILS Function

The localizer needle indicates, by deflection, whether the aircraft is right or left of the localizer centerline, regardless of the position or heading of the aircraft. Rotating the OBS has no effect on the operation of the localizer needle, although it is useful to rotate the OBS to put the LOC inbound course under the course index. When inbound on the front course, or outbound on the back course, the course indication remains directional. (See *Figure 9-38,* aircraft C, D, and E.)

Unless the aircraft has reverse sensing capability and it is in use, when flying inbound on the back course or outbound on the front course, heading corrections to on-course are made opposite the needle deflection. This is commonly described as "flying away from the needle." (See *Figure 9-38*, aircraft A and B.) Back course signals should not be used for an approach unless a back course approach procedure is published for that particular runway and the approach is authorized by ATC.

Once you have reached the localizer centerline, maintain the inbound heading until the CDI moves off center. Drift corrections should be small and reduced proportionately as the course narrows. By the time you reach the OM, your drift correction should be established accurately enough on a well-executed approach to permit completion of the approach, with heading corrections no greater than 2°.

The heaviest demand on pilot technique occurs during descent from the OM to the MM, when you maintain the localizer course, adjust pitch attitude to maintain the proper rate of descent, and adjust power to maintain proper airspeed. Simultaneously, the altimeter must be checked and preparation made for visual transition to land or for a missed approach. You can appreciate the need for accurate instrument interpretation and aircraft control within the ILS as a whole, when you notice the relationship between CDI and glidepath needle indications and aircraft displacement from the localizer and glidepath centerlines.

Deflection of the GS needle indicates the position of the aircraft with respect to the glidepath. When the aircraft is above the glidepath, the needle is deflected downward. When the aircraft is below the glidepath, the needle is deflected upward. *[Figure 9-39]*

ILS Errors

The ILS and its components are subject to certain errors, which are listed below. Localizer and GS signals are subject to the same type of bounce from hard objects as space waves.

1. Reflection. Surface vehicles and even other aircraft flying below 5,000 feet above ground level (AGL) may disturb the signal for aircraft on the approach.
2. False courses. In addition to the desired course, GS facilities inherently produce additional courses at higher vertical angles. The angle of the lowest of these false courses occurs at approximately 9°– 12°. An aircraft flying the LOC/GS course at a constant altitude would observe gyrations of both the GS needle and GS warning flag as the aircraft passed through the various false courses. Getting established on one of these false courses results in either confusion (reversed GS needle indications) or in the need for a very high descent rate. However, if the approach is conducted at the altitudes specified on the appropriate approach chart, these false courses are not encountered.

Marker Beacons

The very low power and directional antenna of the marker beacon transmitter ensures that the signal is not received any distance from the transmitter site. Problems with signal reception are usually caused by the airborne receiver not being turned on or by incorrect receiver sensitivity.

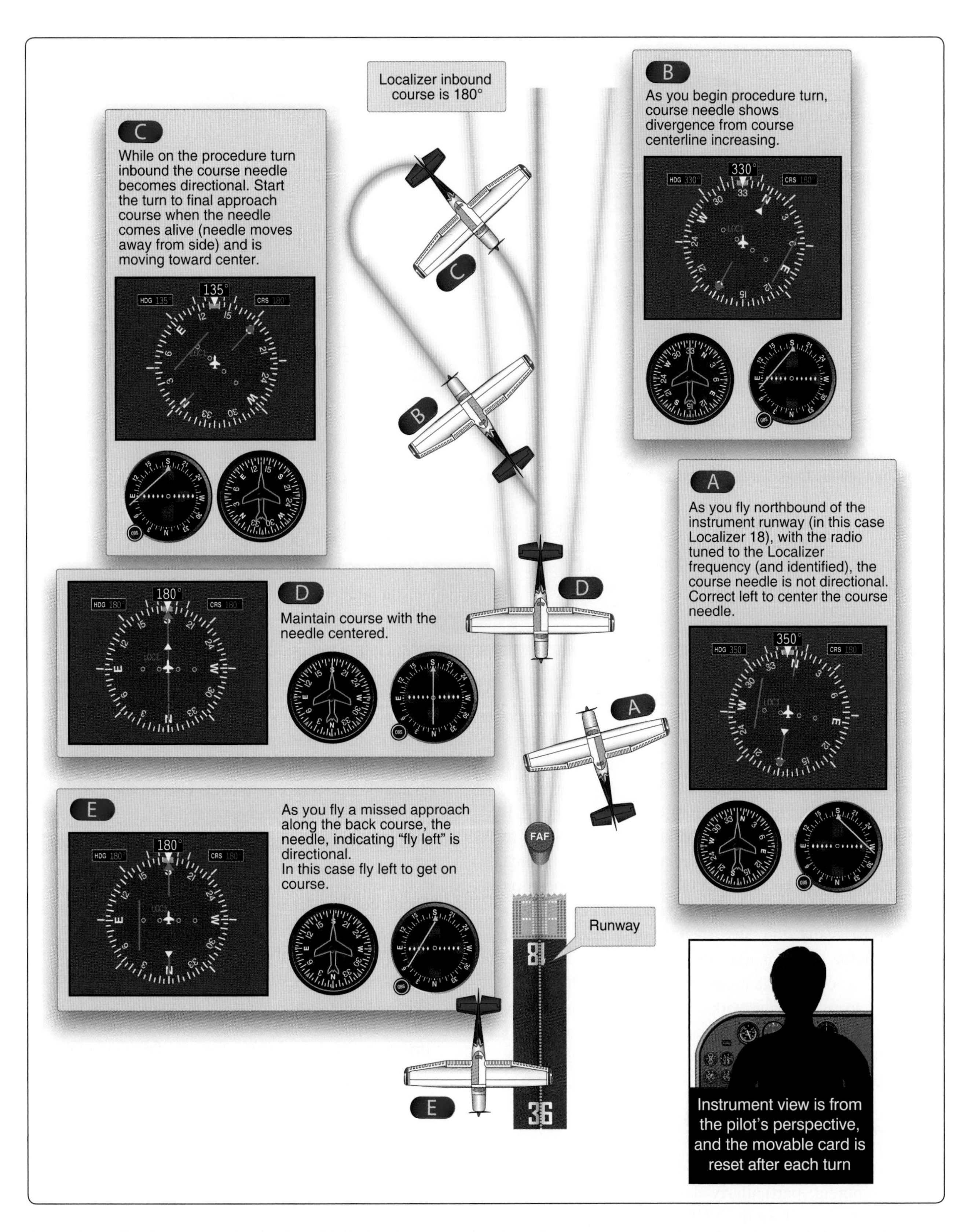

Figure 9-38. *Localizer course indications. To follow indications displayed in the aircraft, start from A and proceed through E.*

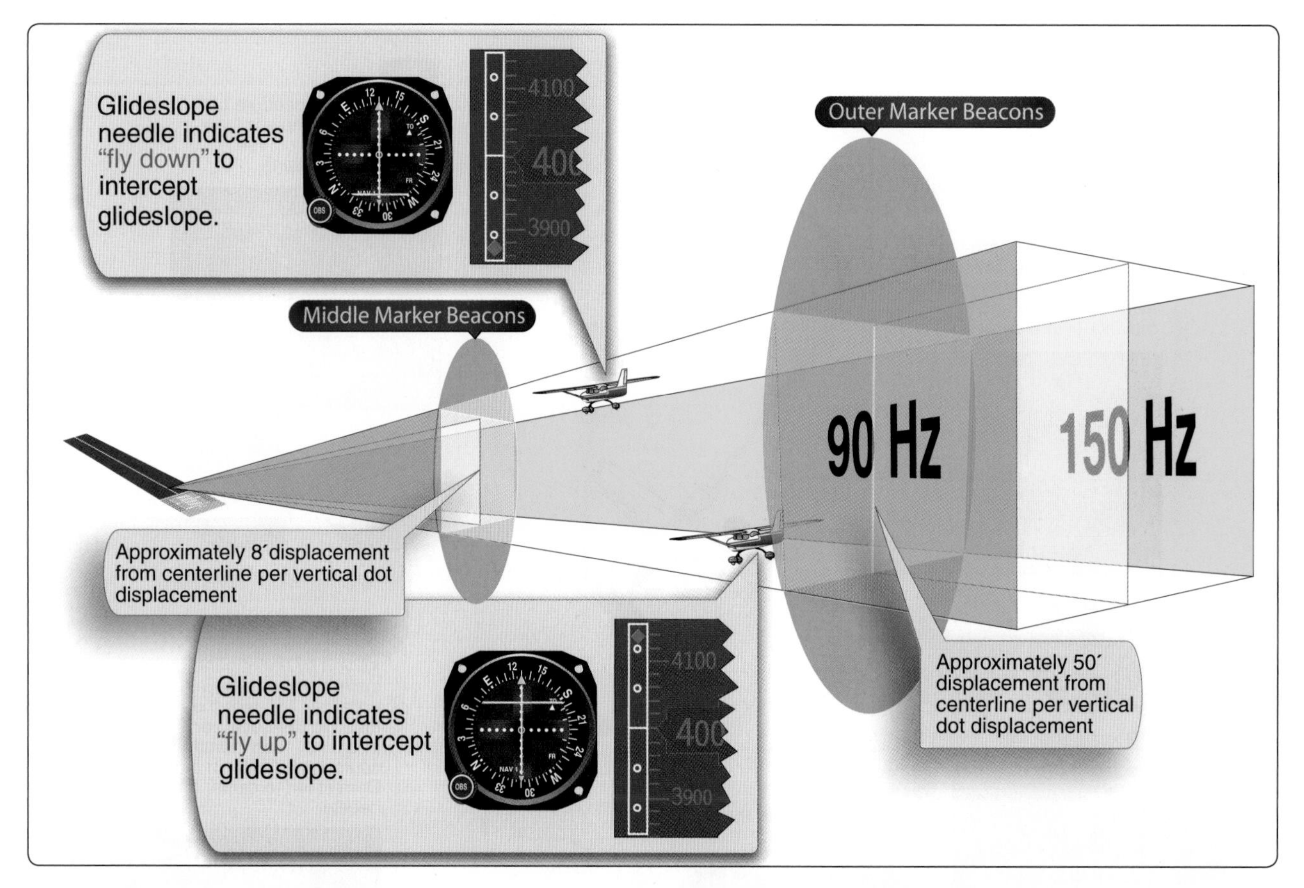

Figure 9-39. *A GS receiver indication and aircraft displacement. An analog system is on the left and the same indication on the Garmin PFD on the right.*

Some marker beacon receivers, to decrease weight and cost, are designed without their own power supply. These units utilize a power source from another radio in the avionics stack, often the ADF. In some aircraft, this requires the ADF to be turned on in order for the marker beacon receiver to function, yet no warning placard is required. Another source of trouble may be the "HIGH/LOW/OFF" three-position switch, which both activates the receiver and selects receiver sensitivity. Usually, the "test" feature only tests to see if the light bulbs in the marker beacon lights are working. Therefore, in some installations, there is no functional way for the pilot to ascertain the marker beacon receiver is actually on except to fly over a marker beacon transmitter and see if a signal is received and indicated (e.g., audibly, and visually via marker beacon lights).

Operational Errors

1. Failure to understand the fundamentals of ILS ground equipment, particularly the differences in course dimensions. Since the VOR receiver is used on the localizer course, the assumption is sometimes made that interception and tracking techniques are identical when tracking localizer courses and VOR radials. Remember that the CDI sensing is sharper and faster on the localizer course.
2. Disorientation during transition to the ILS due to poor planning and reliance on one receiver instead of on all available airborne equipment. Use all the assistance available; a single receiver may fail.
3. Disorientation on the localizer course, due to the first error noted above.
4. Incorrect localizer interception angles. A large interception angle usually results in overshooting and possible disorientation. When intercepting, if possible, turn to the localizer course heading immediately upon the first indication of needle movement. An ADF receiver is an excellent aid to orient you during an ILS approach if there is a locator or NDB on the inbound course.
5. Chasing the CDI and glidepath needles, especially when you have not sufficiently studied the approach before the flight.

Simplified Directional Facility (SDF)

The simplified directional facility (SDF) provides a final approach course similar to the ILS localizer. The SDF course may or may not be aligned with the runway and the course may be wider than a standard ILS localizer, resulting in less precision. Usable off-course indications are limited to 35° either side of the course centerline. Instrument indications in the area between 35° and 90° from the course centerline are not controlled and should be disregarded.

The SDF must provide signals sufficient to allow satisfactory operation of a typical aircraft installation within a sector which extends from the center of the SDF antenna system to distances of 18 NM covering a sector 10° either side of centerline up to an angle 7° above the horizontal. The angle of convergence of the final approach course and the extended runway centerline must not exceed 30°. Pilots should note this angle since the approach course originates at the antenna site, and an approach continued beyond the runway threshold would lead the aircraft to the SDF offset position rather than along the runway centerline.

The course width of the SDF signal emitted from the transmitter is fixed at either 6° or 12°, as necessary, to provide maximum flyability and optimum approach course quality. A three-letter identifier is transmitted in code on the SDF frequency; there is no letter "I" (two dots) transmitted before the station identifier, as there is with the LOC. For example, the identifier for Lebanon, Missouri, SDF is LBO.

Localizer Type Directional Aid (LDA)

The localizer type directional aid (LDA) is of comparable utility and accuracy to a localizer but is not part of a complete ILS. The LDA course width is between 3° and 6° and thus provides a more precise approach course than an SDF installation. Some LDAs are equipped with a GS. The LDA course is not aligned with the runway, but straight-in minimums may be published where the angle between the runway centerline and the LDA course does not exceed 30°. If this angle exceeds 30°, only circling minimums are published. The identifier is three letters preceded by "I" transmitted in code on the LDA frequency. For example, the identifier for Van Nuys, California, LDA is I-BUR.

Microwave Landing System (MLS)

The microwave landing system (MLS) provides precision navigation guidance for exact alignment and descent of aircraft on approach to a runway. It provides azimuth, elevation, and distance. Both lateral and vertical guidance may be displayed on conventional course deviation indicators or incorporated into multipurpose flight deck displays. Range information can be displayed by conventional DME indicators and also incorporated into multipurpose displays. *[Figure 9-40]*

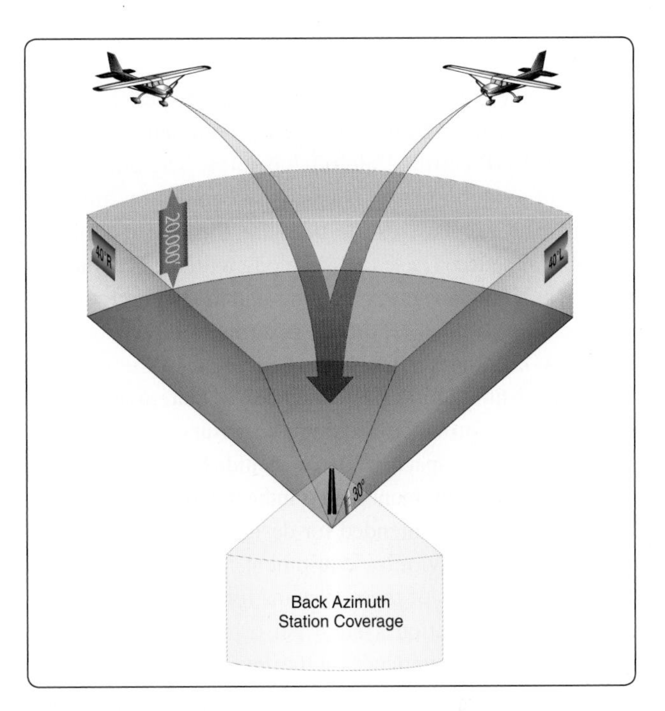

Figure 9-40. *MLS coverage volumes, 3-D representation.*

The system may be divided into five functions, which are approach azimuth, back azimuth, approach elevation, range; and data communications. The standard configuration of MLS ground equipment includes an azimuth station to perform functions as indicated above. In addition to providing azimuth navigation guidance, the station transmits basic data, which consists of information associated directly with the operation of the landing system, as well as advisory data on the performance of the ground equipment.

Approach Azimuth Guidance

The azimuth station transmits MLS angle and data on one of 200 channels within the frequency range of 5031 to 5091 MHz. The equipment is normally located about 1,000 feet beyond the stop end of the runway, but there is considerable flexibility in selecting sites. For example, for heliport operations the azimuth transmitter can be collocated with the elevation transmitter. The azimuth coverage extends laterally at least 40° on either side of the runway centerline in a standard configuration, in elevation up to an angle of 15° and to at least 20,000 feet, and in range to at least 20 NM.

MLS requires separate airborne equipment to receive and process the signals from what is normally installed in general aviation aircraft today. It has data communications capability, and can provide audible information about the condition of the transmitting system and other pertinent data such as weather, runway status, etc. The MLS transmits an audible identifier consisting of four letters beginning with the letter M, in Morse code at a rate of at least six per minute. The MLS system monitors itself and transmits ground-to-air data messages about the system's operational condition. During periods of routine or emergency maintenance, the coded identification is missing from the transmissions. At this time there are only a few systems installed.

Required Navigation Performance

RNP is a navigation system that provides a specified level of accuracy defined by a lateral area of confined airspace in which an RNP-certified aircraft operates. The continuing growth of aviation places increasing demands on airspace capacity and emphasizes the need for the best use of the available airspace. These factors, along with the accuracy of modern aviation navigation systems and the requirement for increased operational efficiency in terms of direct routings and track-keeping accuracy, have resulted in the concept of required navigation performance—a statement of the navigation performance accuracy necessary for operation within a defined airspace. RNP can include both performance and functional requirements and is indicated by the RNP type. These standards are intended for designers, manufacturers, and installers of avionics equipment, as well as service providers and users of these systems for global operations. The minimum aviation system performance specification (MASPS) provides guidance for the development of airspace and operational procedures needed to obtain the benefits of improved navigation capability. *[Figure 9-41]*

The RNP type defines the total system error (TSE) that is allowed in lateral and longitudinal dimensions within a particular airspace. The TSE, which takes account of navigation system errors (NSE), computation errors, display errors and flight technical errors (FTE), must not exceed the specified RNP value for 95 percent of the flight time on any part of any single flight. RNP combines the accuracy standards laid out in the ICAO Manual (Doc 9613) with specific accuracy requirements, as well as functional and performance standards, for the RNAV system to realize a system that can meet future air traffic management requirements. The functional criteria for RNP address the need for the flightpaths of participating aircraft to be both predictable and repeatable to the declared levels of accuracy. More information on RNP is contained in subsequent chapters.

The term RNP is also applied as a descriptor for airspace, routes, and procedures (including departures, arrivals, and IAPs). The descriptor can apply to a unique approach procedure or to a large region of airspace. RNP applies to navigation performance within a designated airspace and includes the capability of both the available infrastructure (navigation aids) and the aircraft.

RNP type is used to specify navigation requirements for the airspace. The following are ICAO RNP Types: RNP-1.0, RNP-4.0, RNP-5.0, and RNP-10.0. The required performance is obtained through a combination of aircraft capability and the level of service provided by the corresponding navigation infrastructure. From a broad perspective:

Aircraft Capability + Level of Service = Access

In this context, aircraft capability refers to the airworthiness certification and operational approval elements (including avionics, maintenance, database, human factors, pilot procedures, training, and other issues). The level of service element refers to the NAS infrastructure, including published routes, signal-in-space performance and availability, and air traffic management. When considered collectively, these elements result in providing access. Access provides the desired benefit (airspace, procedures, routes of flight, etc.).

RNP levels are actual distances from the centerline of the flightpath, which must be maintained for aircraft and obstacle separation. Although additional FAA-recognized RNP levels may be used for specific operations, the United States currently supports three standard RNP levels:

- RNP 0.3 – Approach
- RNP 1.0 – Departure, Terminal
- RNP 2.0 – En route

RNP 0.3 represents a distance of 0.3 NM either side of a specified flightpath centerline. The specific performance that is required on the final approach segment of an instrument approach is an example of this RNP level. At the present time, a 0.3 RNP level is the lowest level used in normal RNAV operations. Specific airlines, using special procedures, are approved to use RNP levels lower than RNP 0.3, but those levels are used only in accordance with their approved operations specifications (OpsSpecs). For aircraft equipment to qualify for a specific RNP type, it must maintain navigational accuracy at least 95 percent of the total flight time.

Flight Management Systems (FMS)

A flight management system (FMS) is not a navigation system in itself. Rather, it is a system that automates the tasks of managing the onboard navigation systems. FMS may perform other onboard management tasks, but this discussion is limited to its navigation function.

FMS is an interface between flight crews and flightdeck systems. FMS can be thought of as a computer with a large database of airport and NAVAID locations and associated data, aircraft performance data, airways, intersections, DPs, and STARs. FMS also has the ability to accept and store numerous user-defined WPs, flight routes consisting of departures, WPs, arrivals, approaches, alternates, etc. FMS can quickly define a desired route from the aircraft's current position to any point in the world, perform flight plan computations, and display the total picture of the flight route to the crew.

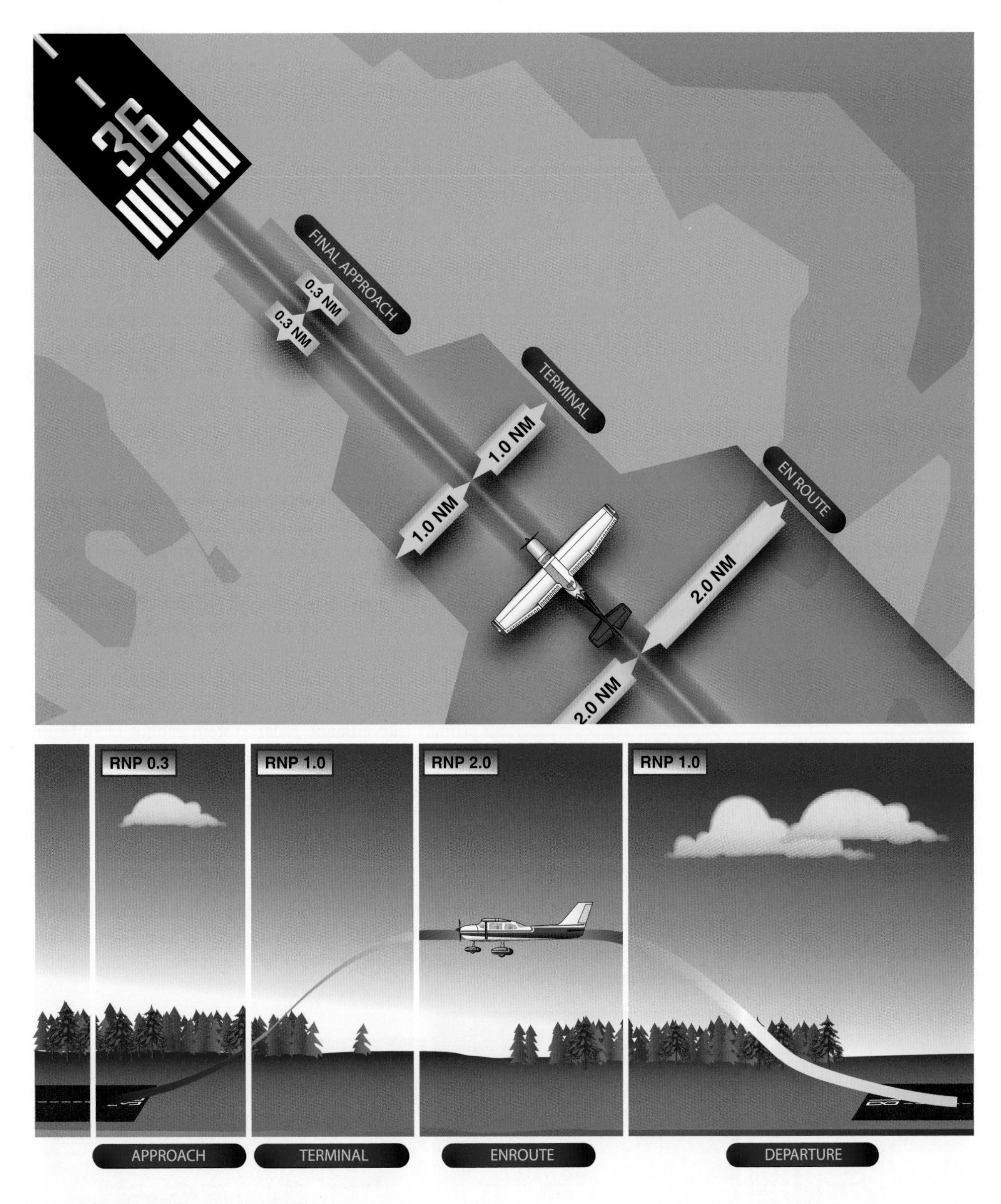

Figure 9-41. *Required navigation performance.*

FMS also has the capability of controlling (selecting) VOR, DME, and LOC NAVAIDs, and then receiving navigational data from them. INS, LORAN, and GPS navigational data may also be accepted by the FMS computer. The FMS may act as the input/output device for the onboard navigation systems, so that it becomes the "go-between" for the crew and the navigation systems.

Function of FMS

At startup, the crew programs the aircraft location, departure runway, DP (if applicable), WPs defining the route, approach procedure, approach to be used, and routing to alternate. This may be entered manually, be in the form of a stored flight plan, or be a flight plan developed in another computer and transferred by disk or electronically to the FMS computer. The crew enters this basic information in the control/display unit (CDU). *[Figure 9-42]*

Once airborne, the FMS computer channels the appropriate NAVAIDs and takes radial/distance information or channels two NAVAIDs, taking the more accurate distance information. FMS then indicates position, track, desired heading, groundspeed, and position relative to desired track. Position information from the FMS updates the INS. In more sophisticated aircraft, the FMS provides inputs to the HSI, RMI, glass flight deck navigation displays, head-up display (HUD), autopilot, and autothrottle systems.

Head-Up Display (HUD)

The HUD is a display system that provides a projection of navigation and air data (airspeed in relation to approach reference speed, altitude, left/right and up/down GS) on a transparent screen between the pilot and the windshield. Other information may be displayed, including a runway target in relation to the nose of the aircraft. This allows the pilot to see the information necessary to make the approach while also being able to see out the windshield, which diminishes the need to shift between looking at the panel to looking outside. Virtually any information desired can be displayed on the HUD if it is available in the aircraft's flight computer and if the display is user definable. *[Figure 9-43]*

Radar Navigation (Ground-Based)

Radar works by transmitting a pulse of RF energy in a specific direction. The return of the echo or bounce of that pulse from a target is precisely timed. From this, the distance traveled by the pulse and its echo is determined and displayed on a radar screen in such a manner that the distance and bearing to this target can be instantly determined. The radar transmitter must be capable of delivering extremely high power levels toward the airspace under surveillance, and the associated radar receiver must be able to detect extremely small signal levels of the returning echoes.

The radar display system provides the controller with a map-like presentation upon which appear all the radar echoes of aircraft within detection range of the radar facility. By means of electronically-generated range marks and azimuth-indicating devices, the controller can locate each radar target with respect to the radar facility, or can locate one radar target with respect to another.

Another device, a video-mapping unit, generates an actual airway or airport map and presents it on the radar display equipment. Using the video-mapping feature, the air traffic controller not only can view the aircraft targets, but can see these targets in relation to runways, navigation aids, and hazardous ground obstructions in the area. Therefore, radar becomes a NAVAID, as well as the most significant means of traffic separation.

In a display presenting perhaps a dozen or more targets, a primary surveillance radar system cannot identify one specific radar target, and it may have difficulty "seeing" a small target at considerable distance—especially if there is a rain shower or thunderstorm between the radar site and the aircraft. This problem is solved with the Air Traffic Control Radar Beacon System (ATCRBS), sometimes called secondary surveillance radar (SSR), which utilizes a transponder in the aircraft. The ground equipment is an interrogating unit, in which the beacon antenna is mounted so it rotates with the surveillance antenna. The interrogating

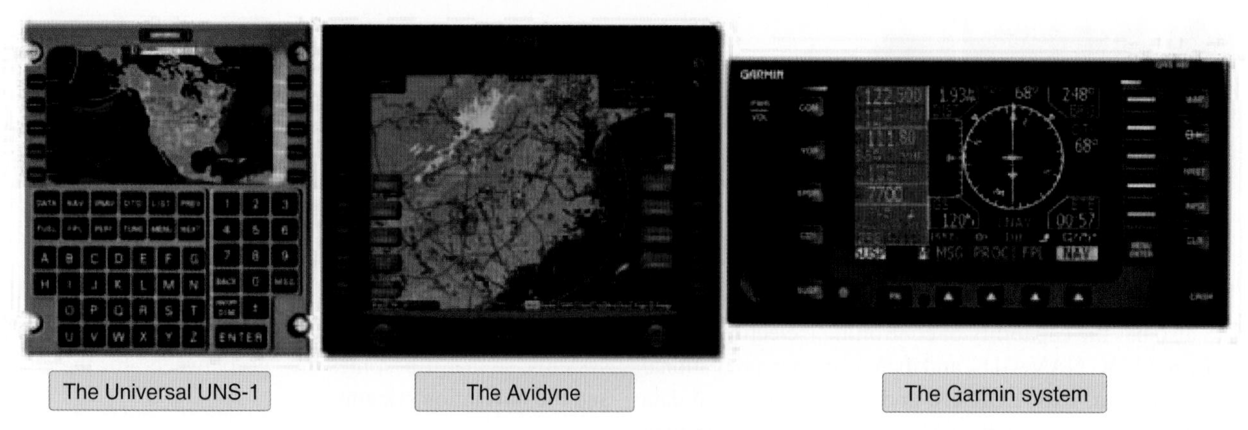

Figure 9-42. *Typical display and control unit(s) in general aviation. The Universal UNS-1 (left) controls and integrates all other systems. The Avidyne (center) and Garmin systems (right) illustrate and are typical of completely integrated systems. Although the Universal CDU is not typically found on smaller general aviation aircraft, the difference in capabilities of the CDUs and stand-alone sytems is diminishing each year.*

Figure 9-43. *Example of a head-up display (top) and a head-down display (bottom). The head-up display presents information in front of the pilot along his/her normal field of view while a head-down display may present information beyond the normal head-up field of view.*

unit transmits a coded pulse sequence that actuates the aircraft transponder. The transponder answers the coded sequence by transmitting a preselected coded sequence back to the ground equipment, providing a strong return signal and positive aircraft identification, as well as other special data such as aircraft altitude.

Functions of Radar Navigation

The radar systems used by ATC are air route surveillance radar (ARSR), airport surveillance radar (ASR), and precision approach radar (PAR) and airport surface detection equipment (ASDE). Surveillance radars scan through 360° of azimuth and present target information on a radar display located in a tower or center. This information is used independently or in conjunction with other navigational aids in the control of air traffic.

ARSR is a long-range radar system designed primarily to cover large areas and provide a display of aircraft while en route between terminal areas. The ARSR enables air route traffic control center (ARTCC) controllers to provide radar service when the aircraft are within the ARSR coverage. In some instances, ARSR may enable ARTCC to provide terminal radar services similar to but usually more limited than those provided by a radar approach control.

ASR is designed to provide relatively short-range coverage in the general vicinity of an airport and to serve as an expeditious means of handling terminal area traffic through observation of precise aircraft locations on a radarscope. Nonprecision instrument approaches are available at airports that have an approved surveillance radar approach procedure. ASR provides radar vectors to the final approach course and then azimuth information to the pilot during the approach. In addition to range (distance) from the runway, the pilot is advised of MDA, when to begin descent, and when the aircraft is at the MDA. If requested, recommended altitudes are furnished each mile while on final.

PAR is designed to be used as a landing aid displaying range, azimuth, and elevation information rather than as an aid for sequencing and spacing aircraft. PAR equipment may be used as a primary landing aid, or it may be used to monitor other types of approaches. Two antennas are used in the PAR array: one scanning a vertical plane and the other scanning horizontally. Since the range is limited to 10 miles, azimuth to 20°, and elevation to 7°, only the final approach area is covered. The controller's scope is divided into two parts. The upper half presents altitude and distance information, and the lower half presents azimuth and distance.

PAR is a system in which a controller provides highly accurate navigational guidance in azimuth and elevation to a pilot. Pilots are given headings to fly to direct them to and keep their aircraft aligned with the extended centerline of the landing runway. They are told to anticipate glidepath interception approximately 10–30 seconds before it occurs and when to start descent. The published decision height (DH) is given only if the pilot requests it. If the aircraft is observed to deviate above or below the glidepath, the pilot is given the relative amount of deviation by use of terms "slightly" or "well" and is expected to adjust the aircraft's rate of descent/ascent to return to the glidepath. Trend information is also issued with respect to the elevation of the aircraft and may be modified by the terms "rapidly" and "slowly" (e.g., "well above glidepath, coming down rapidly").

Range from touchdown is given at least once each mile. If an aircraft is observed by the controller to proceed outside of specified safety zone limits in azimuth and/or elevation and continue to operate outside these prescribed limits, the pilot will be directed to execute a missed approach or to fly a specified course unless the pilot has the runway environment (runway, approach lights, etc.) in sight. Navigational guidance in azimuth and elevation is provided to the pilot until the aircraft reaches the published decision altitude (DA)/DH. Advisory course and glidepath information is furnished by the controller until the aircraft passes over the landing threshold, at which point the pilot is advised of any deviation from the runway centerline. Radar service is automatically terminated upon completion of the approach.

Airport Surface Detection Equipment

Radar equipment is specifically designed to detect all principal features on the surface of an airport, including aircraft and vehicular traffic, and to present the entire image on a radar indicator console in the control tower. It is used to augment visual observation by tower personnel of aircraft and/or vehicular movements on runways and taxiways.

Radar Limitations

1. It is very important for the aviation community to recognize the fact that there are limitations to radar service and that ATC may not always be able to issue traffic advisories concerning aircraft which are not under ATC control and cannot be seen on radar.
2. The characteristics of radio waves are such that they normally travel in a continuous straight line unless they are "bent" by abnormal atmospheric phenomena such as temperature inversions; reflected or attenuated by dense objects such as heavy clouds, precipitation, ground obstacles, mountains, etc.; or screened by high terrain features.
3. Primary radar energy that strikes dense objects is reflected and displayed on the operator's scope, thereby blocking out aircraft at the same range and greatly weakening or completely eliminating the display of targets at a greater range.
4. Relatively low altitude aircraft are not seen if they are screened by mountains or are below the radar beam due to curvature of the Earth.
5. The amount of reflective surface of an aircraft determines the size of the radar return. Therefore, a small light airplane or a sleek jet fighter is more difficult to see on primary radar than a large commercial jet or military bomber.
6. All ARTCC radar in the conterminous United States and many ASR have the capability to interrogate Mode C and display altitude information to the controller from appropriately-equipped aircraft. However, a number of ASR do not have Mode C display capability; therefore, altitude information must be obtained from the pilot.

Chapter 10

IFR Flight

Introduction

This chapter is a discussion of conducting a flight under instrument flight rules (IFR). It also explains the sources for flight planning, the conditions associated with instrument flight, and the procedures used for each phase of IFR flight: departure, en route, and approach. The chapter concludes with an example of an IFR flight that applies many of the procedures discussed in the chapter.

Sources of Flight Planning Information

The following resources are available for a pilot planning a flight conducted under IFR.

National Aeronautical Navigation Products (AeroNav Products) Group publications:

- IFR en route charts
- Area charts
- United States Terminal Procedures Publications (TPP)

The Federal Aviation Administration (FAA) publications:

- Aeronautical Information Manual (AIM)
- Airport/Facility Directory (A/FD)
- Notices to Airmen Publication (NTAP) for flight planning in the National Airspace System (NAS)

Pilots should also consult the Pilot's Operating Handbook/Airplane Flight Manual (POH/AFM) for flight planning information pertinent to the aircraft to be flown.

A review of the contents of all the listed publications helps determine which material should be referenced for each flight. As a pilot becomes more familiar with these publications, the flight planning process becomes quicker and easier.

Aeronautical Information Manual (AIM)

The AIM provides the aviation community with basic flight information and air traffic control (ATC) procedures used in the United States NAS. An international version called the Aeronautical Information Publication contains parallel information, as well as specific information on the international airports used by the international community.

Airport/Facility Directory (A/FD)

The A/FD contains information on airports, communications, and navigation aids (NAVAIDs) pertinent to IFR flight. It also includes very-high frequency omnidirectional range (VOR) receiver checkpoints, flight service station (FSS), weather service telephone numbers, and air route traffic control center (ARTCC) frequencies. Various special notices essential to flight are also included, such as land-and-hold-short operations (LAHSO) data, the civil use of military fields, continuous power facilities, and special flight procedures.

In the major terminal and en route environments, preferred routes have been established to guide pilots in planning their routes of flight, to minimize route changes, and to aid in the orderly management of air traffic using the Federal airways. The A/FD lists both high and low altitude preferred routes.

Notices to Airmen Publication (NTAP)

The NTAP is a publication containing current Notices to Airmen (NOTAMs) that are essential to the safety of flight, as well as supplemental data affecting the other operational publications listed. It also includes current Flight Data Center (FDC) NOTAMs, which are regulatory in nature, issued to establish restrictions to flight or to amend charts or published instrument approach procedures (IAPs).

POH/AFM

The POH/AFM contain operating limitations, performance, normal and emergency procedures, and a variety of other operational information for the respective aircraft. Aircraft manufacturers have done considerable testing to gather and substantiate the information in the aircraft manual. Pilots should refer to it for information relevant to a proposed flight.

IFR Flight Plan

As specified in Title 14 of the Code of Federal Regulations (14 CFR) part 91, no person may operate an aircraft in controlled airspace under IFR unless that person has filed an IFR flight plan. Flight plans may be submitted to the nearest FSS or air traffic control tower (ATCT) either in person, by telephone (1-800-WX-BRIEF), by computer (using the direct user access terminal system (DUATS)), or by radio if no other means are available. Pilots should file IFR flight plans at least 30 minutes prior to estimated time of departure to preclude possible delay in receiving a departure clearance from ATC. The AIM provides guidance for completing and filing FAA Form 7233-1, Flight Plan. These forms are available at flight service stations (FSSs) and are generally found in flight planning rooms at airport terminal buildings. *[Figure 10-1]*

Filing in Flight

IFR flight plans may be filed from the air under various conditions, including:

1. A flight outside controlled airspace before proceeding into IFR conditions in controlled airspace.
2. A visual flight rules (VFR) flight expecting IFR weather conditions en route in controlled airspace.

In either of these situations, the flight plan may be filed with the nearest FSS or directly with the ARTCC. A pilot who files with the FSS submits the information normally entered during preflight filing, except for "point of departure," together with present position and altitude. FSS then relays this information to the ARTCC. The ARTCC then clears the pilot from present position or from a specified navigation fix.

Form Approved OMB No. 2120-0026

U.S. DEPARTMENT OF TRANSPORTATION
FEDERAL AVIATION ADMINISTRATION

FLIGHT PLAN

(FAA USE ONLY) ☐ PILOT BRIEFING ☐ VNR ☐ STOPOVER | TIME STARTED | SPECIALIST INITIALS

1. TYPE	2. AIRCRAFT IDENTIFICATION	3. AIRCRAFT TYPE/ SPECIAL EQUIPMENT	4. TRUE AIRSPEED	5. DEPARTURE POINT	6. DEPARTURE TIME		7. CRUISING ALTITUDE
					PROPOSED (Z)	ACTUAL (Z)	
VFR							
IFR							
DVFR			KTS				

8. ROUTE OF FLIGHT

9. DESTINATION (Name of airport and city)	10. EST. TIME ENROUTE		11. REMARKS
	HOURS	MINUTES	

12. FUEL ON BOARD		13. ALTERNATE AIRPORT(S)	14. PILOT'S NAME, ADDRESS & TELEPHONE NUMBER & AIRCRAFT HOME BASE	15. NUMBER ABOARD
HOURS	MINUTES		17. DESTINATION CONTACT/TELEPHONE (OPTIONAL)	

16. COLOR OF AIRCRAFT

CIVIL AIRCRAFT PILOTS. 14 CFR Part 91 requires you file an IFR flight plan to operate under instrument flight rules in controlled airspace. Failure to file could result in a civil penalty not to exceed $1,000 for each violation (Section 901 of the Federal Aviation Act of 1958, as amended). Filing of a VFR flight plan is recommended as a good operating practice. See also Part 99 for requirements concerning DVFR flight plans.

FAA Form 7233-1 (8-82) CLOSE VFR FLIGHT PLAN WITH ____________ FSS ON ARRIVAL

Figure 10-1. *Flight plan form.*

A pilot who files directly with the ARTCC reports present position and altitude, and submits only the flight plan information normally relayed from the FSS to the ARTCC. Be aware that traffic saturation frequently prevents ARTCC personnel from accepting flight plans by radio. In such cases, a pilot is advised to contact the nearest FSS to file the flight plan.

Cancelling IFR Flight Plans

An IFR flight plan may be cancelled any time a pilot is operating in VFR conditions outside Class A airspace by stating "cancel my IFR flight plan" to the controller or air-to-ground station. After cancelling an IFR flight plan, the pilot should change to the appropriate air-to-ground frequency, transponder code as directed, and VFR altitude/flight level.

ATC separation and information services (including radar services, where applicable) are discontinued when an IFR flight plan is cancelled. If VFR radar advisory service is desired, a pilot must specifically request it. Be aware that other procedures may apply when cancelling an IFR flight plan within areas such as Class C or Class B airspace.

When operating on an IFR flight plan to an airport with an operating control tower, a flight plan is cancelled automatically upon landing. If operating on an IFR flight plan to an airport without an operating control tower, the pilot is responsible for cancelling the flight plan. This can be done by telephone after landing if there is no operating FSS or other means of direct communications with ATC. When there is no FSS or air-to-ground communications are not possible below a certain altitude, a pilot may cancel an IFR flight plan while still airborne and able to communicate with ATC by radio. If using this procedure, be certain the remainder of the flight can be conducted under VFR. It is essential that IFR flight plans be cancelled expeditiously. This allows other IFR traffic to utilize the airspace.

Clearances

An ATC clearance allows an aircraft to proceed under specified traffic conditions within controlled airspace for the purpose of providing separation between known aircraft. A major contributor to runway incursions is lack of communication with ATC and not understanding the instructions that they give. The primary way the pilot and ATC communicate is by voice. The safety and efficiency of taxi operations at airports with operating control towers depend on this communication loop. ATC uses standard phraseology and require readbacks and other responses from the pilot in order to verify that clearances and instructions are understood. In order to complete the communication loop, the controllers must also clearly understand the pilot's readback and other responses. Pilots can help enhance the controller's understanding by responding appropriately and using standard phraseology. Regulatory requirements, the AIM, approved flight training programs, and operational manuals provide information for pilots on standard ATC phraseology and communications requirements.

Examples

A flight filed for a short distance at a relatively low altitude in an area of low traffic density might receive a clearance as follows:

> "Cessna 1230 Alpha, cleared to Doeville airport direct, cruise 5,000."

The term "cruise" in this clearance means a pilot is authorized to fly at any altitude from the minimum IFR altitude up to and including 5,000 feet and may level off at any altitude within this block of airspace. A climb or descent within the block may be made at the pilot's discretion. However, once a pilot reports leaving an altitude within the block, the pilot may not return to that altitude without further ATC clearance.

When ATC issues a cruise clearance in conjunction with an unpublished route, an appropriate crossing altitude is specified to ensure terrain clearance until the aircraft reaches a fix, point, or route where the altitude information is available. The crossing altitude ensures IFR obstruction clearance to the point at which the aircraft enters a segment of a published route or IAP.

Once a flight plan is filed, ATC issues the clearance with appropriate instructions, such as the following:

> "Cessna 1230 Alpha is cleared to Skyline airport via the Crossville 055 radial, Victor 18, maintain 5,000. Clearance void if not off by 1330."

Or a more complex clearance, such as:

> "Cessna 1230 Alpha is cleared to Wichita Mid-continent airport via Victor 77, left turn after takeoff, proceed direct to the Oklahoma City VORTAC. Hold west on the Oklahoma City 277 radial, climb to 5,000 in holding pattern before proceeding on course. Maintain 5,000 to CASHION intersection. Climb to and maintain 7,000. Departure control frequency will be 121.05, Squawk 0412."

Clearance delivery may issue the following "abbreviated clearance" which includes a departure procedure (DP):

> "Cessna 1230 Alpha, cleared to La Guardia as filed, RINGOES 8 departure Phillipsburg transition, maintain 8,000. Departure control frequency will be 120.4, Squawk 0700."

This clearance may be readily copied in shorthand as follows:

> "CAF RNGO8 PSB M80 DPC 120.4 SQ 0700."

The information contained in this DP clearance is abbreviated using clearance shorthand (see appendix 1). The pilot should know the locations of the specified navigation facilities, together with the route and point-to-point time, before accepting the clearance.

The DP enables a pilot to study and understand the details of a departure before filing an IFR flight plan. It provides the information necessary to set up communication and navigation equipment and be ready for departure before requesting an IFR clearance.

Once the clearance is accepted, a pilot is required to comply with ATC instructions. A clearance different from that issued may be requested if the pilot considers another course of action more practicable or if aircraft equipment limitations or other considerations make acceptance of the clearance inadvisable.

A pilot should also request clarification or amendment, as appropriate, any time a clearance is not fully understood or considered unacceptable for safety of flight. The pilot is responsible for requesting an amended clearance if ATC issues a clearance that would cause a pilot to deviate from a rule or regulation or would place the aircraft in jeopardy.

Clearance Separations

ATC provides the pilot on an IFR clearance with separation from other IFR traffic. This separation is provided:

1. Vertically—by assignment of different altitudes.
2. Longitudinally—by controlling time separation between aircraft on the same course.
3. Laterally—by assignment of different flightpaths.
4. By radar—including all of the above.

ATC does not provide separation for an aircraft operating:

1. Outside controlled airspace.
2. On an IFR clearance:
    a) With "VFR-On-Top" authorized instead of a specific assigned altitude.
    b) Specifying climb or descent in "VFR conditions."
    c) At any time in VFR conditions, since uncontrolled VFR flights may be operating in the same airspace.

In addition to heading and altitude assignments, ATC occasionally issues speed adjustments to maintain the required separations. For example:

"Cessna 30 Alpha, slow to 100 knots."

A pilot who receives speed adjustments is expected to maintain that speed plus or minus 10 knots. If for any reason the pilot is not able to accept a speed restriction, the pilot should advise ATC.

At times, ATC may also employ visual separation techniques to keep aircraft safely separated. A pilot who obtains visual contact with another aircraft may be asked to maintain visual separation or to follow the aircraft. For example:

"Cessna 30 Alpha, maintain visual separation with that traffic, climb and maintain 7,000."

The pilot's acceptance of instructions to maintain visual separation or to follow another aircraft is an acknowledgment that the aircraft is maneuvered as necessary to maintain safe separation. It is also an acknowledgment that the pilot accepts the responsibility for wake turbulence avoidance.

In the absence of radar contact, ATC relies on position reports to assist in maintaining proper separation. Using the data transmitted by the pilot, the controller follows the progress of each flight. ATC must correlate the pilots' reports to provide separation; therefore, the accuracy of each pilot's report can affect the progress and safety of every other aircraft operating in the area on an IFR flight plan.

Departure Procedures (DPs)

Instrument departure procedures are preplanned IFR procedures that provide obstruction clearance from the terminal area to the appropriate en route structure and provide the pilot with a way to depart the airport and transition to the en route structure safely. Pilots operating under 14 CFR part 91 are strongly encouraged to file and fly a DP when one is available. *[Figure 10-2]*

There are two types of DPs: Obstacle Departure Procedures (ODP), printed either textually or graphically, and Standard Instrument Departures (SID), always printed graphically. All DPs, either textual or graphic, may be designed using either conventional or area navigation (RNAV) criteria. RNAV procedures have RNAV printed in the title (e.g., SHEAD TWO DEPARTURE (RNAV)).

Obstacle Departure Procedures (ODP)

ODPs provide obstruction clearance via the least onerous route from the terminal area to the appropriate en route structure. ODPs are recommended for obstruction clearance and may be flown without ATC clearance unless an alternate departure procedure (SID or radar vector) has been specifically assigned by ATC. Graphic ODPs have (OBSTACLE) printed in the procedure title (e.g., GEYSR THREE DEPARTURE (OBSTACLE), CROWN ONE DEPARTURE (RNAV)(OBSTACLE)).

Standard Instrument Departures

SIDs are ATC procedures printed for pilot/controller use in graphic form to provide obstruction clearance and a transition from the terminal area to the appropriate en route structure. SIDs are primarily designed for system enhancement and to reduce pilot/controller workload. ATC clearance must be received prior to flying a SID.

ODPs are found in section C of each booklet published regionally by the AeroNav Products, TPP, along with "IFR Take-off Minimums" while SIDs are collocated with the approach procedures for the applicable airport. Additional information on the development of DPs can be found in paragraph 5-2-7 of the AIM. However, the following points are important to remember.

1. The pilot of IFR aircraft operating from locations where DP procedures are effective may expect an ATC clearance containing a DP. The use of a DP requires pilot possession of at least the textual description of the approved DP.
2. If a pilot does not possess a preprinted DP or for any other reason does not wish to use a DP, he or she is expected to advise ATC. Notification may be accomplished by filing "NO DP" in the remarks section of the filed flight plan or by advising ATC.
3. If a DP is accepted in a clearance, a pilot must comply with it.

Radar-Controlled Departures

On IFR departures from airports in congested areas, a pilot normally receives navigational guidance from departure control by radar vector. When a departure is to be vectored immediately following takeoff, the pilot is advised before takeoff of the initial heading to be flown. This information is vital in the event of a loss of two-way radio communications during departure.

The radar departure is normally simple. Following takeoff, contact departure control on the assigned frequency when advised to do so by the control tower. At this time, departure control verifies radar contact and gives headings, altitude, and climb instructions to move an aircraft quickly and safely out of the terminal area. A pilot is expected to fly the assigned headings and altitudes until informed by the controller of the aircraft's position with respect to the route given in

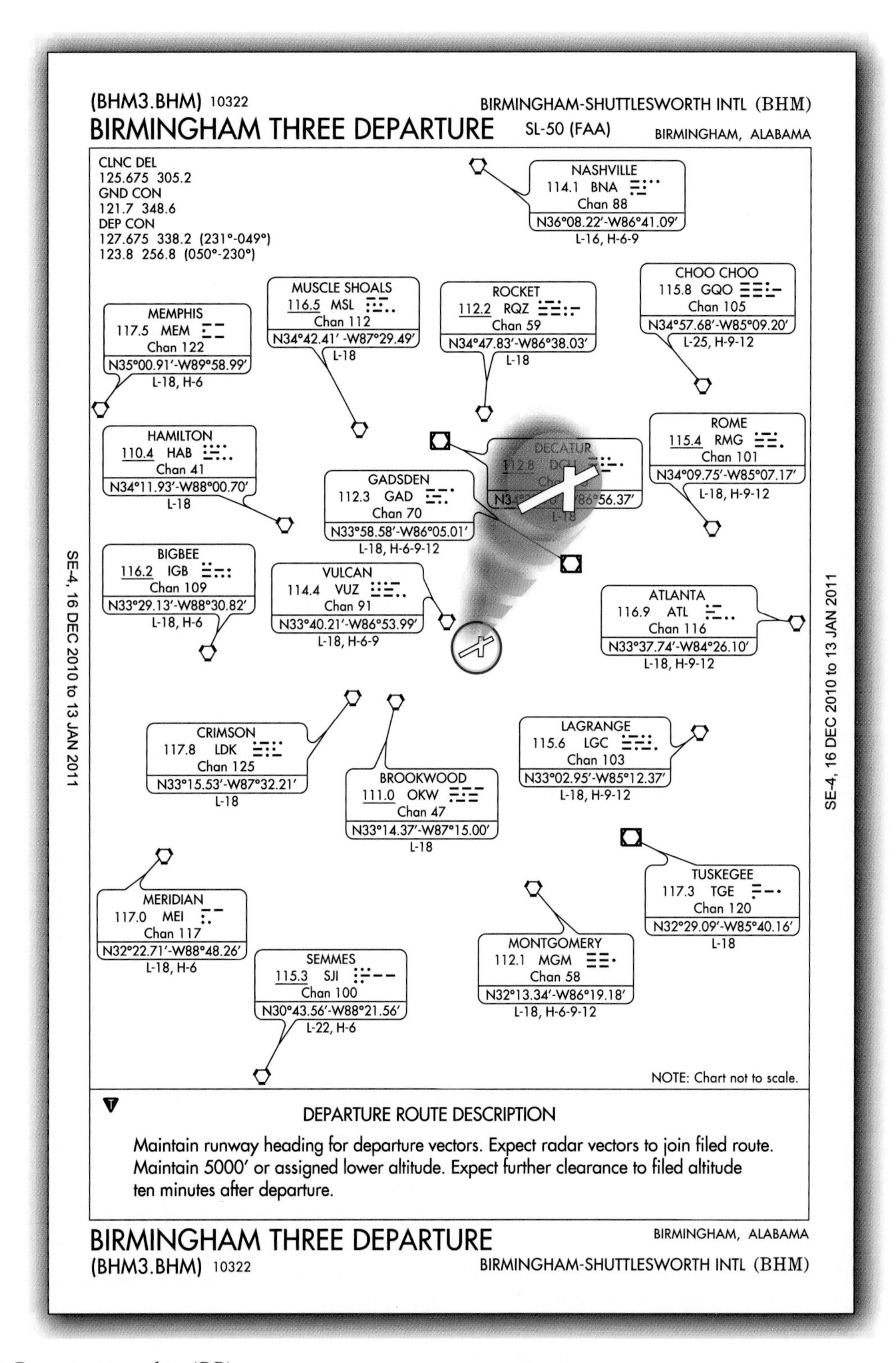

Figure 10-2. *Departure procedure (DP).*

the clearance, whom to contact next, and to "resume own navigation."

Departure control provides vectors to either a navigation facility, or an en route position appropriate to the departure clearance, or transfer to another controller with further radar surveillance capabilities. *[Figure 10-2]*

A radar controlled departure does not relieve the pilot of responsibilities as pilot-in-command. Be prepared before takeoff to conduct navigation according to the ATC clearance with navigation receivers checked and properly tuned. While under radar control, monitor instruments to ensure continuous orientation to the route specified in the clearance and record the time over designated checkpoints.

Departures From Airports Without an Operating Control Tower

When departing from airports that have neither an operating tower nor an FSS, a pilot should telephone the flight plan to the nearest ATC facility at least 30 minutes before the estimated departure time. If weather conditions permit, depart VFR and request IFR clearance as soon as radio contact is established with ATC.

If weather conditions make it undesirable to fly VFR, telephone clearance request. In this case, the controller would probably issue a short-range clearance pending establishment of radio contact and might restrict the departure time to a certain period. For example:

"Clearance void if not off by 0900."

This would authorize departure within the allotted period and permit a pilot to proceed in accordance with the clearance. In the absence of any specific departure instructions, a pilot would be expected to proceed on course via the most direct route.

En Route Procedures

Procedures en route vary according to the proposed route, the traffic environment, and the ATC facilities controlling the flight. Some IFR flights are under radar surveillance and controlled from departure to arrival and others rely entirely on pilot navigation.

Where ATC has no jurisdiction, it does not issue an IFR clearance. It has no control over the flight, nor does the pilot have any assurance of separation from other traffic.

ATC Reports

All pilots are required to report unforecast weather conditions or other information related to safety of flight to ATC. The pilot-in-command of each aircraft operated in controlled airspace under IFR shall report as soon as practical to ATC any malfunctions of navigational, approach, or communication equipment occurring in flight:

1. Loss of VOR, tactical air navigation (TACAN) or automatic direction finder (ADF) receiver capability.
2. Complete or partial loss of instrument landing system (ILS) receiver capability.
3. Impairment of air-to-ground communications capability.

The pilot-in-command shall include within the report (1) aircraft identification, (2) equipment affected, (3) degree to which the pilot to operate under IFR within the ATC system is impaired, and (4) nature and extent of assistance desired from ATC.

Position Reports

Position reports are required over each compulsory reporting point (shown on the chart as a solid triangle) along the route being flown regardless of altitude, including those with a VFR-on-top clearance. Along direct routes, reports are required of all IFR flights over each point used to define the route of flight. Reports at reporting points (shown as an open triangle) are made only when requested by ATC. A pilot should discontinue position reporting over designated reporting points when informed by ATC that the aircraft is in "RADAR CONTACT." Position reporting should be resumed when ATC advises "RADAR CONTACT LOST" or "RADAR SERVICE TERMINATED."

Position reports should include the following items:

1. Identification
2. Position
3. Time
4. Altitude or flight level (include actual altitude or flight level when operating on a clearance specifying VFR-on-top)
5. Type of flight plan (not required in IFR position reports made directly to ARTCCs or approach control)
6. Estimated time of arrival (ETA) and name of next reporting point
7. The name only of the next succeeding reporting point along the route of flight
8. Pertinent remarks

En route position reports are submitted normally to the ARTCC controllers via direct controller-to-pilot communications channels using the appropriate ARTCC frequencies listed on the en route chart.

Whenever an initial contact with a controller is to be followed by a position report, the name of the reporting point should be included in the call-up. This alerts the controller that such information is forthcoming. For example:

"Atlanta Center, Cessna 1230 Alpha at JAILS intersection."

"Cessna 1230 Alpha Atlanta Center."

"Atlanta Center, Cessna 1230 Alpha at JAILS intersection, 5,000, estimating Monroeville at 1730."

Additional Reports

In addition to required position reports, the following reports should be made to ATC without a specific request.

1. At all times:

a) When vacating any previously assigned altitude or flight level for a newly assigned altitude or flight level

b) When an altitude change is made if operating on a clearance specifying VFR-on-top

c) When unable to climb/descend at a rate of at least 500 feet per minute (fpm)

d) When an approach has been missed (Request clearance for specific action (to alternative airport, another approach, etc.))

e) Change in average true airspeed (at cruising altitude) when it varies by 5 percent or 10 knots (whichever is greater) from that filed in the flight plan

f) The time and altitude upon reaching a holding fix or point to which cleared

g) When leaving any assigned holding fix or point

NOTE: The reports in (f) and (g) may be omitted by pilots of aircraft involved in instrument training at military terminal area facilities when radar service is being provided.

h) Any loss in controlled airspace of VOR, TACAN, ADF, low frequency navigation receiver capability, global positioning system (GPS) anomalies while using installed IFR-certified GPS/Global Navigation Satellite Systems (GNSS) receivers, complete or partial loss of ILS receiver capability, or impairment of air/ground communications capability. Reports should include aircraft identification, equipment affected, degree to which the capability to operate under IFR in the ATC system is impaired, and the nature and extent of assistance desired from ATC.

i) Any information relating to the safety of flight.

2. When not in radar contact:

a) When leaving the final approach fix inbound on final approach (nonprecision approach), or when leaving the outer marker or fix used in lieu of the outer marker inbound on final approach (precision approach).

b) A corrected estimate at any time it becomes apparent that an estimate as previously submitted is in error in excess of 3 minutes.

Any pilot who encounters weather conditions that have not been forecast, or hazardous conditions which have been forecast, is expected to forward a report of such weather to ATC.

Planning the Descent and Approach

ATC arrival procedures and flight deck workload are affected by weather conditions, traffic density, aircraft equipment, and radar availability.

When landing at an airport with approach control services and where two or more IAPs are published, information on the type of approach to expect is provided in advance of arrival or vectors are provided to a visual approach. This information is broadcast either on automated terminal information service (ATIS) or by a controller. It is not furnished when the visibility is 3 miles or more and the ceiling is at or above the highest initial approach altitude established for any low altitude IAP for the airport.

The purpose of this information is to help the pilot plan arrival actions; however, it is not an ATC clearance or commitment and is subject to change. Fluctuating weather, shifting winds, blocked runway, etc., are conditions that may result in changes to the approach information previously received. It is important for a pilot to advise ATC immediately if he or she is unable to execute the approach or prefers another type of approach.

If the destination is an airport without an operating control tower and has automated weather data with broadcast capability, the pilot should monitor the automated surface observing system/automated weather observing system (ASOS/AWOS) frequency to ascertain the current weather for the airport. ATC should be advised that weather information has been received and what the pilot's intentions are.

When the approach to be executed has been determined, the pilot should plan for and request a descent to the appropriate altitude prior to the initial approach fix (IAF) or transition route depicted on the IAP. When flying the transition route, a pilot should maintain the last assigned altitude until ATC gives the instructions "cleared for the approach." Lower altitudes can be requested to bring the transition route altitude closer to the required altitude at the initial approach fix. When ATC uses the phrase "at pilot's discretion" in the altitude information of a clearance, the pilot has the option to start a descent at any rate and may level off temporarily at any intermediate altitude. However, once an altitude has been vacated, return to that altitude is not authorized without a clearance. When ATC has not used the term "at pilot's discretion" nor imposed any descent restrictions, initiate descent promptly upon acknowledgment of the clearance.

Descend at an optimum rate (consistent with the operating characteristics of the aircraft) to 1,000 feet above the assigned altitude. Then attempt to descend at a rate of between 500 and 1,500 fpm until the assigned altitude is reached. If at anytime a pilot is unable to maintain a descent rate of at least 500 fpm, advise ATC. Also advise ATC if it is necessary to level off at an intermediate altitude during descent. An exception to this is when leveling off at 10,000 feet mean sea level (MSL) on descent or 2,500 feet above airport elevation (prior to entering a Class B, Class C, or Class D surface area) when required for speed reduction.

Standard Terminal Arrival Routes (STARs)

Standard Terminal Arrival Routes (STARs) (as described in Chapter 1) have been established to simplify clearance delivery procedures for arriving aircraft at certain areas having high density traffic. A STAR serves a purpose parallel to that of a DP for departing traffic. *[Figure 10-3]*

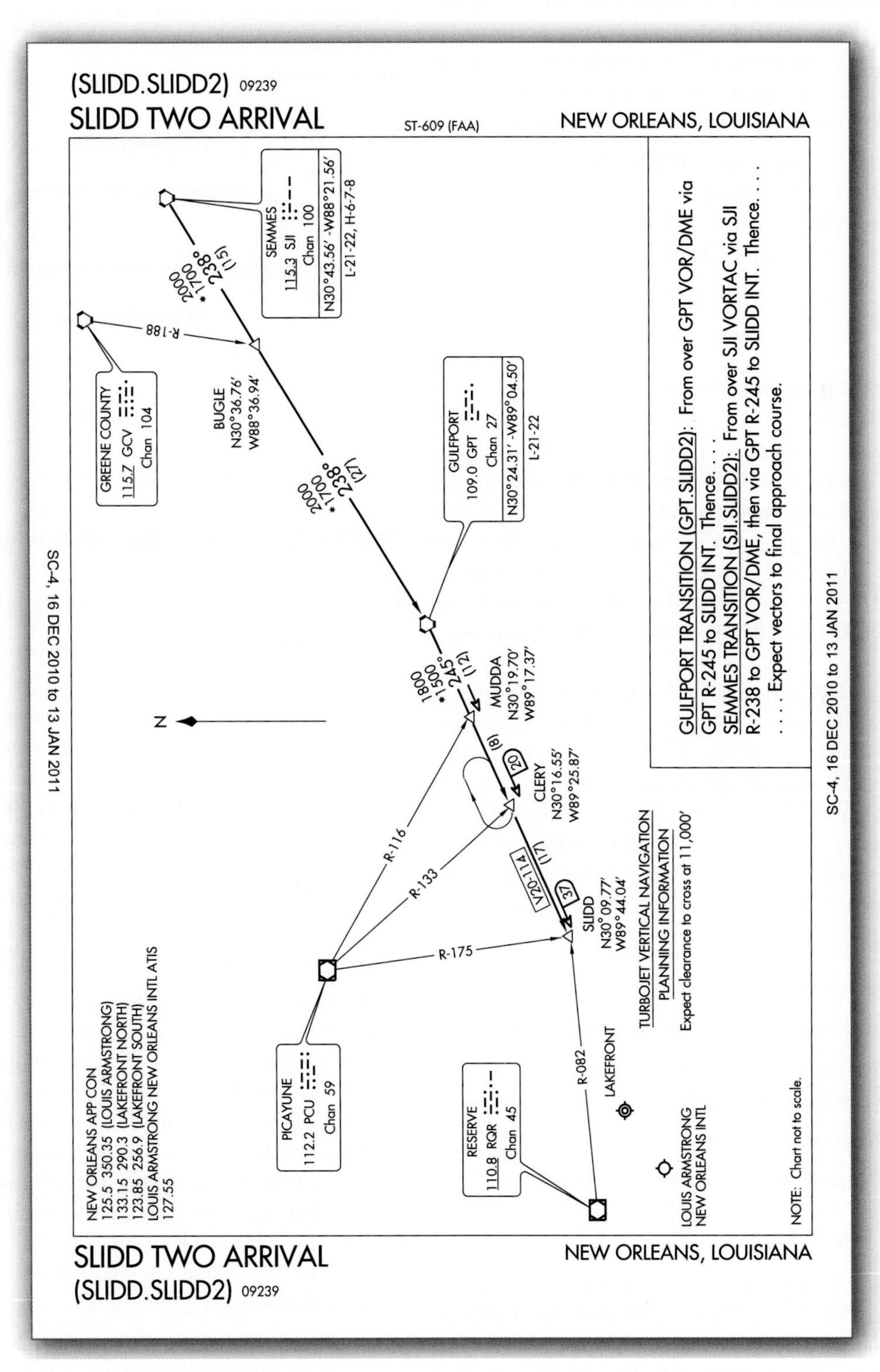

Figure 10-3. *Standard terminal arrival route (STAR).*

The following points regarding STARs are important to remember:

1. All STARs are contained in the Terminal Procedures Publication (TPP), along with the IAP charts for the destination airport. The AIM also describes STAR procedures.
2. If the destination is a location for which STARs have been published, a pilot may be issued a clearance containing a STAR whenever ATC deems it appropriate. To accept the clearance, a pilot must possess at least the approved textual description.
3. It is the pilot's responsibility to either accept or refuse an issued STAR. If a STAR will not or cannot be used, advise ATC by placing "NO STAR" in the remarks section of the filed flight plan or by advising ATC.
4. If a STAR is accepted in a clearance, compliance is mandatory.

Substitutes for Inoperative or Unusable Components

The basic ground components of an ILS are the localizer, glideslope, outer marker, middle marker, and inner marker (when installed). A compass locator or precision radar may be substituted for the outer or middle marker. Distance measuring equipment (DME), VOR, or nondirectional beacon (NDB) fixes authorized in the standard IAP or surveillance radar may be substituted for the outer marker.

Additionally, IFR-certified GPS equipment, operated in accordance with Advisory Circular (AC) 90-94, Guidelines for Using Global Positioning System Equipment for IFR En Route and Terminal Operations and for Nonprecision Instrument Approaches in the United States National Airspace System, may be substituted for ADF and DME equipment, except when flying NDB IAP. Specifically, GPS can be substituted for ADF and DME equipment when:

1. Flying a DME arc;
2. Navigating TO/FROM an NDB;
3. Determining the aircraft position over an NDB;
4. Determining the aircraft position over a fix made up of a crossing NDB bearing;
5. Holding over an NDB;
6. Determining aircraft position over a DME fix.

Holding Procedures

Depending upon traffic and weather conditions, holding may be required. Holding is a predetermined maneuver that keeps aircraft within a specified airspace while awaiting further clearance from ATC. A standard holding pattern uses right turns, and a nonstandard holding pattern uses left turns. The ATC clearance always specifies left turns when a nonstandard pattern is to be flown.

Standard Holding Pattern (No Wind)

In a standard holding pattern with no winds *[Figure 10-4]*, the aircraft follows the specified course inbound to the holding fix, turns 180° to the right, flies a parallel straight course outbound for 1 minute, turns 180° to the right, and flies the inbound course to the fix.

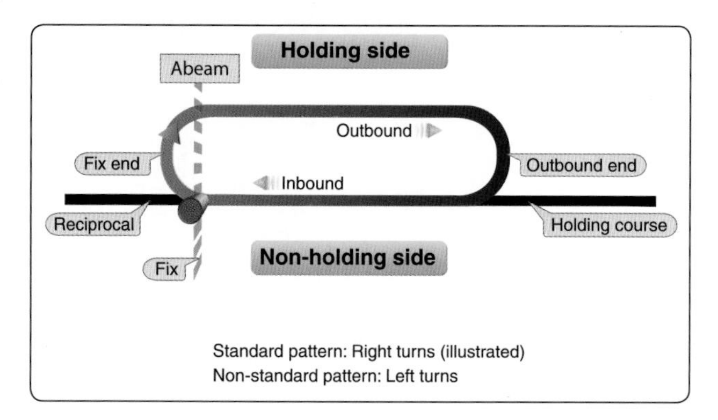

Figure 10-4. *Standard holding pattern—no wind.*

Standard Holding Pattern (With Wind)

A standard symmetrical holding pattern cannot be flown when winds exist. In those situations, the pilot is expected to:

1. Compensate for the effect of a known wind except when turning.
2. Adjust outbound timing to achieve a 1-minute (1½ minutes above 14,000 feet) inbound leg.

Figure 10-5 illustrates the holding track followed with a left crosswind. The effect of wind is counteracted by applying drift corrections to the inbound and outbound legs and by applying time allowances to the outbound leg.

Holding Instructions

If an aircraft arrives at a clearance limit before receiving clearance beyond the fix, ATC expects the pilot to maintain the last assigned altitude and begin holding in accordance with the charted holding pattern. If no holding pattern is charted and holding instructions have not been issued, enter a standard holding pattern on the course on which the aircraft approached the fix and request further clearance as soon as possible. Normally, when no delay is anticipated, ATC issues holding instructions at least 5 minutes before the estimated arrival at the fix. Where a holding pattern is not charted, the ATC clearance specifies the following:

1. Direction of holding from the fix in terms of the eight cardinal compass points (N, NE, E, SE, etc.)

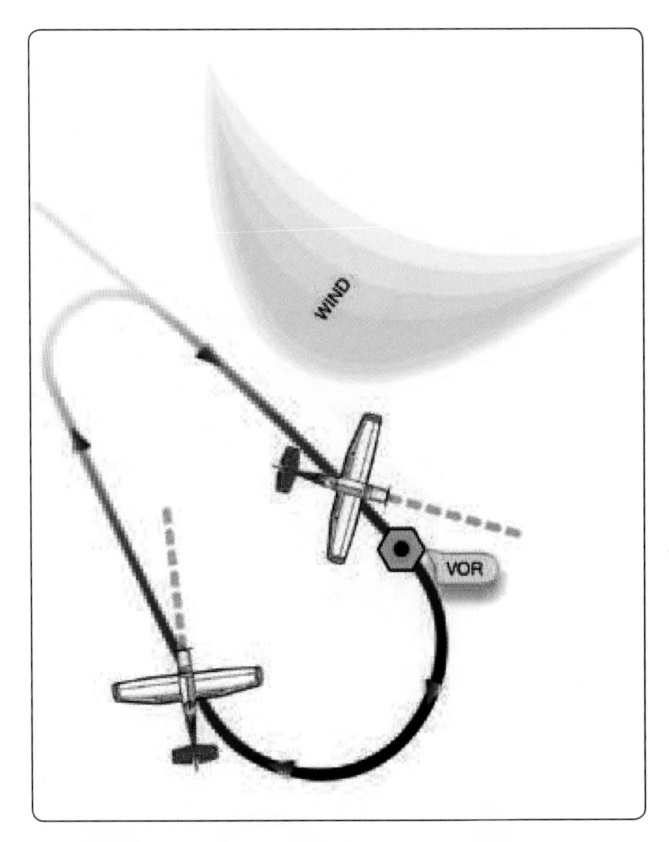

Figure 10-5. *Drift correction in holding pattern.*

2. Holding fix (the fix may be omitted if included at the beginning of the transmission as the clearance limit)
3. Radial, course, bearing, airway, or route on which the aircraft is to hold.
4. Leg length in miles if DME or RNAV is to be used (leg length is specified in minutes on pilot request or if the controller considers it necessary).
5. Direction of turn, if left turns are to be made, because the pilot requests or the controller considers it necessary.
6. Time to expect-further-clearance (EFC) and any pertinent additional delay information.

ATC instructions are also issued whenever:

1. It is determined that a delay will exceed 1 hour.
2. A revised EFC is necessary.
3. In a terminal area having a number of NAVAIDs and approach procedures, a clearance limit may not indicate clearly which approach procedures will be used. On initial contact, or as soon as possible thereafter, approach control advises the pilot of the type of approach to expect.
4. Ceiling and/or visibility is reported as being at or below the highest "circling minimums" established for the airport concerned. ATC transmits a report of current weather conditions and subsequent changes, as necessary.
5. An aircraft is holding while awaiting approach clearance, and the pilot advises ATC that reported weather conditions are below minimums applicable to the operation. In this event, ATC issues suitable instructions to aircraft desiring either to continue holding while awaiting weather improvement or proceed to another airport.

Standard Entry Procedures

The entry procedures given in the AIM evolved from extensive experimentation under a wide range of operational conditions. The standardized procedures should be followed to ensure that an aircraft remains within the boundaries of the prescribed holding airspace.

When a speed reduction is required, start the reduction when 3 minutes or less from the holding fix. Cross the holding fix initially at or below the maximum holding airspeed (MHA). The purpose of the speed reduction is to prevent overshooting the holding airspace limits, especially at locations where adjacent holding patterns are close together.

All aircraft may hold at the following altitudes and maximum holding airspeeds:

Altitude Mean Sea Level (MSL)	*Airspeed (KIAS)*
Up to 6,000 feet	200
6,001 – 14,000 feet	230
14,001 feet and above	265

The following are exceptions to the maximum holding airspeeds:

1. Holding patterns from 6,001 to 14,000 feet may be restricted to a maximum airspeed of 210 knots indicated airspeed (KIAS). This nonstandard pattern is depicted by an icon.
2. Holding patterns may be restricted to a maximum airspeed of 175 KIAS. This nonstandard pattern is depicted by an icon. Holding patterns restricted to 175 KIAS are generally found on IAPs applicable to category A and B aircraft only.
3. Holding patterns at Air Force airfields only—310 KIAS maximum, unless otherwise depicted.
4. Holding patterns at Navy airfields only—230 KIAS maximum, unless otherwise depicted.
5. The pilot of an aircraft unable to comply with maximum airspeed restrictions should notify ATC.

While other entry procedures may enable the aircraft to enter the holding pattern and remain within protected airspace, the parallel, teardrop, and direct entries are the procedures for entry and holding recommended by the FAA. Additionally, paragraph 5-3-7 in the AIM should be reviewed. *[Figure 10-6]*

1. Parallel Procedure. When approaching the holding fix from anywhere in sector (a), fly to the fix. Afterwards, turn to a heading to parallel the holding course outbound. Fly outbound for 1 minute, turn in the direction of the holding pattern through more than 180°, and return to the holding fix or intercept the holding course inbound.
2. Teardrop Procedure. When approaching the holding fix from anywhere in sector (b), the teardrop entry procedure would be to fly to the fix, turn outbound to a heading for a 30° teardrop entry within the pattern (on the holding side) for a period of 1 minute, then turn in the direction of the holding pattern to intercept the inbound holding course.
3. Direct Entry Procedure. When approaching the holding fix from anywhere in sector (c), the direct entry procedure would be to fly directly to the fix and turn to follow the holding pattern.

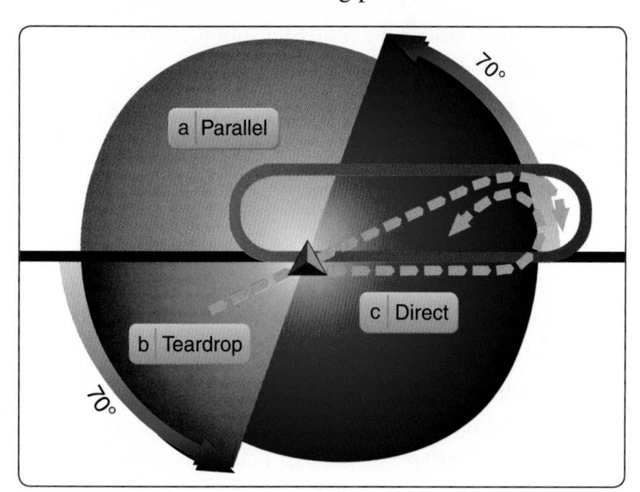

Figure 10-6. *Holding pattern entry procedures.*

A pilot should make all turns during entry and while holding at:

1. 3° per second, or
2. 30° bank angle, or
3. A bank angle provided by a flight director system.

Time Factors

The holding pattern entry time reported to ATC is the initial time of arrival over the fix. Upon entering a holding pattern, the initial outbound leg is flown for 1 minute at or below 14,000 feet MSL, and for 1½ minutes above 14,000 feet MSL. Timing for subsequent outbound legs should be adjusted as necessary to achieve proper inbound leg time. The pilot should begin outbound timing over or abeam the fix, whichever occurs later. If the abeam position cannot be determined, start timing when the turn to outbound is completed. *[Figure 10-7]*

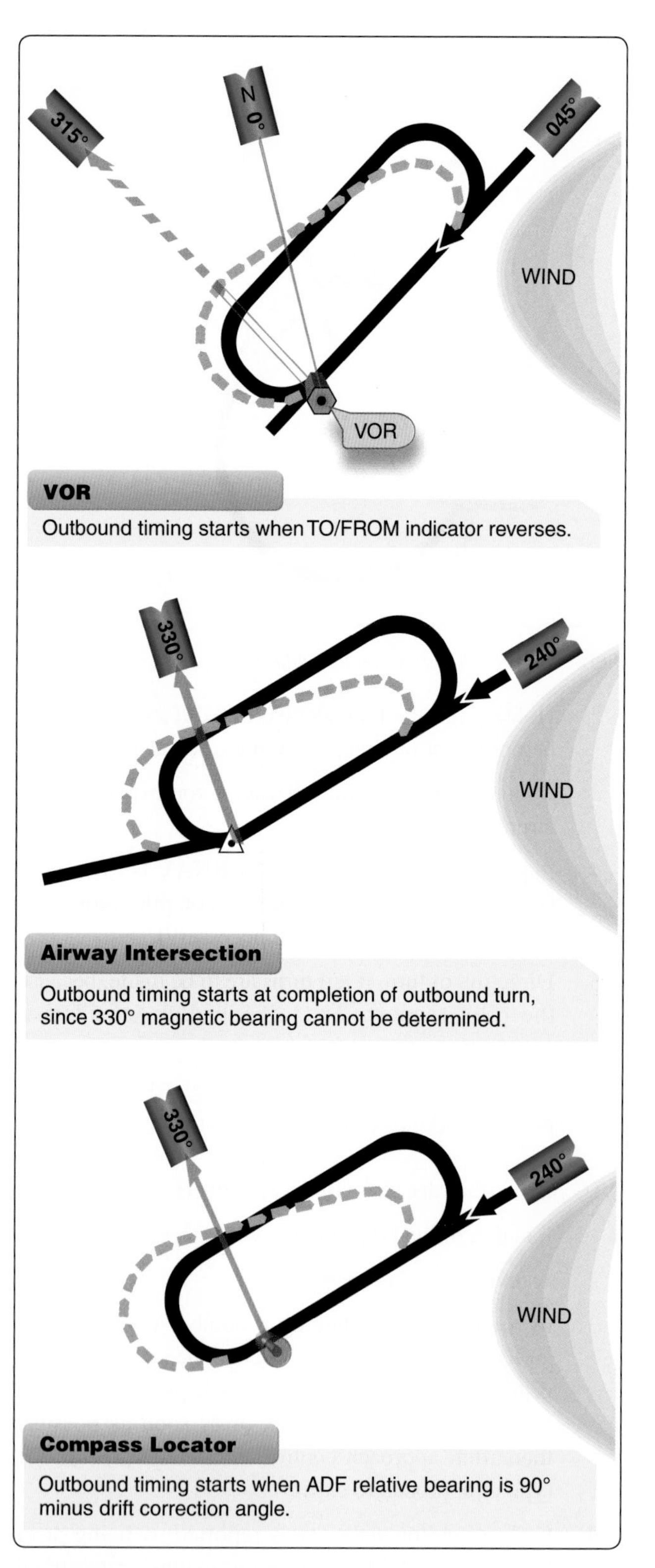

Figure 10-7. *Holding—outbound timing.*

Time leaving the holding fix must be known to ATC before succeeding aircraft can be cleared to the vacated airspace. Leave the holding fix:

1. When ATC issues either further clearance en route or approach clearance;
2. As prescribed in 14 CFR part 91 (for IFR operations; two-way radio communications failure, and responsibility and authority of the pilot-in-command); or
3. After the IFR flight plan has been cancelled, if the aircraft is holding in VFR conditions.

DME Holding

The same entry and holding procedures apply to DME holding, but distances (nautical miles) are used instead of time values. The length of the outbound leg is specified by the controller, and the end of this leg is determined by the DME readout.

Approaches

Compliance With Published Standard Instrument Approach Procedures

Compliance with the approach procedures shown on the approach charts provides necessary navigation guidance information for alignment with the final approach courses, as well as obstruction clearance. Under certain conditions, a course reversal maneuver or procedure turn may be necessary. However, this procedure is not authorized when:

1. The symbol "NoPT" appears on the approach course on the plan view of the approach chart.
2. Radar vectoring is provided to the final approach course.
3. A holding pattern is published in lieu of a procedure turn.
4. Executing a timed approach from a holding fix.
5. Otherwise directed by ATC.

Instrument Approaches to Civil Airports

Unless otherwise authorized, when an instrument letdown to an airport is necessary, the pilot should use a standard IAP prescribed for that airport. IAPs are depicted on IAP charts and are found in the TPP.

ATC approach procedures depend upon the facilities available at the terminal area, the type of instrument approach executed, and the existing weather conditions. The ATC facilities, NAVAIDs, and associated frequencies appropriate to each standard instrument approach are given on the approach chart. Individual charts are published for standard approach procedures associated with the following types of facilities:

1. Nondirectional beacon (NDB)
2. Very-high frequency omnirange (VOR)
3. Very-high frequency omnirange with distance measuring equipment (VORTAC or VOR/DME)
4. Localizer (LOC)
5. Instrument landing system (ILS)
6. Localizer-type directional aid (LDA)
7. Simplified directional facility (SDF)
8. Area navigation (RNAV)
9. Global positioning system (GPS)

An IAP can be flown in one of two ways: as a full approach or with the assistance of radar vectors. When the IAP is flown as a full approach, pilots conduct their own navigation using the routes and altitudes depicted on the instrument approach chart. A full approach allows the pilot to transition from the en route phase, to the instrument approach, and then to a landing with minimal assistance from ATC. This type of procedure may be requested by the pilot but is most often used in areas without radar coverage. A full approach also provides the pilot with a means of completing an instrument approach in the event of a communications failure.

When an approach is flown with the assistance of radar vectors, ATC provides guidance in the form of headings and altitudes, which position the aircraft to intercept the final approach. From this point, the pilot resumes navigation, intercepts the final approach course, and completes the approach using the IAP chart. This is often a more expedient method of flying the approach, as opposed to the full approach, and allows ATC to sequence arriving traffic. A pilot operating in radar contact can generally expect the assistance of radar vectors to the final approach course.

Approach to Airport Without an Operating Control Tower

Figure 10-8 shows an approach procedure at an airport without an operating control tower. When approaching such a facility, the pilot should monitor the AWOS/ASOS if available for the latest weather conditions. When direct communication between the pilot and controller is no longer required, the ARTCC or approach controller issues a clearance for an instrument approach and advises "change to advisory frequency approved." When the aircraft arrives on a "cruise" clearance, ATC does not issue further clearance for approach and landing.

If an approach clearance is required, ATC authorizes the pilots to execute his or her choice of standard instrument approach (if more than one is published for the airport) with the

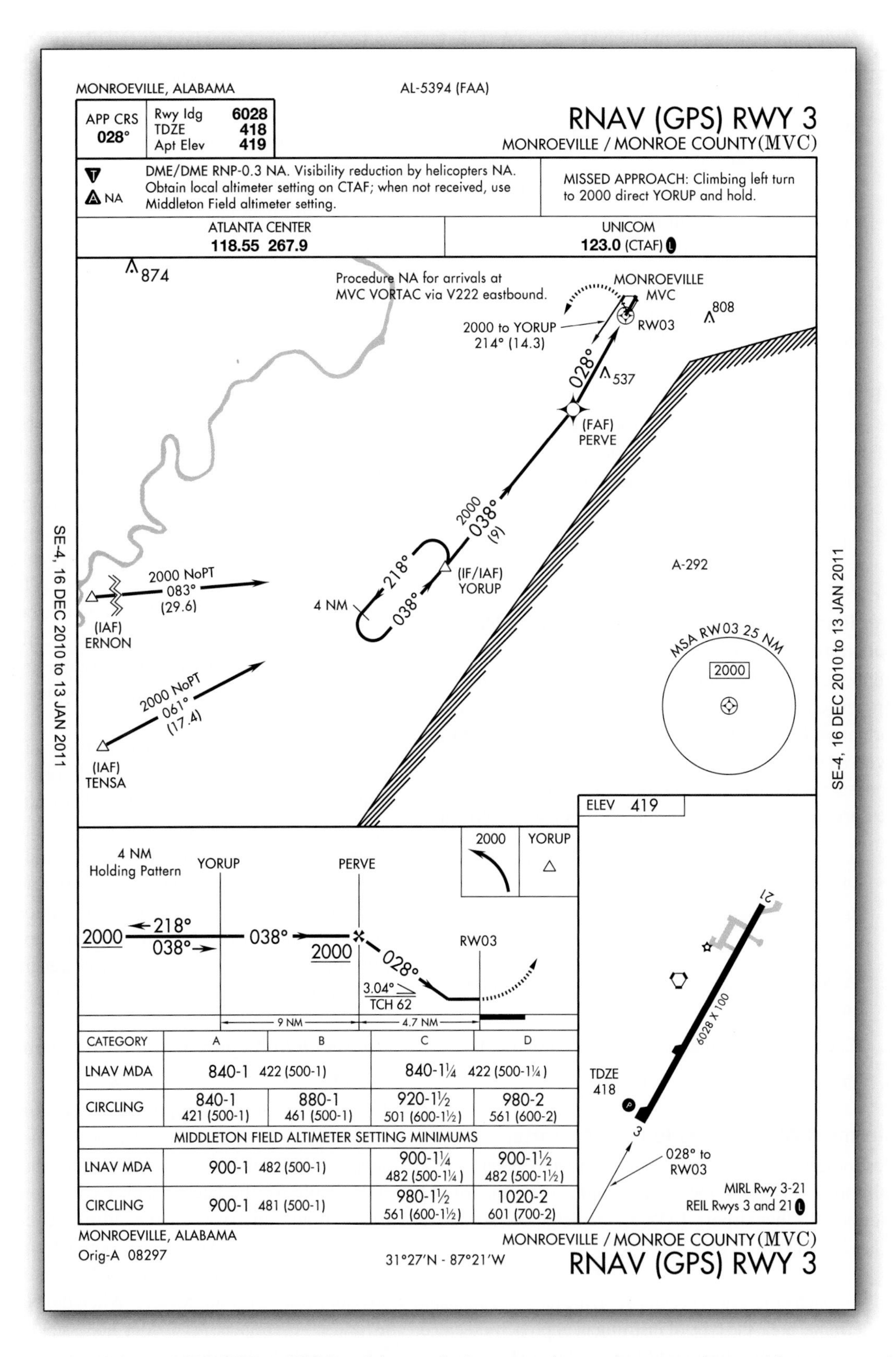

Figure 10-8. *Monroeville, Alabama (MVC) VOR or GPS Rwy 3 Approach: An approach procedure at an airport without an operating control tower.*

phrase "Cleared for the approach" and the communications frequency change required, if any. From this point on, there is no contact with ATC. The pilot is responsible for closing the IFR flight plan before landing, if in VFR conditions, or by telephone after landing.

Unless otherwise authorized by ATC, a pilot is expected to execute the complete IAP shown on the chart.

Approach to Airport With an Operating Tower, With No Approach Control

When an aircraft approaches an airport with an operating control tower, but no approach control, ATC issues a clearance to an approach/outer fix with the appropriate information and instructions as follows:

1. Name of the fix
2. Altitude to be maintained
3. Holding information and expected approach clearance time, if appropriate
4. Instructions regarding further communications, including:
    a) facility to be contacted
    b) time and place of contact
    c) frequency/ies to be used

If ATIS is available, a pilot should monitor that frequency for information such as ceiling, visibility, wind direction and velocity, altimeter setting, instrument approach, and runways in use prior to initial radio contact with the tower. If ATIS is not available, ATC provides weather information from the nearest reporting station.

Approach to an Airport With an Operating Tower, With an Approach Control

Where radar is approved for approach control service, it is used to provide vectors in conjunction with published IAPs. Radar vectors can provide course guidance and expedite traffic to the final approach course of any established IAP. *Figure 10-9* shows an IAP chart with maximum ATC facilities available.

Approach control facilities that provide this radar service operate in the following manner:

1. Arriving aircraft are either cleared to an outer fix most appropriate to the route being flown with vertical separation and, if required, given holding information; or,
2. When radar hand-offs are effected between ARTCC and approach control, or between two approach control facilities, aircraft are cleared to the airport or to a fix so located that the hand-off is completed prior to the time the aircraft reaches the fix.
    a) When the radar hand-offs are utilized, successive arriving flights may be handed off to approach control with radar separation in lieu of vertical separation.
    b) After hand-off to approach control, an aircraft is vectored to the appropriate final approach course.
3. Radar vectors and altitude/flight levels are issued as required for spacing and separating aircraft; do not deviate from the headings issued by approach control.
4. Aircraft are normally informed when it becomes necessary to be vectored across the final approach course for spacing or other reasons. If approach course crossing is imminent and the pilot has not been informed that the aircraft will be vectored across the final approach course, the pilot should query the controller. The pilot is not expected to turn inbound on the final approach course unless an approach clearance has been issued. This clearance is normally issued with the final vector for interception of the final approach course, and the vector enables the pilot to establish the aircraft on the final approach course prior to reaching the final approach fix.
5. Once the aircraft is established inbound on the final approach course, radar separation is maintained with other aircraft, and the pilot is expected to complete the approach using the NAVAID designated in the clearance (ILS, VOR, NDB, GPS, etc.) as the primary means of navigation.
6. After passing the final approach fix inbound, the pilot is expected to proceed direct to the airport and complete the approach or to execute the published missed approach procedure.
7. Radar service is automatically terminated when the landing is completed or when the pilot is instructed to change to advisory frequency at uncontrolled airports, whichever occurs first.

Radar Approaches

With a radar approach, the pilot receives course and altitude guidance from a controller who monitors the progress of the flight with radar. This is an option should the pilot experience an emergency or distress situation.

The only airborne radio equipment required for radar approaches is a functioning radio transmitter and receiver.

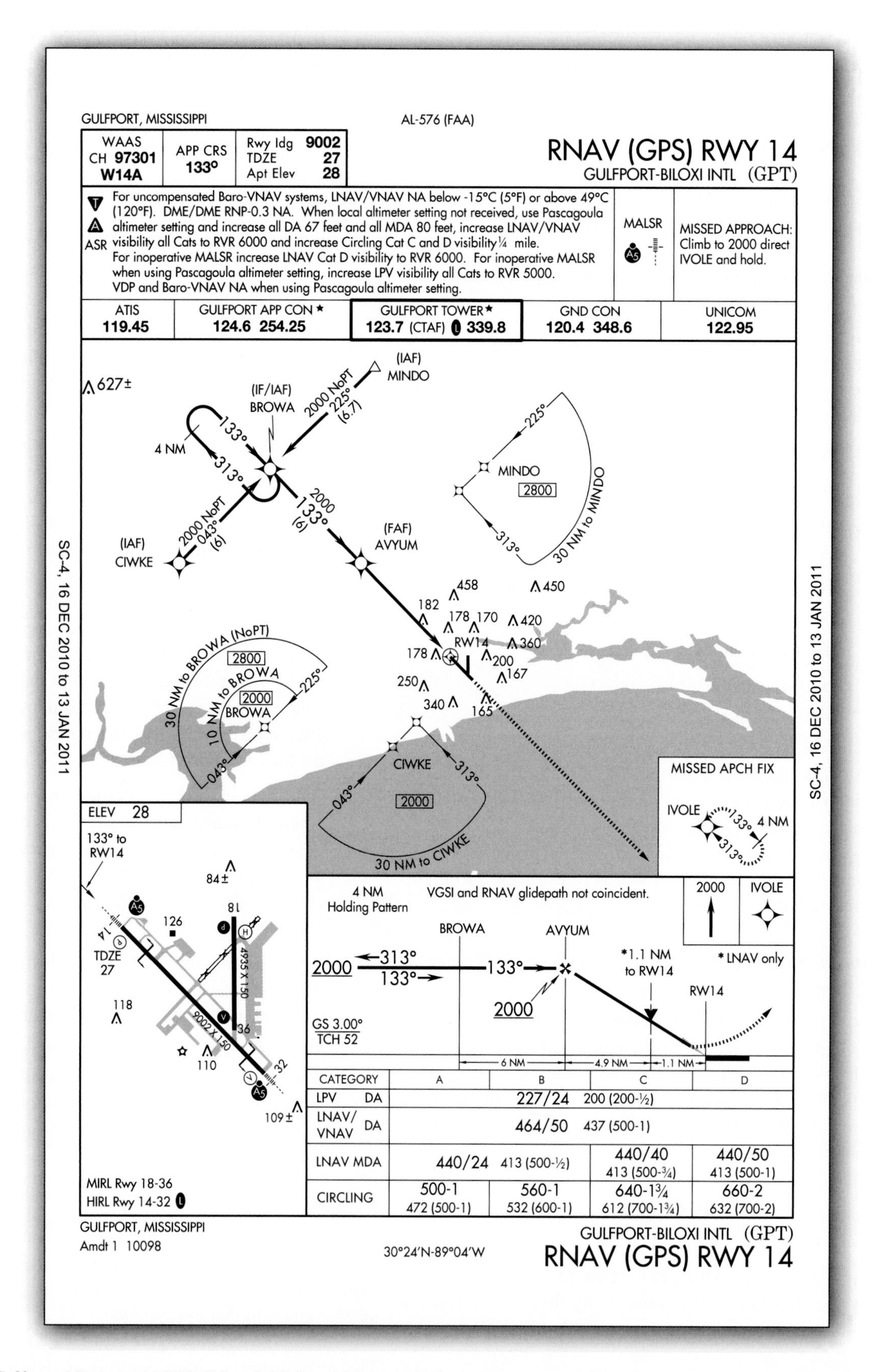

CATEGORY	A	B	C	D
LPV DA	227/24 200 (200-½)			
LNAV/VNAV DA	464/50 437 (500-1)			
LNAV MDA	440/24 413 (500-½)		440/40 413 (500-¾)	440/50 413 (500-1)
CIRCLING	500-1 472 (500-1)	560-1 532 (600-1)	640-1¾ 612 (700-1¾)	660-2 632 (700-2)

Figure 10-9. *Gulfport, Mississippi (GPT) ILS or LOC Rwy 14 Approach: An instrument procedure chart with maximum ATC facilities available.*

The radar controller vectors the aircraft to align it with the runway centerline. The controller continues the vectors to keep the aircraft on course until the pilot can complete the approach and landing by visual reference to the surface. There are two types of radar approaches: Precision (PAR) and Surveillance (ASR).

A radar approach may be given to any aircraft upon request and may be offered to pilots of aircraft in distress or to expedite traffic; however, an ASR might not be approved unless there is an ATC operational requirement or in an unusual or emergency situation. Acceptance of a PAR or ASR by a pilot does not waive the prescribed weather minimums for the airport or for the particular aircraft operator concerned. The decision to make a radar approach when the reported weather is below the established minimums rests with the pilot.

PAR and ASR minimums are published on separate pages in the FAA TPP. *Figure 10-10.*

PAR is one in which a controller provides highly accurate navigational guidance in azimuth and elevation to a pilot.

The controller gives the pilot headings to fly that direct the aircraft to, and keep the aircraft aligned with, the extended centerline of the landing runway. The pilot is told to anticipate glidepath interception approximately 10 to 30 seconds before it occurs and when to start descent. The published decision height (DH) is given only if the pilot requests it. If the aircraft is observed to deviate above or below the glidepath, the pilot is given the relative amount of deviation by use of terms "slightly" or "well" and is expected to adjust the aircraft's rate of descent/ascent to return to the glidepath. Trend information is also issued with respect to the elevation of the aircraft and may be modified by the terms "rapidly" and "slowly" (e.g., "well above glidepath, coming down rapidly").

Range from touchdown is given at least once each mile. If an aircraft is observed by the controller to proceed outside of specified safety zone limits in azimuth and/or elevation and continue to operate outside these prescribed limits, the pilot is directed to execute a missed approach or to fly a specified course unless the pilot has the runway environment (runway, approach lights, etc.) in sight. Navigational guidance in azimuth and elevation is provided to the pilot until the aircraft reaches the published DH. Advisory course and glidepath information is furnished by the controller until the aircraft passes over the landing threshold. At this point, the pilot is advised of any deviation from the runway centerline. Radar service is automatically terminated upon completion of the approach.

ASR is one in which a controller provides navigational guidance in azimuth only.

The controller furnishes the pilot with headings to fly to align the aircraft with the extended centerline of the landing runway. Since the radar information used for a surveillance approach is considerably less precise than that used for a precision approach, the accuracy of the approach is not as great and higher minimums apply. Guidance in elevation is not possible, but the pilot is advised when to commence

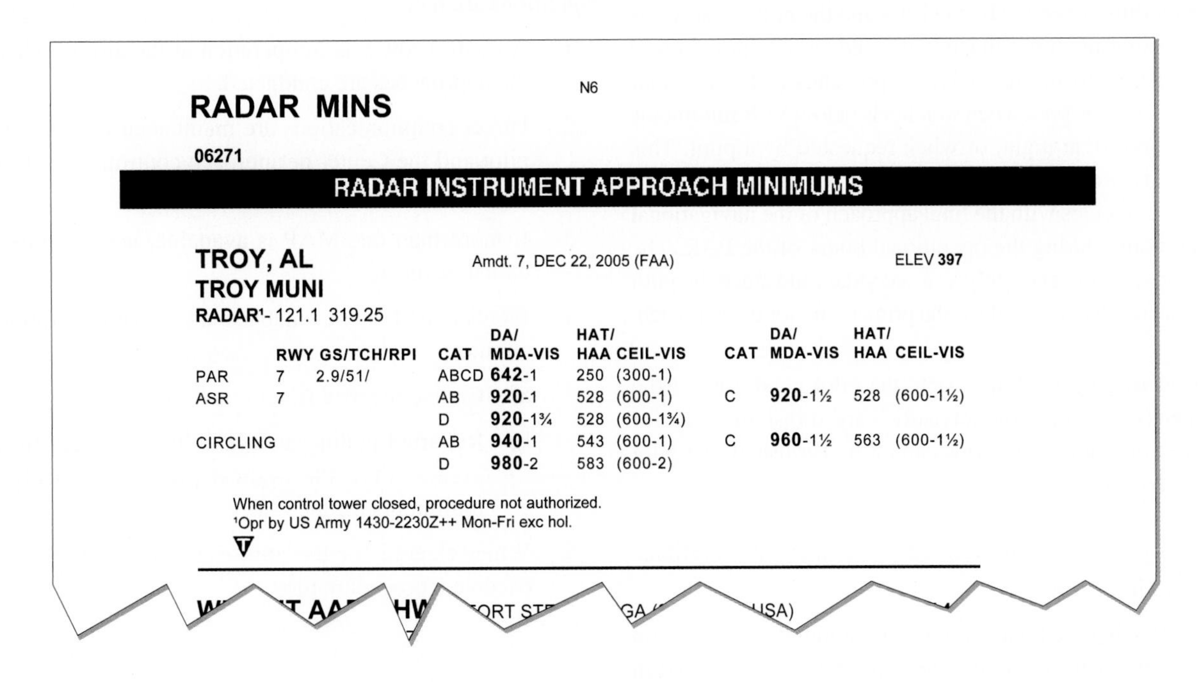

RADAR MINS N6

06271

RADAR INSTRUMENT APPROACH MINIMUMS

TROY, AL Amdt. 7, DEC 22, 2005 (FAA) ELEV **397**

TROY MUNI

RADAR[1]- 121.1 319.25

	RWY	GS/TCH/RPI	CAT	DA/ MDA-VIS	HAT/ HAA	CEIL-VIS	CAT	DA/ MDA-VIS	HAT/ HAA	CEIL-VIS
PAR	7	2.9/51/	ABCD	**642**-1	250	(300-1)				
ASR	7		AB	**920**-1	528	(600-1)	C	**920**-1½	528	(600-1½)
			D	**920**-1¾	528	(600-1¾)				
CIRCLING			AB	**940**-1	543	(600-1)	C	**960**-1½	563	(600-1½)
			D	**980**-2	583	(600-2)				

When control tower closed, procedure not authorized.
[1]Opr by US Army 1430-2230Z++ Mon-Fri exc hol.

Figure 10-10. *Radar instrument approach minimums for Troy, Alabama.*

descent to the Minimum Descent Altitude (MDA) or, if appropriate, to an intermediate step-down fix Minimum Crossing Altitude (MCA) and subsequently to the prescribed MDA. In addition, the pilot is advised of the location of the Missed Approach Point (MAP) prescribed for the procedure and the aircraft's position each mile on final from the runway, airport, heliport, or MAP, as appropriate.

If requested by the pilot, recommended altitudes are issued at each mile, based on the descent gradient established for the procedure, down to the last mile that is at or above the MDA. Normally, navigational guidance is provided until the aircraft reaches the MAP.

Radar service is automatically terminated at the completion of a radar approach.

No-Gyro Approach is available to a pilot under radar control who experiences circumstances wherein the directional gyro or other stabilized compass is inoperative or inaccurate. When this occurs, the pilot should so advise ATC and request a no-gyro vector or approach. The pilot of an aircraft not equipped with a directional gyro or other stabilized compass who desires radar handling may also request a no-gyro vector or approach. The pilot should make all turns at standard rate and should execute the turn immediately upon receipt of instructions. For example, "TURN RIGHT," "STOP TURN." When a surveillance or precision approach is made, the pilot is advised after the aircraft has been turned onto final approach to make turns at half standard rate.

Radar Monitoring of Instrument Approaches

PAR facilities operated by the FAA and the military services at some joint-use (civil and military) and military installations monitor aircraft on instrument approaches and issue radar advisories to the pilot when weather is below VFR minimums (1,000 and 3), at night, or when requested by a pilot. This service is provided only when the PAR Final Approach Course coincides with the final approach of the navigational aid and only during the operational hours of the PAR. The radar advisories serve only as a secondary aid since the pilot has selected the NAVAID as the primary aid for the approach.

Prior to starting final approach, the pilot is advised of the frequency on which the advisories are transmitted. If, for any reason, radar advisories cannot be furnished, the pilot is so advised.

Advisory information, derived from radar observations, includes information on:

1. Passing the final approach fix inbound (nonprecision approach) or passing the outer marker or fix used in lieu of the outer marker inbound (precision approach).
2. Trend advisories with respect to elevation and/or azimuth radar position and movement are provided.
3. If, after repeated advisories, the aircraft proceeds outside the PAR safety limit or if a radical deviation is observed, the pilot is advised to execute a missed approach unless the prescribed visual reference with the surface is established.

Radar service is automatically terminated upon completion of the approach. *[Figure 10-11]*

Timed Approaches From a Holding Fix

Timed approaches from a holding fix are conducted when many aircraft are waiting for an approach clearance. Although the controller does not specifically state "timed approaches are in progress," the assigning of a time to depart the FAF inbound (nonprecision approach), or the outer marker or fix used in lieu of the outer marker inbound (precision approach), indicates that timed approach procedures are being utilized.

In lieu of holding, the controller may use radar vectors to the final approach course to establish a distance between aircraft that ensures the appropriate time sequence between the FAF and outer marker or fix used in lieu of the outer marker and the airport. Each pilot in the approach sequence is given advance notice of the time they should leave the holding point on approach to the airport. When a time to leave the holding point is received, the pilot should adjust the flightpath in order to leave the fix as closely as possible to the designated time.

Timed approaches may be conducted when the following conditions are met:

1. A control tower is in operation at the airport where the approaches are conducted.
2. Direct communications are maintained between the pilot and the Center or approach controller until the pilot is instructed to contact the tower.
3. If more than one MAP is available, none require a course reversal.
4. If only one MAP is available, the following conditions are met:
    a) Course reversal is not required; and
    b) Reported ceiling and visibility are equal to or greater than the highest prescribed circling minimums for the IAP.
5. When cleared for the approach, pilots should not execute a procedure turn.

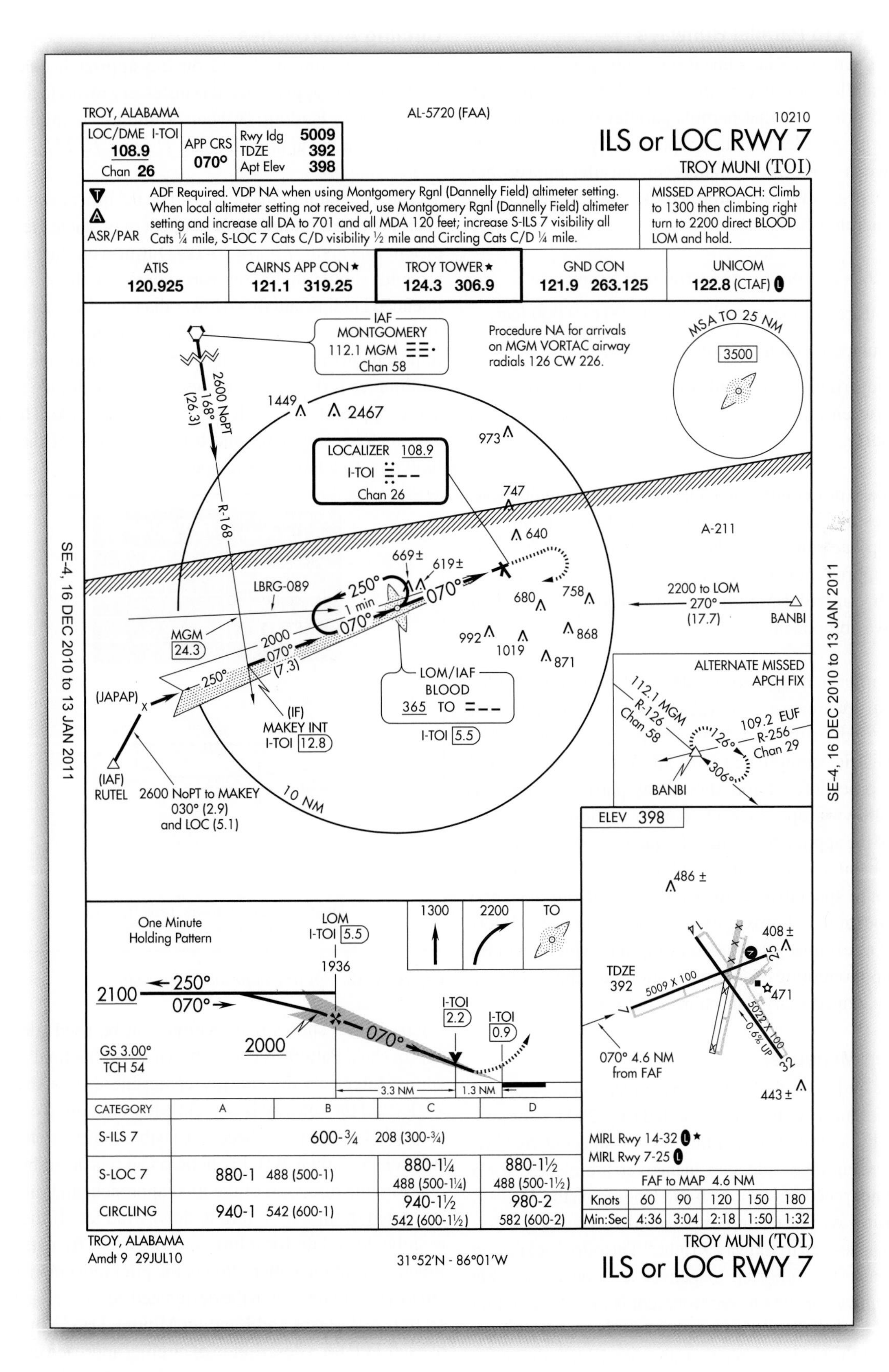

CATEGORY	A	B	C	D
S-ILS 7	600-¾ 208 (300-¾)			
S-LOC 7	880-1 488 (500-1)		880-1¼ 488 (500-1¼)	880-1½ 488 (500-1½)
CIRCLING	940-1 542 (600-1)		940-1½ 542 (600-1½)	980-2 582 (600-2)

FAF to MAP 4.6 NM

Knots	60	90	120	150	180
Min:Sec	4:36	3:04	2:18	1:50	1:32

Figure 10-11. *ILS RWY 7 Troy, Alabama.*

Approaches to Parallel Runways

Procedures permit ILS instrument approach operations to dual or triple parallel runway configurations. A parallel approach is an ATC procedure that permits parallel ILS approach to airports with parallel runways separated by at least 2,500 feet between centerlines. Wherever parallel approaches are in progress, pilots are informed that approaches to both runways are in use.

Simultaneous approaches are permitted to runways:

1. With centerlines separated by 4,300 to 9,000 feet;
2. Equipped with final monitor controllers;
3. Requiring radar monitoring to ensure separation between aircraft on the adjacent parallel approach course.

The approach procedure chart includes the note "simultaneous approaches authorized RWYS 14L and 14R," identifying the appropriate runways. When advised that simultaneous parallel approaches are in progress, pilots must advise approach control immediately of malfunctioning or inoperative components.

Parallel approach operations demand heightened pilot situational awareness. The close proximity of adjacent aircraft conducting simultaneous parallel approaches mandates strict compliance with all ATC clearances and approach procedures. Pilots should pay particular attention to the following approach chart information: name and number of the approach, localizer frequency, inbound course, glideslope intercept altitude, DA/DH, missed approach instructions, special notes/procedures, and the assigned runway location and proximity to adjacent runways. Pilots also need to exercise strict radio discipline, which includes continuous monitoring of communications and the avoidance of lengthy, unnecessary radio transmissions.

Side-Step Maneuver

ATC may authorize a side-step maneuver to either one of two parallel runways that are separated by 1,200 feet or less, followed by a straight-in landing on the adjacent runway. Aircraft executing a side-step maneuver are cleared for a specified nonprecision approach and landing on the adjacent parallel runway. For example, "Cleared ILS runway 7 left approach, side-step to runway 7 right." The pilot is expected to commence the side-step maneuver as soon as possible after the runway or runway environment is in sight. Landing minimums to the adjacent runway are based on nonprecision criteria and therefore higher than the precision minimums to the primary runway, but are normally lower than the published circling minimums.

Circling Approaches

Landing minimums listed on the approach chart under "CIRCLING" apply when it is necessary to circle the airport, maneuver for landing, or when no straight-in minimums are specified on the approach chart. *[Figure 10-11]*

The circling minimums published on the instrument approach chart provide a minimum of 300 feet of obstacle clearance in the circling area. *[Figure 10-12]* During a circling approach, the pilot should maintain visual contact with the runway of intended landing and fly no lower than the circling minimums until positioned to make a final descent for a landing. It is important to remember that circling minimums are only minimums. If the ceiling allows it, fly at an altitude that more nearly approximates VFR traffic pattern altitude. This makes any maneuvering safer and brings the view of the landing runway into a more normal perspective.

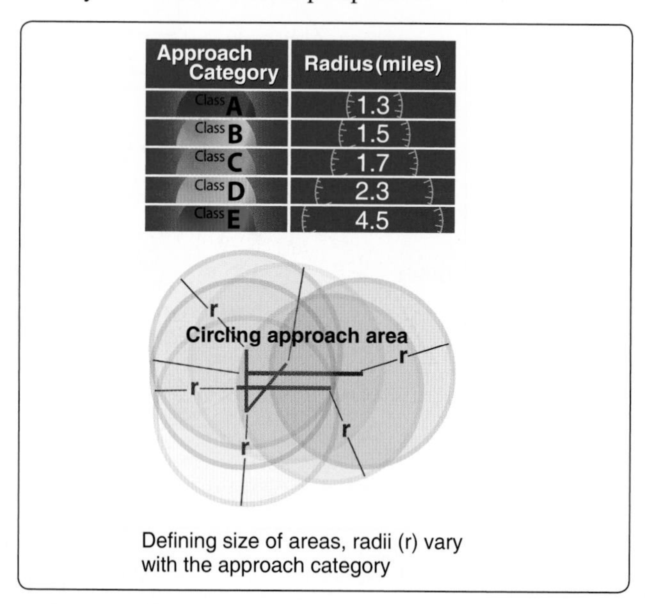

Figure 10-12. *Circling approach area radii.*

Figure 10-13 shows patterns that can be used for circling approaches. Pattern A can be flown when the final approach course intersects the runway centerline at less than a 90° angle, and the runway is in sight early enough to establish a base leg. If the runway becomes visible too late to fly pattern A, circle as shown in B. Fly pattern C if it is desirable to land opposite the direction of the final approach, and the runway is sighted in time for a turn to downwind leg. If the runway is sighted too late for a turn to downwind, fly pattern "D." Regardless of the pattern flown, the pilot must maneuver the aircraft to remain within the designated circling area. Refer to section A ("Terms and Landing Minima Data") in the front of each TPP for a description of circling approach categories. The criteria for determining the pattern to be flown are based on personal flying capabilities and knowledge of the

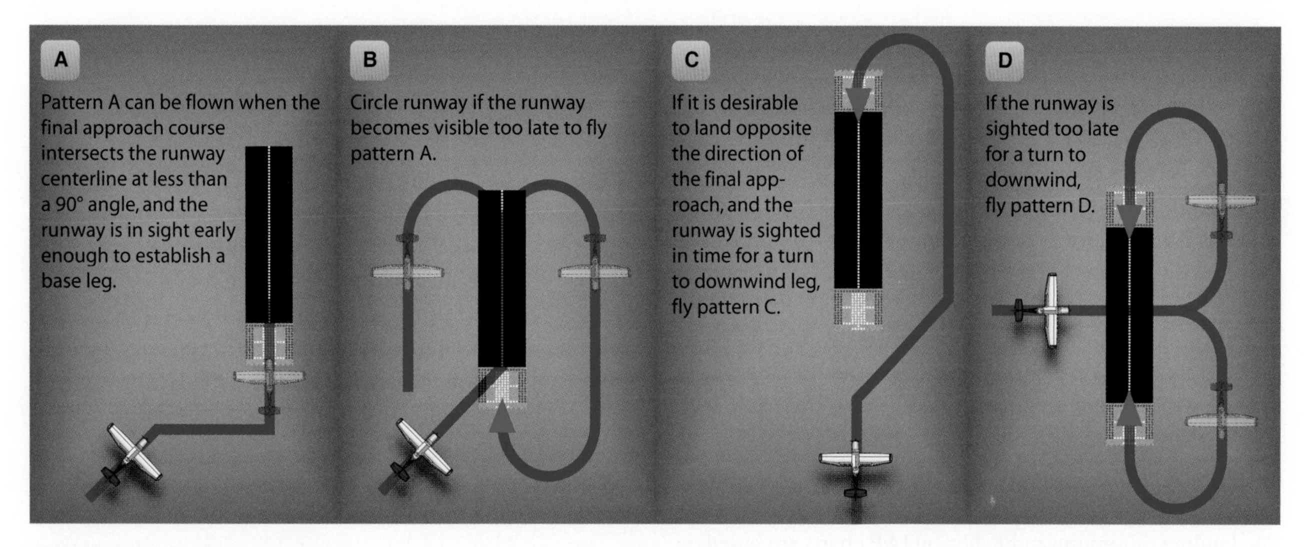
A
Pattern A can be flown when the final approach course intersects the runway centerline at less than a 90° angle, and the runway is in sight early enough to establish a base leg.
B
Circle runway if the runway becomes visible too late to fly pattern A.
C
If it is desirable to land opposite the direction of the final app-roach, and the runway is sighted in time for a turn to downwind leg, fly pattern C.
D
If the runway is sighted too late for a turn to downwind, fly pattern D.

Figure 10-13. *Circling approaches.*

performance characteristics of the aircraft. In each instance, the pilot must consider all factors: airport design, ceiling and visibility, wind direction and velocity, final approach course alignment, distance from the final approach fix to the runway, and ATC instructions.

IAP Minimums

Pilots may not operate an aircraft at any airport below the authorized MDA or continue an approach below the authorized DA/DH unless:

1. The aircraft is continuously in a position from which a descent to a landing on the intended runway can be made at a normal descent rate using normal maneuvers;
2. The flight visibility is not less than that prescribed for the approach procedure being used; and
3. At least one of the following visual references for the intended runway is visible and identifiable to the pilot:
    a) Approach light system
    b) Threshold
    c) Threshold markings
    d) Threshold lights
    e) Runway end identifier lights (REIL)
    f) Visual approach slope indicator (VASI)
    g) Touchdown zone or touchdown zone markings
    h) Touchdown zone lights
    i) Runway or runway markings
    j) Runway lights

Missed Approaches

A MAP is formulated for each published instrument approach and allows the pilot to return to the airway structure while remaining clear of obstacles. The procedure is shown on the approach chart in text and graphic form. Since the execution of a missed approach occurs when the flight deck workload is at a maximum, the procedure should be studied and mastered before beginning the approach.

When a MAP is initiated, a climb pitch attitude should be established while setting climb power. Configure the aircraft for climb, turn to the appropriate heading, advise ATC that a missed approach is being executed, and request further clearances.

If the missed approach is initiated prior to reaching the MAP, unless otherwise cleared by ATC, continue to fly the IAP as specified on the approach chart. Fly to the MAP at or above the MDA or DA/DH before beginning a turn.

If visual reference is lost while circling-to-land from an instrument approach, execute the appropriate MAP. Make the initial climbing turn toward the landing runway and then maneuver to intercept and fly the missed approach course.

Pilots should immediately execute the MAP:

1. Whenever the requirements for operating below DA/DH or MDA are not met when the aircraft is below MDA, or upon arrival at the MAP and at any time after that until touchdown;
2. Whenever an identifiable part of the airport is not visible to the pilot during a circling maneuver at or above MDA; or
3. When so directed by ATC.

Landing

According to 14 CFR part 91, no pilot may land when the flight visibility is less than the visibility prescribed in the standard IAP being used. ATC provides the pilot with the current visibility reports appropriate to the runway in use. This may be in the form of prevailing visibility, runway visual value (RVV), or runway visual range (RVR). However, only the pilot can determine if the flight visibility meets the landing requirements indicated on the approach chart. If the flight visibility meets the minimum prescribed for the approach, then the approach may be continued to a landing. If the flight visibility is less than that prescribed for the approach, then the pilot must execute a missed approach regardless of the reported visibility.

The landing minimums published on IAP charts are based on full operation of all components and visual aids associated with the instrument approach chart being used. Higher minimums are required with inoperative components or visual aids. For example, if the ALSF-1 approach lighting system were inoperative, the visibility minimums for an ILS would need to be increased by one-quarter mile. If more than one component is inoperative, each minimum is raised to the highest minimum required by any single component that is inoperative. ILS glideslope inoperative minimums are published on instrument approach charts as localizer minimums. Consult the "Inoperative Components or Visual Aids Table" (printed on the inside front cover of each TPP) for a complete description of the effect of inoperative components on approach minimums.

Instrument Weather Flying

Flying Experience

The more experience a pilot has in VFR and IFR flight, the more proficient a pilot becomes. VFR experience can be gained by flying in terminal areas with high traffic activity. This type of flying forces the pilot to polish the skill of dividing his or her attention between aircraft control, navigation, communications, and other flight deck duties. IFR experience can be gained through night flying which also promotes both instrument proficiency and confidence. The progression from flying at night under clear, moonlit conditions to flying at night without moonlight, natural horizon, or familiar landmarks teaches a pilot to trust the aircraft instruments with minimal dependence upon what can be seen outside the aircraft. It is a pilot's decision to proceed with an IFR flight or to wait for more acceptable weather conditions.

Recency of Experience

Currency as an instrument pilot is an equally important consideration. No person may act as pilot in command of an aircraft under IFR or in weather conditions less than VFR minimums unless he or she has met the requirements of Part 91. Remember, these are minimum requirements.

Airborne Equipment and Ground Facilities

Regulations specify minimum equipment for filing an IFR flight plan. It is the pilot's responsibility to determine the adequacy of the aircraft and navigation/communication (NAV/COM) equipment for the proposed IFR flight. Performance limitations, accessories, and general condition of the equipment are directly related to the weather, route, altitude, and ground facilities pertinent to the flight, as well as to the flight deck workload.

Weather Conditions

In addition to the weather conditions that might affect a VFR flight, an IFR pilot must consider the effects of other weather phenomena (e.g., thunderstorms, turbulence, icing, and visibility).

Turbulence

Inflight turbulence can range from occasional light bumps to extreme airspeed and altitude variations that make aircraft control difficult. To reduce the risk factors associated with turbulence, pilots must learn methods of avoidance, as well as piloting techniques for dealing with an inadvertent encounter.

Turbulence avoidance begins with a thorough preflight weather briefing. Many reports and forecasts are available to assist the pilot in determining areas of potential turbulence. These include the Severe Weather Warning (WW), SIGMET (WS), Convective SIGMET (WST), AIRMET (WA), Severe Weather Outlook (AC), Center Weather Advisory (CWA), Area Forecast (FA), and Pilot Reports (UA or PIREPs). Since thunderstorms are always indicative of turbulence, areas of known and forecast thunderstorm activity is always of interest to the pilot. In addition, clear air turbulence (CAT) associated with jet streams, strong winds over rough terrain, and fast moving cold fronts are good indicators of turbulence.

Pilots should be alert while in flight for the signposts of turbulence. For example, clouds with vertical development such as cumulus, towering cumulus, and cumulonimbus are indicators of atmospheric instability and possible turbulence. Standing lenticular clouds lack vertical development but indicate strong mountain wave turbulence. While en route, pilots can monitor hazardous inflight weather advisory service (HIWAS) broadcast for updated weather advisories, or contact the nearest FSS or En Route Flight Advisory Service (EFAS) for the latest turbulence-related PIREPs.

To avoid turbulence associated with strong thunderstorms, circumnavigate cells by at least 20 miles. Turbulence may also be present in the clear air above a thunderstorm. To avoid this, fly at least 1,000 feet above the top for every 10 knots of wind at that level, or fly around the storm. Finally, do not underestimate the turbulence beneath a thunderstorm. Never attempt to fly under a thunderstorm. The possible results of turbulence and wind shear under the storm could be disastrous.

When moderate to severe turbulence is encountered, aircraft control is difficult, and a great deal of concentration is required to maintain an instrument scan. *[Figure 10-14]* Pilots should immediately reduce power and slow the aircraft to the recommended turbulence penetration speed as described in the POH/AFM. To minimize the load factor imposed on the aircraft, the wings should be kept level and the aircraft's pitch attitude should be held constant. The aircraft is allowed to fluctuate up and down because maneuvering to maintain a constant altitude only increases the stress on the aircraft. If necessary, the pilot should advise ATC of the fluctuations and request a block altitude clearance. In addition, the power should remain constant at a setting that maintains the recommended turbulence penetration airspeed.

The best source of information on the location and intensity of turbulence are PIREPs. Therefore, pilots are encouraged to familiarize themselves with the turbulence reporting criteria found in the AIM, which also describes the procedure for volunteering PIREPs relating to turbulence.

Structural Icing

The very nature of flight in instrument meteorological conditions (IMC) means operating in visible moisture such as clouds. At the right temperatures, this moisture can freeze on the aircraft, causing increased weight, degraded performance, and unpredictable aerodynamic characteristics. Understanding avoidance and early recognition followed by prompt action are the keys to avoiding this potentially hazardous situation.

Structural icing refers to the accumulation of ice on the exterior of the aircraft and is broken down into three classifications: rime ice, clear ice, and mixed ice. For ice to form, there must be moisture present in the air, and the air must be cooled to a temperature of 0 °C (32 °F) or less. Aerodynamic cooling can lower the surface temperature of an airfoil and cause ice to form on the airframe even though the ambient temperature is slightly above freezing.

Rime ice forms if the droplets are small and freeze immediately when contacting the aircraft surface. This type of ice usually forms on areas such as the leading edges of wings or struts. It has a somewhat rough-looking appearance and a milky-white color.

Figure 10-14. *Maintaining an instrument scan in severe turbulence can be difficult.*

Clear ice is usually formed from larger water droplets or freezing rain that can spread over a surface. This is the most dangerous type of ice since it is clear, hard to see, and can change the shape of the airfoil.

Mixed ice is a mixture of clear ice and rime ice. It has the bad characteristics of both types and can form rapidly. Ice particles become embedded in clear ice, building a very rough accumulation. The table in *Figure 10-15* lists the temperatures at which the various types of ice form.

Outside Air Temperature Range	Icing Type
0 °C to −10 °C	Clear
−10 °C to −15 °C	Mixed clear and rime
−15 °C to −20°C	Rime

Figure 10-15. *Temperature ranges for ice formation.*

Structural icing is a condition that can only get worse. Therefore, during an inadvertent icing encounter, it is important the pilot act to prevent additional ice accumulation. Regardless of the level of anti-ice or deice protection offered by the aircraft, the first course of action should be to leave the area of visible moisture. This might mean descending to an altitude below the cloud bases, climbing to an altitude that is above the cloud tops, or turning to a different course. If this is not possible, then the pilot must move to an altitude where the temperature is above freezing. Pilots should report icing conditions to ATC and request new routing or altitude if icing will be a hazard. Refer to the AIM for information on reporting icing intensities.

Fog

Instrument pilots must learn to anticipate conditions leading to the formation of fog and take appropriate action early in the progress of the flight. Before a flight, close examination of current and forecast weather should alert the pilot to the possibility of fog formation. When fog is a consideration, pilots should plan adequate fuel reserves and alternate landing sites. En route, the pilot must stay alert for fog formation through weather updates from EFAS, ATIS, and ASOS/AWOS sites.

Two conditions lead to the formation of fog. Either the air is cooled to saturation, or sufficient moisture is added to the air until saturation occurs. In either case, fog can form when the temperature/dewpoint spread is 5° or less. Pilots planning to arrive at their destination near dusk with decreasing temperatures should be particularly concerned about the possibility of fog formation.

Volcanic Ash

Volcanic eruptions create volcanic ash clouds containing an abrasive dust that poses a serious safety threat to flight operations. Adding to the danger is the fact that these ash clouds are not easily discernible from ordinary clouds when encountered at some distance from the volcanic eruption.

When an aircraft enters a volcanic ash cloud, dust particles and smoke may become evident in the cabin, often along with the odor of an electrical fire. Inside the volcanic ash cloud, the aircraft may also experience lightning and St. Elmo's fire on the windscreen. The abrasive nature of the volcanic ash can pit the windscreens, thus reducing or eliminating forward visibility. The pitot-static system may become clogged, causing instrument failure. Severe engine damage is probable in both piston and jet-powered aircraft.

Every effort must be made to avoid volcanic ash. Since volcanic ash clouds are carried by the wind, pilots should plan their flights to remain upwind of the ash-producing volcano. Visual detection and airborne radar are not considered a reliable means of avoiding volcanic ash clouds. Pilots witnessing volcanic eruptions or encountering volcanic ash should immediately pass this information along in the form of a pilot report. The National Weather Service (NWS) monitors volcanic eruptions and estimates ash trajectories. This information is passed along to pilots in the form of SIGMETs.

As for many other hazards to flight, the best source of volcanic information comes from PIREPs. Pilots who witness a volcanic eruption or encounter volcanic ash in flight should immediately inform the nearest agency. Volcanic Ash Forecast Transport and Dispersion (VAFTAD) charts are also available; these depict volcanic ash cloud locations in the atmosphere following an eruption and also forecast dispersion of the ash concentrations over 6- and 12-hour time intervals. See AC 00-45, Aviation Weather Services.

Thunderstorms

A thunderstorm packs just about every weather hazard known to aviation into one vicious bundle. Turbulence, hail, rain, snow, lightning, sustained updrafts and downdrafts, and icing conditions are all present in thunderstorms. Do not take off in the face of an approaching thunderstorm or fly an aircraft that is not equipped with thunderstorm detection in clouds or at night in areas of suspected thunderstorm activity. *[Figure 10-16]*

There is no useful correlation between the external visual appearance of thunderstorms and the severity or amount of turbulence or hail within them. All thunderstorms should be considered hazardous, and thunderstorms with tops above 35,000 feet should be considered extremely hazardous.

Figure 10-16. *A thunderstorm packs just about every weather hazard known to aviation into one vicious bundle.*

Weather radar, airborne or ground based, normally reflects the areas of moderate to heavy precipitation (radar does not detect turbulence). The frequency and severity of turbulence generally increases with the radar reflectivity closely associated with the areas of highest liquid water content of the storm. A flightpath through an area of strong or very strong radar echoes separated by 20 to 30 miles or less may not be considered free of severe turbulence.

The probability of lightning strikes occurring to aircraft is greatest when operating at altitudes where temperatures are between –5 ° C and +5 ° C. In addition, an aircraft flying in the clear air near a thunderstorm is also susceptible to lightning strikes. Thunderstorm avoidance is always the best policy.

Wind Shear

Wind shear can be defined as a change in wind speed and/or wind direction in a short distance. It can exist in a horizontal or vertical direction and occasionally in both. Wind shear can occur at all levels of the atmosphere but is of greatest concern during takeoffs and landings. It is typically associated with thunderstorms and low-level temperature inversions; however, the jet stream and weather fronts are also sources of wind shear.

As *Figure 10-17* illustrates, while an aircraft is on an instrument approach, a shear from a tailwind to a headwind causes the airspeed to increase and the nose to pitch up with a corresponding balloon above the glidepath. A shear from a headwind to a tailwind has the opposite effect, and the aircraft will sink below the glidepath.

A headwind shear followed by a tailwind/downdraft shear is particularly dangerous because the pilot has reduced power and lowered the nose in response to the headwind shear. This leaves the aircraft in a nose-low, power-low configuration when the tailwind shear occurs, which makes recovery more difficult, particularly near the ground. This type of wind shear scenario is likely while making an approach in the face of an oncoming thunderstorm. Pilots should be alert for indications of wind shear early in the approach phase and be ready to initiate a missed approach at the first indication. It may be impossible to recover from a wind shear encounter at low altitude.

To inform pilots of hazardous wind shear activity, some airports have installed a Low-Level Wind Shear Alert System (LLWAS) consisting of a centerfield wind indicator and several surrounding boundary-wind indicators. With

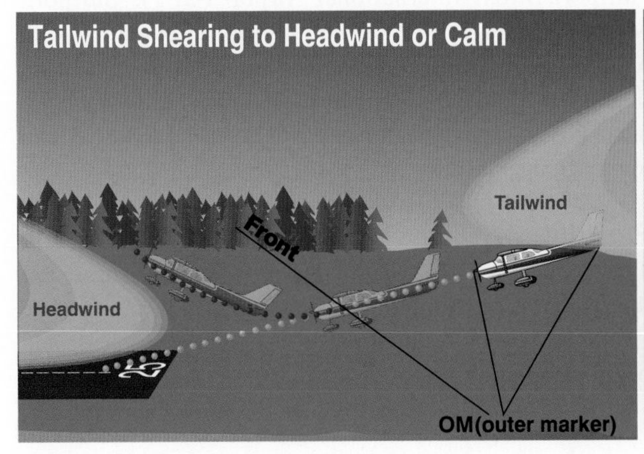

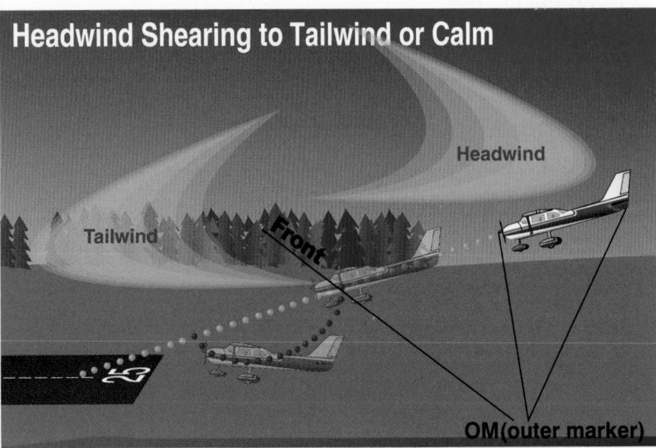

Figure 10-17. *Glideslope deviations due to wind shear encounter.*

this system, controllers are alerted of wind discrepancies (an indicator of wind shear possibility) and provide this information to pilots. A typical wind shear alert issued to a pilot would be:

> "Runway 27 arrival, wind shear alert, 20 knot loss 3 mile final, threshold wind 200 at 15"

In plain language, the controller is advising aircraft arriving on runway 27 that at about 3 miles out they can expect a wind shear condition that will decrease their airspeed by 20 knots and possibly encounter turbulence. Additionally, the airport surface winds for landing runway 27 are reported as 200° at 15 knots.

Pilots encountering wind shear are encouraged to pass along pilot reports. Refer to AIM for additional information on wind shear PIREPs.

VFR-On-Top

Pilots on IFR flight plans operating in VFR weather conditions may request VFR-on-top in lieu of an assigned altitude. This permits them to select an altitude or flight level of their choice (subject to any ATC restrictions).

Pilots desiring to climb through a cloud, haze, smoke, or other meteorological formation and then either cancel their IFR flight plan or operate VFR-on-top may request a climb to VFR-on-top. The ATC authorization contains a top report (or a statement that no top report is available) and a request to report upon reaching VFR-on-top. Additionally, the ATC authorization may contain a clearance limit, routing, and an alternative clearance if VFR-on-top is not reached by a specified altitude.

A pilot on an IFR flight plan, operating in VFR conditions, may request to climb/descend in VFR conditions. When operating in VFR conditions with an ATC authorization to "maintain VFR-on-top/maintain VFR conditions," pilots on IFR flight plans must:

1. Fly at the appropriate VFR altitude as prescribed in 14 CFR part 91.
2. Comply with the VFR visibility and distance-from-cloud criteria in 14 CFR part 91.
3. Comply with IFR applicable to this flight (minimum IFR altitudes, position reporting, radio communications, course to be flown, adherence to ATC clearance, etc.).

Pilots operating on a VFR-on-top clearance should advise ATC before any altitude change to ensure the exchange of accurate traffic information.

ATC authorization to "maintain VFR-on-top" is not intended to restrict pilots to operating only above an obscuring meteorological formation (layer). Rather, it permits operation above, below, between layers, or in areas where there is no meteorological obstruction. It is imperative pilots understand, however, that clearance to operate "VFR-on-top/VFR conditions" does not imply cancellation of the IFR flight plan.

Pilots operating VFR-on-top/VFR conditions may receive traffic information from ATC on other pertinent IFR or VFR aircraft. However, when operating in VFR weather conditions, it is the pilot's responsibility to be vigilant to see and avoid other aircraft.

This clearance must be requested by the pilot on an IFR flight plan. VFR-on-top is not permitted in certain areas, such as Class A airspace. Consequently, IFR flights operating VFR-on-top must avoid such airspace.

VFR Over-The-Top

VFR over-the-top must not be confused with VFR-on-top. VFR-on-top is an IFR clearance that allows the pilot to fly VFR altitudes. VFR over-the-top is strictly a VFR operation in which the pilot maintains VFR cloud clearance requirements while operating on top of an undercast layer. This situation might occur when the departure airport and the destination airport are reporting clear conditions, but a low overcast layer is present in between. The pilot could conduct a VFR departure, fly over the top of the undercast in VFR conditions, then complete a VFR descent and landing at the destination. VFR cloud clearance requirements would be maintained at all times, and an IFR clearance would not be required for any part of the flight.

Conducting an IFR Flight

To illustrate some of the concepts introduced in this chapter, follow along on a typical IFR flight from the Birmingham International Airport (BHM), Birmingham, Alabama to Gulfport-Biloxi International Airport (GPT), Gulfport, Mississippi. *[Figure 10-18]* For this trip, a Cessna 182 with a call sign of N1230A is flown. The aircraft is equipped with dual navigation and communication radios, a transponder, and a GPS system approved for IFR en route, terminal, and approach operations.

Preflight

The success of the flight depends largely upon the thoroughness of the preflight planning. The evening before the flight, pay close attention to the weather forecast and begin planning the flight.

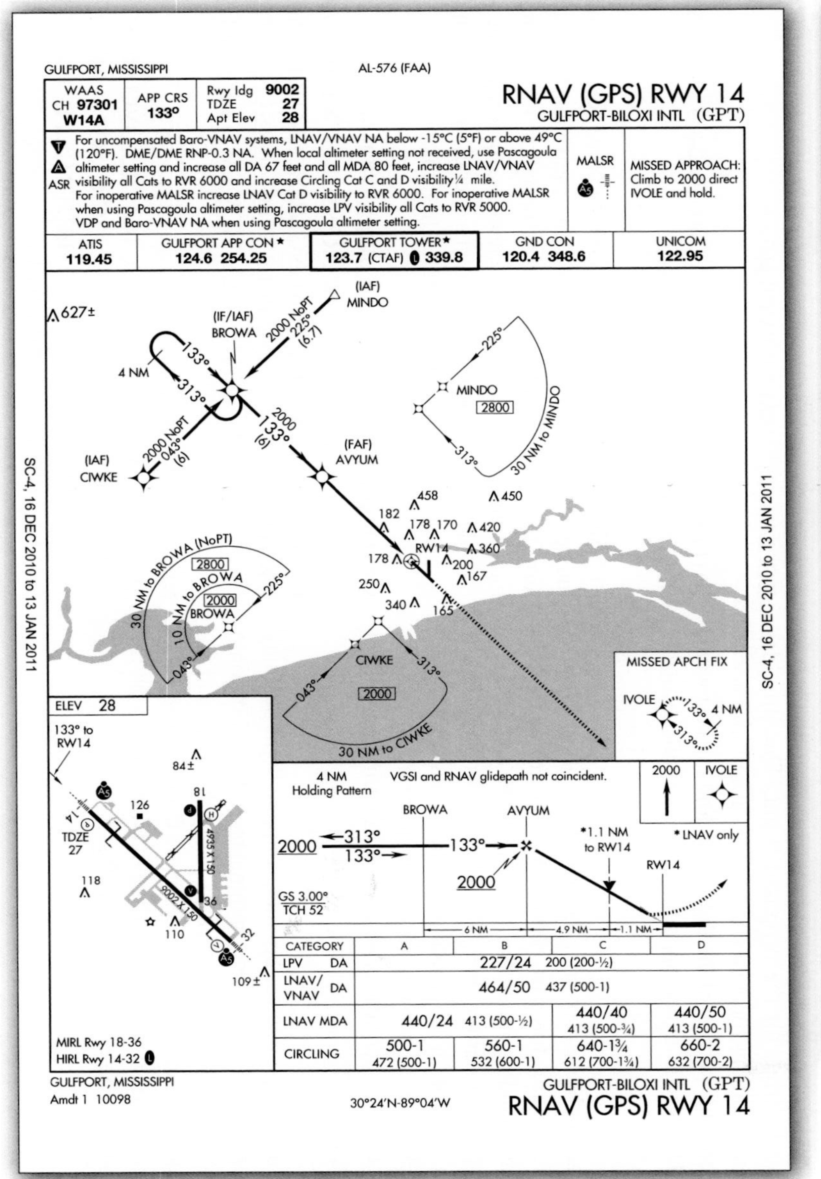

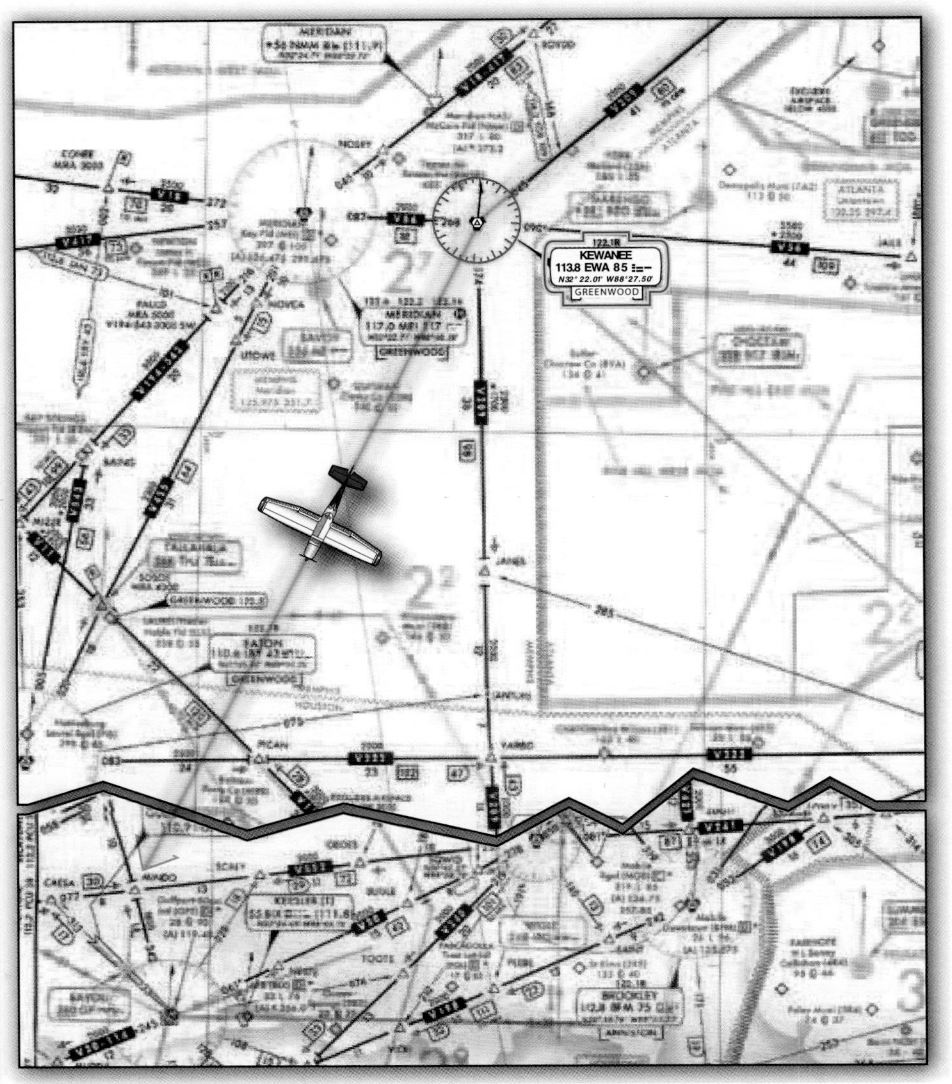

Figure 10-18. *Route planning.*

The Weather Channel indicates a large, low-pressure system has settled in over the Midwest, pulling moisture up from the Gulf of Mexico and causing low ceilings and visibility with little chance for improvement over the next couple of days. To begin planning, gather all the necessary charts and materials, and verify everything is current. This includes en route charts, approach charts, DPs, STAR charts, the GPS database, as well as an A/FD, some navigation logs, and the aircraft's POH/AFM. The charts cover both the departure and arrival airports and any contingency airports that will be needed if the flight cannot be completed as planned. This is also a good time for the pilot to consider recent flight experience, pilot proficiency, fitness, and personal weather minimums to fly this particular flight.

Check the A/FD to become familiar with the departure and arrival airport, and check for any preferred routing between BHM and GPT. Next, review the approach charts and any DP or STAR that pertains to the flight. Finally, review the en route charts for potential routing, paying close attention to the minimum en route and obstacle clearance altitudes.

After this review, select the best option. For this flight, the Birmingham Three Departure *[Figure 10-2]* to Brookwood VORTAC, V 209 to Kewanee VORTAC, direct to Gulfport using GPS would be a logical route. An altitude of 4,000 feet meets all the regulatory requirements and falls well within the performance capabilities of the aircraft.

Next, call 1-800-WX-BRIEF to obtain an outlook-type weather briefing for the proposed flight. This provides forecast conditions for departure and arrival airports, as well as the en route portion of the flight including forecast winds aloft. This also is a good opportunity to check the available NOTAMs.

The weather briefer confirms the predictions of the Weather Channel giving forecast conditions that are at or near minimum landing minimums at both BHM and GPT for the proposed departure time. The briefer provides NOTAM information for GPT indicating that the localizer to runway 32 is scheduled to be out of service and that runway 18/36 is closed until further notice. Also check for temporary flight restrictions (TFRs) along the proposed route.

After receiving a weather briefing, continue flight planning and begin to transfer some preliminary information onto the navigation log, listing each fix along the route and the distances, frequencies, and altitudes. Consolidating this information onto an organized navigation log keeps the workload to a minimum during the flight.

Next, obtain a standard weather briefing online for the proposed route. A check of current conditions indicates low IFR conditions at both the departure airport and the destination, with visibility of one-quarter mile:

SURFACE WEATHER OBSERVATIONS
METAR KBHM 111155Z VRB04KT ¼ SM FG –RA VV004 06/05 A2994 RMK A02 SLP140

METAR KGPT 111156Z 24003KT ¼ SM FG OVC001 08/07 A2962 RMK A02 SLP033

The small temperature/dewpoint spread is causing the low visibility and ceilings. Conditions should improve later in the day as temperatures increase. A check of the terminal forecast confirms this theory:

TERMINAL FORECASTS
TAF KBHM 111156Z 111212 VRB04KT ¼ SM FG VV004 TEMPO1316 ¾ SM OVC004

FM1600 VRB05KT 2SM BR OVC007 TEMPO 1720 3SM DZ BKN009
FM2000 22008KT 3SM –RA OVC015 TEMP 2205 3SM –RA OVC025 FM0500 23013KT P6SM OVC025

FM0800 23013KT P6SM BKN030 PROB40 1012 2SM BR OVC030

TAF KGPT 111153Z 111212 24004KT ¼ SM FG OVC001 BECMG 1317 3SM BR 0VC004

FM1700 24010KT 4SM –RA OVC006 FM0400 24010 5SM SCT080 TEMPO 0612 P6SM SKC

In addition to the terminal forecast, the area forecast also indicates gradual improvement along the route. Since the terminal forecast only provides information for a 5-mile radius around a terminal area, checking the area forecast provides a better understanding of the overall weather picture along the route, as well as potential hazards:

SYNOPSIS AND VFR CLOUDS/WEATHER FORECASTS
SYNOPSIS… AREA OF LOW PRESSURE CNTD OV AL RMNG GENLY STNRY BRNGNG MSTR AND WD SPRD IFR TO E TN. ALF…LOW PRES TROF ACRS CNTR PTN OF THE DFW FA WILL GDLY MOV EWD DURG PD.

NRN LA, AR, NRN MS
SWLY WND THRUT THE PD. 16Z CIG OVC006. SCT –SHRA. OTLK… IFR SRN ½ … CIG SCT – BKN015 TOPS TO FL250 SWLY WND THRUT THE PD. 17Z AGL BKN040. OTLK…MVFR CIG VIS.

LA MS CSTL WTRS
CIG OVC001 – OVC006. TOPS TO FL240. VIS ¼ – ¾ SM FG. SWLY WND. 16Z CIG OVC010 VIS 2 SM BR. OCNL VIS 3-5SM –RN BR OVC009. OTLK…MVFR CIG VIS.

FL
CIG BKN020 TOPS TO FL180. VIS 1–3 SM BR. SWLY WND. 18Z BRK030. OTLK…MVFR CIG.

At this time, there are no SIGMETs or PIREPs reported. However, there are several AIRMETs, one for IFR conditions, one for turbulence that covers the entire route, and another for icing conditions that covers an area just north of the route:

WAUS44 KKCI 111150

DFWS WA 0111150

AIRMET SIERRA FOR IFR VALID UNTIL 111800

AIRMET IFR...OK TX LA AR MS AL FL
TS IMPLY SEV OR GTR TURB SEV ICE LLWS AND IFR CONDS.

NON MSL HGHTS DENOTED BY AGL OR CIG.

A recheck of NOTAMs for Gulfport confirms that the localizer to runway 32 is out of service until further notice and runway 18/36 is closed. If runway 6 is planned for the departure, confirm that the climb restriction for the departure can be met.

GPT 12/006 GPT LOC OS UFN

GPT 12/008 GPT MIRL RWY 18/36 OS UFN

Since the weather is substantially better to the east, Pensacola Regional Airport is a good alternate with current conditions and a forecast of marginal VFR.

METAR KPNS 111150Z 21010Z 3SM BKN014 OVC025 09/03 A2973

TAF KPNS 111152Z 111212 22010KT 3 SM BR OVC020 BECMG 1317 4 SM BR OVC025

FM1700 23010KT 4SM –RA OVC030

FM 0400 25014KT 5SM OVC050 TEMPO1612 P6SM OVC080

If weather minimums are below a pilot's personal minimums, a delay in departure to wait for improved conditions is a good decision. This time can be used to complete the navigation log, which is the next step in planning an IFR flight. *[Figure 10-19]*

Use the POH/AFM to compute a true airspeed, cruise power setting, and fuel burn based on the forecast temperatures aloft and cruising pressure altitude. Also, compute weight-and-balance information and determine takeoff and landing distances. There will be a crosswind if weather conditions require a straight-in landing on runway 14 at GPT. Therefore, compute the landing distance assuming a 10-knot crosswind and determine if the runway length is adequate to allow landing. Determine the estimated flight time and fuel burn using the winds aloft forecast and considering Pensacola Regional Airport as an alternate airport. With full tanks, the flight can be made nonstop with adequate fuel for flight to the destination, alternate, and the reserve requirement.

Next, check the surface analysis chart, which shows where the pressure systems are found. The weather depiction chart shows areas of IFR conditions and can be used to find areas of improving conditions. These charts provide information a pilot needs should a diversion to VFR conditions be required. For this flight, the radar depicts precipitation along the route, and the latest satellite photo confirms what the weather depiction chart showed.

When the navigation log is finished, complete the flight plan in preparation for filing with flight service. *[Figure 10-20]*

Call an FSS for an updated weather briefing. Birmingham INTL airport is now reporting 700 overcast with 3 miles visibility, and Gulfport-Biloxi is now 400 overcast with 2 miles visibility. The alternate, Pensacola Regional Airport, continues to report adequate weather conditions with 2,000 overcast and 3 miles visibility in light rain.

Several pilot reports have been submitted for light icing conditions; however, all the reports are north of the route of flight and correspond to the AIRMET that was issued earlier. No pilot reports have included cloud tops, but the area forecast predicted cloud tops to flight level 240. Since the weather conditions appear to be improving, a flight plan can be filed using the completed form.

Analyze the latest weather minimums to determine if they exceed personal minimums. With the absence of icing reported along the route and steadily rising temperatures,

FLIGHT LOG							
TIME			DISTANCE	FUEL			
TAKE OFF 1600 E	LANDING		TOTAL 228	REQUIRED 51 Gal		AVAILABLE 87 Gal	
ROUTE (Check Point)	IDENT / FREQ	MAG CRSE	LEG / REMAINING	ETE / ATE	ETE / ATE	ALTITUDE / GND SPD	REMARKS
Brookwood	OKW / 111.0	230	31 / 197	+16 /	16:16 /	4000 / 120	3 Gal
Kewanee	EWA / 113.8	225	80 / 117	+40 /	16:56 /	4000 / 120	8 Gal
Mindo	/	195	110 / 17	+54 /	17:50 /	4000 / 125	12 Gal
Appr	/		17 / 0	+08 /	17:58 /	/	2 Gal
	/		/	118 / 1+58	/	/	
Rascagoula Regional	PNS /	085	91 / 0	+35 /	/	3000 / 158	18 Gal
	/		/	/	/	/	

ATIS			
DEPERTURE		ARRIVAL	
INFORMATION		INFORMATION	
CEILING		CEILING	
VISIBILITY		VISIBILITY	
TEMP / DEWPOINT	/	TEMP / DEWPOINT	/
WINDS		WINDS	
ALTIMETER		ALTIMETER	
RWY IN USE		RWY IN USE	
REMARKS		REMARKS	

Figure 10-19. *Navigation log.*

structural icing should not be a problem. Make a note to do an operational check of the pitot heat during preflight and to take evasive action immediately should even light icing conditions be encountered in flight. This may require returning to BHM or landing at an intermediate spot before reaching GPT. The go/no-go decision is constantly reevaluated during the flight.

Once at the airport, conduct a thorough preflight inspection. A quick check of the logbooks indicates all airworthiness requirements have been met to conduct this IFR flight including an altimeter, static, and transponder test within the preceding 24 calendar months. In addition, a log on the clipboard indicates the VOR system has been checked

Form Approved OMB No. 2120-0026

U.S. DEPARTMENT OF TRANSPORTATION
FEDERAL AVIATION ADMINISTRATION

FLIGHT PLAN

(FAA USE ONLY) ☐ PILOT BRIEFING ☐ VNR ☐ STOPOVER | TIME STARTED | SPECIALIST INITIALS

1. TYPE	2. AIRCRAFT IDENTIFICATION	3. AIRCRAFT TYPE/ SPECIAL EQUIPMENT	4. TRUE AIRSPEED	5. DEPARTURE POINT	6. DEPARTURE TIME PROPOSED (Z)	6. DEPARTURE TIME ACTUAL (Z)	7. CRUISING ALTITUDE
VFR / X IFR / DVFR	n/230a	C182/G	140 KTS	BHM	1600		4000

8. ROUTE OF FLIGHT
OKW V209 EWA MINDO

9. DESTINATION (Name of airport and city)	10. EST. TIME ENROUTE HOURS	10. EST. TIME ENROUTE MINUTES	11. REMARKS
Gulfport, Biloxi	1	58	

12. FUEL ON BOARD HOURS	12. FUEL ON BOARD MINUTES	13. ALTERNATE AIRPORT(S)	14. PILOT'S NAME, ADDRESS & TELEPHONE NUMBER & AIRCRAFT HOME BASE	15. NUMBER ABOARD
04	20	PNS	James Jones, 841 Oak St. Gardendail, AL 205-555-6028	2

17. DESTINATION CONTACT/TELEPHONE (OPTIONAL)

16. COLOR OF AIRCRAFT: blue/white

CIVIL AIRCRAFT PILOTS. FAR Part 91 requires you file an IFR flight plan to operate under instrument flight rules in controlled airspace. Failure to file could result in a civil penalty not to exceed $1,000 for each violation (Section 901 of the Federal Aviation Act of 1958, as amended). Filing of a VFR flight plan is recommended as a good operating practice. See also Part 99 for requirements concerning DVFR flight plans.

FAA Form 7233-1 (8-82) CLOSE VFR FLIGHT PLAN WITH ________ FSS ON ARRIVAL

Figure 10-20. *Flight plan form.*

within the preceding 30 days. Turn on the master switch and pitot heat, and quickly check the heating element before it becomes too hot. Then, complete the rest of the walk-around procedure. Since this is a flight in actual IFR conditions, place special emphasis on IFR equipment during the walk-around, including the alternator belt and antennas. After completing the preflight, organize charts, pencils, paper, and navigation log in the flight deck for quick, easy access. This is also the time to enter the planned flight into the GPS.

Departure

After starting the engine, tune in ATIS and copy the information to the navigation log. The conditions remain the same as the updated weather briefing with the ceiling at 700 overcast and visibility at 3 miles. Call clearance delivery to receive a clearance:

> "Clearance Delivery, Cessna 1230A IFR to Gulfport Biloxi with information Kilo, ready to copy."

> "Cessna 1230A is cleared to Gulfport-Biloxi via the Birmingham Three Departure, Brookwood, Victor 209 Kewanee then direct Mindo, Gulfport. Climb and maintain 4,000. Squawk 0321."

Read back the clearance and review the DP. Although a departure frequency was not given in the clearance, note that on the DP, the departure control frequency is listed as 123.8 for the southern sector. Since a departure from runway 24 is anticipated, note the instruction to climb to 2,100 prior to turning. After tuning in the appropriate frequencies and setting up navigation equipment for the departure routing, contact ground control (noting that this is IFR) and receive the following clearance:

> "Cessna 1230A taxi to runway 24 via taxiway Mike."

Read back the clearance and aircraft call sign. After a review of the taxi instructions on the airport diagram, begin to taxi and check the flight instruments for proper indications.

Hold short of runway 24 and complete the before takeoff checklist and engine run-up. Advise the tower when ready for takeoff. The tower gives the following clearance:

> "Cessna 30A cleared for takeoff runway 24. Caution wake turbulence from 737 departing to the northwest."

Taxi into position. Note the time off on the navigation log, verify that the heading indicator and magnetic compass are in agreement, the transponder is in the ALT position, all the necessary lights, equipment, and pitot heat are on. Start the takeoff roll. To avoid the 737's wake turbulence, make note of its lift off point and take off prior to that point.

En Route

After departure, climb straight ahead to 2,100 feet as directed by the Birmingham Three Departure. While continuing a climb to the assigned altitude of 4,000 feet, the following instructions are received from the tower:

"Cessna 30A contact Departure."

Acknowledge the clearance and contact departure on the frequency designated by the DP. State the present altitude so the departure controller can check the encoded altitude against indicated altitude:

"Birmingham Departure Cessna 1230A climbing through 2,700 heading 240."

Departure replies:

"Cessna 30A proceed direct to Brookwood and resume own navigation. Contact Atlanta Center on 134.05."

Acknowledge the clearance, contact Atlanta Center and proceed direct to Brookwood VORTAC, using the IFR-approved GPS equipment. En route to Kewanee, VORTAC Atlanta Center issues the following instructions:

"Cessna 1230A contact Memphis Center on 125.975."

Acknowledge the instructions and contact Memphis Center with aircraft ID and present altitude. Memphis Center acknowledges contact:

"Cessna 1230A, Meridian altimeter is 29.87. Traffic at your 2 o'clock and 6 miles is a King Air at 5,000 climbing to 12,000."

Even when on an IFR flight plan, pilots are still responsible for seeing and avoiding other aircraft. Acknowledge the call from Memphis Center and inform them of negative contact with traffic due to IMC.

"Roger, altimeter setting 29.87. Cessna 1230A is in IMC negative contact with traffic."

Continue the flight, and at each fix note the arrival time on the navigation log to monitor progress.

To get an update of the weather at the destination and issue a pilot report, contact the FSS servicing the area. To find the nearest FSS, locate a nearby VOR and check above the VOR information box for a frequency. In this case, the nearest VOR is Kewanee VORTAC which lists a receive-only frequency of 122.1 for Greenwood FSS. Request a frequency change from Memphis and then attempt to contact Greenwood on 122.1 while listening over the Kewanee VORTAC frequency of 113.8:

"Greenwood Radio Cessna 1230A receiving on frequency 113.8, over."

"Cessna 30A, this is Greenwood, go ahead."

"Greenwood Radio, Cessna 30A is currently 30 miles south of the Kewanee VORTAC at 4,000 feet en route to Gulfport. Requesting an update of en route conditions and current weather at GPT, as well as PNS."

"Cessna 30A, Greenwood Radio, current weather at Gulfport is 400 overcast with 3 miles visibility in light rain. The winds are from 140 at 7 and the altimeter is 29.86. Weather across your route is generally IFR in light rain with ceilings ranging from 300 to 1,000 overcast with visibilities between 1 and 3 miles. Pensacola weather is much better with ceilings now at 2,500 and visibility 6 miles. Checking current NOTAMs at GPT shows the localizer out of service and runway 18/36 closed."

"Roger, Cessna 30A copies the weather. I have a PIREP when you are ready to copy."

"Cessna 30A go ahead with your PIREP."

"Cessna 30A is a Cessna 182 located on the Kewanee 195° radial at 30 miles level at 4,000 feet. I am currently in IMC conditions with a smooth ride. Outside air temperature is plus 1° Celsius. Negative icing."

"Cessna 30A thank you for the PIREP."

With the weather check and PIREP complete, return to Memphis Center:

"Memphis Center, Cessna 1230A is back on your frequency."

"Cessna 1230A, Memphis Center, roger, contact Houston Center now on frequency 126.8."

"Roger, contact Houston Center frequency 126.8, Cessna 1230A."

"Houston Center, Cessna 1230A level at 4,000 feet."

"Cessna 30A, Houston Center area altimeter 29.88."

Arrival

40 miles north of Gulfport, tune in ATIS on number two communication radio. The report reveals there has been no change in the weather and ATIS is advertising ILS runway 14 as the active approach.

Houston Center completes a hand off to Gulfport approach control with instructions to contact approach:

"Gulfport Approach, Cessna 1230A level 4,000 feet with information TANGO. Request GPS Runway 14 approach."

"Cessna 30A, Gulfport Approach, descend and maintain 3,000 feet."

"Descend to 3,000, Cessna 30A."

Begin a descent to 3,000 and configure your navigation radios for the approach. The GPS automatically changes from the en route mode to the terminal mode. This change affects the sensitivity of the CDI. Tune in the VORTAC frequency of 109.0 on the number one navigation radio and set in the final approach course of 133° on the OBS. This setup helps with situational awareness should the GPS lose signal.

"Cessna 30A your position is 7 miles from MINDO, maintain 3,000 feet until MINDO, cleared for the GPS runway 14 approach."

Read back the clearance and concentrate on flying the aircraft. At MINDO descend to 2,000 as depicted on the approach chart. At BROWA turn to the final approach course of 133°. Just outside the Final Approach Way Point (FAWP) AVYUM, the GPS changes to the approach mode and the CDI becomes even more sensitive. Gulfport approach control issues instructions to contact Gulfport tower:

"Cessna 30A contact Tower on 123.7."

"123.7, Cessna 30A."

"Tower, Cessna 1230A outside AVYUM on the GPS runway 14."

"Cessna 30A Gulfport Tower, the ceiling is now 600 overcast and the visibility is 4 miles."

"Cleared to land runway 14, Cessna 30A."

Continue the approach, complete the appropriate checklists, cross AVYUM, and begin the final descent. At 700 feet MSL visual contact with the airport is possible. Slow the aircraft and configure it to allow a normal descent to landing. As touch down is completed, Gulfport Tower gives further instructions:

"Cessna 30A turn left at taxiway Bravo and contact ground on 120.4."

"Roger, Cessna 30A."

Taxi clear of the runway and complete the appropriate checklists. The tower automatically cancels the IFR flight plan.

Chapter 11

Emergency Operations

Introduction

Changing weather conditions, air traffic control (ATC), the aircraft, and the pilot are all variables that make instrument flying an unpredictable and challenging operation. The safety of the flight depends upon the pilot's ability to manage these variables while maintaining positive aircraft control and adequate situational awareness. This chapter discusses the recognition and suggested remedies for such abnormal and emergency events related to unforecasted, adverse weather; aircraft system malfunctions; communication/navigation system malfunctions; and loss of situational awareness.

Unforecast Adverse Weather

Inadvertent Thunderstorm Encounter

A pilot should avoid flying through a thunderstorm of any intensity. However, certain conditions may be present that could lead to an inadvertent thunderstorm encounter. For example, flying in areas where thunderstorms are embedded in large cloud masses may make thunderstorm avoidance difficult, even when the aircraft is equipped with thunderstorm detection equipment. Therefore, pilots must be prepared to deal with an inadvertent thunderstorm penetration. At the very least, a thunderstorm encounter subjects the aircraft to turbulence that could be severe. The pilot and passengers should tighten seat belts and shoulder harnesses, and secure any loose items in the cabin.

As with any emergency, the first order of business during an inadvertent thunderstorm encounter must be to fly the aircraft. The pilot workload is heavy; therefore, increased concentration is necessary to maintain an instrument scan. If a pilot inadvertently enters a thunderstorm, it is better to maintain a course straight through the thunderstorm rather than turning around. A straight course minimizes the amount of time in the thunderstorm, and turning maneuvers only increase structural stress on the aircraft.

Reduce power to a setting that maintains a speed at the recommended turbulence penetration speed as described in the Pilot's Operating Handbook/Airplane Flight Manual (POH/AFM), and try to minimize additional power adjustments. Concentrate on maintaining a level attitude while allowing airspeed and altitude to fluctuate. Similarly, if using the autopilot, disengage the altitude hold and speed hold modes, as they only increase the aircraft's maneuvering—thereby increasing structural stress.

During a thunderstorm encounter, the potential for icing also exists. As soon as possible, turn on anti-icing/deicing equipment and carburetor heat, if equipped. Icing can be rapid at any altitude and may lead to power failure and/or loss of airspeed indication.

Lightning is also present in a thunderstorm and can temporarily blind a pilot. To reduce this risk, turn up flight deck lights to the highest intensity, concentrate on the flight instruments, and resist the urge to look outside.

Inadvertent Icing Encounter

Because icing is unpredictable in nature, pilots may find themselves in icing conditions even though they have done everything practicable to avoid it. In order to stay alert to this possibility while operating in visible moisture, pilots should monitor the outside air temperature (OAT).

The effects of ice on aircraft are cumulative—thrust is reduced, drag increases, lift lessens, and weight increases. The results are an increase in stall speed and a deterioration of aircraft performance. In extreme cases, two to three inches of ice can form on the leading edge of the airfoil in less than 5 minutes. It takes only ½ inch of ice to reduce the lifting power of some aircraft by 50 percent and increases the frictional drag by an equal percentage.

A pilot can expect icing when flying in visible precipitation, such as rain or cloud droplets, and the temperature is between +02 and –10° Celsius. When icing is detected, a pilot should do one of two things, particularly if the aircraft is not equipped with deicing equipment: leave the area of precipitation or go to an altitude where the temperature is above freezing. This "warmer" altitude may not always be a lower altitude. Proper preflight action includes obtaining information on the freezing level and the above-freezing levels in precipitation areas.

If neither option is available, consider an immediate landing at the nearest suitable airport. Even if the aircraft is equipped with anti-icing/deicing equipment, it is not designed to allow aircraft to operate indefinitely in icing conditions. Anti-icing/deicing equipment gives a pilot more time to get out of the icing conditions. Report icing to ATC and request new routing or altitude. Be sure to report the type of aircraft, and use the following terms when reporting icing to ATC:

1. Trace. Ice becomes perceptible. Rate of accumulation is slightly greater than sublimation. Anti-icing/deicing equipment is not utilized unless encountered for an extended period of time (over 1 hour).
2. Light. The rate of accumulation may create a problem if flight is prolonged in this environment (over 1 hour). Occasional use of anti-icing/deicing equipment removes/prevents accumulation. It does not present a problem if anti-icing/deicing equipment is used.
3. Moderate. The rate of accumulation is such that even short encounters become potentially hazardous and use of anti-icing/deicing equipment or flight diversion is necessary.
4. Severe. The rate of accumulation is such that anti-icing/deicing equipment fails to reduce or control the hazard. Immediate flight diversion is necessary.

Early ice detection is critical and is particularly difficult during night flight. Use a flashlight to check for ice accumulation on the wings. At the first indication of ice accumulation, take action to get out of the icing conditions. Refer to the POH/AFM for the proper use of anti-icing/deicing equipment.

Precipitation Static

Precipitation static, often referred to as P-static, occurs when accumulated static electricity is discharged from the extremities of the aircraft. This discharge has the potential to create problems for the instrument pilot. These problems range from the serious, such as erroneous magnetic compass readings and the complete loss of very high frequency (VHF) communications to the annoyance of high-pitched audio squealing and St. Elmo's fire. *[Figure 11-1]*

Precipitation static is caused when an aircraft encounters airborne particles during flight (e.g., rain or snow) and develops a negative charge. It can also result from atmospheric electric fields in thunderstorm clouds. When a significant negative voltage level is reached, the aircraft discharges it, which can create electrical disturbances. This electrical discharge builds with time as the aircraft flies in precipitation. It is usually encountered in rain, but snow can cause the same effect. As the static buildup increases, the effectiveness of both communication and navigation systems decreases to the point of potential unusability.

To reduce the problems associated with P-static, the pilot should ensure the aircraft's static wicks are properly maintained and accounted for. Broken or missing static wicks should be replaced before an instrument flight. *[Figure 11-2]*

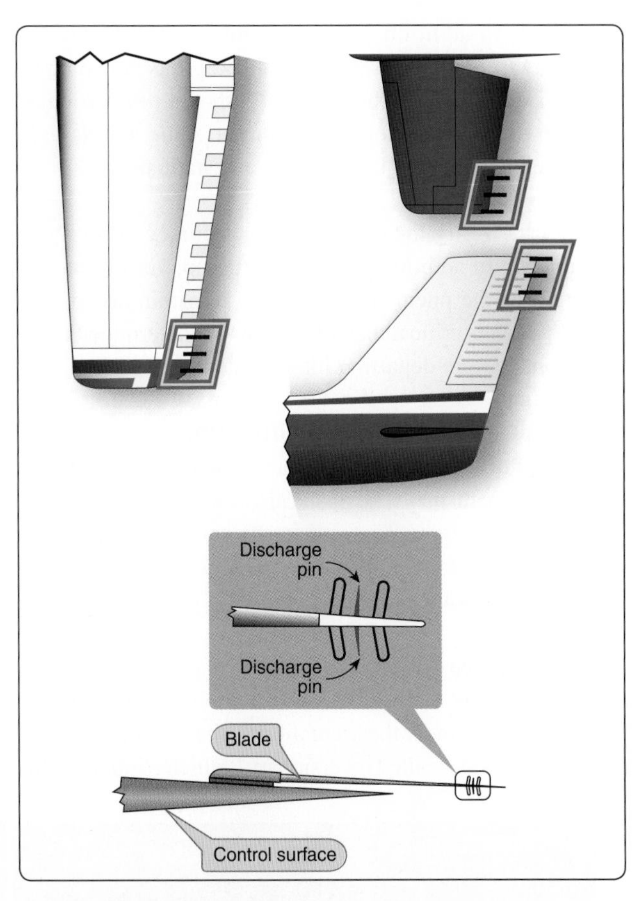

Figure 11-2. *One example of a static wick installed on aircraft control surface to bleed off static charges built up during flight. This prevents static buildup and St. Elmo's fire by allowing the static electricity to dissipate harmlessly.*

Aircraft System Malfunctions

Preventing aircraft system malfunctions that might lead to an inflight emergency begins with a thorough preflight

Figure 11-1. *St. Elmo's Fire is harmless but may affect both communication and navigation radios, especially the lower frequencies such as those used on the automatic direction finding (ADF).*

inspection. In addition to those items normally checked prior to a visual flight rules (VFR) flight, pilots intending to fly under instrument flight rules (IFR) should pay particular attention to the alternator belt, antennas, static wicks, anti-icing/deicing equipment, pitot tube, and static ports.

During taxi, verify the operation and accuracy of all flight instruments. In addition, during the run-up, verify that the operation of the pneumatic system(s) is within acceptable parameters. It is critical that all systems are determined to be operational before departing into IFR conditions.

Electronic Flight Display Malfunction

When a pilot becomes familiar and comfortable with the new electronic displays, he or she also tends to become more reliant on the system. The system then becomes a primary source of navigation and data acquisition instead of the supplementary source of data as initially intended.

Complete reliance on the moving map for navigation becomes a problem during a failure of one, more, or all of the flight display screens. Under these conditions, the systems revert to a composite mode (called reversionary), which eliminates the moving map display and combines the primary flight display (PFD) with the engine indicating system. *[Figure 11-3]* If a pilot has relied on the display for navigation information and situational awareness, he or she lacks any concept of critical data such as the aircraft's position, the nearest airport, or proximity to other aircraft.

The electronic flight display (EFD) is a supplementary source of navigation data and does not replace en route charts. To maintain situational awareness, a pilot must follow the flight on the en route chart while monitoring the PFD. It is important for the pilot to know the location of the closest airport as well as surrounding traffic relative to the location of his or her aircraft. This information becomes critical should the EFD fail.

For the pilot who utilizes the electronic database as a substitute for the Airport/Facilities Directory (A/FD), screen failure or loss of electrical power can mean the pilot is no longer able to access airport information. Once the pilot loses the ability to call up airport information, aeronautical decision-making (ADM) is compromised.

Figure 11-3. *G1000 PFD display in normal mode and in the reversionary mode activated upon system failure.*

Alternator/Generator Failure

Depending upon the aircraft being flown, an alternator failure is indicated in different ways. Some aircraft use an ammeter that indicates the state of charge or discharge of the battery. *[Figure 11-4]* A positive indication on the ammeter indicates a charge condition; a negative indication reveals a discharge condition. Other aircraft use a loadmeter to indicate the load being carried by the alternator. *[Figure 11-4]*

Sometimes an indicator light is also installed in the aircraft to alert the pilot to an alternator failure. On some aircraft, such as the Cessna 172, the light is located on the lower left side making it difficult to see its illumination if charts are open. Ensure that these safety indicators are visible during flight.

When a loss of the electrical charging system is experienced, the pilot has approximately 40 minutes of battery life remaining before the system fails entirely. The time mentioned is an approximation and should not be relied upon as specific to all aircraft. In addition, the battery charge that exists in a battery may not be full, altering the time available before electrical exhaustion occurs. At no time should a pilot consider continuing a flight once the electrical charging system has failed. Land at the nearest suitable airport.

Techniques for Electrical Usage

Master Battery Switch

One technique for conserving the main battery charge is to fly the aircraft to the airport of intended landing while operating with minimal power. If a two-position battery master/alternator rocker switch is installed, it can be utilized to isolate the main battery from the electrical system and conserve power. *[Figure 11-5]*

Figure 11-5. *Double rocker switch seen on many aircraft.*

Operating on the Main Battery

While en route to the airport of intended landing, reduce the electrical load as much as practical. Turn off all unnecessary electrical items, such as duplicate radios, non-essential lighting, etc. If unable to turn off radios, lights, etc., manually, consider pulling circuit breakers to isolate those pieces of equipment from the electrical system. Maximum time of useful voltage may be between 30 and 40 minutes and is influenced by many factors, that degrade the useful time.

Loss of Alternator/Generator for Electronic Flight Instrumentation

With the increase in electrical components being installed in modern technically advanced aircraft, the power supply and the charging system need increased attention and

Figure 11-4. *Ammeter (left) and loadmeter (right).*

understanding. Traditional round dial aircraft do not rely as heavily on electrical power for the primary six-pack instrumentation. Modern EFDs utilize the electrical system to power the Attitude Heading Reference System (AHRS), air data computer (ADC), engine indicating system (EIS), etc. A loss of an alternator or generator was considered an abnormality in traditionally-equipped aircraft; however, a failure of this magnitude is considered an emergency in technically advanced aircraft.

Due to the increased demand for electrical power, it is necessary for manufacturers to install a standby battery in conjunction with the primary battery. The standby battery is held in reserve and kept charged in case of a failure of the charging system and a subsequent exhaustion of the main battery. The standby battery is brought online when the main battery voltage is depleted to a specific value, approximately 19 volts. Generally, the standby battery switch must be in the ARM position for this to occur but pilots should refer to the aircraft flight manual (AFM) for specifics on an aircraft's electrical system. The standby battery powers the essential bus and allows the PFD to be utilized.

The essential bus usually powers the following components:

1. AHRS (Attitude and Heading Reference System)
2. ADC (Air Data Computer)
3. PFD (Primary Flight Display)
4. Navigation Radio #1
5. Communication Radio #1
6. Standby Indicator Light

Techniques for Electrical Usage

Standby Battery

One technique for conserving the main battery charge is to fly the aircraft to the airport of intended landing while using the standby battery. A two-position battery master/ alternator rocker switch is installed on most aircraft with EFDs, which can be utilized to isolate the main battery from the electrical system. By switching the MASTER side off, the battery is taken offline and the standby battery comes online to power the essential bus. However, the standby battery switch must be in the ARM position for this to occur. *[Figure 11-6]* Utilization of the standby battery first reserves the main battery for use when approaching to land. With this technique, electrical power may be available for the use of flaps, gear, lights, etc. Do not rely on any power to be available after the standby battery has exhausted itself. Once the charging system has failed, flight with a powered electrical system is not guaranteed.

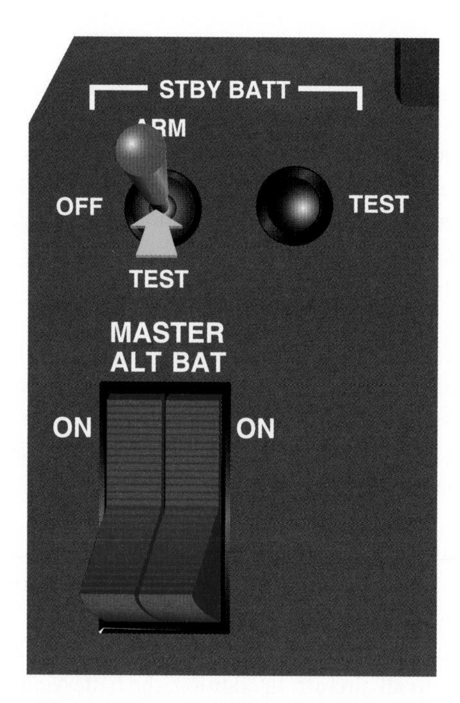

Figure 11-6. *Note the double rocker switch and the standby battery switch in this aircraft. The standby battery must be armed to work correctly; arming should be done prior to departure.*

Operating on the Main Battery

While en route to the airport of intended landing, reduce the electrical load as much as practical. Turn off all unnecessary electrical items, such as duplicate radios, non-essential lighting, etc. If unable to turn off radios, lights, etc., manually, consider pulling circuit breakers to isolate those pieces of equipment from the electrical system. Keep in mind that once the standby battery has exhausted its charge, the flight deck may become very dark depending on what time of day the failure occurs. The priority during this emergency situation is landing the aircraft as soon as possible without jeopardizing safety.

A standby attitude indicator, altimeter, airspeed indicator (ASI) and magnetic compass are installed in each aircraft for use when the PFD instrumentation is unavailable. *[Figure 11-7]* These would be the only instruments left available to the pilot. Navigation would be limited to pilotage and dead reckoning unless a hand-held transceiver with a global positioning system (GPS)/navigation function is onboard.

Once an alternator failure has been detected, the pilot must reduce the electrical load on the battery and land as soon as practical. Depending upon the electrical load and condition of the battery, there may be sufficient power available for 45 minutes of flight—or for only a matter of minutes. Pilots should also know which systems on the aircraft are electric and

Figure 11-7. *Emergency instrumentation available to the pilot on electronic flight instrumented aircraft.*

those that continue to operate without electrical power. Pilots can attempt to troubleshoot alternator failure by following the established alternator failure procedure published in the POH/AFM. If the alternator cannot be reset, advise ATC of the situation and inform them of the impending electrical failure.

Analog Instrument Failure

A warning indicator, or an inconsistency between indications on the attitude indicator and the supporting performance instruments, usually identifies system or instrument failure. Aircraft control must be maintained while identifying the failed component(s). Expedite the cross-check and include all flight instruments. The problem may be individual instrument failure or a system failure affecting multiple instruments.

One method of identification involves an immediate comparison of the attitude indicator with the rate-of-turn indicator and vertical speed indicator (VSI). Along with providing pitch-and-bank information, this technique compares the static system with the suction or pressure system and the electrical system. Identify the failed component(s) and use the remaining functional instruments to maintain aircraft control.

Attempt to restore the inoperative component(s) by checking the appropriate power source, changing to a backup or alternate system, and resetting the instrument if possible. Covering the failed instrument(s) may enhance a pilot's ability to maintain aircraft control and navigate the aircraft. Usually, the next step is to advise ATC of the problem and, if necessary, declare an emergency before the situation deteriorates beyond the pilot's ability to recover.

Pneumatic System Failure

One possible cause of instrument failure is a loss of the suction or pressure source. This pressure or suction is supplied by a vacuum pump mechanically driven off the engine. Occasionally these pumps fail, leaving the pilot with inoperative attitude and heading indicators.

Figure 11-8 illustrates inoperative vacuum driven attitude and heading indicators that can fail progressively. As the gyroscopes slow down, they may wander, which, if connected to the autopilot and/or flight director, can cause incorrect movement or erroneous indications. In *Figure 11-8,* the aircraft is actually level and at 2,000 feet mean sea level (MSL). It is not in a turn to the left which the pilot may misinterpret if he or she fails to see the off or failed flags. If that occurs, the pilot may transform a normally benign situation into a hazardous situation. Again, good decision-making by the pilot only occurs after a careful analysis of systems.

Many small aircraft are not equipped with a warning system for vacuum failure; therefore, the pilot should monitor the

Figure 11-8. *Vacuum failure.*

system's vacuum/pressure gauge. This can be a hazardous situation with the potential to lead the unsuspecting pilot into a dangerous unusual attitude that would require a partial panel recovery. It is important that pilots practice instrument flight without reference to the attitude and heading indicators in preparation for such a failure.

Pitot/Static System Failure

A pitot or static system failure can also cause erratic and unreliable instrument indications. When a static system problem occurs, it affects the ASI, altimeter, and the VSI. In most aircraft, provisions have been made for the pilot to select an alternate static source. Check the POH/AFM for the location and operation of the alternate static source. In the absence of an alternate static source, in an unpressurized aircraft, the pilot could break the glass on the VSI. The VSI is not required for instrument flight, and breaking the glass provides the altimeter and the ASI a source of static pressure. This procedure could cause additional instrument errors.

Communication/Navigation System Malfunction

Avionics equipment has become very reliable, and the likelihood of a complete communications failure is remote. However, each IFR flight should be planned and executed in anticipation of a two-way radio failure. At any given point during a flight, the pilot must know exactly what route to fly, what altitude to fly, and when to continue beyond a clearance limit. Title 14 of the Code of Federal Regulations (14 CFR) part 91 describes the procedures to be followed in case of a two-way radio communications failure. If operating in VFR conditions at the time of the failure, the pilot should continue the flight under VFR and land as soon as practicable. If the failure occurs in IFR conditions, or if VFR conditions cannot be maintained, the pilot must continue the flight:

1. Along the route assigned in the last ATC clearance received;
2. If being radar vectored, by the direct route from the point of radio failure to the fix, route, or airway specified in the vector clearance;
3. In the absence of an assigned route, by the route that ATC has advised may be expected in a further clearance; or
4. In the absence of an assigned route or a route that ATC has advised may be expected in a further clearance, by the route filed in the flight plan.

The pilot should maintain the highest of the following altitudes or flight levels for the route segment being flown:

1. The altitude or flight level assigned in the last ATC clearance received;
2. The minimum altitude (converted, if appropriate, to minimum flight level as prescribed in 14 CFR, part 91 for IFR operations); or
3. The altitude or flight level ATC has advised may be expected in a further clearance.

In addition to route and altitude, the pilot must also plan the progress of the flight to leave the clearance limit.

1. When the clearance limit is a fix from which an approach begins, commence descent or descent and approach as close as possible to the expect-further-clearance time if one has been received. If an expect-further-clearance time has not been received, commence descent or descent and approach as close as possible to the estimated time of arrival as calculated from the filed or amended (with ATC) estimated time en route.
2. If the clearance limit is not a fix from which an approach begins, leave the clearance limit at the expect-further-clearance time if one has been received. If no expect-further-clearance time has been received, leave the clearance limit upon arrival over it, and proceed to a fix from which an approach begins and commence descent or descent and approach as close as possible to the estimated time of arrival as calculated from the filed or amended (with ATC) estimated time en route. *[Figure 11-8]*

While following these procedures, set the transponder to code 7600, and use all means possible to reestablish two-way radio communication with ATC. This includes monitoring navigational aids (NAVAIDs), attempting radio contact with other aircraft, and attempting contact with a nearby flight service station (FSS).

GPS Nearest Airport Function

Procedures for accessing the nearest airport information vary by the type of display installed in an aircraft. Pilots can obtain information relative to the nearest airport by using the PFD, multi-function display (MFD), or the nearest function on the GPS receiver. The following examples are based on a popular system. Pilots should become familiar with the operational characteristics of the equipment to be used.

Nearest Airports Using the PFD

With the advancements in electronic databases, diverting to alternate airports has become easier. Simply by pressing a soft key on the PFD, pilots can access information for up to 25 of the nearest airports that meet the criteria set in the systems configuration page. *[Figure 11-9]* Pilots are able to specify what airports are acceptable for their aircraft requirements based on landing surface and length of runway.

When the text box opens, the flashing cursor is located over the nearest airport that meets the criteria set in the auxiliary setup page as shown in *Figure 11-10*. Scrolling through the 25 airports is accomplished by turning the outer FMS knob, which is located on the lower right corner of the display screen. Turning the FMS knob clockwise moves the blinking cursor to the next closest airport. By continuing to turn the knob, the pilot is able to scroll through all 25 nearest airports. Each airport box contains the information illustrated in

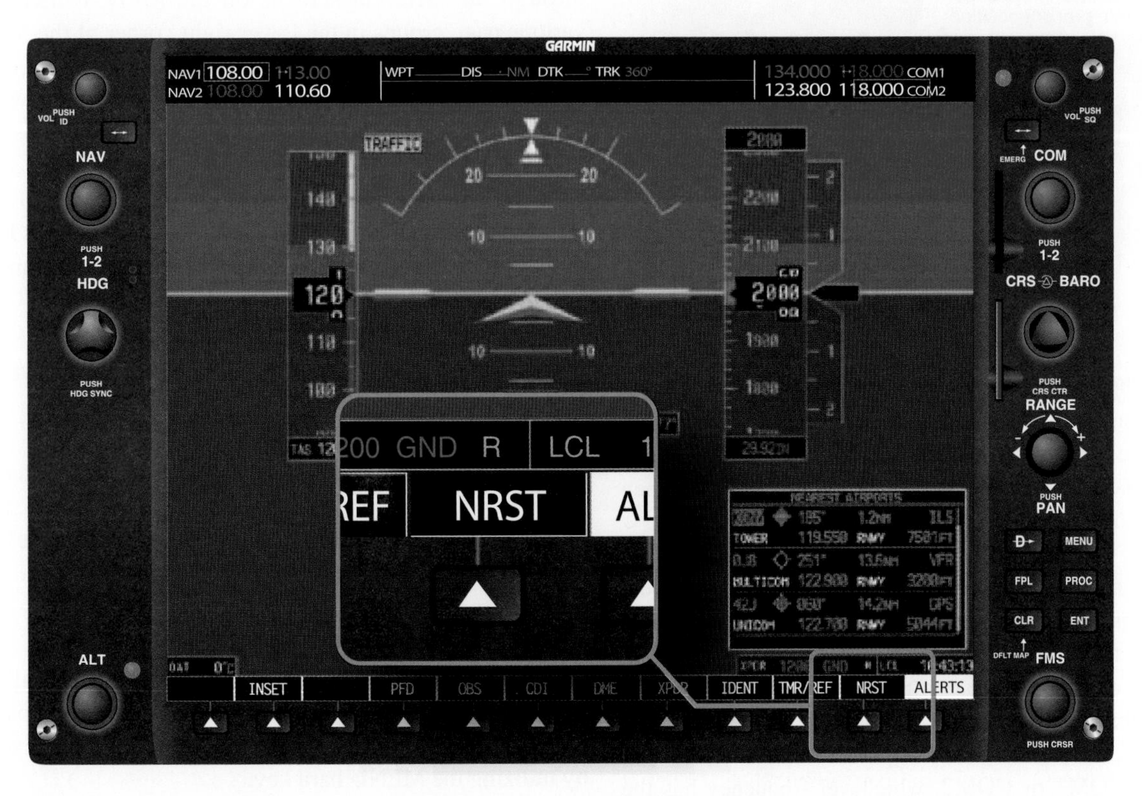

Figure 11-9. *The default soft key menu that is displayed on the PFD contains a "NRST" (Nearest Airport) soft key. Pressing this soft key opens a text box that displays the nearest 25 airports.*

Figure 11-10. *An enlargement of the box shown in the lower right of Figure 11-9. Note that KGNV would be flashing.*

Figure 11-11, which the pilot can utilize to determine which airport best suits their individual needs.

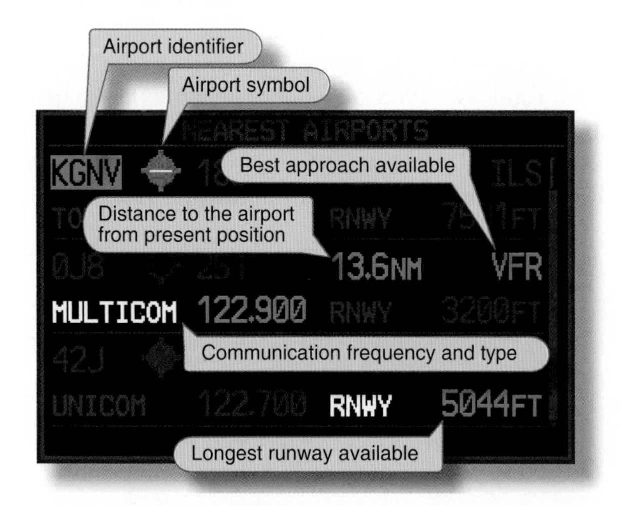

Figure 11-11. *Information shown on the nearest airport page.*

Additional Information for a Specific Airport

In addition to the information that is presented on the first screen, the pilot can view additional information as shown in *Figure 11-12* by highlighting the airport identifier and then pressing the enter key.

From this menu or the previous default nearest airport screen, the pilot is able to activate the Direct-To function, which provides a direct GPS course to the airport. In addition, the pilot can auto-tune communication frequencies by highlighting the appropriate frequency and then pressing the enter key. The frequency is placed in the stand-by box of either COM1 or COM2, whichever frequency has the cyan box around it.

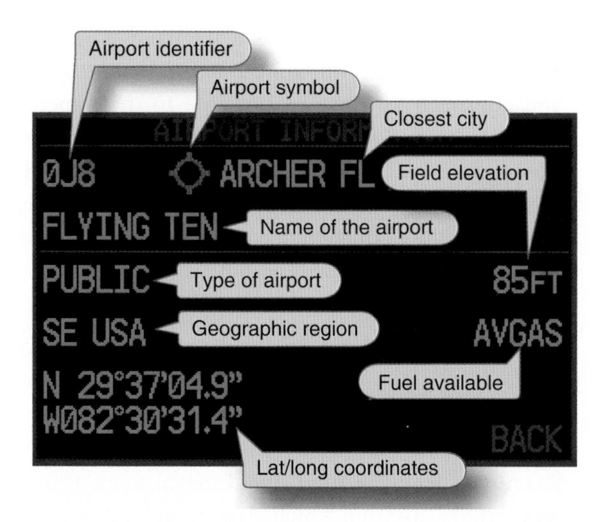

Figure 11-12. *Information shown on the additional information page that will aid the pilot in making a more informed decision about which airport to choose when diverting.*

Nearest Airports Using the MFD

A second way to determine the nearest airport is by referencing the NRST Page Group located on the MFD. This method provides additional information to the pilot; however, it may require additional steps to view. *[Figure 11-13]*

Navigating the MFD Page Groups

Most display systems are designed for ease of navigation through the different screens on the MFD. Notice the various page groups in the lower right corner of the MFD screen. Navigation through these four page groups is accomplished by turning the outer FMS knob clockwise. *[Figure 11-14]*

Within each page group are specific pages that provide additional information pertaining to that specific group. Once the desired page group and page is selected, the MFD remains in that configuration until the page is changed or the CLR button is depressed for more than 2 seconds. Holding the CLR button returns the display to the default moving map page.

Nearest Airport Page Group

The nearest airport page contains specific areas of interest for the airport selected. *[Figure 11-15]* The pilot is furnished information regarding runways, frequencies, and types of approaches available.

Nearest Airports Page Soft Keys

Figure 11-16 illustrates four specific soft keys that allow the pilot to access independent windows of the airport page. Selection of each of these windows can also be accomplished by utilizing the MENU hard key.

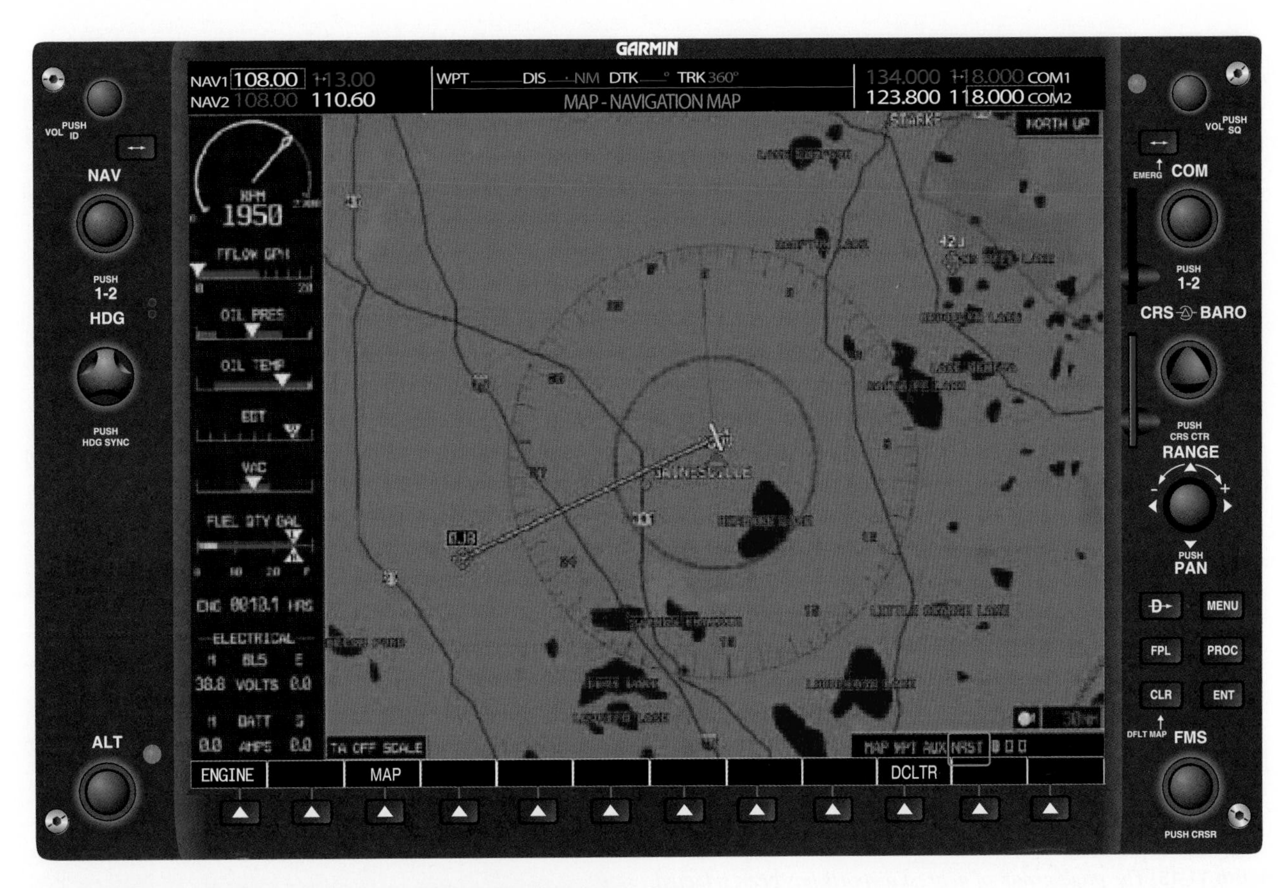

Figure 11-13. *The MFD is another means of viewing the nearest airports.*

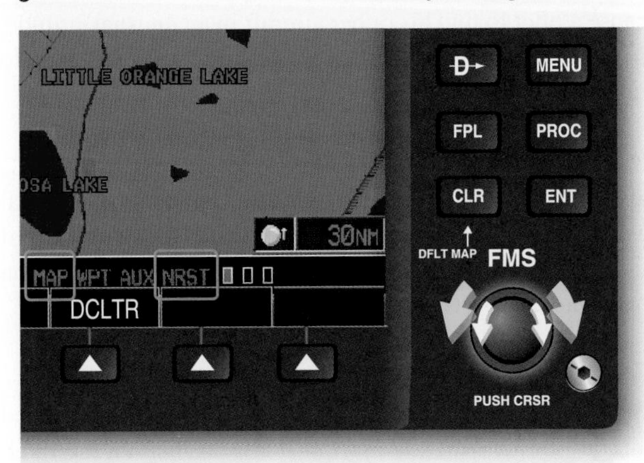

Figure 11-14. *Page groups. As the FMS outer knob is rotated, the current page group is indicated by highlighting the specific group indicator. Notice that the MAP page group is highlighted.*

The soft keys and functions are as follows: Scroll through each section with the cursor, then press enter to accept the selection.

1. APT. Allows the user access to scroll through the 25 nearest airports. The white arrow indicates which airport is selected. The INFORMATION window is slaved to the white arrow. The INFORMATION window decodes the airport identifier. Scroll through the 25 airports by turning the outer FMS knob.
2. RNWY. Moves the cursor into the Runways section and allows the user to scroll through the available runways at a specific airport that is selected in conjunction with the APT soft key. A green arrow indicates additional runways to view.
3. FREQ. Moves the cursor into the Frequencies section and allows the pilot to highlight and auto-tune the frequency into the selected standby box.
4. APR. Moves the cursor into the Approach section and allows the pilot to review approaches and load them into the flight plan. When the APR soft key is selected, an additional soft key appears. The LD APR (Load Approach) soft key must be pressed once the desired instrument approach procedure has been highlighted. Once the soft key is pressed, the screen changes to the PROC Page Group. From this page, the pilot is able to choose the desired approach, the transition, and choose the option to activate the approach or just load it into the flight plan.

Situational Awareness

Situational awareness is not simply a mental picture of aircraft location; rather, it is an overall assessment of each element of the environment and how it affects a flight. On one end of the situational awareness spectrum is a pilot who

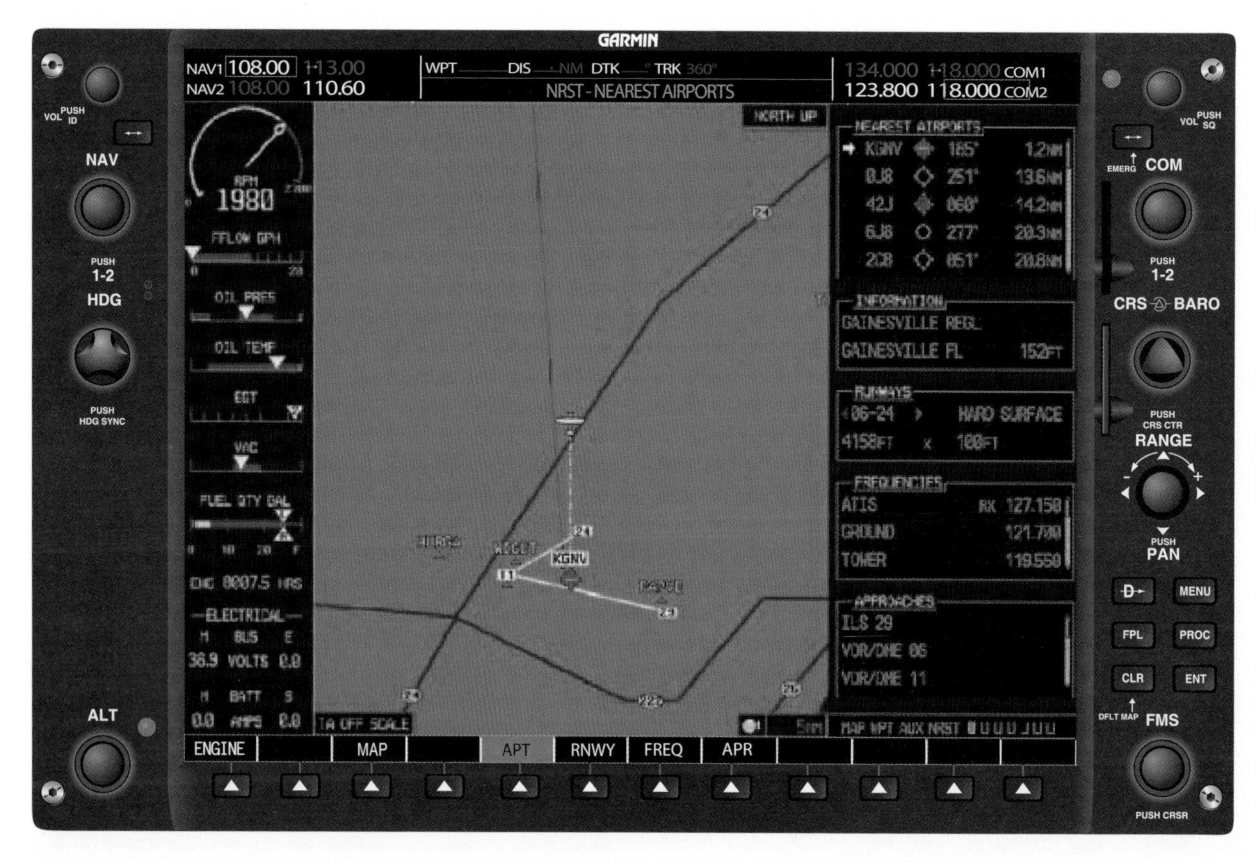

Figure 11-15. *The page group of nearest airports has been selected.*

is knowledgeable of every aspect of the flight; consequently, this pilot's decision-making is proactive. With good situational awareness, this pilot is able to make decisions well ahead of time and evaluate several different options. On the other end of the situational awareness spectrum is a pilot who is missing important pieces of the puzzle: "I knew exactly where I was when I ran out of fuel." Consequently, this pilot's decision-making is reactive. With poor situational awareness, a pilot lacks a vision of future events and is forced to make decisions quickly, often with limited options.

During a typical IFR flight, a pilot operates at varying levels of situational awareness. For example, a pilot may be cruising to his or her destination with a high level of situational awareness when ATC issues an unexpected standard terminal arrival route (STAR). Since the pilot was not expecting the STAR and is not familiar with it, situational awareness is lowered. However, after becoming familiar with the STAR and resuming normal navigation, the pilot returns to a higher level of situational awareness.

Factors that reduce situational awareness include: distractions, unusual or unexpected events, complacency, high workload, unfamiliar situations, and inoperative equipment. In some situations, a loss of situational awareness may be beyond a pilot's control. For example, a pneumatic system failure and associated loss of the attitude and heading indicators could cause a pilot to find his or her aircraft in an unusual attitude. In this situation, established procedures must be used to regain situational awareness.

Pilots should be alert to a loss of situational awareness anytime they are in a reactive mindset. To regain situational awareness, reassess the situation and seek additional information from other sources, such as the navigation instruments or ATC.

Traffic Avoidance

EFDs have the capability of displaying transponder-equipped aircraft on the MFD, as well as the inset map on the PFD. However, due to the limitations of the systems, not all traffic is displayed. Some traffic information service (TIS) units display only eight intruding targets within the service volume. The normal service volume has altitude limitations of 3,500 feet below the aircraft to 3,500 feet above the aircraft. The lateral limitation is 7 NM. *[Figure 11-17]* Pilots unfamiliar with the limitations of the system may rely on the aural warnings to alert them to approaching traffic.

In addition to an outside visual scan of traffic, a pilot should incorporate any traffic information electronically displayed, such as TIS. This innovation in traffic alerting reinforces and adds synergy to the ability to see and avoid. However, it is an aid and not a replacement for the responsibilities of the

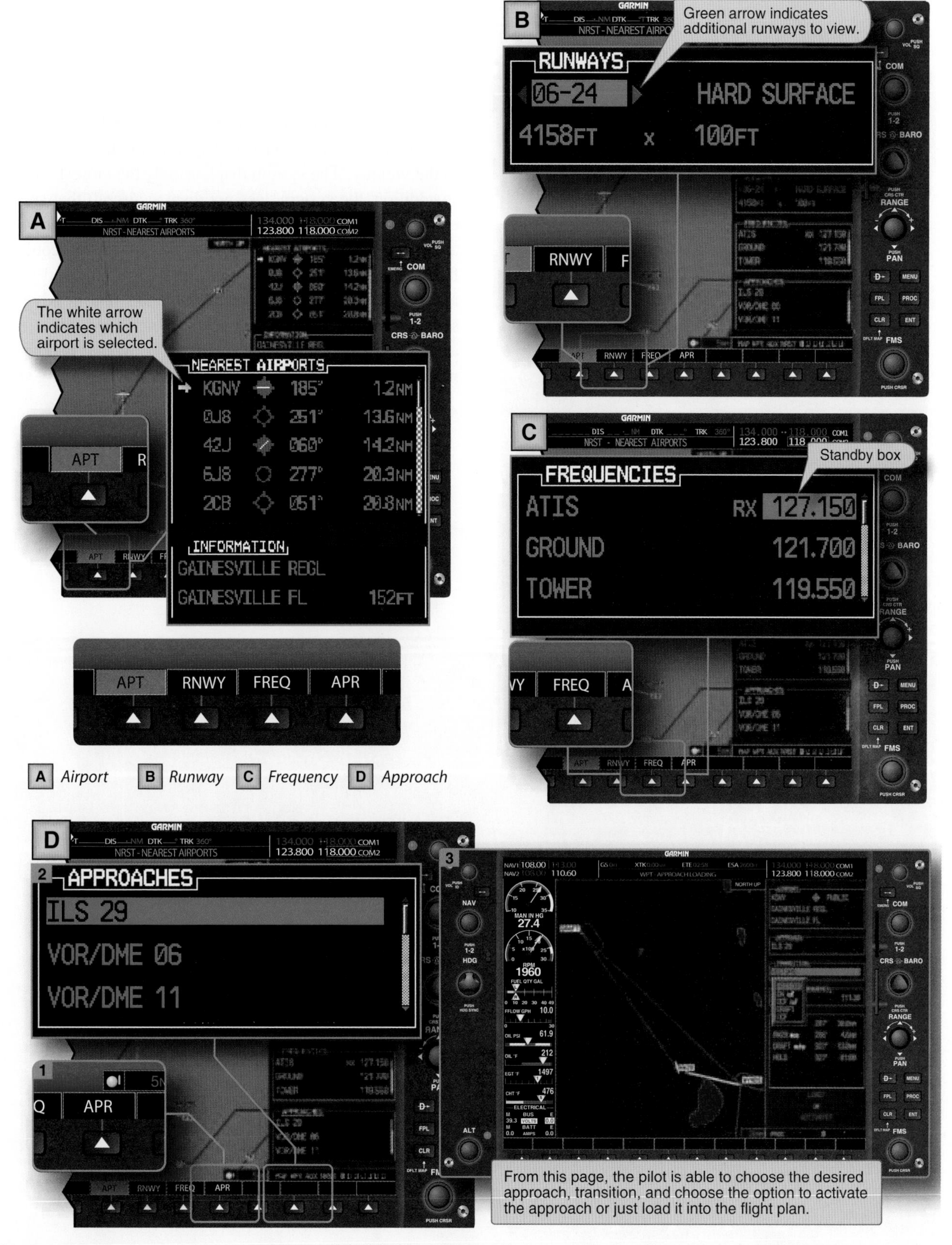

Figure 11-16. *The four soft keys at the bottom of the MFD are airport (A), runway (B), frequency (C), and approach (D).*

pilot. Systems such as TIS provide a visual representation of nearby traffic and displays a symbol on the moving map display with relative information about altitude, vertical trends, and direction of flight. *[Figure 11-18]*

It is important to remember that most systems display only a specific maximum number of targets allowed. Therefore, it does not mean that the targets displayed are the only aircraft in the vicinity. The system displays only the closest aircraft. In addition, the system does not display aircraft that are not equipped with transponders. The display may not show any aircraft; however, a Piper Cub with no transponder could be flying in the area. TIS coverage can be sporadic and is not available in some areas of the United States. Traffic advisory software is to be utilized only for increased situational awareness and not the sole means of traffic avoidance. There is no substitute for a good visual scan of the surrounding sky.

Figure 11-17. *The area surrounding the aircraft for coverage using TIS.*

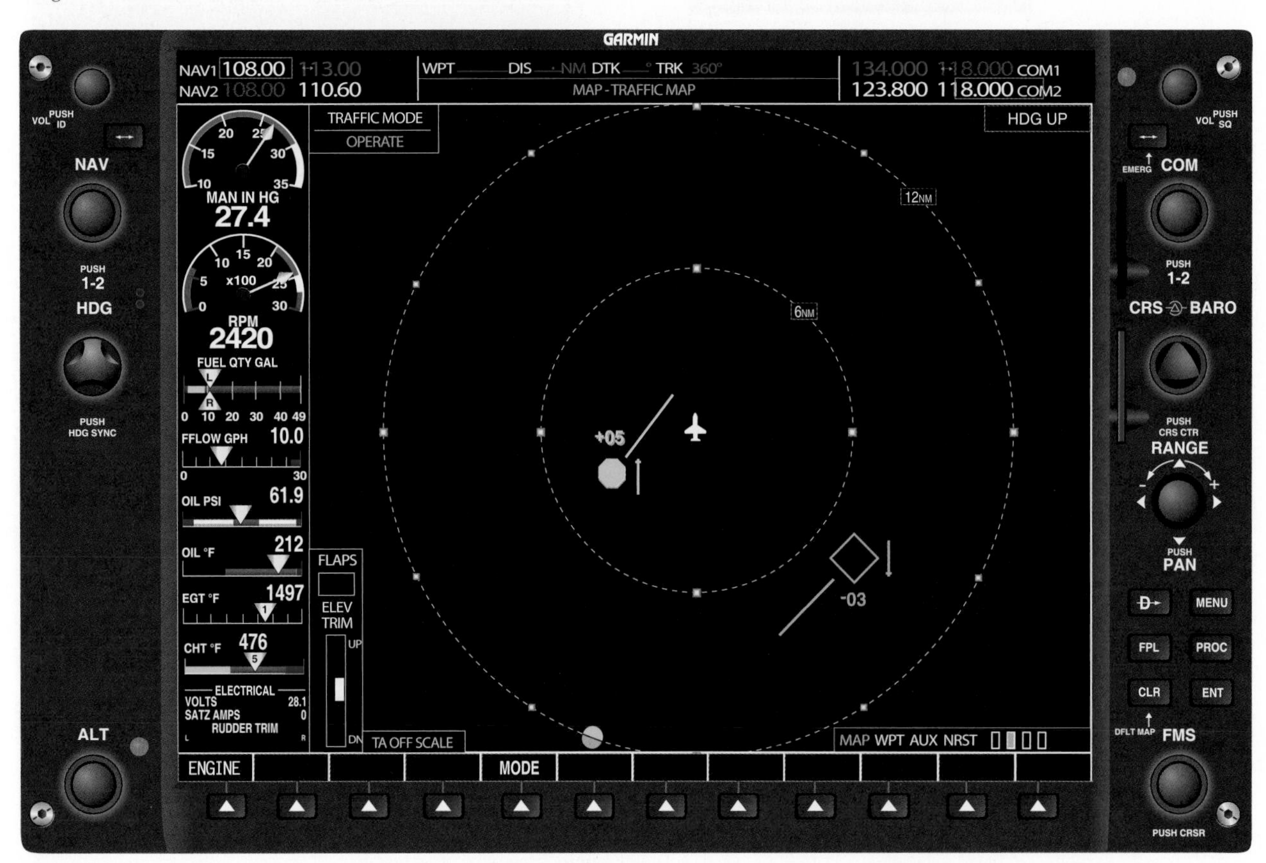

Figure 11-18. *A typical display on aircraft MFD when using TIS.*

Appendix A

Clearance Shorthand

The following shorthand system is recommended by the Federal Aviation Administration (FAA). Applicants for the instrument rating may use any shorthand system, in any language, which ensures accurate compliance with air traffic control (ATC) instructions. No shorthand system is required by regulation and no knowledge of shorthand is required for the FAA Knowledge Test; however, because of the vital need for reliable communication between the pilot and controller, clearance information should be unmistakably clear.

The following symbols and contractions represent words and phrases frequently used in clearances. Most are used regularly by ATC personnel. By practicing this shorthand, omitting the parenthetical words, you will be able to copy long clearances as fast as they are read.

Example: CAF M→ RH RV V18 ↑40 SQ 0700 DPC 120.4
Cleared as filed, maintain runway heading for radar vector to Victor 18, climb to 4,000, squawk 0700, departure control frequency is 120.4.

Words and Phrases	**Shorthand**
Above	ABV
Above (Altitude, Hundreds of Feet)	$\underline{70}$
Adjust speed to 250 knots	250 K
Advise	ADZ
After (Passing)	<
Airway (Designation)	V26
Airport	A
Alternate Instructions	( )
Altitude 6,000–17,000	60-170
And	&
Approach	AP
Approach Control	APC
Area Navigation	RNAV
Arriving	↓
At	@
At or Above	$\underline{\uparrow}$
At or Below	$\overline{\downarrow}$
(ATC) Advises	CA
(ATC) Clears or Cleared	C
(ATC) Requests	CR
Back Course	BC
Bearing	BR
Before (Reaching, Passing)	>
Below	BLO
Below (Altitude, Hundreds of Feet)	$\overline{70}$
Center	CTR
Clearance Void if Not Off By (Time)	v<
Cleared as Filed	CAF
Cleared to Airport	A
Cleared to Climb/Descend at Pilot's Discretion	PD
Cleared to Cross	X
Cleared to Depart From the Fix	D
Cleared to the Fix	F
Cleared to Hold and Instructions Issued	H
Cleared to Land	L
Cleared to the Outer Marker	O
Climb to (Altitude, Hundreds of Feet)	↑70
Contact Approach	CT
Contact (Denver) Approach Control	(den
Contact (Denver) Center	(DEN
Course	CRS
Cross	X
Cruise	→
Delay Indefinite	DLI
Depart (Direction, if Specified)	T→( )
Departure Control	DPC
Descend To (Altitude, Hundreds of Feet)	↓70
Direct	DR
Direction (Bound)	
Eastbound	EB
Westbound	WB
Northbound	NB
Southbound	SB
Inbound	IB
Outbound	OB
DME Fix (Mile)	(21)
Each	EA
Enter Control Area	△
Estimated Time of Arrival	ETA
Expect	EX
Expect-Further-Clearance	EFC

Fan Marker	FM
Final	F
Final Approach	FA
Flight Level	FL
Flight Planned Route	FPR
For Further Clearance	FFC
For Further Headings	FFH
From	FM
Ground	GND
GPS Approach	GPS
Heading	HDG
Hold (Direction)	H-W
Holding Pattern	⬭
ILS Approach	ILS
Increase Speed 30 Knots	+30 K
Initial Approach	I
Instrument Departure Procedure	DP
Intersection	XN
Join or Intercept Airway/Jet Route/Track or Course	⩾
Left Turn After Takeoff	↰
Locator Outer Marker	LOM
Magnetic	M
Maintain	~~M~~→
Maintain VFR Conditions On Top	<u>VFR</u>
Middle Compass Locator	ML
Middle Marker	MM
Missed Approach	MA
Nondirectional Beacon Approach	NDB
Out of (Leave) Control Area	△↗
Outer Marker	OM
Over (Station)	<u>OKC</u>
On Course	OC
Precision Approach Radar	PAR
Procedure Turn	PT
Radar Vector	RV
Radial (080° Radial)	080R
Reduce Speed 20 Knots	-20 K
Remain This Frequency	RTF
Remain Well to Left Side	LS
Remain Well to Right Side	RS
Report Crossing	RX
Report Departing	RD
Report Leaving	RL
Report on Course	R-CRS
Report Over	RO
Report Passing	RP
Report Reaching	RR
Report Starting Procedure Turn	RSPT
Reverse Course	RC
Right Turn After Takeoff	↱
Runway Heading	RH
Runway (Number)	RY18
Squawk	SQ
Standby	STBY
Straight-in Approach	SI
Surveillance Radar Approach	ASR
Takeoff (Direction)	T→N
Tower	Z
Turn Left	TL
Turn Right	TR
Until	/
Until Advised (By)	UA
Until Further Advised	UFA
VFR Conditions On Top	OTP
Via	VIA
Victor (Airway Number)	V14
Visual Approach	VA
VOR	⊙
VOR Approach	VR
VORTAC	Ⓣ
While in Control Area	△

Appendix B

Instrument Training Lesson Guide

Introduction

Flight instructors may use this guide in the development of lesson plans. The lessons are arranged in a logical learning sequence and use the building-block technique. Each lesson includes ground training appropriate to the flight portion of the lesson. It is vitally important that the flight instructor brief the student on the objective of the lesson and how it will be accomplished. Debriefing the student's performance is also necessary to motivate further progress. To ensure steady progress, student pilots should master the objective of each lesson before advancing to the next lesson. Lessons should be arranged to take advantage of each student's knowledge and skills.

Flight instructors must monitor progress closely during training to guide student pilots in how to properly divide their attention. The importance of this division of attention or "cross-check" cannot be overemphasized. Cross-check and proper instrument interpretation are essential components of "attitude instrument flying" that enables student pilots to accurately visualize the aircraft's attitude at all times.

When possible, each lesson should incorporate radio communications, basic navigation, and emergency procedures so the student pilot is exposed to the entire IFR experience with each flight. Cross-reference the Instrument Training Lesson Guide with this handbook and the Instrument Practical Test Standards for a comprehensive instrument rating training program.

Lesson 1—Ground and flight evaluation of student's knowledge and performance

Aircraft systems
Aircraft performance
Preflight planning
Use of checklists
Basic flight maneuvers
Radio communications procedures
Navigation systems

Lesson 2—Preflight preparation and flight by reference to instruments

Ground Training

Instrument system preflight procedures
Attitude instrument flying
Fundamental instrument skills
Instrument cross-check techniques

Flight Training

Aircraft and instrument preflight inspection
Use of checklists
Fundamental instrument skills
Basic flight maneuvers
Instrument approach (demonstrated)
Postflight procedures

Lesson 3—Flight instruments and human factors

Ground Training

Human factors
Flight instruments and systems
Aircraft systems
Navigation instruments and systems

Flight Training

Aircraft and instrument preflight inspection
Radio communications
Checklist procedures
Attitude instrument flying
Fundamental instrument skills
Basic flight maneuvers
Spatial disorientation demonstration
Navigation systems
Postflight procedures

Lesson 4—Attitude instrument flying

Ground Training

Human factors
Flight instruments and systems

Aircraft systems
Navigation instruments and systems
Attitude instrument flying
Fundamental instrument skills
Basic flight maneuvers

Flight Training

Aircraft and instrument preflight inspection
Checklist procedures
Radio communications
Attitude instrument flying
Fundamental instrument skills
Basic flight maneuvers
Spatial disorientation
Navigation
Postflight procedures

Lesson 5—Aerodynamic factors and basic flight maneuvers

Ground Training

Basic aerodynamic factors
Basic instrument flight patterns
Emergency procedures

Flight Training

Aircraft and instrument preflight inspection
Checklist procedures
Radio communications
Basic instrument flight patterns
Emergency procedures
Navigation
Postflight procedures

Lesson 6—Partial panel operations

Ground Training

ATC system
Flight instruments
Partial panel operations

Flight Training

Aircraft and instrument preflight inspection
Checklist procedures
Radio communications
Basic instrument flight patterns
Emergency procedures
Partial panel practice
Navigation
Postflight procedures

Lesson 7—Recovery from unusual attitudes

Ground Training

Attitude instrument flying
ATC system
NAS overview

Flight Training

Preflight
Aircraft and instrument preflight inspection
Checklist procedures
Radio communications
Instrument takeoff
Navigation
Partial panel practice
Recovery from unusual attitudes
Postflight procedures

Lesson 8—Navigation systems

Ground Training

ATC clearances
Departure procedures
IFR en route charts

Flight Training

Aircraft and instrument preflight inspection
Checklist procedures
Radio communications
Intercepting and tracking
Holding
Postflight procedures

Lesson 9—Review and practice

Ground Training

Aerodynamic factors
Flight instruments and systems
Attitude instrument flying
Navigation systems
NAS
ATC
Emergency procedures

Flight Training

Aircraft and instrument preflight inspection
Checklist procedures
Radio communications
Review and practice as determined by the flight instructor

Instrument takeoff
Radio communications
Navigation systems
Emergency procedures
Postflight procedures

Lessons 10 through 19—Orientation, intercepting, tracking, and holding using each navigation system installed in the aircraft

Ground Training

Preflight planning
Navigation systems
NAS
ATC
Emergencies

Flight Training

Aircraft and instrument preflight inspection
Checklist procedures
Radio communications
Departure procedures
En route navigation
Terminal operations
Partial panel operation
Instrument approach
Missed approach
Approach to a landing
Postflight procedures

Lessons 20 and 21—Cross-country flights

Ground Training

Preflight planning
Aircraft performance
Navigation systems
NAS
ATC
Emergencies

Flight Training

Emergency procedures
Partial panel operation
Aircraft and instrument preflight inspection
Checklist procedures
Radio communications
Departure procedures
En route navigation
Terminal operations
Instrument approach
Missed approach
Approach to a landing
Postflight procedures

Lessons 22 and 23—Review and practice

Ground Training

Human factors
Aerodynamic factors
Flight instruments and systems
Attitude instrument flying
Basic flight maneuvers
Navigation systems
NAS
ATC
Emergency operations

Flight Training

Aircraft and instrument preflight inspection
Checklist procedures
Radio communications
Review and practice as determined by the flight instructor
Instrument takeoff
Partial panel operations
Unusual attitude recoveries
Radio communications
Navigation systems
Emergency procedures
Postflight procedures

Lessons 24 and subsequent—Practical test preparation

Ground Training

Title 14 of the Code of Federal Regulations (14 CFR) parts 61, 71, 91, 95, and 97
Instrument Flying Handbook
Practical test standards
Administrative requirements
Equipment requirements
Applicant's requirements

Flight Training

Review and practice until the student can consistently perform all required tasks in accordance with the appropriate practical test standards.

NOTE: It is the recommending instructor's responsibility to ensure that the applicant meets 14 CFR part 61 requirements and is prepared for the practical test, including: training, knowledge, experience, and the appropriate instructor endorsements.

Glossary

Absolute accuracy. The ability to determine present position in space independently, and is most often used by pilots.

Absolute altitude. The actual distance between an aircraft and the terrain over which it is flying.

Absolute pressure. Pressure measured from the reference of zero pressure, or a vacuum.

A.C. Alternating current.

Acceleration error. A magnetic compass error apparent when the aircraft accelerates while flying on an easterly or westerly heading, causing the compass card to rotate toward North.

Accelerometer. A part of an inertial navigation system (INS) that accurately measures the force of acceleration in one direction.

ADF. See automatic direction finder.

ADI. See attitude director indicator.

ADM. See aeronautical decision-making.

ADS–B. See automatic dependent surveillance–broadcast.

Adverse yaw. A flight condition at the beginning of a turn in which the nose of the aircraft starts to move in the direction opposite the direction the turn is being made, caused by the induced drag produced by the downward-deflected aileron holding back the wing as it begins to rise.

Aeronautical decision-making (ADM). A systematic approach to the mental process used by pilots to consistently determine the best course of action in response to a given set of circumstances.

A/FD. See Airport/Facility Directory.

Agonic line. An irregular imaginary line across the surface of the Earth along which the magnetic and geographic poles are in alignment, and along which there is no magnetic variation.

Aircraft approach category. A performance grouping of aircraft based on a speed of 1.3 times the stall speed in the landing configuration at maximum gross landing weight.

Air data computer (ADC). An aircraft computer that receives and processes pitot pressure, static pressure, and temperature to calculate very precise altitude, indicated airspeed, true airspeed, and air temperature.

AIRMET. Inflight weather advisory issued as an amendment to the area forecast, concerning weather phenomena of operational interest to all aircraft and that is potentially hazardous to aircraft with limited capability due to lack of equipment, instrumentation, or pilot qualifications.

Airport diagram. The section of an instrument approach procedure chart that shows a detailed diagram of the airport. This diagram includes surface features and airport configuration information.

Airport/Facility Directory (A/FD). An FAA publication containing information on all airports, communications, and NAVAIDs.

Airport surface detection equipment (ASDE). Radar equipment specifically designed to detect all principal features and traffic on the surface of an airport, presenting the entire image on the control tower console; used to augment visual observation by tower personnel of aircraft and/or vehicular movements on runways and taxiways.

Airport surveillance radar (ASR). Approach control radar used to detect and display an aircraft's position in the terminal area.

Airport surveillance radar approach. An instrument approach in which ATC issues instructions for pilot compliance based on aircraft position in relation to the final approach course and the distance from the end of the runway as displayed on the controller's radar scope.

Air route surveillance radar (ARSR). Air route traffic control center (ARTCC) radar used primarily to detect and display an aircraft's position while en route between terminal areas.

Air route traffic control center (ARTCC). Provides ATC service to aircraft operating on IFR flight plans within controlled airspace and principally during the en route phase of flight.

Airspeed indicator. A differential pressure gauge that measures the dynamic pressure of the air through which the aircraft is flying. Displays the craft's airspeed, typically in knots, to the pilot.

Air traffic control radar beacon system (ATCRBS). Sometimes called secondary surveillance radar (SSR), which utilizes a transponder in the aircraft. The ground equipment is an interrogating unit, in which the beacon antenna is mounted so it rotates with the surveillance antenna. The interrogating unit transmits a coded pulse sequence that actuates the aircraft transponder. The transponder answers the coded sequence by transmitting a preselected coded sequence back to the ground equipment, providing a strong return signal and positive aircraft identification, as well as other special data.

Airway. An airway is based on a centerline that extends from one navigation aid or intersection to another navigation aid (or through several navigation aids or intersections); used to establish a known route for en route procedures between terminal areas.

Alert area. An area in which there is a high volume of pilot training or an unusual type of aeronautical activity.

Almanac data. Information the global positioning system (GPS) receiver can obtain from one satellite which describes the approximate orbital positioning of all satellites in the constellation. This information is necessary for the GPS receiver to know what satellites to look for in the sky at a given time.

ALS. See approach lighting system.

Alternate airport. An airport designated in an IFR flight plan, providing a suitable destination if a landing at the intended airport becomes inadvisable.

Alternate static source valve. A valve in the instrument static air system that supplies reference air pressure to the altimeter, airspeed indicator, and vertical speed indicator if the normal static pickup should become clogged or iced over.

Altimeter setting. Station pressure (the barometric pressure at the location the reading is taken) which has been corrected for the height of the station above sea level.

AME. See aviation medical examiner.

Amendment status. The circulation date and revision number of an instrument approach procedure, printed above the procedure identification.

Ammeter. An instrument installed in series with an electrical load used to measure the amount of current flowing through the load.

Aneroid. The sensitive component in an altimeter or barometer that measures the absolute pressure of the air. It is a sealed, flat capsule made of thin disks of corrugated metal soldered together and evacuated by pumping all of the air out of it.

Aneroid barometer. An instrument that measures the absolute pressure of the atmosphere by balancing the weight of the air above it against the spring action of the aneroid.

Angle of attack. The acute angle formed between the chord line of an airfoil and the direction of the air striking the airfoil.

Anti-ice. Preventing the accumulation of ice on an aircraft structure via a system designed for that purpose.

Approach lighting system (ALS). Provides lights that will penetrate the atmosphere far enough from touchdown to give directional, distance, and glide path information for safe transition from instrument to visual flight.

Area chart. Part of the low-altitude en route chart series, this chart furnishes terminal data at a larger scale for congested areas.

Area navigation (RNAV). Allows a pilot to fly a selected course to a predetermined point without the need to overfly ground-based navigation facilities, by using waypoints.

ARSR. See air route surveillance radar.

ARTCC. See air route traffic control center.

ASDE. See airport surface detection equipment.

ASOS. See automated surface observing station.

ASR. See airport surveillance radar.

ATC. Air Traffic Control.

ATCRBS. See air traffic control radar beacon system.

ATIS. See automatic terminal information service.

Atmospheric propagation delay. A bending of the electromagnetic (EM) wave from the satellite that creates an error in the GPS system.

Attitude and heading reference systems (AHRS). System composed of three-axis sensors that provide heading, attitude, and yaw information for aircraft. AHRS are designed to replace traditional mechanical gyroscopic flight instruments and provide superior reliability and accuracy.

Attitude director indicator (ADI). An aircraft attitude indicator that incorporates flight command bars to provide pitch and roll commands.

Attitude indicator. The foundation for all instrument flight, this instrument reflects the airplane's attitude in relation to the horizon.

Attitude instrument flying. Controlling the aircraft by reference to the instruments rather than by outside visual cues.

Autokinesis. Nighttime visual illusion that a stationary light is moving, which becomes apparent after several seconds of staring at the light.

Automated Weather Observing System (AWOS). Automated weather reporting system consisting of various sensors, a processor, a computer-generated voice subsystem, and a transmitter to broadcast weather data.

Automated Surface Observing Station (ASOS). Weather reporting system which provides surface observations every minute via digitized voice broadcasts and printed reports.

Automatic dependent surveillance–broadcast (ADS-B). A device used in aircraft that repeatedly broadcasts a message that includes position (such as latitude, longitude, and altitude), velocity, and possibly other information.

Automatic direction finder (ADF). Electronic navigation equipment that operates in the low- and medium-frequency bands. Used in conjunction with the ground-based nondirectional beacon (NDB), the instrument displays the number of degrees clockwise from the nose of the aircraft to the station being received.

Automatic terminal information service (ATIS). The continuous broadcast of recorded non-control information in selected terminal areas. Its purpose is to improve controller effectiveness and relieve frequency congestion by automating repetitive transmission of essential but routine information.

Aviation medical examiner (AME). A physician with training in aviation medicine designated by the Civil Aerospace Medical Institute (CAMI).

AWOS. See automated weather observing system.

Azimuth card. A card that may be set, gyroscopically controlled, or driven by a remote compass.

Back course (BC). The reciprocal of the localizer course for an ILS. When flying a back-course approach, an aircraft approaches the instrument runway from the end at which the localizer antennas are installed.

Baro-aiding. A method of augmenting the GPS integrity solution by using a non-satellite input source. To ensure that baro-aiding is available, the current altimeter setting must be entered as described in the operating manual.

Barometric scale. A scale on the dial of an altimeter to which the pilot sets the barometric pressure level from which the altitude shown by the pointers is measured.

BC. See back course.

Block altitude. A block of altitudes assigned by ATC to allow altitude deviations; for example, "Maintain block altitude 9 to 11 thousand."

Cage. The black markings on the ball instrument indicating its neutral position.

Calibrated. The instrument indication compared with a standard value to determine the accuracy of the instrument.

Calibrated orifice. A hole of specific diameter used to delay the pressure change in the case of a vertical speed indicator.

Calibrated airspeed. The speed at which the aircraft is moving through the air, found by correcting IAS for instrument and position errors.

CAS. Calibrated airspeed.

CDI. Course deviation indicator.

Changeover point (COP). A point along the route or airway segment between two adjacent navigation facilities or waypoints where changeover in navigation guidance should occur.

Circling approach. A maneuver initiated by the pilot to align the aircraft with a runway for landing when a straight-in landing from an instrument approach is not possible or is not desirable.

Class A airspace. Airspace from 18,000 feet MSL up to and including FL 600, including the airspace overlying the waters within 12 NM of the coast of the 48 contiguous states and Alaska; and designated international airspace beyond 12 NM of the coast of the 48 contiguous states and Alaska within areas of domestic radio navigational signal or ATC radar coverage, and within which domestic procedures are applied.

Class B airspace. Airspace from the surface to 10,000 feet MSL surrounding the nation's busiest airports in terms of IFR operations or passenger numbers. The configuration of each Class B airspace is individually tailored and consists of a surface area and two or more layers, and is designed to contain all published instrument procedures once an aircraft enters the airspace. For all aircraft, an ATC clearance is required to operate in the area, and aircraft so cleared receive separation services within the airspace.

Class C airspace. Airspace from the surface to 4,000 feet above the airport elevation (charted in MSL) surrounding those airports having an operational control tower, serviced by radar approach control, and having a certain number of IFR operations or passenger numbers. Although the configuration of each Class C airspace area is individually tailored, the airspace usually consists of a 5 NM radius core surface area that extends from the surface up to 4,000 feet above the airport elevation, and a 10 NM radius shelf area that extends from 1,200 feet to 4,000 feet above the airport elevation.

Class D airspace. Airspace from the surface to 2,500 feet above the airport elevation (charted in MSL) surrounding those airports that have an operational control tower. The configuration of each Class D airspace area is individually tailored, and when instrument procedures are published, the airspace is normally designed to contain the procedures.

Class E airspace. Airspace that is not Class A, Class B, Class C, or Class D, and is controlled airspace.

Class G airspace. Airspace that is uncontrolled, except when associated with a temporary control tower, and has not been designated as Class A, Class B, Class C, Class D, or Class E airspace.

Clean configuration. A configuration in which all flight control surfaces have been placed to create minimum drag. In most aircraft this means flaps and gear retracted.

Clearance. ATC permission for an aircraft to proceed under specified traffic conditions within controlled airspace, for the purpose of providing separation between known aircraft.

Clearance delivery. Control tower position responsible for transmitting departure clearances to IFR flights.

Clearance limit. The fix, point, or location to which an aircraft is cleared when issued an air traffic clearance.

Clearance on request. An IFR clearance not yet received after filing a flight plan.

Clearance void time. Used by ATC, the time at which the departure clearance is automatically canceled if takeoff has not been made. The pilot must obtain a new clearance or cancel the IFR flight plan if not off by the specified time.

Clear ice. Glossy, clear, or translucent ice formed by the relatively slow freezing of large, supercooled water droplets.

Compass course. A true course corrected for variation and deviation errors.

Compass locator. A low-power, low- or medium-frequency (L/MF) radio beacon installed at the site of the outer or middle marker of an ILS.

Compass rose. A small circle graduated in 360° increments, printed on navigational charts to show the amount of compass variation at different locations, or on instruments to indicate direction.

Computer navigation fix. A point used to define a navigation track for an airborne computer system such as GPS or FMS.

Concentric rings. Dashed-line circles depicted in the plan view of IAP charts, outside of the reference circle, that show en route and feeder facilities.

Cone of confusion. A cone-shaped volume of airspace directly above a VOR station where no signal is received, causing the CDI to fluctuate.

Control and performance. A method of attitude instrument flying in which one instrument is used for making attitude changes, and the other instruments are used to monitor the progress of the change.

Control display unit. A display interfaced with the master computer, providing the pilot with a single control point for all navigations systems, thereby reducing the number of required flight deck panels.

Controlled airspace. An airspace of defined dimensions within which ATC service is provided to IFR and VFR flights in accordance with the airspace classification. It includes Class A, Class B, Class C, Class D, and Class E airspace.

Control pressures. The amount of physical exertion on the control column necessary to achieve the desired attitude.

Convective weather. Unstable, rising air found in cumiliform clouds.

Convective SIGMET. Weather advisory concerning convective weather significant to the safety of all aircraft, including thunderstorms, hail, and tornadoes.

Coordinated flight. Flight with a minimum disturbance of the forces maintaining equilibrium, established via effective control use.

COP. See changeover point.

Coriolis illusion. The illusion of rotation or movement in an entirely different axis, caused by an abrupt head movement, while in a prolonged constant rate turn that has ceased stimulating the brain's motion sensing system.

Crew resource management (CRM). The effective use of all available resources—human, hardware, and information.

Critical areas. Areas where disturbances to the ILS localizer and glide slope courses may occur when surface vehicles or aircraft operate near the localizer or glide slope antennas.

CRM. See crew resource management.

Cross-check. The first fundamental skill of instrument flight, also known as "scan," the continuous and logical observation of instruments for attitude and performance information.

Cruise clearance. An ATC clearance issued to allow a pilot to conduct flight at any altitude from the minimum IFR altitude up to and including the altitude specified in the clearance. Also authorizes a pilot to proceed to and make an approach at the destination airport.

Current induction. An electrical current being induced into, or generated in, any conductor that is crossed by lines of flux from any magnet.

DA. See decision altitude.

D.C. Direct current.

Dark adaptation. Physical and chemical adjustments of the eye that make vision possible in relative darkness.

Deceleration error. A magnetic compass error that occurs when the aircraft decelerates while flying on an easterly or westerly heading, causing the compass card to rotate toward South.

Decision altitude (DA). A specified altitude in the precision approach, charted in feet MSL, at which a missed approach must be initiated if the required visual reference to continue the approach has not been established.

Decision height (DH). A specified altitude in the precision approach, charted in height above threshold elevation, at which a decision must be made either to continue the approach or to execute a missed approach.

Deice. The act of removing ice accumulation from an aircraft structure.

Density altitude. Pressure altitude corrected for nonstandard temperature. Density altitude is used in computing the performance of an aircraft and its engines.

Departure procedure (DP). Preplanned IFR ATC departure, published for pilot use, in textual and graphic format.

Deviation. A magnetic compass error caused by local magnetic fields within the aircraft. Deviation error is different on each heading.

DGPS. Differential global positioning system.

DH. See decision height.

Differential Global Positioning System (DGPS). A system that improves the accuracy of Global Navigation Satellite Systems (GNSS) by measuring changes in variables to provide satellite positioning corrections.

Direct indication. The true and instantaneous reflection of aircraft pitch-and-bank attitude by the miniature aircraft, relative to the horizon bar of the attitude indicator.

Direct User Access Terminal System (DUATS). A system that provides current FAA weather and flight plan filing services to certified civil pilots, via personal computer, modem, or telephone access to the system. Pilots can request specific types of weather briefings and other pertinent data for planned flights.

Distance circle. See reference circle.

Distance measuring equipment (DME). A pulse-type electronic navigation system that shows the pilot, by an instrument-panel indication, the number of nautical miles between the aircraft and a ground station or waypoint.

DME. See distance measuring equipment.

DME arc. A flight track that is a constant distance from the station or waypoint.

DOD. Department of Defense.

Doghouse. A turn-and-slip indicator dial mark in the shape of a doghouse.

Domestic Reduced Vertical Separation Minimum (DRVSM). Additional flight levels between FL 290 and FL 410 to provide operational, traffic, and airspace efficiency.

Double gimbal. A type of mount used for the gyro in an attitude instrument. The axes of the two gimbals are at right angles to the spin axis of the gyro, allowing free motion in two planes around the gyro.

DP. See departure procedure.

Drag. The net aerodynamic force parallel to the relative wind, usually the sum of two components: induced drag and parasite drag.

Drag curve. The curve created when plotting induced drag and parasite drag.

DUATS. See direct user access terminal system.

Duplex. Transmitting on one frequency and receiving on a separate frequency.

Eddy currents. Current induced in a metal cup or disc when it is crossed by lines of flux from a moving magnet.

EFAS. See En Route Flight Advisory Service.

EFC. See expect-further-clearance.

Electronic flight display (EFD). For the purpose of standardization, any flight instrument display that uses LCD or other image-producing system (Cathode Ray Tube [CRT], etc.)

Elevator illusion. The sensation of being in a climb or descent, caused by the kind of abrupt vertical accelerations that result from up- or downdrafts.

Emergency. A distress or urgent condition.

Emphasis error. The result of giving too much attention to a particular instrument during the cross-check, instead of relying on a combination of instruments necessary for attitude and performance information.

EM wave. Electromagnetic wave.

Encoding altimeter. A special type of pressure altimeter used to send a signal to the air traffic controller on the ground, showing the pressure altitude the aircraft is flying.

En route facilities ring. Depicted in the plan view of IAP charts, a circle which designates NAVAIDs, fixes, and intersections that are part of the en route low altitude airway structure.

En Route Flight Advisory Service (EFAS). An en route weather-only AFSS service.

En route high-altitude charts. Aeronautical charts for en route instrument navigation at or above 18,000 feet MSL.

En route low-altitude charts. Aeronautical charts for en route IFR navigation below 18,000 feet MSL.

Equivalent airspeed. Airspeed equivalent to CAS in standard atmosphere at sea level. As the airspeed and pressure altitude increase, the CAS becomes higher than it should be, and a correction for compression must be subtracted from the CAS.

Expect-further-clearance (EFC). The time a pilot can expect to receive clearance beyond a clearance limit.

FAA. Federal Aviation Administration.

FAF. See final approach fix.

False horizon. Inaccurate visual information for aligning the aircraft, caused by various natural and geometric formations that disorient the pilot from the actual horizon.

Federal airways. Class E airspace areas that extend upward from 1,200 feet to, but not including, 18,000 feet MSL, unless otherwise specified.

Feeder facilities. Used by ATC to direct aircraft to intervening fixes between the en route structure and the initial approach fix.

Final approach. Part of an instrument approach procedure in which alignment and descent for landing are accomplished.

Final approach fix (FAF). The fix from which the IFR final approach to an airport is executed, and which identifies the beginning of the final approach segment. An FAF is designated on government charts by a Maltese cross symbol for nonprecision approaches, and a lightning bolt symbol for precision approaches.

Fixating. Staring at a single instrument, thereby interrupting the cross-check process.

FL. See flight level.

Flight configurations. Adjusting the aircraft control surfaces (including flaps and landing gear) in a manner that will achieve a specified attitude.

Flight director indicator (FDI). One of the major components of a flight director system, it provides steering commands that the pilot (or the autopilot, if coupled) follows.

Flight level (FL). A measure of altitude (in hundreds of feet) used by aircraft flying above 18,000 feet with the altimeter set at 29.92" Hg.

Flight management system (FMS). Provides pilot and crew with highly accurate and automatic long-range navigation capability, blending available inputs from long- and short-range sensors.

Flightpath. The line, course, or track along which an aircraft is flying or is intended to be flown.

Flight patterns. Basic maneuvers, flown by reference to the instruments rather than outside visual cues, for the purpose of practicing basic attitude flying. The patterns simulate maneuvers encountered on instrument flights such as holding patterns, procedure turns, and approaches.

Flight strips. Paper strips containing instrument flight information, used by ATC when processing flight plans.

FMS. See flight management system.

Form drag. The drag created because of the shape of a component or the aircraft.

Fundamental skills. Pilot skills of instrument cross-check, instrument interpretation, and aircraft control.

Glideslope (GS). Part of the ILS that projects a radio beam upward at an angle of approximately 3° from the approach end of an instrument runway. The glideslope provides vertical guidance to aircraft on the final approach course for the aircraft to follow when making an ILS approach along the localizer path.

Glideslope intercept altitude. The minimum altitude of an intermediate approach segment prescribed for a precision approach that ensures obstacle clearance.

Global landing system (GLS). An instrument approach with lateral and vertical guidance with integrity limits (similar to barometric vertical navigation (BRO VNAV).

Global navigation satellite systems (GNSS). Satellite navigation systems that provide autonomous geo-spatial positioning with global coverage. It allows small electronic receivers to determine their location (longitude, latitude, and altitude) to within a few meters using time signals transmitted along a line of sight by radio from satellites.

GNSS. See global navigation satellite systems.

Global positioning system (GPS). Navigation system that uses satellite rather than ground-based transmitters for location information.

Goniometer. As used in radio frequency (RF) antenna systems, a direction-sensing device consisting of two fixed loops of wire oriented 90° from each other, which separately sense received signal strength and send those signals to two rotors (also oriented 90°) in the sealed direction-indicating instrument. The rotors are attached to the direction-indicating needle of the instrument and rotated by a small motor until minimum magnetic field is sensed near the rotors.

GPS. See global positioning system.

GPS Approach Overlay Program. An authorization for pilots to use GPS avionics under IFR for flying designated existing nonprecision instrument approach procedures, with the exception of LOC, LDA, and SDF procedures.

Graveyard spiral. The illusion of the cessation of a turn while still in a prolonged, coordinated, constant rate turn, which can lead a disoriented pilot to a loss of control of the aircraft.

Great circle route. The shortest distance across the surface of a sphere (the Earth) between two points on the surface.

Ground proximity warning system (GPWS). A system designed to determine an aircraft's clearance above the Earth and provides limited predictability about aircraft position relative to rising terrain.

Groundspeed. Speed over the ground, either closing speed to the station or waypoint, or speed over the ground in whatever direction the aircraft is going at the moment, depending upon the navigation system used.

GS. See glide slope.

GWPS. See ground proximity warning system.

HAA. See height above airport.

HAL. See height above landing.

HAT. See height above touchdown elevation.

Hazardous attitudes. Five aeronautical decision-making attitudes that may contribute to poor pilot judgment: antiauthority, impulsivity, invulnerability, machismo, and resignation.

Hazardous Inflight Weather Advisory Service (HIWAS). Service providing recorded weather forecasts broadcast to airborne pilots over selected VORs.

Head-up display (HUD). A special type of flight viewing screen that allows the pilot to watch the flight instruments and other data while looking through the windshield of the aircraft for other traffic, the approach lights, or the runway.

Height above airport (HAA). The height of the MDA above the published airport elevation.

Height above landing (HAL). The height above a designated helicopter landing area used for helicopter instrument approach procedures.

Height above touchdown elevation (HAT). The DA/DH or MDA above the highest runway elevation in the touchdown zone (first 3,000 feet of the runway).

HF. High frequency.

Hg. Abbreviation for mercury, from the Latin hydrargyrum.

HIWAS. See Hazardous Inflight Weather Advisory Service.

Holding. A predetermined maneuver that keeps aircraft within a specified airspace while awaiting further clearance from ATC.

Holding pattern. A racetrack pattern, involving two turns and two legs, used to keep an aircraft within a prescribed airspace with respect to a geographic fix. A standard pattern uses right turns; nonstandard patterns use left turns.

Homing. Flying the aircraft on any heading required to keep the needle pointing to the 0° relative bearing position.

Horizontal situation indicator (HSI). A flight navigation instrument that combines the heading indicator with a CDI, in order to provide the pilot with better situational awareness of location with respect to the courseline.

HSI. See horizontal situation indicator.

HUD. See head-up display.

Human factors. A multidisciplinary field encompassing the behavioral and social sciences, engineering, and physiology, to consider the variables that influence individual and crew performance for the purpose of optimizing human performance and reducing errors.

Hypoxia. A state of oxygen deficiency in the body sufficient to impair functions of the brain and other organs.

IAF. See initial approach fix.

IAP. See instrument approach procedures.

IAS. See indicated airspeed.

ICAO. See International Civil Aviation Organization.

Ident. Air Traffic Control request for a pilot to push the button on the transponder to identify return on the controller's scope.

IFR. See instrument flight rules.

ILS. See instrument landing system.

ILS categories. Categories of instrument approach procedures allowed at airports equipped with the following types of instrument landing systems:

> **ILS Category I:** Provides for approach to a height above touchdown of not less than 200 feet, and with runway visual range of not less than 1,800 feet.
>
> **ILS Category II:** Provides for approach to a height above touchdown of not less than 100 feet and with runway visual range of not less than 1,200 feet.
>
> **ILS Category IIIA:** Provides for approach without a decision height minimum and with runway visual range of not less than 700 feet.
>
> **ILS Category IIIB:** Provides for approach without a decision height minimum and with runway visual range of not less than 150 feet.
>
> **ILS Category IIIC:** Provides for approach without a decision height minimum and without runway visual range minimum.

IMC. See instrument meteorological conditions.

Indicated airspeed (IAS). Shown on the dial of the instrument airspeed indicator on an aircraft. Directly related to calibrated airspeed (CAS), IAS includes instrument errors and position error.

Indirect indication. A reflection of aircraft pitch-and-bank attitude by the instruments other than the attitude indicator.

Induced drag. Drag caused by the same factors that produce lift; its amount varies inversely with airspeed. As airspeed decreases, the angle of attack must increase, in turn increasing induced drag.

Induction icing. A type of ice in the induction system that reduces the amount of air available for combustion. The most commonly found induction icing is carburetor icing.

Inertial navigation system (INS). A computer-based navigation system that tracks the movement of an aircraft via signals produced by onboard accelerometers. The initial location of the aircraft is entered into the computer, and all subsequent movement of the aircraft is sensed and used to keep the position updated. An INS does not require any inputs from outside signals.

Initial approach fix (IAF). The fix depicted on IAP charts where the instrument approach procedure (IAP) begins unless otherwise authorized by ATC.

Inoperative components. Higher minimums are prescribed when the specified visual aids are not functioning; this information is listed in the Inoperative Components Table found in the United States Terminal Procedures Publications.

INS. See inertial navigation system.

Instantaneous vertical speed indicator (IVSI). Assists in interpretation by instantaneously indicating the rate of climb or descent at a given moment with little or no lag as displayed in a vertical speed indicator (VSI).

Instrument approach procedures (IAP). A series of predetermined maneuvers for the orderly transfer of an aircraft under IFR from the beginning of the initial approach to a landing or to a point from which a landing may be made visually.

Instrument flight rules (IFR). Rules and regulations established by the Federal Aviation Administration to govern flight under conditions in which flight by outside visual reference is not safe. IFR flight depends upon flying by reference to instruments in the flight deck, and navigation is accomplished by reference to electronic signals.

Instrument landing system (ILS). An electronic system that provides both horizontal and vertical guidance to a specific runway, used to execute a precision instrument approach procedure.

Instrument meteorological conditions (IMC). Meteorological conditions expressed in terms of visibility, distance from clouds, and ceiling less than the minimums specified for visual meteorological conditions, requiring operations to be conducted under IFR.

Instrument takeoff. Using the instruments rather than outside visual cues to maintain runway heading and execute a safe takeoff.

Interference drag. Drag generated by the collision of airstreams creating eddy currents, turbulence, or restrictions to smooth flow.

International Civil Aviation Organization (ICAO). The United Nations agency for developing the principles and techniques of international air navigation, and fostering planning and development of international civil air transport.

International standard atmosphere (IAS). A model of standard variation of pressure and temperature.

Inversion illusion. The feeling that the aircraft is tumbling backwards, caused by an abrupt change from climb to straight-and-level flight while in situations lacking visual reference.

Inverter. A solid-state electronic device that converts D.C. into A.C. current of the proper voltage and frequency to operate A.C. gyro instruments.

Isogonic lines. Lines drawn across aeronautical charts to connect points having the same magnetic variation.

IVSI. See instantaneous vertical speed indicator.

Jet route. A route designated to serve flight operations from 18,000 feet MSL up to and including FL 450.

Jet stream. A high-velocity narrow stream of winds, usually found near the upper limit of the troposphere, which flows generally from west to east.

KIAS. Knots indicated airspeed.

Kollsman window. A barometric scale window of a sensitive altimeter used to adjust the altitude for the altimeter setting.

LAAS. See local area augmentation system.

Lag. The delay that occurs before an instrument needle attains a stable indication.

Land as soon as possible. ATC instruction to pilot. Land without delay at the nearest suitable area, such as an open field, at which a safe approach and landing is assured.

Land as soon as practical. ATC instruction to pilot. The landing site and duration of flight are at the discretion of the pilot. Extended flight beyond the nearest approved landing area is not recommended.

Land immediately. ATC instruction to pilot. The urgency of the landing is paramount. The primary consideration is to ensure the survival of the occupants. Landing in trees, water, or other unsafe areas should be considered only as a last resort.

LDA. See localizer-type directional aid.

Lead radial. The radial at which the turn from the DME arc to the inbound course is started.

Leans, the. A physical sensation caused by an abrupt correction of a banked attitude entered too slowly to stimulate the motion sensing system in the inner ear. The abrupt correction can create the illusion of banking in the opposite direction.

Lift. A component of the total aerodynamic force on an airfoil and acts perpendicular to the relative wind.

Lines of flux. Invisible lines of magnetic force passing between the poles of a magnet.

L/MF. See low or medium frequency.

LMM. See locator middle marker.

Load factor. The ratio of a specified load to the total weight of the aircraft. The specified load is expressed in terms of any of the following: aerodynamic forces, inertial forces, or ground or water reactions.

Loadmeter. A type of ammeter installed between the generator output and the main bus in an aircraft electrical system.

LOC. See localizer.

Local area augmentation system (LAAS). A differential global positioning system (DGPS) that improves the accuracy of the system by determining position error from the GPS satellites, then transmitting the error, or corrective factors, to the airborne GPS receiver.

Localizer (LOC). The portion of an ILS that gives left/right guidance information down the centerline of the instrument runway for final approach.

Localizer-type directional aid (LDA). A NAVAID used for nonprecision instrument approaches with utility and accuracy comparable to a localizer but which is not a part of a complete ILS and is not aligned with the runway. Some LDAs are equipped with a glide slope.

Locator middle marker (LMM). Nondirectional radio beacon (NDB) compass locator, collocated with a middle marker (MM).

Locator outer marker (LOM). NDB compass locator, collocated with an outer marker (OM).

LOM. See locator outer marker.

Long range navigation (LORAN). An electronic navigational system by which hyperbolic lines of position are determined by measuring the difference in the time of reception of synchronized pulse signals from two fixed transmitters. LORAN A operates in the 1750 to 1950 kHz frequency band. LORAN C and D operate in the 100 to 110 kHz frequency band.

LORAN. See long range navigation.

Low or medium frequency. A frequency range between 190–535 kHz with the medium frequency above 300 kHz. Generally associated with nondirectional beacons transmitting a continuous carrier with either a 400 or 1,020 Hz modulation.

Lubber line. The reference line used in a magnetic compass or heading indicator.

MAA. See maximum authorized altitude.

Mach number. The ratio of the true airspeed of the aircraft to the speed of sound in the same atmospheric conditions, named in honor of Ernst Mach, late 19th century physicist.

Mach meter. The instrument that displays the ratio of the speed of sound to the true airspeed an aircraft is flying.

Magnetic bearing (MB). The direction to or from a radio transmitting station measured relative to magnetic north.

Magnetic heading (MH). The direction an aircraft is pointed with respect to magnetic north.

Mandatory altitude. An altitude depicted on an instrument approach chart with the altitude value both underscored and overscored. Aircraft are required to maintain altitude at the depicted value.

Mandatory block altitude. An altitude depicted on an instrument approach chart with two underscored and overscored altitude values between which aircraft are required to maintain altitude.

MAP. See missed approach point.

Margin identification. The top and bottom areas on an instrument approach chart that depict information about the procedure, including airport location and procedure identification.

Marker beacon. A low-powered transmitter that directs its signal upward in a small, fan-shaped pattern. Used along the flight path when approaching an airport for landing, marker beacons indicate both aurally and visually when the aircraft is directly over the facility.

Maximum altitude. An altitude depicted on an instrument approach chart with overscored altitude value at which or below aircraft are required to maintain altitude.

Maximum authorized altitude (MAA). A published altitude representing the maximum usable altitude or flight level for an airspace structure or route segment.

MB. See magnetic bearing.

MCA. See minimum crossing altitude.

MDA. See minimum descent altitude.

MEA. See minimum en route altitude.

Mean sea level. The average height of the surface of the sea at a particular location for all stages of the tide over a 19-year period.

MFD. See multi-function display.

MH. See magnetic heading.

MHz. Megahertz.

Microwave landing system (MLS). A precision instrument approach system operating in the microwave spectrum which normally consists of an azimuth station, elevation station, and precision distance measuring equipment.

Mileage breakdown. A fix indicating a course change that appears on the chart as an "x" at a break between two segments of a federal airway.

Military operations area (MOA). Airspace established for the purpose of separating certain military training activities from IFR traffic.

Military training route (MTR). Airspace of defined vertical and lateral dimensions established for the conduct of military training at airspeeds in excess of 250 knots indicated airspeed (KIAS).

Minimum altitude. An altitude depicted on an instrument approach chart with the altitude value underscored. Aircraft are required to maintain altitude at or above the depicted value.

Minimum crossing altitude (MCA). The lowest allowed altitude at certain fixes an aircraft must cross when proceeding in the direction of a higher minimum en route altitude (MEA).

Minimum descent altitude (MDA). The lowest altitude (in feet MSL) to which descent is authorized on final approach, or during circle-to-land maneuvering in execution of a nonprecision approach.

Minimum en route altitude (MEA). The lowest published altitude between radio fixes that ensures acceptable navigational signal coverage and meets obstacle clearance requirements between those fixes.

Minimum obstruction clearance altitude (MOCA). The lowest published altitude in effect between radio fixes on VOR airways, off-airway routes, or route segments, which meets obstacle clearance requirements for the entire route segment and which ensures acceptable navigational signal coverage only within 25 statute (22 nautical) miles of a VOR.

Minimum reception altitude (MRA). The lowest altitude at which an airway intersection can be determined.

Minimum safe altitude (MSA). The minimum altitude depicted on approach charts which provides at least 1,000 feet of obstacle clearance for emergency use within a specified distance from the listed navigation facility.

Minimum vectoring altitude (MVA). An IFR altitude lower than the minimum en route altitude (MEA) that provides terrain and obstacle clearance.

Minimums section. The area on an IAP chart that displays the lowest altitude and visibility requirements for the approach.

Missed approach. A maneuver conducted by a pilot when an instrument approach cannot be completed to a landing.

Missed approach point (MAP). A point prescribed in each instrument approach at which a missed approach procedure shall be executed if the required visual reference has not been established.

Mixed ice. A mixture of clear ice and rime ice.

MLS. See microwave landing system.

MM. Middle marker.

MOA. See military operations area.

MOCA. See minimum obstruction clearance altitude.

Mode C. Altitude reporting transponder mode.

MRA. See minimum reception altitude.

MSA. See minimum safe altitude.

MSL. See mean sea level.

MTR. See military training route.

Multi-function display (MFD). Small screen (CRT or LCD) in an aircraft that can be used to display information to the pilot in numerous configurable ways. Often an MFD will be used in concert with a Primary Flight Display.

MVA. See minimum vectoring altitude.

NACG. See National Aeronautical Charting Group.

NAS. See National Airspace System.

National Airspace System (NAS). The common network of United States airspace—air navigation facilities, equipment and services, airports or landing areas; aeronautical charts, information and services; rules, regulations and procedures, technical information; and manpower and material.

National Aeronautical Charting Group (NACG). A Federal agency operating under the FAA, responsible for publishing charts such as the terminal procedures and en route charts.

National Route Program (NRP). A set of rules and procedures designed to increase the flexibility of user flight planning within published guidelines.

National Security Area (NSA). Areas consisting of airspace of defined vertical and lateral dimensions established at locations where there is a requirement for increased security and safety of ground facilities. Pilots are requested to voluntarily avoid flying through the depicted NSA. When it is necessary to provide a greater level of security and safety, flight in NSAs may be temporarily prohibited. Regulatory prohibitions are disseminated via NOTAMs.

National Transportation Safety Board (NTSB). A United States Government independent organization responsible for investigations of accidents involving aviation, highways, waterways, pipelines, and railroads in the United States. NTSB is charged by congress to investigate every civil aviation accident in the United States.

NAVAID. Naviagtional aid.

NAV/COM. Navigation and communication radio.

NDB. See nondirectional radio beacon.

NM. Nautical mile.

NOAA. National Oceanic and Atmospheric Administration.

No-gyro approach. A radar approach that may be used in case of a malfunctioning gyro-compass or directional gyro. Instead of providing the pilot with headings to be flown, the controller observes the radar track and issues control instructions "turn right/left" or "stop turn," as appropriate.

Nondirectional radio beacon (NDB). A ground-based radio transmitter that transmits radio energy in all directions.

Nonprecision approach. A standard instrument approach procedure in which only horizontal guidance is provided.

No procedure turn (NoPT). Term used with the appropriate course and altitude to denote that the procedure turn is not required.

NoPT. See no procedure turn.

Notice to Airmen (NOTAM). A notice filed with an aviation authority to alert aircraft pilots of any hazards en route or at a specific location. The authority in turn provides means of disseminating relevant NOTAMs to pilots.

NRP. See National Route Program.

NSA. See National Security Area.

NTSB. See National Transportation Safety Board.

NWS. National Weather Service.

Obstacle departure procedures (ODP). Obstacle clearance protection provided to aircraft in instrument meteorological conditions (IMC).

ODP. See obstacle departure procedures.

OM. Outer marker.

Omission error. The failure to anticipate significant instrument indications following attitude changes; for example, concentrating on pitch control while forgetting about heading or roll information, resulting in erratic control of heading and bank.

Optical illusion. A misleading visual image. For the purpose of this handbook, the term refers to the brain's misinterpretation of features on the ground associated with landing, which causes a pilot to misread the spatial relationships between the aircraft and the runway.

Orientation. Awareness of the position of the aircraft and of oneself in relation to a specific reference point.

Otolith organ. An inner ear organ that detects linear acceleration and gravity orientation.

Outer marker. A marker beacon at or near the glide slope intercept altitude of an ILS approach. It is normally located four to seven miles from the runway threshold on the extended centerline of the runway.

Overcontrolling. Using more movement in the control column than is necessary to achieve the desired pitch-and bank condition.

Overpower. To use more power than required for the purpose of achieving a faster rate of airspeed change.

P-static. See precipitation static.

PAPI. See precision approach path indicator.

PAR. See precision approach radar.

Parasite drag. Drag caused by the friction of air moving over the aircraft structure; its amount varies directly with the airspeed.

PFD. See primary flight display.

PIC. See pilot-in-command.

Pilot-in-command (PIC). The pilot responsible for the operation and safety of an aircraft.

Pilot report (PIREP). Report of meteorological phenomena encountered by aircraft.

Pilot's Operating Handbook/Airplane Flight Manual (POH/AFM). FAA-approved documents published by the airframe manufacturer that list the operating conditions for a particular model of aircraft.

PIREP. See pilot report.

Pitot pressure. Ram air pressure used to measure airspeed.

Pitot-static head. A combination pickup used to sample pitot pressure and static air pressure.

Plan view. The overhead view of an approach procedure on an instrument approach chart. The plan view depicts the routes that guide the pilot from the en route segments to the IAF.

POH/AFM. See Pilot's Operating Handbook/Airplane Flight Manual.

Point-in-space approach. A type of helicopter instrument approach procedure to a missed approach point more than 2,600 feet from an associated helicopter landing area.

Position error. Error in the indication of the altimeter, ASI, and VSI caused by the air at the static system entrance not being absolutely still.

Position report. A report over a known location as transmitted by an aircraft to ATC.

Precession. The characteristic of a gyroscope that causes an applied force to be felt, not at the point of application, but 90° from that point in the direction of rotation.

Precipitation static (P-static). A form of radio interference caused by rain, snow, or dust particles hitting the antenna and inducing a small radio-frequency voltage into it.

Precision approach. A standard instrument approach procedure in which both vertical and horizontal guidance is provided.

Precision approach path indicator (PAPI). A system of lights similar to the VASI, but consisting of one row of lights in two- or four-light systems. A pilot on the correct glide slope will see two white lights and two red lights. See VASI.

Precision approach radar (PAR). A type of radar used at an airport to guide an aircraft through the final stages of landing, providing horizontal and vertical guidance. The radar operator directs the pilot to change heading or adjust the descent rate to keep the aircraft on a path that allows it to touch down at the correct spot on the runway.

Precision runway monitor (PRM). System allows simultaneous, independent Instrument Flight Rules (IFR) approaches at airports with closely spaced parallel runways.

Preferred IFR routes. Routes established in the major terminal and en route environments to increase system efficiency and capacity. IFR clearances are issued based on these routes, listed in the A/FD except when severe weather avoidance procedures or other factors dictate otherwise.

Pressure altitude. Altitude above the standard 29.92" Hg plane.

Prevailing visibility. The greatest horizontal visibility equaled or exceeded throughout at least half the horizon circle (which is not necessarily continuous).

Primary and supporting. A method of attitude instrument flying using the instrument that provides the most direct indication of attitude and performance.

Primary flight display (PFD). A display that provides increased situational awareness to the pilot by replacing the traditional six instruments used for instrument flight with an easy-to-scan display that provides the horizon, airspeed, altitude, vertical speed, trend, trim, rate of turn among other key relevant indications.

PRM. See precision runway monitor.

Procedure turn. A maneuver prescribed when it is necessary to reverse direction to establish an aircraft on the intermediate approach segment or final approach course.

Profile view. Side view of an IAP chart illustrating the vertical approach path altitudes, headings, distances, and fixes.

Prohibited area. Designated airspace within which flight of aircraft is prohibited.

Propeller/rotor modulation error. Certain propeller RPM settings or helicopter rotor speeds can cause the VOR course deviation indicator (CDI) to fluctuate as much as ±6°. Slight changes to the RPM setting will normally smooth out this roughness.

Rabbit, the. High-intensity flasher system installed at many large airports. The flashers consist of a series of brilliant blue-white bursts of light flashing in sequence along the approach lights, giving the effect of a ball of light traveling towards the runway.

Radar. Radio Detection And Ranging.

Radar approach. The controller provides vectors while monitoring the progress of the flight with radar, guiding the pilot through the descent to the airport/heliport or to a specific runway.

Radials. The courses oriented from a station.

Radio or radar altimeter. An electronic altimeter that determines the height of an aircraft above the terrain by measuring the time needed for a pulse of radio-frequency energy to travel from the aircraft to the ground and return.

Radio frequency (RF). A term that refers to alternating current (AC) having characteristics such that, if the current is input to antenna, an electromagnetic (EM) field is generated suitable for wireless broadcasting and/or communications.

Radio magnetic indicator (RMI). An electronic navigation instrument that combines a magnetic compass card with two bearing pointers (typically). Generally, one pointer is for the ADF while the other is for an RNAV or VOR navigation system. The pointers are typically different colors and of different widths for ease of identification. Sometimes a function switch is provided to allow the #2 pointer to be slaved to either a VOR or RNAV system. The card of the RMI acts as a gyro-stabilized magnetic compass (usually corrected for north via a flux valve) and shows the magnetic heading the aircraft is flying.

Radio wave. An electromagnetic wave (EM wave) with frequency characteristics useful for radio transmission.

RAIM. See receiver autonomous integrity monitoring.

Random RNAV routes. Direct routes, based on area navigation capability, between waypoints defined in terms of latitude/longitude coordinates, degree-distance fixes, or offsets from established routes/airways at a specified distance and direction.

Ranging signals. Transmitted from the GPS satellite, these allow the aircraft's receiver to determine range (distance) from each satellite.

RB. See relative bearing.
RBI. See relative bearing indicator.

RCO. See remote communications outlet.

Receiver autonomous integrity monitoring (RAIM). A system used to verify the usability of the received GPS signals and warns the pilot of any malfunction in the navigation system. This system is required for IFR-certified GPS units.

Recommended altitude. An altitude depicted on an instrument approach chart with the altitude value neither underscored nor overscored. The depicted value is an advisory value.

Receiver-transmitter (RT). A system that permits selection of a unique channel or frequency whereupon a signal (typically communication) can be transmitted and received.

Reduced vertical separation minimum (RVSM). Reduces the vertical separation between flight level (FL) 290–410 from 2,000 feet to 1,000 feet and makes six additional FLs available for operation. Also see DRVSM.

Reference circle (also, distance circle). The circle depicted in the plan view of an IAP chart that typically has a 10 NM radius, within which chart the elements are drawn to scale.

Regions of command. The "regions of normal and reversed command" refers to the relationship between speed and the power required to maintain or change that speed in flight.

REIL. See runway end identifier lights.

Relative bearing (RB). The angular difference between the aircraft heading and the direction to the station, measured clockwise from the nose of the aircraft.

Relative bearing indicator (RBI). Also known as the fixed-card ADF, zero is always indicated at the top of the instrument and the needle indicates the relative bearing to the station.

Relative wind. Direction of the airflow produced by an object moving through the air. The relative wind for an airplane in flight flows in a direction parallel with and opposite to the direction of flight; therefore, the actual flight path of the airplane determines the direction of the relative wind.

Remote communications outlet (RCO). An unmanned communications facility that is remotely controlled by air traffic personnel.

Required navigation performance (RNP). A specified level of accuracy defined by a lateral area of confined airspace in which an RNP-certified aircraft operates.

Restricted area. Airspace designated under 14 CFR part 73 within which the flight of aircraft, while not wholly prohibited, is subject to restriction.

Reverse sensing. The VOR needle appearing to indicate the reverse of normal operation.

RF. Radio frequency.

Rhodopsin. The photosensitive pigments that initiate the visual response in the rods of the eye.

Rigidity. The characteristic of a gyroscope that prevents its axis of rotation tilting as the Earth rotates.

Rime ice. Rough, milky, opaque ice formed by the instantaneous freezing of small supercooled water droplets.

Risk. The future impact of a hazard that is not eliminated or controlled.

RMI. See radio magnetic indicator.

RNAV. See area navigation.

RNP. See required navigation performance.

Runway end identifier lights (REIL). A pair of synchronized flashing lights, located laterally on each side of the runway threshold, providing rapid and positive identification of the approach end of a runway.

Runway visibility value (RVV). The visibility determined for a particular runway by a transmissometer.

Runway visual range (RVR). The instrumentally derived horizontal distance a pilot should be able to see down the runway from the approach end, based on either the sighting of high-intensity runway lights, or the visual contrast of other objects.

RVR. See runway visual range.

RVV. See runway visibility value.

St. Elmo's Fire. A corona discharge which lights up the aircraft surface areas where maximum static discharge occurs.

Satellite ephemeris data. Data broadcast by the GPS satellite containing very accurate orbital data for that satellite, atmospheric propagation data, and satellite clock error data.

Scan. The first fundamental skill of instrument flight, also known as "cross-check;" the continuous and logical observation of instruments for attitude and performance information.

SDF. See simplified directional facility.

Selective availability. A satellite technology permitting the Department of Defense (DOD) to create, in the interest of national security, a significant clock and ephemeris error in the satellites, resulting in a navigation error.

Semicircular canal. An inner ear organ that detects angular acceleration of the body.

Sensitive altimeter. A form of multipointer pneumatic altimeter with an adjustable barometric scale that allows the reference pressure to be set to any desired level.

SIDS. See standard instrument departure procedures.

SIGMET. The acronym for Significant Meteorological information. A weather advisory issued concerning weather significant to the safety of all aircraft.

Signal-to-noise ratio. An indication of signal strength received compared to background noise, which is a measure of how adequate the received signal is.

Simplex. Transmission and reception on the same frequency.

Simplified directional facility (SDF). A NAVAID used for nonprecision instrument approaches. The final approach course is similar to that of an ILS localizer; however, the SDF course may be offset from the runway, generally not more than 3°, and the course may be wider than the localizer, resulting in a lower degree of accuracy.

Single-pilot resource management (SRM). The ability for crew or pilot to manage all resources effectively to ensure the outcome of the flight is successful.

Situational awareness. Pilot knowledge of where the aircraft is in regard to location, air traffic control, weather, regulations, aircraft status, and other factors that may affect flight.

Skidding turn. An uncoordinated turn in which the rate of turn is too great for the angle of bank, pulling the aircraft to the outside of the turn.

Skin friction drag. Drag generated between air molecules and the solid surface of the aircraft.

Slant range. The horizontal distance from the aircraft antenna to the ground station, due to line-of-sight transmission of the DME signal.

Slaved compass. A system whereby the heading gyro is "slaved to," or continuously corrected to bring its direction readings into agreement with a remotely located magnetic direction sensing device (usually this is a flux valve or flux gate compass).

Slipping turn. An uncoordinated turn in which the aircraft is banked too much for the rate of turn, so the horizontal lift component is greater than the centrifugal force, pulling the aircraft toward the inside of the turn.

Small airplane. An airplane of 12,500 pounds or less maximum certificated takeoff weight.

Somatogravic illusion. The misperception of being in a nose-up or nose-down attitude, caused by a rapid acceleration or deceleration while in flight situations that lack visual reference.

Spatial disorientation. The state of confusion due to misleading information being sent to the brain from various sensory organs, resulting in a lack of awareness of the aircraft position in relation to a specific reference point.

Special use airspace. Airspace in which flight activities are subject to restrictions that can create limitations on the mixed use of airspace. Consists of prohibited, restricted, warning, military operations, and alert areas.

SRM. See single-pilot resource management.

SSR. See secondary surveillance radar.

SSV. See standard service volume.

Standard holding pattern. A holding pattern in which all turns are made to the right.

Standard instrument departure procedures (SIDS). Published procedures to expedite clearance delivery and to facilitate transition between takeoff and en route operations.

Standard rate turn. A turn in which an aircraft changes its direction at a rate of 3° per second. The turn indicators are typically 2 minute or 4 minute instruments. In a 2 minute instrument, if the needle is one needle width either side of the center alignment mark, the turn is 3° per second and the turn takes 2 minutes to execute a 360° turn. In a 4 minute instrument, the same turn takes two widths deflection of the needle to achieve 3° per second. The 4 minute turn instrument is usually found on high performance aircraft.

Standard service volume (SSV). Defines the limits of the volume of airspace which the VOR serves.

Standard terminal arrival route (STAR). A preplanned IFR ATC arrival procedure published for pilot use in graphic and/or textual form.

STAR. See standard terminal arrival route.

Static longitudinal stability. The aerodynamic pitching moments required to return the aircraft to the equilibrium angle of attack.

Static pressure. Pressure of air that is still, or not moving, measured perpendicular to the surface of the aircraft.

Steep turns. In instrument flight, any turn greater than standard rate; in visual flight, anything greater than a 45° bank.

Stepdown fix. The point after which additional descent is permitted within a segment of an IAP.

Strapdown system. An INS in which the accelerometers and gyros are permanently "strapped down" or aligned with the three axes of the aircraft.

Stress. The body's response to demands placed upon it.

Structural icing. The accumulation of ice on the exterior of the aircraft.

Suction relief valve. A relief valve in an instrument vacuum system required to maintain the correct low pressure inside the instrument case for the proper operation of the gyros.

Synchro. A device used to transmit indications of angular movement or position from one location to another.

Synthetic vision. A realistic display depiction of the aircraft in relation to terrain and flight path.

TAA. See terminal arrival area.

TACAN. See tactical air navigation.

Tactical air navigation (TACAN). An electronic navigation system used by military aircraft, providing both distance and direction information.

TAWS. See terrain awareness and warning system.

TCAS. See traffic alert collision avoidance system.

TCH. See threshold crossing height.

TDZE. See touchdown zone elevation.

TEC. See Tower En Route Control.

Technique. The manner in which procedures are executed.

Temporary flight restriction (TFR). Restriction to flight imposed in order to:

1. Protect persons and property in the air or on the surface from an existing or imminent flight associated hazard;
2. Provide a safe environment for the operation of disaster relief aircraft;
3. Prevent an unsafe congestion of sightseeing aircraft above an incident;
4. Protect the President, Vice President, or other public figures; and,
5. Provide a safe environment for space agency operations.

Pilots are expected to check appropriate NOTAMs during flight planning when conducting flight in an area where a temporary flight restriction is in effect.

Tension. Maintaining an excessively strong grip on the control column, usually resulting in an overcontrolled situation.

Terminal Instrument Approach Procedure (TERP). Prescribes standardized methods for use in designing instrument flight procedures.

Terminal arrival area (TAA). A procedure to provide a new transition method for arriving aircraft equipped with FMS and/or GPS navigational equipment. The TAA contains a "T" structure that normally provides a NoPT for aircraft using the approach.

TERP. See terminal instrument approach procedure.

Terrain Awareness and Warning System (TAWS). A timed-based system that provides information concerning potential hazards with fixed objects by using GPS positioning and a database of terrain and obstructions to provide true predictability of the upcoming terrain and obstacles.

TFR. See temporary flight restriction.

Threshold crossing height (TCH). The theoretical height above the runway threshold at which the aircraft's glide slope antenna would be if the aircraft maintains the trajectory established by the mean ILS glide slope or MLS glide path.

Thrust (aerodynamic force). The forward aerodynamic force produced by a propeller, fan, or turbojet engine as it forces a mass of air to the rear, behind the aircraft.

Time and speed table. A table depicted on an instrument approach procedure chart that identifies the distance from the FAF to the MAP, and provides the time required to transit that distance based on various groundspeeds.

Timed turn. A turn in which the clock and the turn coordinator are used to change heading a definite number of degrees in a given time.

TIS. See traffic information service.

Title 14 of the Code of Federal Regulations (14 CFR). The federal aviation regulations governing the operation of aircraft, airways, and airmen.

Touchdown zone elevation (TDZE). The highest elevation in the first 3,000 feet of the landing surface, TDZE is indicated on the instrument approach procedure chart when straight-in landing minimums are authorized.

Tower En Route Control (TEC). The control of IFR en route traffic within delegated airspace between two or more adjacent approach control facilities, designed to expedite traffic and reduce control and pilot communication requirements.

TPP. See United States Terminal Procedures Publication.

Tracking. Flying a heading that will maintain the desired track to or from the station regardless of crosswind conditions.

Traffic Alert Collision Avoidance System (TCAS). An airborne system developed by the FAA that operates independently from the ground-based Air Traffic Control system. Designed to increase flight deck awareness of proximate aircraft and to serve as a "last line of defense" for the prevention of mid-air collisions.

Traffic information service (TIS). A ground-based service providing information to the flight deck via data link using the S-mode transponder and altitude encoder to improve the safety and efficiency of "see and avoid" flight through an automatic display that informs the pilot of nearby traffic.

Transcribed Weather Broadcast (TWEB). Meteorological and aeronautical data recorded on tapes and broadcast over selected NAVAIDs. Generally, the broadcast contains route-oriented data with specially prepared NWS forecasts, inflight advisories, and winds aloft. It also includes selected current information such as weather reports (METAR/SPECI), NOTAMs, and special notices.

Transponder. The airborne portion of the ATC radar beacon system.

Transponder code. One of 4,096 four-digit discrete codes ATC assigns to distinguish between aircraft.

Trend. Immediate indication of the direction of aircraft movement, as shown on instruments.

Trim. Adjusting the aerodynamic forces on the control surfaces so that the aircraft maintains the set attitude without any control input.

TWEB. See Transcribed Weather Broadcast.

True airspeed. Actual airspeed, determined by applying a correction for pressure altitude and temperature to the CAS.

UHF. See ultra-high frequency.

Ultra-high frequency (UHF). The range of electromagnetic frequencies between 962 MHz and 1213 MHz.

Uncaging. Unlocking the gimbals of a gyroscopic instrument, making it susceptible to damage by abrupt flight maneuvers or rough handling.

Underpower. Using less power than required for the purpose of achieving a faster rate of airspeed change.

United States Terminal Procedures Publication (TPP). Booklets published in regional format by the NACO that include DPs, STARs, IAPs, and other information pertinent to IFR flight.

Unusual attitude. An unintentional, unanticipated, or extreme aircraft attitude.

User-defined waypoints. Waypoint location and other data which may be input by the user, this is the only GPS database information that may be altered (edited) by the user.

Variation. Compass error caused by the difference in the physical locations of the magnetic north pole and the geographic north pole.

VASI. See visual approach slope indicator.

VDP. See visual descent point.

Vectoring. Navigational guidance by assigning headings.

Venturi tube. A specially shaped tube attached to the outside of an aircraft to produce suction to allow proper operation of gyro instruments.

Vertical speed indicator (VSI). A rate-of-pressure change instrument that gives an indication of any deviation from a constant pressure level.

Very-high frequency (VHF). A band of radio frequencies falling between 30 and 300 MHz.

Very-high frequency omnidirectional range (VOR). Electronic navigation equipment in which the flight deck instrument identifies the radial or line from the VOR station, measured in degrees clockwise from magnetic north, along which the aircraft is located.

Vestibule. The central cavity of the bony labyrinth of the ear, or the parts of the membranous labyrinth that it contains.

VFR. See visual flight rules.

VFR-on-top. ATC authorization for an IFR aircraft to operate in VFR conditions at any appropriate VFR altitude.

VFR over-the-top. A VFR operation in which an aircraft operates in VFR conditions on top of an undercast.

Victor airways. Airways based on a centerline that extends from one VOR or VORTAC navigation aid or intersection, to another navigation aid (or through several navigation aids or intersections); used to establish a known route for en route procedures between terminal areas.

Visual approach slope indicator (VASI). A visual aid of lights arranged to provide descent guidance information during the approach to the runway. A pilot on the correct glide slope will see red lights over white lights.

Visual descent point (VDP). A defined point on the final approach course of a nonprecision straight-in approach procedure from which normal descent from the MDA to the runway touchdown point may be commenced, provided the runway environment is clearly visible to the pilot.

Visual flight rules (VFR). Flight rules adopted by the FAA governing aircraft flight using visual references. VFR operations specify the amount of ceiling and the visibility the pilot must have in order to operate according to these rules. When the weather conditions are such that the pilot can not operate according to VFR, he or she must use instrument flight rules (IFR).

Visual meteorological conditions (VMC). Meteorological conditions expressed in terms of visibility, distance from cloud, and ceiling meeting or exceeding the minimums specified for VFR.

VMC. See visual meteorological conditions.

VOR. See very-high frequency omnidirectional range.

VORTAC. A facility consisting of two components, VOR and TACAN, which provides three individual services: VOR azimuth, TACAN azimuth, and TACAN distance (DME) at one site.

VOR test facility (VOT). A ground facility which emits a test signal to check VOR receiver accuracy. Some VOTs are available to the user while airborne, while others are limited to ground use only.

VOT. See VOR test facility.

VSI. See vertical speed indicator.

WAAS. See wide area augmentation system.

Warning area. An area containing hazards to any aircraft not participating in the activities being conducted in the area. Warning areas may contain intensive military training, gunnery exercises, or special weapons testing.

Waypoint. A designated geographical location used for route definition or progress-reporting purposes and is defined in terms of latitude/longitude coordinates.

WCA. See wind correction angle.

Weather and radar processor (WARP). A device that provides real-time, accurate, predictive and strategic weather information presented in an integrated manner in the National Airspace System (NAS).

Weight. The force exerted by an aircraft from the pull of gravity.

Wide area augmentation system (WAAS). A differential global positioning system (DGPS) that improves the accuracy of the system by determining position error from the GPS satellites, then transmitting the error, or corrective factors, to the airborne GPS receiver.

Wind correction angle (WCA). The angle between the desired track and the heading of the aircraft necessary to keep the aircraft tracking over the desired track.

Wind shear. A change in wind speed and/or wind direction in a short distance.

Work. A measurement of force used to produce movement.

Zone of confusion. Volume of space above the station where a lack of adequate navigation signal directly above the VOR station causes the needle to deviate.

Index

A

M

N

O

P

R

S

T

U

V

W